How PRODUCTS Are MADE

How PRODUCTS Are MADE

An Illustrated Guide to

Product Manufacturing

Volume 7

Deirdre S. Blanchfield, Associate Editor

GALE GROUP

THOMSON LEARNING

Detroit • New York • San Diego • San Francisco
Boston • New Haven, Conn. • Waterville, Maine
London • Munich

How Products Are Made, volume 7

GALE GROUP STAFF

Deirdre S. Blanchfield, *Associate Editor*
Christine B. Jeryan, *Managing Editor*

Stacey L. Blachford, *Editor*
Kate Kretschmann, *Assistant Editor*

Mary Beth Trimper, *Manager, Composition and electronic prepress*
Evi Seoud, *Assistant Manager, Composition and electronic prepress*
Rhonda Williams, *Buyer*

Michelle DiMercurio, *Art Director*
Bernadette M. Gornie, *Page Designer*
Mike Logusz, *Graphic Artist*

Barbara J. Yarrow, *Manager, Imaging and Multimedia Content*
Robyn V. Young, *Project Manager, Imaging and Multimedia Content*
Dean Dauphinais, *Senior Editor, Imaging and Multimedia Content*
Kelly A. Quin, *Editor, Imaging and Multimedia Content*
Leitha Etheridge-Sims, Mary K. Grimes, David G. Oblender, *Image Catalogers*
Randy Bassett, *Imaging Supervisor*
Robert Duncan, *Senior Imaging Specialist*
Dan Newell, *Imaging Specialist*
Christine O'Bryan, *Graphic Specialist*

Maria Franklin, *Permissions Manager*
Shalice Shah-Caldwell, *Permissions Specialist*

ISBN 0-7876-3643-6
ISSN 1072-5091

Printed in the United States of America
10 9 8 7 6 5

Gale Group and Design is a trademark used herein under license.

Contents

Introduction vii

Contributors ix

Acknowledgments xi

Air Purifier 1

Amber 6

Aneroid Barometer 11

Artificial Turf 15

Bank Vault 20

Bicycle Seat 25

Binocular 30

Bird Cage 33

Birdseed 37

Brandy 41

Breath Alcohol Tester 46

Bullet 51

Candy Cane 57

Cash Register 62

Catheter 66

Condensed Soup 69

Corset 73

Cyclotron 77

Diving Bell 82

Drain Cleaner 87

Dry Ice 91

EEG Machine 95

Electric Guitar 100

Electric Tea Kettle 106

Epilation Device 110

External Defibrillator 114

Fabric Softener 120

Feather Duster 125

Felt 130

Fluoride Treatment 135

Fountain Pen 139

Furnace 144

Gas Lantern 147

Glucometer Test Kit 151

Goalie Mask 156

Grenade 160

Guillotine 163

Hair Dryer 168

Hairspray 172

Handcuffs 177

Headstone 180

Ice-Resurfacing Machine . .184

Insulin 187

Ironing Board 194

Juice Box 198

Kerosene 203

Laser Pointer 207

Lawn Sprinkler 211

Lighter 215

Manhole Cover 220

Maracas 224

Microphone 229

Mop 233

Movie Projector 237

Night Scope 244

Oxygen Tank 249

Pacifier 253

Paper Clip 257

Pizza 260

Pop Up Book 263

Punching Bag 266

Pyrex 270

Radio 275

Radio Collar 279

Revolving Door 282

Rolling Pin 286

Rubik's Cube 290

Scale 295

Ship In A Bottle 298

Shrapnel Shell 304

Sleeping Pill 309

Smoked Ham 314

Speedometer 317

Spinning Wheel 321

Spork 325

Spray Paint 329

Stereo Speaker 333

Stereoptic Viewer 336

Stirling Cycle Engine 342

Straight Pin 348

Sushi Roll 352

Suture 356

Swimming Pool 361

Swimsuit 365

Table 369

Tattoo 374

Teflon 378

Telephone Booth 382

Tiara 385

Titanium 391

Toaster 395

Tuxedo 399

Typewriter 404

Unicycle 410

Vending Machine 414

Videotape 420

Vinegar 425

Windmill 429

Windshield Wiper 434

X-Ray Glasses 439

Index 443

Introduction

About the Series

Welcome to *How Products Are Made: An Illustrated Guide to Product Manufacturing*. This series provides information on the manufacture of a variety of items, from everyday household products to heavy machinery to sophisticated electronic equipment. You will find step-by-step descriptions of processes, simple explanations of technical terms and concepts, and clear, easy-to-follow illustrations.

Each volume of *How Products Are Made* covers a broad range of manufacturing areas: food, clothing, electronics, transportation, machinery, instruments, sporting goods, and more. Some are intermediate goods sold to manufacturers of other products, while others are retail goods sold directly to consumers. You will find items made from a variety of materials, including products such as precious metals and minerals that are not "made" so much as they are extracted and refined.

Organization

Every volume in this series is comprised of many individual entries, each covering a single product. Although each entry focuses on the product's manufacturing process, it also provides a wealth of other information: who invented the product or how it has developed, how it works, what materials are used, how it is designed, quality control procedures, byproducts generated during its manufacture, future applications, and books and periodical articles containing more information.

To make it easier for you to find what you're looking for, the entries are broken up into standard sections. Among the sections you will find are the following:

- Background
- History
- Raw Materials
- Design
- The Manufacturing Process

- Quality Control
- Byproducts/Waste
- The Future
- Where To Learn More

Every entry is accompanied by illustrations. Uncomplicated and easy to understand, these illustrations may follow the step-by-step description of the manufacturing process found in the text, highlight a certain aspect of the manufacturing process, or illustrate how the product works.

A cumulative subject index of important terms, processes, materials, and people is found at the end of the book. Bold faced volume and page numbers in the index refer to main entries in the present or previous volumes.

About this Volume

This volume contains essays on 100 products, arranged alphabetically, and 15 special boxed sections, describing interesting historical developments or biographies of individuals related to a product. Photographs are also included. Bold faced terms found in main entries direct the user to the topical essay of the same name.

Contributors/Advisor

The entries in this volume were written by a skilled team of technical writers and engineers, often in cooperation with manufacturers and industry associations. The advisor for this volume was David L. Wells, PhD, CMfgE, a long time member of the Society of Manufacturing Engineers (SME) and Professor and Chair of the Industrial and Manufacturing Engineering Department at North Dakota State University.

Suggestions

Your questions, comments, and suggestions for future products are welcome. Please send all such correspondence to:

The Editor
How Products Are Made
Gale Group, Inc.
27500 Drake Rd.
Farmington Hills, MI 48331-3535

Contributors

Margaret Alic

Nancy EV Bryk

Andrew Dawson

Steve Guerriero

O. Harold Boutilier

Dan Harvey

Gillian S. Holmes

Carrie Lystila

Bonny McClain

Mary McNulty

Mya Nelson

Jeff Rains

Jeff Roberts

Perry Romanowski

Kathy Saporito

Randy Schueller

Ernst Sibberson

David L. Wells

Angela Woodward

Acknowledgments

The editor would like to thank the following individuals, companies, and associations for providing assistance with Volume 7 of *How Products Are Made*:

Artificial Turf: Patrick Dubeau, FieldTurf Inc., Montreal, Quebec. **Birdcage:** Guy Cone, Quality Cage Company, Portland, Oregon; Willis Kurtz, Safegaurd, New Holland, Pennsylvania. **Diving Bell:** Gregory D. Frye, Museum of Man in the Sea, Panama City Beach, Florida. **Drain Cleaner:** Dennis West, Rooto Corporation. **Dry Ice:** Robert Foster, Cold Jet, Inc., Loveland, Ohio. **Felt:** Mark Pryne, Affco, New Windsor, New York. **Gas Lantern:** Richard Long, Coleman Company, Wichita, Kansas. **Headstone:** Linda Mathiasen, Cold Spring Granite Company, Cold Spring, Minnesota; John J. Spaulding, Association for Gravestone Studies, Greenfield, Massachusetts. **Ironing Board:** Joseph Deppen, Home Products International, Chicago, Illinois. **Juice Box:** Jeff Lofton, SIG Combibloc Inc., Columbus, Ohio; Amy Brower, Tetra Pak, Inc., Denton, Texas. **Lawn Sprinkler:** Mike Simpson, L. R. Nelson Company, Peoria, Illinois. **Laser Pointer:** Whitney Kim, Lasermate; Jim Heusinger, Berea Hard Woods. **Microphone:** Heil Sound Ltd., Fairview Heights, Illinois. **Night Vision:** Laurel Holder, ITT Night Vision, Roanoke, Virginia. **Pacifier:** Paul Dailey, Children's Medical Ventures, Inc., Norwell, Massachusetts. **Pop Up Book:** Greg Witt, Leo Paper Company, Seattle, Washington; Sarah Ketcherside, Candlewick Press, New York, New York. **Revolving Door:** O. Harold Boutilier, Sierra Automatic Doors, Inc., Ontario, California. **Spinning Wheel:** Gord Lendrum, Lendrum Spinning Wheels, Odessa, Ontario. **Table:** Roger Shinn, Westview Products, Dallas, Oregon. **Tuxedo:** Barry Cohen, Hartz and Company, Frederick, Maryland. **Vending Machine:** John Turner, Dixie-Narco, Inc., Williston, South Carolina; Kari Klassen and James Radant, Automatic Products International, Ltd.

Photographs appearing in Volume 7 of *How Products Are Made* were received from the following sources:

AP/Wide World Photos. Reproduced by permission: **Artificial Turf.** Houston Astrodome, photograph; **Brandy.** Prohibition demonstration, photograph; **Electric Guitar.** Hendrix, Jimi, playing guitar, photograph; **External Defibrillator.** Kouwenhoven, William Bennett, photograph; **Goalie Mask.** Rheaume, Manon, photograph.

Archive Photos Inc. Reproduced by permission: **Michrophone.** Morrison, Jim, photograph.

Corbis Corporation. Reproduced by permission: **Bank Vault.** Dillinger, John Herbert (wanted poster); **Punching Bag.** Frazier, Joe, throwing punch at Muhammad Ali, photograph.

Corning Consumer Products Company. Reproduced by permission: **Pyrex.** Corning cookbook, photograph.

The Library of Congress: **Cyclotron.** Lawrence, Ernest Orlando, photograph; **Guillotine.** Louis XVI, King, painting; **Insulin.** Banting, Frederick Grant, photograph; **Movie Projector.** Lumiere, Louis, photograph; **Stereo Speaker.** Morita, Akio, photograph.

All line art illustrations in this volume were created by **Electronic Illustrators Group (EIG)** of Morgan Hill, California.

Air Purifier

Background

Air purifiers evolved in response to people's reactions to allergens like pollen, animal dander, dust, and mold spores. Reactions (sneezing, runny nose, scratchy eyes, and even more severe consequences such as asthma attacks) are the result of antigens found in the home. These antigens are major triggers of asthma, and there are more than 17 million asthmatics in the United States alone. Air purifiers remove a portion of these particles, thus reducing allergic-type responses.

Due to their extremely small size, allergens are able to pass through a standard vacuum cleaner bag and redistribute into the air where they stay for days. Even a single microgram of cat allergens is enough to invoke an allergic response in most of the six to 10 million Americans who are allergic to cats. Other airborne particles—such as bacteria and viruses—can cause illnesses, some of which are fatal. There are many reasons—allergies, asthma, fatal illnesses—that millions of air purifiers are sold in the United States every year.

There are two common types of air purifiers that can remove some or all of the disease and allergy-causing particles in the air: mechanical filters—the most effective are classified as High Efficiency Particulate Air filters (HEPA filters)—and electrostatic precipitators.

HEPA filters are made out of very fine glass threads with a diameter of less than 1 micron (a micron is 0.00004 in, 0.001 mm). By comparison, a human hair has a diameter of about 75 microns (0.003 in, 0.07 mm). The fine glass threads are tangled together and compressed to form a filter mat. Because the individual threads are so microscopic, most of the mat consists of air. The openings in the mat are very small, generally less than 0.5 micron (0.00002 in, 0.0005 mm). HEPA filters will collect particles down to 0.3 microns (0.00001 in, 0.0003 mm) in diameter. Even though the filter may only be 0.10 in (2.5 mm) wide, it would consist of 2,500 layers of glass threads.

Electrostatic precipitators rely on electrostatic forces to remove particles from the air. They work by creating a cloud of free electrons through which dust particles are forced to pass. As the dust particles pass through the plasma, they become charged, making them easy to collect. Electrostatic precipitators can collect particles down to a diameter of 0.01 microns (0.00001 mm).

Neither HEPA filters nor electrostatic precipitators can remove volatile organic compounds from the air, therefore do nothing to reduce odors. For this reason, most air purifiers are equipped with a pre- or post-filter composed of activated carbon. Activated carbon is produced by heating a carbon source (coconut shells, old tires, bones, etc.) at very high temperatures in the absence of oxygen, a process also known as pyrolysis or destructive distillation. Pyrolysis separates the pure carbon from the other materials contained in the raw material. The pure carbon is then exposed to steam at 1,500°F (800°C). The high temperature steam activates the carbon. The activation process forms millions of cracks in the carbon grains. These cracks have diameters of about 0.002 microns (0.000002 mm). Because there are so many cracks, the activation process provides the carbon with an enormous surface area per weight—about

The current generation of HEPA filters can only remove particles down to 0.3 microns (0.00001 in, 0.0003 mm) in diameter, while it is believed that particles down to 0.1 microns (0.0001 mm) in diameter can cause mechanical damage to lung tissue.

6.5 acres/oz (1,000 m²/g). The millions of cracks provide locations where organic compounds can be adsorbed. In addition, the surface of the carbon carries a residual electrical charge that attracts non-polar chemicals (chemicals that do not have separated positive and negative charges) to it. Activated carbon is very effective at adsorbing odor producing compounds.

History

Air purity has been a concern as long as human beings have lived in groups. One of the reasons that hunter-gatherers are nomadic is that they periodically need to move away from their garbage dumps and latrines. In A.D. 61, the Roman philosopher Seneca complained about the miasma of chimney smoke that constantly hung over Rome. In 1306, King Edward I of England banned the burning of coal in London due to the heavy pollutants left in the air.

The Industrial Revolution of the eighteenth and nineteenth century only worsened the problem. Burning coal to produce electricity and fuel trains produced a dark cloud of smoke over every major center of industry in the world and covered entire cities with soot. To deal with this problem, engineers built higher smoke stacks to move airborne waste further away from the source. Regardless of how high the stacks got, the people down wind complained about the ashes and the acid gases from coal combustion (the source of acid rain) destroying their crops. Air pollution took another turn for the worse after World War II when automobiles became the primary means of transportation in the industrialized world. Automobile smog has provided Los Angeles with the worst air quality in the world.

Raw Materials

The materials that go into both HEPA filters and electrostatic precipitators are: a case made out of plastic, an electric fan to induce air flow through the filter, the filter itself, and control switches to control the speed of the fan and turn the air purifier on and off. The HEPA filters are made of borosilicate glass fibers or plastic fibers (e.g., polypropylene) bound together with up to 5% acrylic binder (the same compound that binds latex

paint to a house). Electrostatic precipitators generate ions by running extremely high positive direct current voltages through steel wires set between grounded steel charging plates. Cases are almost universally made from plastic, usually high-impact polystyrene, polyvinyl chloride, high-density polyethylene, or polypropylene. Most air purifiers are also usually equipped with a post-filter composed of activated carbon.

Design

HEPA filters are designed based on the size of particles to be removed and the required air flow rate. The finer the pores in the HEPA material, the finer the particles removed from the air. However, collecting finer particles means the filter material will clog sooner and need replacing on a more frequent basis. The designer will specify the diameter of the glass fibers and the mat density of the filter fabric that fixes the filter pore size. HEPA filters can contain binders that provide additional strength, but this also produces a filter that clogs sooner.

Design of an electrostatic precipitator is considerably more complex. Home electrostatic precipitators usually are designed to have two components, an ionizing component (where the electron cloud is created) and a collecting component (where the charged dust particles are pulled out of the air). The collecting component consists of a series of parallel steel plates—half are grounded and half carry a positive direct current voltage—thus alternate plates are either positively or negatively charged. The ionizing unit consists of thin wires strung between a separate set of grounded steel ionizing plates parallel to, but set in front of, the collector plates. The thin wires carry a very high positive voltage direct current (up to 25,000 volts in a home air purifier). The positive charges in the wires induce a flow of electrons between the wires and the adjacent ionizing plates. Because there is a very high voltage on the wire, electrons are pushed toward it by an acceleration of around 1,000 times the acceleration of gravity, which accelerates the electrons to very high velocities. For example, as a particle of dust mite excrement floats past the wire, the high-speed electrons collide with the electrons in the molecules of the particle, knock-

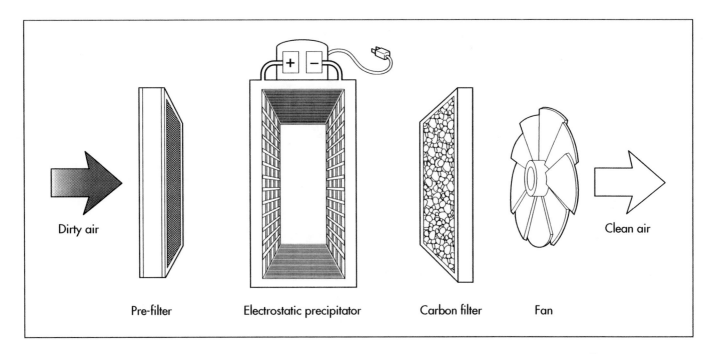

Dirty air

Pre-filter   Electrostatic precipitator   Carbon filter   Fan   Clean air

ing some of them free. As these molecules lose electrons, they take on a positive charge and are thus attracted to the negatively-charged collector plate. The designer must select a voltage high enough to produce sufficient numbers of electrons to ionize the particles passing through the precipitator, and space the collector plates close enough together so that the ionized dust particles will be captured on the plates before the precipitator fan can pull them completely through the air purifier.

The Manufacturing Process

The case

1 Pellets of the raw material (high-impact polystyrene, polyvinyl chloride, high-density polyethylene, or polypropylene) are fed into a hopper and heated to the melting point, 300–590°F (150–310°C).

2 The molten plastic is injected under high pressure into a mold of the case. The mold is usually made from tool steel by a highly skilled mold maker. Vents in the mold allow the entrained air to escape as the plastic enters. The mold designer must assure that the mold fills evenly with plastic and that all of the entrained air is allowed to escape, otherwise the final part might contain small air bubbles or even voids.

3 Water is forced through channels built into the mold to transfer heat from the molten plastic into the environment. Once the part is sufficiently cool, which can take up to two minutes, the mold opens. Hydraulically-operated pins push the part out of the open mold into a receiving bin.

The fan

4 An electric fan is used to pull air through the air purifier. The fan is usually purchased from a small-parts supplier. The fan consists of a small electric motor with metal fan blades attached to the motor's power take-off. The fan blades are usually spot welded to a collar, which is slipped onto the power take-off and bolted in place.

5 The fan is usually attached to the case with steel screws.

HEPA filters

1 The glass fibers that make up a HEPA filter are created by passing molten glass or plastic through very fine pores in a spinning nozzle. The resulting glass fibers cool and harden almost instantly because of their tiny diameters.

2 The spinning nozzle moves back and forth (causing the glass fibers to form a web) above a moving conveyor belt onto which the fibers are collected. The speed of

An example of an electrostatic precipitator and its components.

the conveyor belt determines the thickness of the filter material—a slow conveyor belt allows more glass fibers to build up on the belt.

3 The melting and cooling of the fiber produces some bonding of the fibers. As the conveyor progresses, a latex binder is sprayed onto the fabric to provide additional strength. The fabric can be any width up to the practical size of the machinery and can be cut down to the size specified by the customer before the fabric is taken up on rollers.

4 Once the HEPA mats are formed, they are folded into an accordion pattern in an automatic folder. The accordion pattern allows up to 50 ft² (5 m²) of filter material to be enclosed in a small space.

5 The accordion-shaped filter is then enclosed in a filter case, usually consisting of an open wire grid. The purpose of the filter case is to support the filter.

Electrostatic precipitators

1 The electrostatic precipitator collection system is manufactured by enclosing steel plates into a plastic casing, often by hand assembly. The plates are arranged parallel to each other in the case.

2 Wires are then connected to alternate plates through which the high voltage positive direct current will be applied to the plates. The other plates are grounded.

3 The ionizing unit is constructed by running small diameter wires in front of the collector plates.

4 A voltage transformer, which is used to convert 115 volt household alternating current into high voltage direct current, is fixed to the precipitator case. This voltage is run to both the positively charged collector plates and the ionizing wires.

The activated carbon filter

1 The activated carbon filter (for odor reduction) usually consists of carbon-impregnated cloth or foam. This is manufactured by infusing the raw material with powdered activated carbon.

2 The carbon filter is then wrapped around the inside or outside of the HEPA filter,

or stretched in a frame at either the inlet or outlet of the electrostatic precipitator.

Assembly

3 There are very few components in an air purifier. For this reason, they are usually bench assembled. In bench assembly, moving conveyors bring the individual components or sub-assemblies (e.g., the fan already attached to the case) to a bench where a person then hand assembles them. In a typical HEPA air purifier, there may only be five components that require assembly: casing, fan, particulate filter, carbon filter, and the on/off switch.

Quality Control

Filter efficiency is the most important quality control test for air purifiers. The American Society for Testing and Materials (ASTM) publishes quality control tests that filters must meet before they can be used in certain applications or be marketed as HEPA filters (e.g., ASTM-F50: Standard Practices for Continuous Sizing and Counting of Airborne Particles in Dust Controlled Areas and Cleanrooms Using Instruments Capable of Detecting Single Submicrometer and Larger Particles). The United States Department of Defense has promulgated a standard in which dioctylphthalate (DOP) particles are blown through a filter. To pass, the filter must remove 99.97% of the influent DOP.

Byproducts/Waste

The byproducts of manufacturing include the non-carbon materials that are distilled from the manufacture of activated carbon, specification filter material, and excess material that must be discarded in the production of HEPA filters. Most of the other manufacturing wastes, plastic runners from the injection machines and excess sheet metal, can be recycled.

Additional wastes are produced during the operation of air filters. The ions produced by electrostatic precipitators interact with oxygen in the air to produce ozone. At high concentrations, ozone is poisonous. The ozone levels produced in a home electrostatic precipitator are unlikely to reach dangerous levels, but some people are sensitive to even low levels of ozone. The collector plates in an electrostatic precipitator need to be cleaned periodically.

HEPA filters have limited lifetimes, depending on the amount of air that is filtered through them and the amount of particulates in the air. Most manufacturers recommend that they be replaced every few years. The used filters cannot be recycled and thus end up in landfills.

Activated carbon can be recycled, but the cost of handling the small amount of carbon contained in a home air purifier would be prohibitive. Generally, it also ends up in landfills after it is used completely.

The Future

As scientists learn more about environmental pollutants and their impact on human health, the need to provide cleaner air in homes and offices will only grow. The current generation of HEPA filters can only remove particles down to 0.3 microns (0.00001 in, 0.0003 mm) in diameter while it is believed that particles down to 0.1 microns (0.0001 mm) in diameter can cause mechanical damage to lung tissue. Viruses can be as small as 0.02 microns (0.00002 mm) in diameter. Clearly, there is still progress that can be made in controlling indoor air pollution. The current direction of technology is toward ever finer filter materials. The new standard in filtration is the ULPA filter, which stands for Ultra Low Penetrating Air. An ULPA filter is required to be able to remove particles down to 0.12 microns (0.00012 mm) in diameter, about one third of the diameter of the smallest particle a HEPA filter can remove.

Where to Learn More

Books

Cooper, David C., and F. C. Alley. *Air Pollution Control: A Design Approach.* Prospect Heights, IL: Waveland Press, Inc., 1994.

Godish, Thad. *Indoor Environmental Quality.* New York: Lewis Publishers, 1999.

Mycock, John C., et al. *Handbook of Air Pollution Control Engineering and Technology.* New York: Lewis Publishers, 1995.

Periodicals

Christiansen, S. C., et al. "Exposure and Sensitization to Environmental Allergen of Predominantly Hispanic Children with Asthma in San Diego's Inner City." *Journal of Allergy and Clinical Immunology* (August 1996): 288–294.

Other

"Electrostatic Precipitation." *Tin Works, Inc.* 3 June 2001. <http://www.tinworks.com>.

"Filter Media." *Mac Equipment.* 3 June 2001. <http://www.macequipment.com>.

"Plastic: Injection Molding." *Industrial Designers Society of America.* 17 June 2001. <http://www.idsa-mp.org/proc/plastic/injection/injection_process.htm>.

Jeff Raines

Amber

For thousands of years, amber has been carved and worked into beads, jewelry, and other types of ornamentation.

Background

Although considered a gem, amber is a wholly-organic material derived from the resin of extinct species of trees. In the dense forests of the Middle Cretaceous and Tertiary periods, between 10 and 100 million years ago, these resin-bearing trees fell and were carried by rivers to coastal regions. There, the trees and their resins became covered with sediment, and over millions of years the resin hardened into amber. Although many amber deposits remain in ocean residue, geological events often repositioned the amber elsewhere.

For thousands of years, amber has been carved and worked into beads, jewelry, and other types of ornamentation. However, today amber is valued primarily for the astounding array of fossils preserved inside. As sticky resin was exuded by the trees, animals, minerals, and plant materials were trapped in it. As the resin hardened, these fossils—called inclusions—were perfectly preserved, providing modern scientists with invaluable information about extinct species.

Unlike other types of fossils, amber fossils are three-dimensional, with life-like colors and patterns. Even the internal structures of cells may be intact. Often, insects were caught by the resin in active poses, along with their predators, prey, and internal and external parasites. Previously-unknown genera of fossilized insects have been discovered in amber. Intact frogs and lizards, snake skins, bird feathers, hair and bones of mammals, and various plant materials have been preserved in amber. In some cases, deoxyribonucleic acid (DNA) can be extracted from the fossilized organisms and compared with that of its modern-day counterparts.

History

Amber has been a highly-valued material since earliest times. Worked amber dating back to 11,000 B.C. has been found at archeological sites in England. Amber was widely believed to have magical healing powers. It was used to make varnish as long ago as 250 B.C., and powdered amber was valued as incense. Amber was also traded throughout the world. By identifying the type of amber used in ancient artifacts, scholars can determine the geographical source of the amber and draw conclusions about early trade routes.

In about 600 B.C., the Greek philosopher Thales rubbed amber with silk, causing it to attract dust and feathers. This static electricity was believed to be a unique property of amber until the sixteenth century, when English scientist William Gilbert demonstrated that it was characteristic of numerous materials. He called it electrification, after *elektron*, the Greek word for amber.

In the Western Hemisphere, the Aztecs and Mayans carved amber and burned it as incense. The Taino Indians of the island of Hispaniola offered gifts of amber to Christopher Columbus.

The decorative use of amber culminated in 1712 with the completion of an entire banquet room made of amber panels constructed for King Frederick I of Prussia. In the nineteenth century amber attained new significance when German scientists began studying the fossils imbedded in it.

Raw Materials

Resins are complex substances that include oily compounds called terpenes. Over time, some terpenes evaporate while others condense and become cross-linked to each other, forming hard polymers. However, different species of trees produce different types and amounts of resins. The exact structure and composition of amber depends on the makeup of the original tree resin, the age of the amber, the environment in which it was deposited, and the thermal conditions and geological forces to which it was exposed. Thus, even amber obtained from similar locations may vary in chemical structure and physical characteristics.

Types of amber

Although deposits of amber occur throughout the world, amber from the coast of the Baltic Sea is the best-known. It is called succinite amber because it contains a substantial amount of succinic acid. Most Baltic amber came from pine tree resin. Amber that lacks succinic acid is classified as retinite amber.

Amber from Mexico and the Dominican Republic began forming 20–30 million years ago from the resins of extinct species of *Hymenaea* or algarrobo trees. These flowering trees thrived in the canopy of extensive tropical rain forests. They produced copious amounts of resin that eventually hardened into amber. Torrential rains washed the amber to deltas where it was covered with silt. As sea levels changed, the amber settled on the sea floor and the sediment over it hardened into rock. Later, mountain formation pushed up the rocks.

Design

Physical characteristics

Many components of amber are similar to those of modern resins. However the cross-linking of these compounds makes the amber hard, with a high melting point and low solubility. Amber has a hardness of 2–3 on Mohs's scale, the standard for minerals and gems. On this scale, talc is 1 and diamond is 10. Amber softens at 302°F (150°C) and melts at 482–662°F (250–350°C). With a specific gravity of 1.05–1.12, amber is only slightly more dense than water. It will not completely dissolve in organic solvents.

Amber usually occurs as small irregular masses, nodules, or droplets. Although it can be many different colors, it is most often pale to golden yellow or orange and can be fluorescent. After a few years of exposure to light and air, amber often turns dark red and develops numerous cracks on the surface. Some amber is translucent or even transparent. However, trapped air bubbles can cause amber to be cloudy or opaque. Amber is a poor conductor of heat and large changes in temperature can cause it to fracture.

The Manufacturing Process

Amber is extracted in different ways, depending on its location. Baltic amber washes up along the shores of the Baltic Sea and as far away as Denmark, Norway, and England. The largest deposits of North American amber are found on the surface of open-pit clay mines in Arkansas. In New Jersey, Cretaceous amber is dug from the sand and clay of abandoned pit mines. It is screened, washed, and examined for inclusions. In Asia, amber is found in coal mines. Until the mid-twentieth century, highly-prized amber was mined from deep pits in northern Burma (now Myanmar).

Mining and washing

1 Drops or blocks of Baltic amber are mined from open pits of 40–60 million-year-old glauconite sand. Glauconite is a hydrated potassium-iron silicate mineral and these deposits are called "blue earth" because of their blue-green color. After the surface has been cleared, the blue earth is dug out with steam shovels and dredges. It is poured through grates at a washing plant, where streams of water are used to separate the amber from the sand. In the early twentieth century, up to one million lb (450,000 kg) of amber per year were extracted from the blue earth layer of the Samland Peninsula in the eastern Baltic.

Mexican and Dominican amber may be exposed by landslides on steep mountain slopes and extracted with picks and shovels. It also is mined from pits dug deep into the ground. Much Dominican amber is mined from narrow tunnels carved as far as 600 ft (183 m) into the sides of mountains. Water is baled or pumped out of the tunnels and the miners crawl through, chisel at the rock, and pick out the exposed amber. Dominican

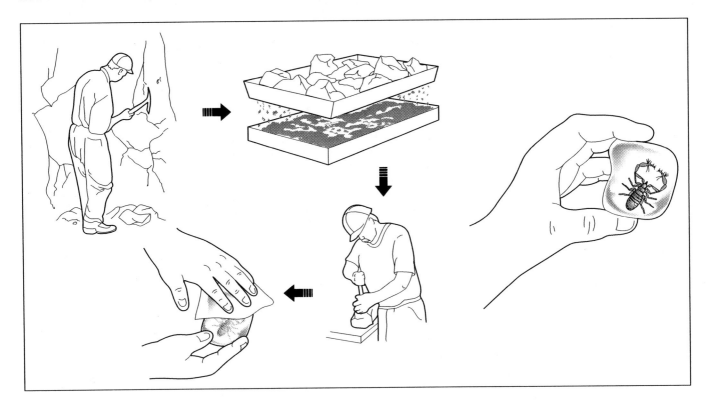

Baltic amber is mined and clarified to obtain a clear view of the inclusion.

amber is washed by the miner, sorted by size, and examined for inclusions.

Clarifying and coloring

2 Large trapped air bubbles result in a foamy or frothy type of amber. Microscopic bubbles result in bony or osseous amber that looks like dried bone. Very cloudy amber is called bastard. Amber is clarified by heating in rapeseed oil. The oil penetrates bubbles near the surface and reduces the cloudiness, making even bony or bastard amber more transparent. Amber also may be clarified by heating under pressure with nitrogen and then baking in an oven. Clarification darkens the amber and produces disc-like stress marks, called "sun spangles." Amber may be stained green or reddish. Mexican and Dominican amber is usually clear and transparent and does not need clarifying.

Cutting and reshaping

3 For jewelry or carving, amber usually is worked by hand, with a jeweler's saw and fine-toothed files. It is wet-sanded with 320-grit cloth and finished with a 400- or 600-grit wet-sanding cloth. It can be drilled with dry steel drills, using a low speed and slight pressure, to prevent heating and cracking.

4 To obtain a clear view of inclusions, one end of an amber piece may be chipped off. Amber with inclusions may be cut or reshaped for examination of the biological specimen or to separate two specimens. Cutting is done with a jeweler's hand saw or, for larger pieces, with a high-speed trim saw with a diamond blade, at speeds up to 4,200 rpm.

5 Reshaping is done with various grades of sandpaper. Rough edges from the saw blade may be smoothed with 200- and 400-grit paper, by hand or with a belt sander equipped with a water cooling system, to remove dust and prevent overheating and fracturing or glazing.

Polishing

6 Amber for jewelry is polished with tin oxide or cerium oxide, using a leather buff, felt wheel or pad, or chamois board. Periodic polishing with a silicone-based wax restores shine and decreases evaporation and surface oxidation.

Dominican amber is polished with a sander, following the natural contours. Surface oxidation of Dominican amber diminishes the fluorescence and the blue, green, or purple color. Removing the outer layer and repolishing restores the fluorescence. Repolishing may be done by hand or with a cotton buffing wheel, using dental polishing compound, an abrasive for plastics, or other fine neutral-colored polishing compounds. A final hand polishing removes the polishing compound.

Cretaceous amber more than 65 million years old is very brittle and fractured. After several years of exposure, it is prone to disintegration. Encasing Cretaceous amber in a synthetic resin helps to preserve it.

Pressing

7 For producing gems, small clear pieces of amber are softened and fused in a vacuum with steam at 400°F (204°C) or above.

8 The pieces are pressed through a fine steel sieve or mesh, mixed, and hardened into blocks. This pressed amber is called ambroid or amberoid and may contain bubbles that have elongated under the heat and pressure. Sometimes modern insect inclusions are inserted into pressed amber and the ambroid may be dyed, usually dark red.

Other processing

9 Small pieces of poor quality amber, including about 90% of Baltic amber, are distilled in huge, dry iron retorts. About 60% is recovered as amber colophony, a high-grade varnish. Another 15–20% becomes amber oil, used in medicines, casting, and the highest grade of varnish. About 2% of the products are distilled acids, such as succinic acid, that are used for medicines and varnishes.

Quality Control

Harder and, presumably, older amber is usually considered to be of higher quality. Since mining costs are 28% higher than the value of raw amber, its value is based primarily on its inclusions or on its eventual processing into jewelry and art objects. Therefore, amber is graded according to its size and beauty, as well as the presence and type of inclusions.

Imitation amber with fake inclusions has been produced for at least 600 years. Fresh resins, synthetic polystyrenes, Bakelite, epoxy resins, celluloid, colored glass, plastics, and polyesters all have been used for imitation amber. However, true amber can be distinguished by its hardness, melting temperature, lack of solubility, fluorescence, specific gravity, refractive index (measure of the degree that it bends light), and odor on burning. Sometimes true amber is fractured, a cavity is carved in it for an inclusion embedded in fresh resin, and the piece is resealed.

Byproducts/Waste

About 90% of the world's extractable amber is located in the Kaliningrad region of Russia on the Baltic Sea. There, amber mining and processing has caused widespread environmental degradation. More than 100 million tons of waste has been discharged into the Baltic from the Palmnicken (Yantarny) mine over the past century. This insoluble waste causes high turbidity in the Baltic Sea. The waters of the pollution-sensitive Baltic take 25–30 years to renew themselves.

The Future

The easily-extracted, top layers of Baltic amber were exhausted by the mid-1800s. However it is estimated that over 180,000 tons of amber remain in the Yantarny mine in Kaliningrad. At the current rate of extraction, amber could be mined there for another 300 years. In addition, mining has resumed in Myanmar and the high-quality Burmese amber is being sold to museums.

Although the process of amber formation from tree resin continues, it takes millions of years for the resin to harden into amber. As amber deposits are depleted by mining, and resin-bearing trees are cut or burned rather than allowed to fossilize, the supply of raw amber will continue to dwindle.

Where to Learn More

Books

Anderson, K. B., and J. C. Crelling, eds. *Amber, Resinite, and Fossil Resins.* Washington, DC: American Chemical Society, 1995.

Grimaldi, David A. *Amber: Window to the Past.* New York: Harry N. Abrams, Inc. and the American Museum of Natural History, 1996.

Poinar Jr., George, and Roberta Poinar. *The Amber Forest: A Reconstruction of a Vanished World.* Princeton: Princeton University Press, 1999.

Other

"Amber Trade and the Environment in the Kaliningrad Oblast." *TED Case Studies.* 27 July 2001. <http://gurukul.ucc.american.edu/ted/amber.htm>.

Brost, Leif. "Amber: A Fossilized Tree Resin." *The Swedish Amber Museum Home Page.* 27 July 2001. <http://www.brost.se/eng/education/facts.html>.

Margaret Alic

Aneroid Barometer

Background

Earth's atmosphere weighs about 6.5×10^{21} (5.98×10^{24}). Spread out across Earth's entire surface area, it exerts an air (barometric) pressure of about 14.7 pounds per square inch (psi) (101 kilopascals [kPa]) at sea level. While that is the average, the actual barometric pressure varies greatly from place to place and from one moment to the next. The barometric pressure at the summit of Mount Everest, is one third of the barometric pressure at sea level. The greatest barometric pressure extremes ever recorded at sea level were 15.7 psi (108 kPa) during a very cold winter in Siberia and 13.5 psi (87 kPa) recorded in the eye of a Pacific Ocean typhoon. Barometric pressure differences are important because they are the basic creators of weather.

The sun is the major factor in causing pressure variations in the atmosphere. Hot equatorial air rises and flows north. As it moves Coriolis forces in the northern hemisphere bend it to the west in the tropics and to the east in the temperate zones, setting up cells of clockwise and counterclockwise atmospheric flow. The changing atmospheric pressures that come with these flows can be used to predict the weather. In fact, prior to the advent of the radio, the only tool sailors had to predict the weather was the barometer, which told them which way the air pressure was changing. A rising barometric pressure was a sign of improving weather. A falling barometer was a sign to batten down the hatches and hope for the best.

History

Many people do not realize that atmospheric pressure exists since it cannot be felt. Its existence was discovered by the Italian scientist Evangelista Torricelli. Torricelli made his discovery during an attempt to help silver miners, who were having trouble keeping their mines dry. The only pump available to the miners were suction pumps, which could only raise water 32 ft (9.8 m). Torricelli deduced the reason the pump could not raise water more than that was because the weight of the atmosphere was only heavy enough to support a column of water 32 ft (9.8 m) high. Torricelli's insight was that if a see-saw were arranged such that half of it was under a vacuum and half of it was under atmospheric pressure, 32 ft (9.8 m) of water would have to be placed on the vacuum side of the see-saw to balance the atmospheric pressure acting on the other side. The miners' pumps were like a see-saw trying to balance more than 32 ft (9.8 m) of water.

To test his theory, Torricelli took a glass tube about 4 ft (1.2 m) long, sealed it at one end, and filled it with mercury. Holding his thumb over the open end, he upended the tube into a bowl of mercury. His theory was that, since mercury is 13.5 times more dense than water, barometric pressure would only be high enough to support a column of mercury 2.4 ft (0.73 m) high (the maximum height the suction pumps could pull water divided by 13.5). In actuality, the atmosphere supported a column of mercury 2.5 ft (0.76 m) high. The extra distance was because the vacuum at the top of the glass tube was almost perfect—Torricelli was also the first person to create a vacuum—and the seals in the miners' pumps were not. It is not clear who noticed that barometers could be used to forecast the weather, though it is possible it was Ferdinand dei Medici, Grand Duke of Tuscany.

Many people do not realize that atmospheric pressure exists since it cannot be felt. Its existence was discovered by the Italian scientist Evangelista Torricelli.

While mercury barometers, even to this day, are the most accurate barometers, they are not without drawbacks. Trying to read a mercury barometer on board a ship caught in a hurricane is not easy. The idea for a mercury-free barometer (an aneroid barometer) first occurred to Gottfried Leibniz (co-inventor of calculus) around 1700. Metallurgy was not sufficiently advanced in 1700 to realize Leibniz's idea. The French inventor Lucien Vidie developed the first practical aneroid barometer in 1843. Aneroid barometers are the most common barometers in use today. They are the circular, brass, clock-like instruments with a sweep indicator pointing to the current barometric pressure. They are commonly seen in weather stations and on board boats. Aneroid barometers function by measuring the expansion and contraction of a hollow metal capsule.

Raw Materials

The only components of a mercury barometer are glass and mercury. Aneroid barometers, on the other hand, are very complex machines similar to fine watches. The aneroid capsule, which is the device that moves with changes in air pressure, is made from an alloy of beryllium and copper. The movements are made from stainless steel (e.g., AISI 304L) with jeweled bearings (synthetic rubies or sapphires). Jewels are used in the bearings because they have very low frictional resistance. Barometer cases can be made out of anything, but are usually made out of brass (a mixture of copper and zinc). There are many types of brass. One of the most common is "clockbrass," a mixture of 65% copper and 35% lead. Barometer dials can be made out of anything: aluminum, steel, brass, or paper.

Design

Product design for an aneroid barometer involves a careful analysis of the contracting and expanding properties of the aneroid capsule, design of the temperature compensation system, and mechanical design of the linkage between the aneroid capsule and the sweep indicator.

The aneroid capsule is very thin, hollow, and usually shaped like a bellows. Most of the air is removed from the capsule so that the contraction and expansion of the capsule

is strictly a function of the elasticity of the capsule and any of its supporting springs. Leaving air in the capsule would induce non-linearity into the capsule response. As the capsule contracted, if there were any air left, the air pressure in the capsule would rise, which would make further compression of the capsule harder. The barometer designer calculates how much the aneroid capsule will expand or contract under the expected range of pressures the barometer will be subjected to. Based on these movements, the designer specifies the linkages that will translate the movement of the capsule into the movement of a sweep indicator on the barometer face.

The aneroid barometer is sensitive to temperature variations both because the capsule and its linkages will expand or contract as the temperature changes and also because the capsule's elastic properties (how much the capsule will deflect under changes in outside pressure) also change with temperature. There are several ways to compensate for temperature-induced movements of the barometer components. One of the more elegant solutions involves the use of a bimetallic strip. A bimetallic strip consists of two flat pieces of metal, made of different types of elements or alloys, welded back to back. Because the temperature changes in the bimetallic strip and capsule are predictable, the bimetallic strip can be used to compensate for the capsule movements. As temperatures change, the two components of the bimetallic strip try to expand by different amounts. This causes the bimetallic strip to bend toward the component with the smaller coefficient of expansion. This bending motion can be used to shift the indicator hand or compress the aneroid capsule to compensate for the temperature change.

The linkage between the aneroid capsule and the sweep indicator is almost as complex as the movement of a fine Swiss watch. In fact, a quality barometer linkage incorporates many of the same components. The linkage's purpose is to translate the tiny horizontal motion of an expanding bellows (a few thousands of an inch or centimeter) into the sweep motion of an indicator arm. The required magnification of the capsule movement can be accomplished using levers. A see-saw is a form of lever. The very end of the see-saw moves through a much greater arc than a point near the pivot. By arranging for the aneroid capsule to push or pull on a point near the pivot of a see-saw-like lever, the movement of the capsule is greatly magnified at the far end of the lever. Any non-linearity of the capsule movement can be compensated for using a *fusee*, pronounced FU-say. A fusee, which was invented by Leonardo da Vinci, is a spiral-cut pulley shaped like a cone. At the zero point of the barometer, the end of the see-saw lever is connected to the middle of the fusee by a chain. As the aneroid capsule compresses, the fusee rotates, shifting the chain down to a smaller diameter. What this accomplishes is that as the aneroid capsule hardens under compression, a smaller movement of the chain can produce the same movement of the sweep indicator.

The Manufacturing Process

The case

1 A fine barometer case can be cast out of brass, bronze, or steel or carved out of wood. Less expensive cases can be stamped out of steel or aluminum and then plated with a decorative finish. A casting is made by pouring molten metal into a mold and allowing the metal to harden.

2 After the metal has hardened, the mold is shaken off the case. Stamping involves pressing a flat piece of metal between two dies at high pressures.

3 The case is finished by removing any excess metal left over from the casting process, grinding down any rough edges, and then polishing the case to a bright finish.

4 Some cases are then varnished or coated with a clear plastic to prevent tarnishing.

The aneroid capsule

5 Thin sheets of copper/beryllium metal (around 0.002 in [0.05 mm] thick) are stamped into the two halves of the aneroid capsule. The stamping dies are designed to leave a knife edge mating surface where the two halves will be joined.

6 The individual aneroid components are electron-beam welded. Electron beam welding requires that a concentrated stream

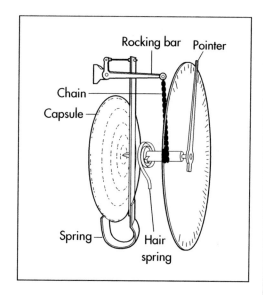

An aneroid barometer.

of electrons be generated and focused on the joint to be welded. As the electrons collide with the part, the kinetic energy of the collision creates heat resulting in the fusion or melting of the two pieces to be joined. Electron beam welding can only be performed in a vacuum (because the air molecules would intercept the electron beam) which is very convenient as the aneroid capsule must also be free of air. Electron beam welding is performed by automated robot welding machines because a human welder could not provide the degree of accuracy needed to join the parts without damaging them.

The linkages

7 High quality linkages consisting of levers, chains, sector racks, pinions, fusee, etc., are machined from tool steel. Machining means that stock bars of metal are ground and cut to shape the final piece. Automated milling machinery can easily produce linkage parts to a tolerance of 0.0001 in (0.0025 mm).

The temperature compensator

8 The temperature compensator, in quality barometers, is usually a bimetallic strip. The bimetallic strip is fixed by welding or riveting one end of the strip to the barometer case. Welding involves partially melting both the casing and the bimetallic strip so that the two pieces flow together and are joined.

Final assembly

9 The finished product is bench-assembled. Because production runs are gener-

ally very small, there is very little automation in the final assembly process. An assembler attaches the movement and temperature compensator to the case.

10 The face of the barometer is placed over the central pin of the movement.

11 The barometer indicator is then fixed to the central pin with a cotter pin or screw.

12 A glass plate is placed over the barometer face and a bezel is screwed onto the barometer to hold the glass in place. In many barometers, the face plate has a hole cut in the center of it so that the movement can be seen.

Quality Control

Quality control requires that the completed barometer be tested under different atmospheric conditions. All aneroid barometers come with a zeroing screw to adjust the initial position of the sweep indicator to be the same barometric pressure as that of a very precise standard barometer kept at the factory. The new barometer is then subjected to varying barometric pressures to assess how accurately it can record the actual pressure. Barometers that cannot meet the required factory tolerances, which vary from manufacturer to manufacturer, have their movements replaced.

Byproducts/Waste

Mercury barometers contain the highly-toxic heavy metal that gives them their name. However, many localities and some states have banned the use of mercury in thermometers, barometers, and blood pressure recording devices. It is only a matter of time before the mercury barometer disappears from common use. Wastes generated during aneroid barometer manufacturing are limited to minor amounts of metal from the linkage machining. Casting wastes from the barometer cases are usually immediately recycled at the casting house.

The Future

The future of the barometer is a digital version. By placing parallel steel plates inside the aneroid capsule and running a current across them, the distance between the two plates can be determined as it is proportional to the capacitance of the plates (capacitance is a measure of the amount of electric charge that can be stored on the plate). As the aneroid capsule shrinks and expands, the capacitance of the two plates changes, providing a measure of the change in atmospheric pressure driving the change in plate position. This obviates the need for jeweled bearings, fusee, and machined linkages, but produces an instrument with all the charm of a digital watch. However, with the insatiable need of the weather service supercomputers for data, the future will inevitably bring a huge number of very inexpensive barometers and thermometers stationed throughout the world and connected through the world wide web.

Where to Learn More

Books

Barry, Roger G., and Richard J. Chorley. *Atmosphere, Weather, and Climate*. 6th ed. New York: Routledge, 1998.

Middleton, W. E. Knowles. *The History of the Barometer*. Baltimore: The Johns Hopkins Press, 1964.

Other

Accuweather Web Page. 20 September 2001. <http://www.accuweather.com>.

Jeff Raines

Artificial Turf

Background

Artificial turf is a surfacing material used to imitate grass. It is generally used in areas where grass cannot grow, or in areas where grass maintenance is impossible or undesired. Artificial turf is used mainly in sports stadiums and arenas, but can also be found on playgrounds and in other spaces.

Artificial turf has been manufactured since the early 1960s, and was originally produced by Chemstrand Company (later renamed Monsanto Textiles Company). It is produced using manufacturing processes similar to those used in the carpet industry. Since the 1960s, the product has been improved through new designs and better materials. The newest synthetic turf products have been chemically treated to be resistant to ultraviolet rays, and the materials have been improved to be more wear-resistant, less abrasive, and, for some applications, more similar to natural grass.

History

In the early 1950s, the tufting process was invented. A large number of needles insert filaments of fiber into a fabric backing. Then a flexible adhesive like polyurethane or polyvinyl chloride is used to bind the fibers to the backing. This is the procedure used for the majority of residential and commercial carpets. A tufting machine can produce a length of carpet that is 15 ft (4.6 m) wide and more than 3 ft (1 m) long in one minute.

In the early 1960s, the Ford Foundation, as part of its mission to advance human achievement, asked science and industry to develop synthetic playing surfaces for urban spaces. They hoped to give urban children year-round play areas with better play quality and more uses than the traditional concrete, asphalt, and compacted soil of small urban playgrounds. In 1964, the first installation of the new playing surface called Chemgrass was installed at Moses Brown School in Providence, Rhode Island.

In 1966, artificial turf was first used in professional major-league sports and gained its most famous brand name when the Astrodome was opened in Houston, Texas. By the first game of the 1966 season, artificial turf was installed, and the brand name Chemgrass was changed to AstroTurf. (Although the name AstroTurf is used as a common name for all types of artificial turf, the name is more accurately used only for the products of the AstroTurf Manufacturing Company.)

Artificial turf also found its way into the applications for which it was originally conceived, and artificial turf was installed at many inner-city playgrounds. Some schools and recreation centers took advantage of artificial turf's properties to convert building roofs into "grassy" play areas.

After the success of the Astrodome installation, the artificial turf market expanded with other manufacturers entering the field, most notably the 3M (Minnesota Mining and Manufacturing) Company with its version known as Tartan Turf. The widespread acceptance of artificial turf also led to the boom in closed and domed stadium construction around the world.

In the early 1970s, artificial turf came under scrutiny due to safety and quality concerns. Some installations, often those done by the number of companies that sprang up to cash

Artificial turf has been manufactured since the early 1960s, and was originally produced by Chemstrand Company (later renamed Monsanto Textiles Company).

The Houston Astrodome.

Dubbed "The Eighth Wonder of the World," the Houston Astrodome opened April 9, 1965 for the first major-league baseball game ever played indoors. Americans hailed the massive $48.9-million concrete, steel, and plastic structure as a historic engineering feat. A rigid dome shielded the 150,000-ft² (13,935 m²) playing field of natural grass from the Texas heat, wind, and rain. The Astrodome was the world's first permanently covered stadium.

The roof—642 ft (196 m) in diameter and constructed on the principles of American architect Buckminster Fuller's geodesic dome—contained 4,596 rectangular panes of Lucite, an acrylic material designed to allow the sun to shine through without casting shadows. Still, the Houston Astros baseball team soon complained that the resulting glare made it difficult to catch fly balls. Stadium officials tinted the Lucite gray, but the tint was not good for the grass, which turned a sickly shade of brown. As a result, when the team took to the field for the 1966 season, their spikes dug into another revolutionary baseball first: synthetic grass. Today, AstroTurf—as the material was called—blankets more than 500 sports arenas in 32 countries.

The Astrodome underwent $60 million worth of renovations to increase its seating capacity in 1989. As the years went on, new technology developed making this "Eighth Wonder" outdated. The Astros played their last game at the Astrodome on October 9, 1999 before moving to Enron Field. The same year, the Houston Oilers relocated to Tennessee and were renamed the Tennessee Titans. Despite these losses, the Astrodome still hosts over 300 events a year.

leg driven along the unyielding surface of artificial turf is more likely to be injured. Since artificial turf does not have the same cooling effects as natural turf, surface temperatures can be 30° warmer above the artificial surfaces. Baseball players claimed that a ball would bounce harder and in less predictable ways, and some soccer players claimed that the artificial surface makes the ball roll faster, directly affecting the game. However, the National Football League and the Stanford Research Institute declared in 1974 that artificial turf was not a health hazard to professional football players, and its use continued to spread.

In the 1990s, biological turf began to make a comeback when a marketing of nostalgia in professional sport resulted in the re-emergence of outdoor stadiums. Many universities—responding to the nostalgia, advances in grass biology, and the fears about increased risk of injury on artificial turf—began to reinstall natural turf systems. However, natural turf systems continue to require sunlight and maintenance (mowing, watering, fertilizing, aerating), and the surface may deteriorate in heavy rain. Artificial turf offers a surface that is nearly maintenance-free, does not require sunlight, and has a drainage system. Recent developments in the artificial turf industry are new systems that have simulated blades of grass supported by an infill material so the "grass" does not compact. The resulting product is closer to the look and feel of grass than the older, rug-like systems. Because of these factors, artificial turf will probably continue to be a turf surface option for communities, schools, and professional sports teams.

Raw Materials

The quality of the raw materials is crucial to the performance of turf systems. Almost anything used as a carpet backing has been used for the backing material, from jute to plastic to polyester. High quality artificial turf uses polyester tire cord for the backing.

The fibers that make up the blades of "grass" are made of nylon or polypropylene and can be manufactured in different ways. The nylon blades can be produced in thin sheets that are cut into strips or extruded through molds to produce fibers with a round or oval cross-section. The extruded

in on the trend, began to deteriorate. The turf would wear too quickly, seams would come apart, and the top layer would soon degrade from exposure to sunlight. Athletes and team doctors began to complain about the artificial surfaces, and blamed the turf for friction burns and blisters. Natural turf yields to the force of a blow, but an arm or product results in blades that feel and act more like biological grass.

Cushioning systems are made from rubber compounds or from polyester foam. Rubber tires are sometimes used in the composition of the rubber base, and some of the materials used in backing can come from plastic or rubber recycling programs. The thread used to sew the pads together and also the top fabric panels has to meet the same criteria of strength, color retention, and durability as the rest of the system. Care and experience must also be applied to the selection of the adhesives used to bond all the components together.

The Manufacturing Process

The "grass" part of a turf system is made with the same tufting techniques used in the manufacture of carpets.

1 The first step is to blend the proprietary ingredients together in a hopper. Dyes and chemicals are added to give the turf its traditional green color and to protect it from the ultraviolet rays from the sun.

2 After the batch has been thoroughly blended, it is fed into a large steel mixer. The batch is automatically mixed until it has a thick, taffy-like consistency.

3 The thickened liquid is then fed into an extruder, and exits in a long, thin strand of material.

4 The strands are placed on a carding machine and spun into a loose rope. The loose ropes are pulled, straightened, and woven into yarn. The nylon yarn is then wound onto large spools.

5 The yarn is then heated to set the twisted shaped.

6 Next, the yarn is taken to a tufting machine. The yarn is put on a bar with skewers (a reel) behind the tufting machine. It is then fed through a tube leading to the tufting needle. The needle pierces the primary backing of the turf and pushes the yarn into the loop. A looper, or flat hook, seizes and release the loop of nylon while the needle pulls back up; the backing is shifted forward and the needle once more pierces the

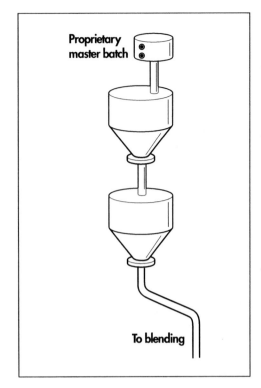

Proprietary
master batch

To blending

How the ingredients of artificial turf are blended.

backing further on. This process is carried out by several hundred needles, and several hundred rows of stitches are carried out per minute. The nylon yarn is now a carpet of artificial turf.

7 The artificial turf carpet is now rolled under a dispenser that spreads a coating of latex onto the underside of the turf. At the same time, a strong secondary backing is also coated with latex. Both of these are then rolled onto a marriage roller, which forms them into a sandwich and seals them together.

8 The artificial turf is then placed under heat lamps to cure the latex.

9 The turf is fed through a machine that clips off any tufts that rise above its uniform surface.

10 Then the turf is rolled into large lengths and packaged. The rolls are then shipped to the wholesaler.

Installation

Artificial turf installation and maintenance is as important as its construction.

1 The base of the installation, which is either concrete or compacted soil, must be leveled by a bulldozer and then smoothed by

A profile of artificial turf.

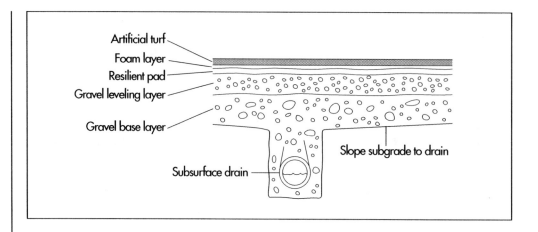

a steam roller. Uneven surfaces will still be evident once the turf is supplied.

2 For outdoor applications, intricate drainage systems must be installed, since the underlying surface can absorb little, if any, rainwater.

3 Turf systems can be either filled or un-filled. A filled system is designed so that once it is installed, a material such as crumbled cork, rubber pellets, or sand (or a mixture) is spread over the turf and raked down in between the fibers. The material helps support the blades of fiber, and also provides a surface with some give, that feels more like the soil under a natural grass surface. Filled systems have some limitations, however. Filling material like cork may break down or the filling material can become contaminated with dirt and become compacted. In either case the blades are no longer supported. Maintenance may require removing and replacing all of the fill.

Quality Control

Because of the high use of artificial turf and the constant scrutiny by professional athletes, new products must undergo a number of tests as they are being developed. In 1994, the American Society for Testing and Materials (ASTM) published a list of standard methods for the testing of synthetic turf systems. It contains over two dozen tests for the properties of turf systems.

As part of ASTM's testing, the backing fabric is tested for strength. The force it takes to separate the individual tufts or blades is also measured. In tufted turf, this test usually measures the strength of the adhesive in-volved. To test how resistant the turf is to abrasion, the ASTM recommends testing the fabric by running it under an abrasive head made of spring steel, while another ASTM test measures how abrasive the turf will be to the players. The ASTM also has tests that measure the shock absorbency of the turf system, and there are also tests to see how well the turf stands up during the course of a game or even prolonged tournament play.

Several quality checks are performed during the manufacturing process, as well. For example, according to AstroTurf Incorporated, the following quality checks are performed: 19 checks for the raw materials, eight checks for extrusion, six checks for unfinished fabric, and 14 checks for finished fabric.

Byproducts/Waste

Defected artificial turf batches are discarded as are nylon yarn that is damaged. Completed turf is generally recycled, but not reused as artificial turf. The earth that is cleared from the installation site is transported to a landfill and discarded. Older turf that has been worn down is typically recycled.

The arguments about the environmental impact of artificial versus biological turf continue. Both create large amount of water run-off, adding to sewage problems. Chemical processes are used in the manufacture of raw materials for artificial turf, but most biological grass in stadium applications requires chemicals in the form of fertilizer and pesticides for maintenance.

The Future

The engineering and design of both artificial and biological turf systems are constantly improving. As new stadiums are built, the owners and architects strive to give a more old-fashioned feel to the structures, which usually means no dome or a dome that allows the use of biological turf.

Recent installations of artificial turf have included new advancements that serve both economic and environmental needs. Large holding tanks are built beneath outdoor installations. The water that runs off the surface is held in the tanks, and used later for watering practice fields or nearby lawns.

Another recent development has been a hybrid of filled turf and biological grass. Once artificial turf is installed, it is filled not with rubber or sand, but with soil. Grass seed is then planted in the soil, nurtured and grown to a height above that of the artificial turf. The resulting combination combines the feel, look, and comfort of biological turf with the resilience and resistance to tearing and divots of artificial turf. Of course, it also requires all the maintenance of both systems, and it is not suitable for most indoor applications.

Where to Learn More

Books

Schmidt. *Natural and Artificial Playing Fields: Characteristics and Safety Features.* Portland: Book News, Inc., 1990.

Other

"Manufacturing Information." *AstroTurf Web Page.* December 2001. <http://www.astroturf.com>.

Wilson, Nicholas. *A Comparison of Filled Artificial Turf with Conventional Alternatives.* Portland: 2000.

Steven Guerriero

Bank Vault

Some nineteenth and early twentieth century vaults were built so well that today they are almost impossible to destroy. Buildings have been renovated around them.

Background

A bank vault is a secure space where money, valuables, records, and documents can be stored. Vaults protect their contents with armored walls and a tightly fashioned door closed with a complex lock. Vault technology developed in a type of arms race with bank robbers. As burglars came up with new ways to break into vaults, vault makers found innovative ways to foil them. Modern vaults may be armed with a wide array of alarms and anti-theft devices. Some nineteenth and early twentieth century vaults were built so well that today they are almost impossible to destroy. Buildings have been renovated around them. A restaurant in a restored bank building even features a dining area inside the indestructible vault. These older vaults were typically made with steel-reinforced concrete. The walls were usually at least 1 ft (0.31 m) thick, and the door itself was typically 3.5 ft (1.1 m) thick. Total weight ran into the hundreds of tons. Today vaults are made with thinner, lighter materials that, while still secure, are easier to dismantle than their earlier counterparts.

History

The need for secure storage stretches far back in time. The earliest known locks were made by the Egyptians. Ancient Romans used a more sophisticated locking system, called warded locks. Warded locks had special notches and grooves that made picking them more difficult. Lock technology advanced independently in ancient India, Russia, and China, where the combination lock is thought to have originated. In the United States, most banks relied on small iron safes fitted with a key lock up until the middle of the nineteenth century. After the Gold Rush of 1849, unsuccessful prospectors turned to robbing banks. The prospectors would often break into the bank using a pickaxe and hammer. The safe was usually small enough that the thief could get it out a window, and take it to a secluded spot to break it open.

Banks demanded more protection and safe makers responded by designing larger, heavier safes. Safes with a key lock were still vulnerable through the key hole, and bank robbers soon learned to blast off the door by pouring explosives in this opening. In 1861, inventor Linus Yale Jr. introduced the modern combination lock. Bankers quickly adopted Yale's lock for their safes, but bank robbers came up with several ways to get past the new invention. It was possible to use force to punch the combination lock through the door. Other experienced burglars learned to drill holes into the lock case and use mirrors to view the slots in the combination wheels inside the mechanism. A more direct approach was to simply kidnap the bank manager and force him to reveal the combination.

After the inventions of the combination lock, James Sargent—an employee of Yale—developed the "theftproof lock." This was a combination lock that worked on a timer. The vault or safe door could only be opened after a set number of hours had passed, thus a kidnapped bank employee could not open the lock in the middle of the night even under force. Time locks became widespread at banks in the 1870s. This reduced the kidnappings, but set bank robbers to work again at prying or blasting open vaults. Thieves developed tools for forcing open a tiny crack between the vault door and frame. As the crack widened, the thieves levered the door open or poured in gunpowder and blasted it off. Vault makers responded with a series of stair-stepped grooves in the door frame so the door could not be levered open. Unfortunately, these grooves proved ideal for a new weapon: liquid nitroglycerin. Professional bank robbers learned to boil dynamite in a kettle of water and skim the nitroglycerin off the top. They could drip this volatile liquid into the door grooves and destroy the door. Vault makers subsequently redesigned their doors so they closed with a thick, smooth, tapered plug. The plug fit so tightly that there was no room for the nitroglycerin.

By the 1920s, most banks avoided using safes and instead turned to gigantic, heavy vaults with walls and doors several feet thick. These were meant to withstand not only robbers but also angry mobs and natural disasters. Despite the new security measures, these vaults were still vulnerable to yet another new invention, the cutting torch. Burning oxygen and acetylene gas at about 6,000°F (3,315°C), the torch could easily cut through steel. It was in use as early as 1907, but became wide spread with World War I. Robbers used cutting torches in over 200 bank robberies in 1924 alone. Manufacturers learned to sandwich a copper alloy into vault doors. If heated, the copper alloy melted and flowed. As soon as the burglar removed the heat, the copper resolidified, sealing the hole. After this design improvement, bank burglaries fell off and were far less common at the end of the 1920s than at the beginning of the decade.

Technology continues in the race with bank robbers, coming up with new devices such as heat sensors, motion detectors, and alarms. Bank robbers have in turn developed even more technological tools to find ways around these systems. Although the number of bank robberies has been cut dramatically, they are still attempted.

Materials used in vaults and vault doors have also changed as well. The earlier vaults had steel doors, but because these could easily be cut by torches, different materials were tried. Massive cast iron doors had more resistance to acetylene torches than

A wanted poster of John Dillinger.

Born June 22, 1903 in Indianapolis, John Dillinger was raised by his sister and stepmother. In 1924 he was arrested for attempted robbery and sentenced to 10–20 years in prison. Confinement trained Dillinger as a criminal and leaving prison in 1933, he carried a map of prospective robbery sites. In three weeks Dillinger robbed 10 banks in five states. Known as "Gentleman Johnnie," he was pleasant and often flirtatious during robberies. The press played Dillinger up as a brilliant, daring, likeable individual, beating the banks that foreclosed on helpless debtors.

Other criminals joined Dillinger, forming the Dillinger Gang. Banks were cased and robberies were timed. The heist was abandoned—no matter what—after a certain amount of time had passed. Getaways were also planned precisely: street lights were timed with back roads and alternate routes noted in the plans. Often, the gangsters didn't race out of town, they casually motored through back roads. FBI director, J. Edgar Hoover increased the reward for Dillinger and issued agents shoot to kill. The Dillinger pursuit was the largest manhunt in the country's history.

In 1934, Dillinger's friend—Anna Sage—agreed to betray Dillinger. On July 22, 1934, Sage, Dillinger, and Dillinger's girlfriend Polly Hamilton attended a movie on Chicago's North Side where Federal agents gunned Dillinger down. Following his death, the FBI press announced Dillinger was shot after resisting arrest and attempting to draw a pistol. Other members of Dillinger's gang were incarcerated or killed in shootouts with police. The reign of John Dillinger and his gang had come to an end. However, Dillinger's exploits earned him a lasting place in American crime history.

steel. The modern preferred vault door material is actually the same concrete as used in the vault wall panels. It is usually clad in steel for cosmetic reasons.

Raw Materials

Vault walls and doors are comprised mainly of concrete, steel rods for reinforcement, and proprietary additives to give the concrete even more strength.

Design

Bank vaults are built as custom orders. The vault is usually the first aspect of a new bank building to be designed and built. The manufacturing process begins with the design if the vault, and the rest of the bank is built around it. The vault manufacturer consults with the customer to determine factors such as the total vault size, desired shape, and location of the door. After the customer signs off on the design, the manufacturer configures the equipment to make the vault panels and door. The customer usually orders the vault to be delivered and installed. That is, the vault manufacturer not only makes the vault parts, but brings the parts to the construction site and puts them together.

Bank vaults are typically made with steel-reinforced concrete. This material was not substantially different from that used in construction work. It relied on its immense thickness for strength. An ordinary vault from the middle of the century might have been 18 in (45.72 cm) thick and was quite heavy and difficult to remove or remodel around. Modern bank vaults are now typically made of modular concrete panels using a special proprietary blend of concrete and additives for extreme strength. The concrete has been engineered for maximum crush resistance. A panel of this material, though only 3 in (7.62 cm) thick, may be up to 10 times as strong as an 18 in-thick (45.72-cm) panel of regular formula cement.

The Manufacturing Process

The panels

1 The first step in the process is to mold the wall panels. Unlike regular concrete used in construction, the concrete for bank vaults is so thick that it cannot be poured. The consistency of concrete is measured by its "slump." Vault concrete is said to have zero slump. It also sets very quickly, drying in only six to 12 hours, instead of the three to

four days needed for most concrete. Workers dump the concrete mix into the panel molds.

2 Next, a network of reinforcing steel rods are manually placed into the damp mix.

3 Then the molds are vibrated for several hours. The vibration settles the material and eliminates air pockets.

4 The edges are smoothed with a trowel, and the concrete is allowed to harden.

5 Workers unmold the product and place the panels on a truck for transport to the customer's construction site.

The door

6 The vault door is also molded of special concrete used to make the panels, but it can be made in several ways. The door mold differs from the panel molds because there is a hole for the lock and the door will be clad in stainless steel. Some manufacturers use the steel cladding as the mold and pour the concrete directly into it. Other manufacturers use a regular mold and screw the steel on after the panel is dry.

The lock

7 The lock for a modern bank vault is usually a dual-control combination lock, meaning it takes two people to open it. This lock is connected to a time lock that can be set so the combination lock will not open until the pre-set number of hours has passed. This is still the "theftproof" lock system that Sargent invented in the late nineteenth century. Such locks are manufactured by only a few companies worldwide. The locking system is supplied already assembled to the vault manufacturer.

Installation

8 The finished vault panels, door, and lock assembly are trucked to the bank construction site. The vault manufacturer's workers then place the panels enclosed in steel at the designated spots and weld them together. The vault manufacturer may also supply an alarm system, which is installed at the same time. While older vaults were armed with multiple weapons against burglars, such as blasts of steam or teargas, this is rarely found in modern vaults. Instead the vault door and interior might be wired with

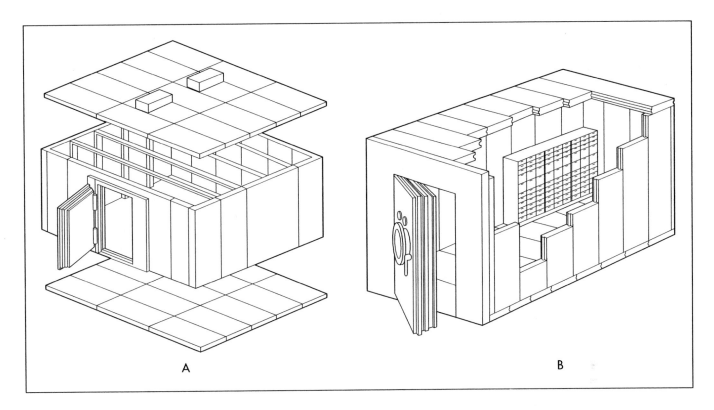

A

B

a listening device that picks up unusual or unusually frequent sounds. The vault may also be surveyed with a camera and an alarm will be hooked up to alert local police if the door or lock is tampered with.

Quality Control

Quality control for the vault industry is overseen by Underwriters Laboratory, Inc. (UL), in Northbrook, Illinois. Until 1991, the United States government also regulated the vault industry. The government set minimum standards for the thickness of vault walls, but advances in concrete technology made thickness an arbitrary measure of strength. Thin panels of new materials were far stronger than the thicker, poured concrete walls. Now the effectiveness of the vault is measured by how well it performs against a mock break-in. Manufacturers strive to make products that repel attacks for a certain number of minutes. A UL Class 1 vault is guaranteed to withstand a break-in attempt for 30 minutes, a Class 2 for 60 minutes, and a Class 3 for 120 minutes. UL's workers attack sample vault walls and doors with equipment that is likely a burglar could carry into a bank and use. This usually includes torches and demolition hammers. If the UL worker can make a hole of at least 6 × 16 in (15.24 × 40.64 cm) in less than the set time, that particular part has failed the test. Manufacturers also do their own testing designing a new product to make sure it is likely to succeed in UL trials.

Byproducts/Waste

The manufacturing process itself has no unusual waste or byproducts, but getting rid of old bank vaults can be a problem. Newer, modular bank vaults can be moved if a bank closes or relocates. They can also be enlarged if the bank's needs change. Older bank vaults are quite difficult to demolish. If an old bank building is to be renovated for another use, in most cases a specialty contractor has to be called in to demolish the vault. A vault's demolition requires massive wrecking equipment and may take months of work at a large expense. At least one company in the United States refurbishes old vault doors that are then resold.

The Future

Bank vault technology changed rapidly in the 1980s and 1990s with the development of improved concrete material. Bank burglaries are also no longer the substantial problem they were in the late nineteenth century up through the 1930s, but vault makers continue to alter their products to counter new break-in methods.

At issue in the twenty-first century is a powerful tool called a "burning bar" or "thermic torch." Burning liquid oxygen ignited by a oxyacetylene torch, this bar burns much hotter than an acetylene torch, getting up to 6,602–8,006°F (3,650–4,430°C). The torch makes a series of small holes that can eventually be linked to form a gap. In the future, the vault manufacturing industry will likely come up with a means to combat the burning bar. Then perhaps criminals will find a more powerful tool, and the industry will change its products again. Vault manufacturers work closely with the banking industry and law enforcement in order to keep up with these advances in burglary.

Where to Learn More

Books

Steele, Sean P. *Heists: Swindles, Stickups, and Robberies that Shocked the World.* New York: Metrobooks, 1995.

Tchudi, Stephen. *Lock & Key.* New York: Charles Scribner's Sons, 1993.

Periodicals

Chiles, James R. "Age-Old Battle to Keep Safes Safe from 'Creepers, Soup Men and Yeggs.'" *Smithsonian* (July 1984): 35–44.

Merrick, Amy. "Immovable Objects, If They're Bank Vaults, Make Nice Restaurants." *Wall Street Journal* (5 February 2001): A1.

Angela Woodward

Bicycle Seat

Background

The bicycle seat, sometimes known as a saddle, is the part of the bicycle on which the rider sits while operating the machine. Generally made from hard plastic and covered with a thin layer of foam and an easily-cleaned cover, the seat is nearly identical on a bicycle whether it was made for a man, woman, or child. Manufacturers refer to this conventional design as a single platform seat (a one-piece seat mounted on a single shaft or post with a sizable horn in front). This conventional design is curved with a bulge in the center of the seat.

This saddle-style bicycle seat is by far the most popular style seen today. Those who order custom bicycles for racing or special sports often receive their machines with seats that may be slightly modified for special sports use. Many owners then replace the seats with special seats with fancy fabrics or custom detailing. The bicycle seat is similar to the motorcycle seat, and some companies often manufacture both.

The bicycle seat is a fairly simple commodity to manufacture as it contains only a few key components such as the platform, bumpers, screws, bolts, rods for support, and fabric to cover the seat. Very few conventional, saddle-type bicycle seats are manufactured in the United States. Rather, they are often produced in sub-assemblies in eastern countries, with the sub-assemblies shipped to the United States. A bicycle manufacturer or distributor then completes the bicycle seat by fully assembling the seat or completing special detailing. Typical sub-assemblies might include the basic seat foam or the rod which supports the seat.

History

The history of the bicycle seat is tied to the development of the bicycle in terms of efficiency of movement and comfort for the rider. When the first bicycle, the *draisine*, appeared in 1818, the seat was simple and unsophisticated, hardly more than a wooden plank. The *penny farthing* bicycle, or the high-wheel bicycle with a small wheel in the back, must have had a terribly uncomfortable seat as the rider put all his weight on the seat and pedaled hard to move the bicycle forward with a primitive gear system. As bicycles became more sophisticated the manufacturers' primary concerns revolved around perfecting the gear system to move the bicycle forward more easily. Safety bicycles, so-called because the wheels were of the same size and the rider did not risk falling over the large wheel of the high wheeler, were extraordinarily popular for both men and women by 1890.

Refinements to the bicycle seat in the early twentieth century were minimal, including the addition of comfortable padding on the convex saddle. More recent refinements to the conventional saddle seat include making the seat cheaper to manufacture by having them made in Taiwan or China, and finding materials to produce a lightweight seat. Some conventional seats have been modified to respond to the concerns regarding reduced blood flow to the genitalia by adding gel to the seat. Only the radical re-designs, which often include two separate lobes for the buttocks and eliminate the horn in front, completely ignore conventional seat design.

Raw Materials

Most conventional-style bicycle saddles are composed of three or four materials. These

The first bicycle, the draisine, appeared in 1818. Its seat was simple and unsophisticated, hardly more than a wooden plank.

include a rigid seat of a molded, nylon-based plastic. The seat is then covered with some sort of padding. Most seat manufacturers prefer to use closed cell foam. Closed-cell foam is a form of latex foam into which a blowing agent has been incorporated in order to force gas to escape during vulcanization (the process by which latex is chemically treated in order to give it elasticity, strength, and stability). The liberation of the gases forms small closed cells, rendering a foam that is nonabsorbent and durable (desirable qualities for a bicycle saddle that may get wet from rain or damp from human perspiration). The plastic base and closed cell foam are then covered with any one of a number of materials, including vinyl, leather, fabric, kelvar, rubber, nylon, or heavy fabric such as canvas. Spray adhesives are used to affix the cover to the foam. These can vary by manufacturer, but effective spray adhesives used with foam include neoprene or urethane based adhesives. Some saddles are constructed with hollow metal tubing or rods extruding so that they may easily attach to the bicycle frame. The metals used in these rods vary and may be either stainless steel or titanium, a lightweight, high-strength/low-corrosion metal.

Design

The saddle seat has changed little in recent years. Simple re-design has included adoption of fancy fabrics that glitter or are decorated with fancy embroidery. Racers or sport enthusiasts have pushed specialty saddle manufacturers to develop rubber fabrics that grip or wick away moisture from the buttocks.

Some small, specialty bicycle seat firms have completely re-designed the saddle in recent years for two reasons. First, seats on bicycles have changed dramatically as the design of bicycles has reflected the physical needs of biking enthusiasts who use the machines for vigorous exercise. For example, downhill bicycle riders prefer a bigger saddle while *roadies* or road bikers prefer sleeker seats. Bicyclists engaged in these sporting pursuits demanded these changes based on practicality. Second, the conventional seat has been re-designed to relieve pressure on the genitalia during bicycling activity.

The conventional single-platform saddle is viewed as unhealthful as well as uncomfort-

able by some cyclists. Many articles appeared in bicycling periodicals criticizing the conventional seats asserting that their design could lead to sterility, impotence, and perhaps even testicular cancer. There is, in fact, some concern that the conventional seat puts undue pressure on the perineum, the area between the rectum and the genitalia, perhaps causing blood flow to be restricted to the area during intensive biking exercise. This occurs because the artery near the perineum thickens as a natural defense mechanism, constricting the inside of the wall and reducing blood flow to the area. The reduction lasts far longer than the bicycle trip; it is asserted by some that impotency may be the result of the reduced blood flow.

As a result, there is a booming market for the non-conventional bicycle seat. These specialty seats are designed to take pressure off the perineum and distribute it broadly on the butt cheeks. This primarily entails eliminating the long horn with its curved spine on the front of the seat that makes contact with the perineum. Most of these unconventional seat manufacturers run small operations and have not yet penetrated the market significantly.

The Manufacturing Process

Few, if any, conventional bicycle seats have all components manufactured in the United States. One prominent company that provides many such bicycle seats to consumers says that American companies tend to get parts from other countries where labor is cheap, and the American companies simply assemble the components. Whether these components are made in this country or not, conventional seats are composed of a hard shell, foam, a seat cover, and a metal rod that attaches the seat to a bicycle frame.

1 The hard plastic shell forms the foundation of the bicycle seat. The contour of the saddle is rendered in a metal mold; this saddle may be longer or broader according to the needs of the seat distributor. Whatever their configuration, these seats are injection molded. Injection molding occurs when a plastic resin is made molten and then rammed and forced through a gate into a cooled mold. The resin solidifies in the mold, the mold is unclamped, and the plastic shell is forced out of the mold using some sort of ejector such as a

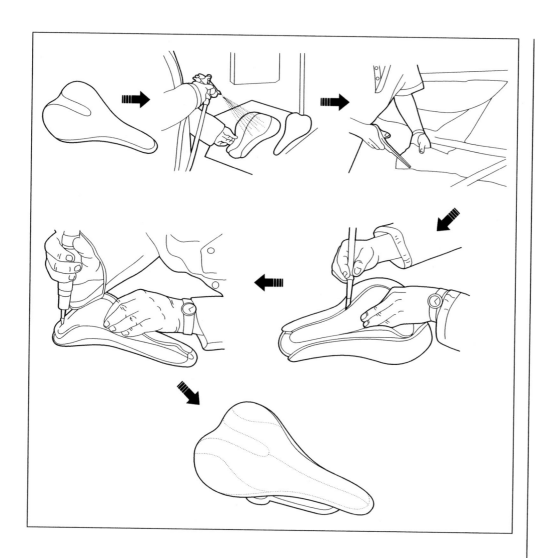

pin. Several injections can be produced per minute. Runners that attach parts of the plastic injection molded pieces (simply pieces of waste) may be knocked off, gathered, and melted down for later use.

2 Next, padding is glued to the plastic shell. The padding is a densely-packed, closed-cell foam that provides some comfort for the rider. The foam is cut using heavy blades along the contours of the shell and moves down and around the edges of the shell. The foam is attached to the plastic shell using a spray adhesive applied using an air compressor and spray gun. These foam seats are applied to the shell by hand, one seat at a time. The adhesive application operator makes sure that the spray adhesive is applied evenly to the sides and parts underneath the seat to ensure that the cover will properly adhere.

3 The cover, sometimes referred to as the topsheet, is cut out by hand using heavy

hand-held scissors. Specialized saddles of leather or other materials such as kevlar, gripping rubber, or metalicized fabrics are also cut out by hand. Those that will be decorated with stitching are sewn on industrial grade machines.

4 The topsheet is now affixed to the foam-covered base. This process entails wrapping the cover down over the seat, around the sides, and sticking to the bottom of the seat. This topsheet is carefully attached by hand using the spray adhesive once again. Wrapped edges are rolled tight to provide a good fit and prevent separation from the foundation. After the topsheet is attached to the base using the glue, it is also stapled to the base to ensure a permanent, smooth fit.

5 Plastic bumpers are then screwed onto the nose (front), the back, and underside of the seat. These bumpers cover the gluing and stapling of the topsheet to the base, giv-

ing the seat a finished look. The bumpers are attached using a hand-held automatic screw gun.

6 Many seats come with a hollow metal rod so that the seat may be dropped into a bicycle frame using the attached rod. In the event that the saddle includes such a rod, it must be cut, configured, and attached to the seat itself. Some companies use titanium rods that are shipped to their plant in 10 ft (3 m) lengths. The rods are heated, cut into smaller sections using a heavy machine saw, and the sections are bent into the desired configuration using molds. These configured rods are then put into a tumbler that literally tumbles the rod using polished rocks in an enclosed cylinder. (This process is used to round off edges of pebbles and give them a smooth shine as well.) This process shines up the rails. Fifty bent rods may be placed in a tumbler at one time.

7 The rods must then be forced into an opening on the plastic shell. In order to do this, the shell (complete with foam and cover) and the rods are put into a machine that applies pressure with the assistance of the operator, forcing the rod into the seat using this pressure. The rails are thus popped into the seat foundation. The seat is now complete and ready for packing and shipping.

Quality Control

Quality control primarily revolves around the successful injection molding of the seat base. The machine for this purpose is carefully monitored to ensure a successful operation and that the mold reflects the desired seat configuration. Products from the operation are visually inspected at regular intervals. However, specifications for its molding must be monitored and assessed to ensure there is no undue variation. There is no question that during the entire manual process of making a saddle the expert ensures a superior fit. Those with topsheets cut incorrectly will not wrap under the seat and pose a problem with fraying. An improperly fit or glued topsheet will rip or tear.

The materials that form the important components of the seat must be of high quality to ensure a good product. The closed-cell foam must be up to specifications desired by the manufacturer; inferior foam will disintegrate with pressure and moisture. Also of great concern is that the material for the metal rail to which the seat is attached will not wear out with average use. One manufacturer used an inferior grade stainless steel rod for this purpose and metal failure proved very costly.

Byproducts/Waste

For those that make plastic injection-molded seats, runners left over from the molding process are gathered and melted for use in future injection moldings. In small operations that make specialty saddle seats the waste of materials is kept to a minimum. Excess cut foam is not reusable and discarded, as is excess fabric from the specialty seats. These specialty cover fabrics may be quite costly, and as they are cut out by hand, the operators are careful to cut out the seats to maximize use of the fabrics. Spray adhesives are carefully used and controlled using a pneumatic machine for their dispensing, greatly eliminating adhesive overspray.

The Future

Bicycle seat manufacturers are curious to see what kind of impact the radically-designed unconventional bicycle seats have on the conventional seat market. Some have addressed the medical concerns voiced about the single-platform seat through small modifications because complete re-tooling is very costly. Some bicyclists believe the medical concerns are unfounded and dispute them. Specialty seats are not put on mass-produced bicycles made for American consumption. That leaves the responsibility for purchasing the conventional seat to the consumer. Eliminating the old seat and replacing it anew with the specialty seat can be costly. In the 1990s many small companies were founded to design, prototype, and manufacture new healthful bicycle seats. These firms are vocal about the perceived problems with conventional bicycle seats and hope they can offer a comfortable and healthful alternative to the seat issue. However, many bicyclists avoid the specialty seat, preferring a well-padded conventional, extremely lightweight saddle to radical new designs.

Where to Learn More

Books

Bijker, Wiebe E. *Of Bicycles, Bakelites, and Bulbs.* Cambridge: MIT Press, 1995.

Bryk, Nancy E. Villa. *American Dress Pattern Catalogs.* New York: Dover Publications, 1989.

Other

"Bicycle." *Encyclopedia Britannica CD Edition.* Encyclopedia Britannica, Inc.: 1994–1998.

Interview with Jeff Dixon, owner of Spongy Wonder Manufacturing Corporation. New Brunswick, Canada. August 2001.

Interview with SDG U.S.A. Management. Santa Ana, CA. September 2001.

"The Manufacturing Process." *SDG U.S.A. Web Page.* December 2001. <http://www.sdgusa.com/process.html>.

Nancy EV Bryk

Binocular

Man has been experimenting with glass since its advent sometime around 3500 B.C.

Background

Modern binoculars consist of two barrel chambers with an objective lens, eyepiece, and a pair of prisms inside. The prisms reflect and lengthen the light, while the objective lenses enhance and magnify images due to stereoscopic vision.

History

Man has been experimenting with glass since its advent sometime around 3500 B.C. These experiments soon became known for their ocular implications. The designs of early optical instruments, like the telescope, were not recorded. It is assumed that these instruments were studied and perfected by Galileo Galilei. Early binoculars were actually called binocular telescopes, and are thought to be based on Galileo's discoveries and designs of prisms.

Early telescopic lenses were full of bubbles and other imperfections. They were also slightly green due to the iron content in the glass. Polishing techniques were crude, and although lenses were of good quality in the center, the peripheral shape was poor resulting in a restricted aperture. As telescopes were improved, binoculars evolved. The first patent application for binocular telescopes was filed early in the seventeenth century by Jan Lippershey in present day Holland. Lippershey primarily used quartz crystal, which is hard to manipulate. The first hand-held binocular originated in 1702 with Johann Zahn's small binocular of two tubes with a lithe connection.

A patent application submitted in 1854 by Ignatio Porro began the use of the modern prism binocular called the Porro prism erecting system. This optical system consisted of an objective lens and ocular lens (eyepiece) with two facing, right angle prisms arranged to invert and correct the orientation of the image. The two most commonly used prism systems are the porro prism and the roof prism design. The roof system uses prisms positioned one over the other resulting in a more compact design.

An other major breakthrough occurred in 1894 when Carl Zeiss, a German optical specialist, developed binoculars with convex lenses and delta prisms to correct the inverted image. In a porro design, the light is bent in a "Z" shape before reaching the eye, allowing the distance between the eyepiece and the objective lens to be compacted. This enables the size and weight of binoculars to be reduced.

Reductions in the weight of the binoculars occurred with the use of aluminum or polycarbonate housings instead of the heavier metal alloys used in pre-civil war binoculars. Performance of smaller and larger binoculars has improved with the introduction of coatings to render the lenses non-reflective and reduce the amount of scattered light. The quality of prisms has also improved over the years, resulting in a reduction of the bubbling effect of optical glass. In the early 1970s, nitrogen filled, waterproof binoculars were developed. A decade later the arrival of infrared transmitters capable of seeing in the dark further transformed binocular technology. Variable magnification models were also developed allowing the user to adjust the level of magnification.

Raw Materials

Early binocular models had brass housing covers and were relatively heavy and expen-

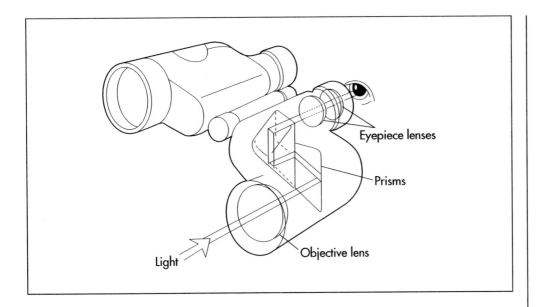

Eyepiece lenses

Prisms

Objective lens

Light

sive to produce. Subsequent leather or hard rubber covers were replaced in Germany during the World War I by a cover of black lacquered cardboard. Galvanized steel replaced the heavier brass in the housing covers. In the 1930s, nearly all of the metal parts of the service glasses were made of aluminum to save brass and reduce the weight.

Modern-day binocular tubes are primarily made out of aluminum coated with silicon or a leather-like material called gutta-percha. The lenses and prisms are made from glass and coated with an anti-reflective coating.

Design

With the exception of the optical glass and some rubber seals, the majority of binocular component parts can be manufactured using a Computer Assisted Design and Manufacturing (CAD/CAM) system that downloads the designs to a variety of Computer Numerically Controlled (CNC) devices (multi-axis mill turn and milling machines as well as vertical and horizontal machining centers, lathes, etc.). Using CAD software provides both drawing, dimensioning, and visualization capabilities. These lead to improvements in the binoculars final design.

The Manufacturing Process

1 The lens material is poured into a lens mold, which has a spherical curved bottom. This results in a lens that is about 4 in (10.2 cm) in diameter and 1–1.5 in (2.4–3.8 cm) thick.

2 The lenses are then removed from the molds and cut into specific pieces using a diamond saw to create the optical lenses.

3 The lenses are placed into the grinding machine and polished.

4 After they have been carefully machined, the lenses are anodized to reduce reflections in vacuum tanks. The more coatings applied, the less light absorbed.

5 The ocular lenses (nearest the eyes) are also molded and carefully polished by auto-polish machines after which they are centered on diamond turning machines and finally cleaned by running through several different solvents in automated machinery.

6 The objective lenses, those furthest from the eyes, are molded and then polished with polishing machines.

7 These components are then manually assembled into a die cast body, which is often made from aluminum.

8 Using a technique called physical vapor deposition, the optics are placed into a "plasma machine" and coated with dielectric coatings. The coatings are essential for high performance.

9 The optics are then inspected and tested for clarity and defects using lasers in specially designed particulate free rooms.

10 Next, the rod shaped prisms are cut by lasers into three-sided shapes depending on the type of prism being manufactured (i.e., roof prisms or porro prisms).

11 The prisms are coated with dielectric materials (metal oxides) by physical vapor deposition inside a vacuum chamber.

12 When all these components are assembled on a belt assembly line, the final assembly station collimates the binocular by hand, making the left side exactly parallel to the right, so only one image will be seen at a time.

13 The binocular housing is then covered with a substance called gutta-percha, which looks like leather but is more durable and flexible. This covering is applied by hand using an adhesive and may be coated with a protective rubber covering.

14 On the assembly line bare metal housing covers are covered with plastic or rubber.

15 The prisms are placed by hand inside the binocular casing and manually screwed in place.

16 The objective lenses are held in place by a metal or plastic ring and the eyepiece is fitted with a rubber eyecap.

17 The focusing lenses are placed in the housing with screws mounted by hand.

18 Waterproof binoculars must have o-rings at every orifice, be purged with nitrogen (injected through a seal), and sealed. The final step would be the packing of binoculars in cases with neck straps, most cases today being of a canvas-like material.

Quality Control

Binoculars that have been hermetically sealed (waterproof) and nitrogen charged (fogproof) are tested underwater. Most binoculars will withstand water immersion at 16.4 (5 m) for five minutes. Both barrels of a binocular need to be optically parallel for the image to merge into one perfect circle and are carefully checked for alignment.

Byproducts/Waste

Lenses and prisms that have defects such as scratches or cracks are either discarded and melted down to be molded again, or they are recycled. If the casing is damaged during production, it is also either remolded or recycled.

The Future

Binoculars continue to advance with new technology. Their ability to see further with better focusing techniques enables the consumer to use the product for a wider variety of tasks. Binoculars are now tending to use the same stabilizing method used in video cameras that automatically stabilizes the prism system so that the image remains steady to the viewer. Some binoculars are also coming equipped with night scope vision. This would enable the consumer to see objects that are far away even at night. Technological advancements are continually made on these specialty binoculars, which are primarily used by the military or for surveillance.

Where to Learn More

Books

Bell, Louis. *The Telescope.* McGraw-Hill Book Company, Inc., 1922.

Von Rohr, Moritz. *Die Binokularen Instrumente.* Berlin: Springer, 1920.

Other

The United States Patent Office Web Page. November 2001. <http://www.uspto.gov/patft>.

Van Helden, Albert. *The Telescope.* 1995. November 2001. <http://es.rice.edu/ES/humsoc/Galileo/Things/telescope.html>.

Bonny McClain

Bird Cage

Background

Bird cages are homes for domesticated birds. Birds require a house in which they can fly and have some freedom but still ensures they do not fly away. Bird cages are constructed to be large enough to accommodate the motion and daily activities of domesticated birds. Cages are generally constructed of wire mesh. Some manufacturers flatten the mesh and others leave the wire round just as it is obtained from the manufacturer. Cages must be constructed with mesh carefully welded in a grid that will not permit a bird to put his or her head through the mesh and strangle. The mesh is generally 1.5×1 in (3.8×2.5 cm) in grid. Even larger birds such as parrots are rarely put into cages with mesh larger than 1×1 in (2.5×2.5 cm).

The design of bird cages is varied. Some cages hold one or two small domesticated birds and are rectangular or square. Polygonal cages are popular and can be quite decorative. Some cages have a plastic or metal tray that fits underneath a mesh cage without a bottom so that cleaning the cage only entails detaching the tray. Others have seed catching trays that are far wider than the cage so that the tray catches all stray seeds dropped by the bird. Still other cages are made specifically to breed birds and are of a very different configuration. These bird-breeding cages are quite wide with a divider in the middle that is removed when the birds in each half of the cage have gotten used to the presence of the other. Then breeding begins. Bird-breeding cages are often made to the specifications of breeders and are designed after observation and feedback from the breeders.

History

The history of the bird cage is tied to the adoption of birds as pets. Birds were caged for their beauty and mystery nearly four centuries ago in ancient Egypt. Doves and parrots were favorites of the Egyptians and are depicted in hieroglyphics. The Mynah has been considered a sacred bird in India for at least 2,000 years as well. The birds were pulled through the streets on oxen, likely in crude cages to ensure they would not escape. It is difficult to determine what some of these cages may have been made of, perhaps wooden twigs, rope mesh, reeds, or bamboo. Some say that Alexander the Great was given a parakeet by one of his generals and the Alexandrine parakeet was named in his honor. Ancient Romans kept and held birds as well and it was considered the duty of a slave to care for the domesticated animal. By the Middle Ages only the wealthy kept caged birds.

When Western traders brought back spices and textiles from the Far East, they also brought the exotic birds as pets. Birds were much beloved pets in the American Colonies. Bamboo and wooden cages were seen in many kitchens in the New World, generally hung near open windows. By the Victorian era the bird was considered even more than a pet. The decorative cage was seen as an important ornamentation within the Victorian parlor. Wire cages were far more effective and long-lasting cages than more ephemeral cane, bamboo, or wood, and wire mesh cages are much preferred in the twentieth century.

Raw Materials

The essential components of the bird cage are very simple. Most American-made cages

Birds were caged for their beauty and mystery nearly four centuries ago in ancient Egypt.

are made from wire mesh. These rolls generally come in varying widths and are keyed to the widths of the panels to be cut from them. These rolls may be between 100–200 ft (30–61 m) in length and are very heavy.

Bird cage makers do not manufacture the mesh; instead, they purchase it from wire mesh or fencing manufacturers. A quality bird cage must be made from precisely configured mesh or the cage will not be stable or safe for the bird. As the panels and sides are cut directly from the mesh, it must be evenly spaced or the cut panels will not meet precisely at the edges. American manufacturers seek suppliers who can deliver tolerance welded wire meaning that the mesh is produced within minute tolerances, generally assumed to vary less than 0.13 in (3.2 mm) within a 10 ft (305 cm) length. The width of the mesh varies greatly depending on which species of bird the cage is designed to accommodate. Larger birds can have a slightly larger mesh, but manufacturers are careful to keep the mesh fairly tight.

Additionally, most American bird cages are made from galvanized steel as this is a material that sturdy and inexpensive, thus keeping the cost of the cage reasonable. Some bird breeders prefer a stainless steel cage. However, stainless steel mesh is nearly five times the cost of galvanized steel mesh and stainless steel cages can cost in excess of two hundred dollars.

The Manufacturing Process

1 The components of the standard American-made rectangular birdcage includes four sides of a rectangular cage, a door, a top, and a plastic or metal tray that serves as a bottom for the cage. Huge rolls of wire mesh of varying widths enter the factory as different widths are used for different parts of the cage. The large rolls, up to 200 ft (61 m) in length, are constructed of true metal wire with rounded edges. Some companies prefer to offer bird cages that do not have round wire but have flattened mesh. In this cage, the mesh rolls are automatically fed into a machine called a cold roller press that flattens the wire mesh without using heat, only pressure. Various widths of rolls are flattened in this fashion; smaller widths used for doors or specialized parts of the cage are thus all flattened using the press.

2 The flattened wire mesh roll is then automatically fed into another machine in which computer-controlled cutting edges come down on the roll at periodic intervals and cut out the side panels of the four-sided cage. Another feed is used to cut out the top (from a different width of wire) as well as the door or any other special piece put into the designed cage. (All of the blades are specially programmed to cut out the correct dimensions of the special panels.)

3 Next, the panels which have been cut out of the wire mesh must be physically separated from the wire mesh. The mesh, with panels still intact, is fed automatically into another machine, and the machine uses a hydraulic punch to remove the panel from the mesh. This happens with all the components of the cage cut out of mesh.

4 The door panels of the cage must be reinforced with another brace or wire around them. Thus, the door panels are automatically soldered with another band of wire along the outside of the door. This strong wire addition leaves no rough edges of cut gauge, strengthens the door in order to provide a secure door for the bird, and provides a stable, consistent edge by which the door can be hinged to the cage body.

5 The components are now all cut. The door must be connected to the front panel. The door, a bit larger than the opening in the front panel, is attached to the panels with ferrules. An air gun is put up to the door and the side panel, and an open clip encircles the door panel and the front panel. At least two ferrules are attached to the door and the front panel and thus serve as hinges.

6 Most bird owners prefer that their pets do not sit in cages with a mesh bottom as it is uncomfortable for the bird to constantly perch on wire. Furthermore, a fully enclosed cage is very difficult to clean. Instead, most bird owners prefer cages with open bottoms into which they fit a metal or plastic tray that can be pulled out for cleaning. A plastic tray is made from plastic pellets fed into a hopper and melted. The liquid is then injected into a mold and cooled. The finished tray is then ejected. A steel tray is made from

thin metal sheets placed in a die. A hydraulic punch is released and forces the metal into the shape of the die.

7 Many American bird cages are shipped flattened as shipping costs are too great to ship them fully constructed. Instead, simple metal clips are enclosed in plastic packaging in order for the retailer or the consumer to assemble and secure the cage at point-of-sale or in the home. Thus, the bird cage is now complete and ready for packaging into cartons.

Quality Control

Quality control is very important in the manufacture of bird cages. Every operator on the manufacturing line looks carefully for sharp edges or stubs in the panels that may hurt the bird. If burrs are found in any of the panels these must be ground down using a machine sander. Additionally, the cage must be secure so that the bird does not slip out of any gaps in the cage should the panels not line up evenly. Poor registration of the sides and/or top is primarily the result of poorly-configured mesh wire. It is imperative that the raw material used in the construction of the bird cage be very carefully inspected upon receipt. Most manufacturers only use materials that are certified meaning they are guaranteed to be made to specifications.

Some manufacturers are very careful not to purchase mesh that is full of drips resulting from the galvanizing process. These drips are heavy in zinc, used in the plating process. Some bird breeders and veterinarians believe that excessive zinc can lead to *zinc toxosis* in which the bird is poisoned by an overabundance of zinc ingested through nipping at the zinc galvanizing from the mesh cage. Excessive globs are generally removed from the mesh for this reason if they are found upon inspection. A batch of wire may be rejected if too many such drips are found upon inspection.

Byproducts/Waste

There is a fair amount of waste that results from punching out the mesh panels of the sides, top, and particularly the doors. This metal waste is gathered up after the panels are punched out, dropped into a dumpster and recycled into new wire.

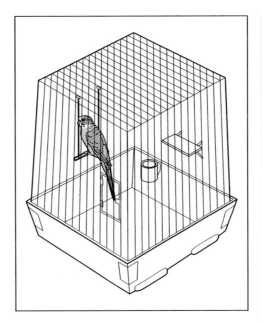

A standard bird cage.

The Future

The future for American bird cage manufacturers is a bit cloudy at present. Many bird cages offered in pet stores and large discount houses are made out of the country where labor costs are about one-tenth of labor costs in the United States. The result is that the foreign product is often far cheaper than the American product. Many manufacturers and consumers agree that the imported products are far less sturdy and may not pay attention to issues such as excessive galvanization which may be harmful to birds. American manufacturers have dwindled in the last several years and many are wary with their ability to compete with cheap imports.

Where to Learn More

Books

Garrett, Elisabeth Donaghy. *At Home: The American Family Home 1750–1870.* New York: Harry N. Abrams, 1998.

Other

"Bird." *Encyclopedia Britannica CD Edition.* Encyclopedia Britannica, Inc., 1994–1998.

Dale, Steve. "The History of Birds as Pets." *The Pet Project Web Page.* December 2001. <http://ecol.webpoint.com/pets/birdhist. htm>.

Oral interview with Willis Kurtz, President of Safeguard. New Holland, PA. October 2001.

Oral interview with Guy Cone, President of Quality Cage Company. Portland, OR. October 2001.

Nancy EV Bryk

Birdseed

Background

Birdseed is a mixture of seeds, nuts, fruits, and vegetables provided to birds for sustenance. It is produced in a two-stage process that involves preparing the component ingredients then combining them in a mixing kettle.

Statistics show that the United States is a nation of bird lovers. In fact, feeding and watching birds has become one of America's favorite pastimes. According to the United States Fish and Wildlife Service (USFWS) one third of the United States population feeds wild birds. In a survey done by the American Pet Products Manufacturers Association (APPMA) it is reported that 6.9 million households in the United States have birds as pets. This fascination with birds has led to a birdseed industry that dispenses over 500,000 tons of birdseed per year.

History

Humans have a long history of interaction with the avian world. As far back as the Egyptian pharaohs and the ancient Romans, people captured and kept birds for both aesthetic and practical reasons. Certainly, modern society continues to depend on birds for food, entertainment, and companionship. In light of this, then, it is not surprising that the United States Congress enacted the Wild Bird Conservation Act of 1992 in an effort to preserve exotic and endangered birds around the world.

Even with such a rich history of interaction between humans and birds, birdseed manufacturing did not get its start until the middle part of the nineteenth century. Many of today's American birdseed manufacturers have similar roots; they were small town agricultural grain companies with retail stores. In the 1940s, Simon Wagner of Wagner Brothers Feed Company and Bill Engler Sr. of Knauf & Tesch (the company is now known as Kaytee, one of the largest pet food manufacturers in the country) collaborated on creating a market for wild birdseed. Up to that point such a market had never existed. Creating birdseed was a natural extension for these feed companies considering they were already making products that contained the same ingredients. Wagner and Engler were able to establish a market for their birdseed relatively quickly. Wagner attributes the early success to the growth of suburbia after World War II, which led to new homeowners' interest in their yards and the animals that visited them.

Raw Materials

The most commonly used birdseed ingredients are sunflowers, corn, millet, fruits (such as raisins and cherries), and peanuts. Many of these crops come from Nebraska, Kansas, North Dakota, and South Dakota.

One of the primary ingredients in birdseed is sunflower seeds. These are four sided, flat seeds that are about 0.25 in (0.64 cm) and 0.13 in (0.32 cm) wide. The seed coat is black with gray stripes. The outer coat protects the inner kernel which is composed of about 20% protein and 30% lipids. It also contains a significant level of iron and fiber. Other seeds that may be used include cottonseed, pumpkin, safflower, hemp, or palm kernel seeds.

Cereal grains are another type of ingredient used in birdseed compositions. Of these corn is one of the most important. About

800 kernels are produced for each ear of corn harvested. An average center kernel measures about 0.15×0.31 in (4×8 mm) thick and 0.5 in (12 mm) long. Corn kernels are made up of about 60% starch and 4% oil. Millet is another grain used. It is smaller than corn with a length of about 0.15 in (4 mm) and a width of 0.11 in (3 mm). It contains about 11% protein, 3% fat, and 8% fiber. Its small size makes it an ideal birdseed ingredient, especially for smaller birds.

Peanuts are a groundnut grown on an upright plant. Its flowers are fertilized above ground and then are pushed into the ground to develop the seeding pods. A pod will contain anywhere from one to three nuts. Peanuts are good components for birdseed because they contain over 25% protein. Additionally, they have about 50% oil which increases their taste appeal for birds. Manufacturers must be particularly careful of any birdseed mixes that contain peanuts because peanuts can harbor the pergillus mold. This mold can do serious liver damage to birds.

Fruits are another component material added to birdseed mixtures. They have a high sugar content that makes them appealing to certain bird species. The most commonly used fruits are cherries and raisins. Cherries are pitted before they are used and then dehydrated. Raisins are produced from grapes using a drying process. They can be either sun-dried or physically dried by forced-air dryers.

In general, pet bird food is a more complex mixture including exotic nuts and fruits. This is because pet birds get all their nutrition from the bird feed that is given to them by their owners; unlike wild birds that have access to and utilize other food sources. Another interesting additive to some wild birdseed is ground hot peppers. It turns out that birds do not mind the hot pepper taste but squirrels have a distinct aversion to it. This prevents squirrels from eating birdseed laid out for wild birds. Other ingredients such as algae extract can be added to improve the tone and color of a bird's feathers.

The Manufacturing Process

Making birdseed is a relatively simple manufacturing process. The first phase of production involves the procurement of seeds, grains, and fruit that make up the various mixes from processors. The second phase involves blending and packaging these materials then shipping them to consumers.

Seed procurement

1 The process of producing birdseed begins with the procurement of raw materials. This is an important phase of production because pure and fresh ingredients are crucial to the quality of the end product. Birdseed manufacturers purchase their raw materials from processors who obtain their grain from the actual growers or grain brokers.

2 The processors clean the component seeds to get them to 98% purity. Once the raw materials are in the plant, they must be cleaned further. The cleaning process consists of sorting unwanted debris and waste products from the birdseed ingredients. Typically the raw materials are put through a three-step air cleaning system that sorts the quality foodstuff from the waste (also known as Chilton). The air sorter separates the lighter debris such as sunflower hulls and stems from the raw materials used in the birdseed mixes. Many manufacturers use the Chilton to make other animal feed such as pellets for companion birds. Additional waste products may be sold to local farmers for use in their animal feed. Dirt and rocks can also be removed with a similar process. Processors may also provide drying services and some seed treatments to prevent fungal growth.

3 Once the seed is cleaned and treated, it is filled into bags, drums, or even trucks and delivered to the birdseed manufacturer within one week. Larger manufacturers are often geographically located near processing companies that clean and store the crops used to make birdseed.

4 Next, the cleaned seeds are blended. Recipes vary depending on the mix, but all ingredients are automatically measured and blended in large, stainless steel containers. In an effort to keep the mixes blended and consistent from package to package, the ingredients are continually mixed until they are deposited into their appropriate packaging. Occasionally, scents, oils (such as anise or orange oil), or colors are added to the blends to enhance consumer appeal.

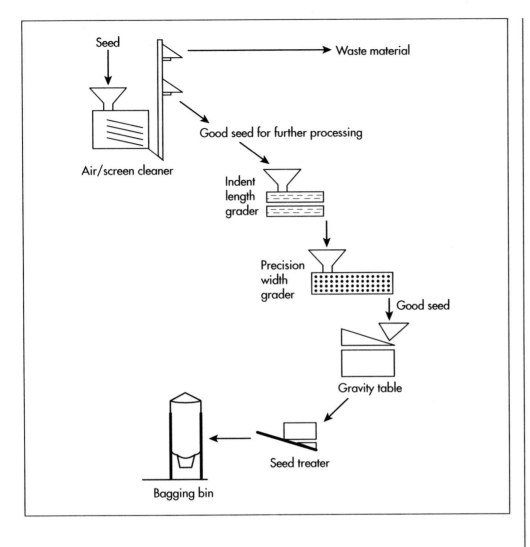

Seed

Waste material

Air/screen cleaner

Good seed for further processing

Indent length grader

Precision width grader

Good seed

Gravity table

Seed treater

Bagging bin

Packaging

5 Proper packaging is crucial to the quality of the birdseed because rodents, mites, or other pests can contaminate seeds which in turn can be spread to birds. Packaging equipment weighs and fills pre-made packages composed of a variety of materials from paper to polyester films. The packaging provides a physical barrier between the feed and potential pests. As a general rule, companion birdseed typically gets more barrier packaging than wild birdseed.

6 To minimize infestations by any pests that may have inadvertently gotten inside the bags after packaging, each bag is flushed with nitrogen. The moisture and oil content of the grain also has an impact on the potential for infestations. Seeds with high oil and moisture content, like sunflowers, require more treatment to prevent infestations or rancidity. Another tactic for cutting down on infestation is to use airtight bins to house the birdseed ingredients prior to it being processed.

Quality Control

To ensure that a high quality product is made, birdseed manufacturers visually inspect the raw materials and finished products during each phase of production. Quality control begins with the raw material growers. For example, the plants are frequently examined while they are growing to make sure they are free from disease and growing properly. If diseased plants are found, they may be isolated and removed before harvest. When the birdseed manufacturers receive component raw materials inspections begin immediately. These ingredients are subjected to a variety of laboratory tests to ensure that specifications related to seed size, nutritional value, and microorganism contamination are met. After batches of birdseed are mixed, the finished product is

checked to see that the correct proportions of ingredients were put in each batch.

The Future

Perhaps the biggest issue facing birdseed manufacturers in regards to pet birds is the trend in avian nutrition to move away from birdseed as the main source of nutrition toward pelleted food. Pelleted food is touted as being "complete nutrition" and, according to manufacturers, no additional supplementation is necessary. Manufacturers recommend using birdseed as a treat or behavior modification tool instead of the bird's mainstay. The advantage of pelleted food over birdseed is that it provides a bird with all the necessary vitamins and nutrients required for optimum health. Of course, the drawback to pelleted food is that some birds refuse to eat it. Birdseed is inherently deficient in some important nutrients like protein and some minerals and high in fat (up to 50% fat in some seed). In the future, birdseed manufacturers will try to formulate more mixtures with a more complete nutritional profile.

Another trend in the industry is to make the industry as a whole more proactive and homogeneous. Overall, the bird feed industry is relatively fragmented with cottage-based industries to massive corporations all producing birdseed. As a result, different states have different regulations and expectations for quality control. Industry trade associations have been attempting to deal with the quality control issue on many levels including an attempt to standardize regulations and expectations from state-to-state. Addi-

tionally, efforts are underway to educate the entire industry on issues affecting it, such as noxious weed control in crops.

Where to Learn More

Books

Alderton, David. *The Cage Bird Question and Answer Manual.* Barron's Publishing, 2000.

Armstrong, Holly, et al. *Gourmet Bird Food Recipes.* San Leandro, CA: Bristol Publishing, 2001.

Gallerstein, Gary A. *The Complete Bird Owner's Handbook.* Macmillan Publishing, 1994.

Periodicals

Allen, Carolyn. "We Have Roots: Early History of the Wild Bird Seed Marketplace." *Birding Business* (Summer 2001).

Rouhi, A. M. "Chili Pepper Studies Paying Off With Hot Birdseed and Better Analgesics. *Chemical and Engineering News* (4 March 1996).

Other

American Pet Products Manufacturers Association Web Page. December 2001. <http://www.appma.org>.

Kaytee Products Web Page. December 2001. <http://www.kaytee.com>.

Sandy De Lisle & Perry Romanowski

Brandy

Background

The name brandy comes from the Dutch word *brandewijn*, meaning "burnt wine." The name is apt as most brandies are made by applying heat, originally from open flames, to wine. The heat drives out and concentrates the alcohol naturally present in the wine. Because alcohol has a lower boiling point (172°F, 78°C) than water (212°F, 100°C), it can be boiled off while the water portion of the wine remains in the still. Heating a liquid to separate components with different boiling points is called heat distillation. While brandies are usually made from wine or other fermented fruit juices, it can be distilled from any liquid that contains sugar. All that is required is that the liquid be allowed to ferment and that the resulting mildly-alcoholic product not be heated past the boiling point of water. The low-boiling point liquids distilled from wine include almost all of the alcohol, a small amount of water, and many of the wine's organic chemicals. It is these chemicals that give brandy its taste and aroma.

Almost every people have their own national brandy, many of which are not made from wine: grappa in Italy is made from grape skins, slivivitz in Poland is made from plums, shochu in Japan is made from rice, and bourbon in the United States is made from corn. Beer brandy is better known as Scotch whiskey. It is universally acknowledged that the finest brandies are the French cognacs that are distilled from wine.

Brandies are easy to manufacture. A fermented liquid is boiled at a temperature between the boiling point of ethyl alcohol and the boiling point of water. The resulting vapors are collected and cooled. The cooled vapors contain most of the alcohol from the original liquid along with some of its water. To drive out more of the water, always saving the alcohol, the distillation process can be repeated several times depending on the alcohol content desired. This process is used to produce both fine and mass-produced brandy, though the final products are dramatically different.

History

It is unknown when people discovered that food could be converted to alcohol through fermentation. It appears that the discovery of fermentation occurred simultaneously with the rise of the first civilizations, which may not be a coincidence. At about the same time that people in Europe discovered that apple and grape juice—both containing fructose—would ferment into hard cider and wine, people in the Middle East discovered that grains—which contain maltose—would naturally ferment into beer, and people in Asia discovered that horse milk—containing lactose—would naturally ferment into airag. The first distilled liquor may in fact have been horse milk brandy, with the alcohol separated from fermented horses' milk by freezing out the water during the harsh Mongolian winter.

It is also not known when it was discovered that the alcohol in fermented liquids could be concentrated by heat distillation. Distilled spirits were made in India as long ago as 800 B.C. The Arabic scientist Jabir ibn Hayyan, known as Geber in the West, described distillation in detail in the eighth century. Regardless of its origin, alcohol was immensely important in the ancient world. In Latin, brandy is known as *aqua vitae*, which translates as "water of life." The French still refer

Almost every people have their own national brandy, many of which are not made from wine: grappa in Italy is made from grape skins, slivivitz in Poland is made from plums, shochu in Japan is made from rice, and bourbon in the United States is made from corn.

A demonstration against Prohibition.

The Eighteenth Amendment made it a crime to make, sell, transport, import, or export liquor. It is the only amendment to be repealed by another (the Twenty-first). The Prohibition era (1920–1933) had been a long time in coming. From the mid-nineteenth century through the beginning of World War I, a growing movement demanded a prohibition on alcohol. When members of Congress finally bowed to pressure from prohibition supporters and passed a constitutional amendment, many did so under the belief that it would not be endorsed by the states. In fact, a clause was added to make it more likely not be sanctioned: if three-quarters of the states did not ratify the amendment before seven years had expired, it would be deemed inoperative.

The amendment was passed by Congress in December 1917 and ratified by three-quarters of the states by January 1919. The popularity of the amendment disappeared soon after it was put into effect. The Volstead Act of 1919 banned beer and wine, something few people had anticipated, and in the minds of many Prohibition became a mistake. Crime rose as gangsters took advantage of the ban on alcohol by making huge profits in bootlegging and smuggling. When Franklin D. Roosevelt campaigned for president in 1932, he called for the repeal of Prohibition. His opponent, President Herbert Hoover, called it "an experiment noble in motive." Roosevelt won the election and his Democratic party won control of the government. Within months the Eighteenth Amendment was repealed.

to brandy as *eau de vie* meaning exactly the same thing. The word whiskey comes from the Gaelic phrase *uisge beatha* also meaning water of life. People in the Middle Ages attributed magical, medicinal properties to distilled spirits, recommending it as a cure for almost every health problem.

Raw Materials

The raw materials used in brandy production are liquids that contain any form of sugar. French brandies are made from the wine of the St. Émillion, Colombard (or Folle Blanche) grapes. However, anything that will ferment can be distilled and turned into a brandy. Grapes, apples, blackberries, sugar cane, honey, milk, rice, wheat, corn, potatoes, and rye are all commonly fermented and distilled. In a time of shortage, desperate people will substitute anything to have access to alcohol. During World War II, people in London made wine out of cabbage leaves and carrot peels, which they subsequently distilled to produce what must have been a truly vile form of brandy.

Heat, used to warm the stills, is the other main raw material required for brandy production. In France, the stills are usually heated with natural gas. During the Middle Ages it would have required about 20 ft^4 of wood (0.6 m^4) to produce 25 gal (100 l) of brandy.

The Manufacturing Process

The fine brandy maker's objective is to capture the alcohol and agreeable aromas of the underlying fruit, and leave all of the off-tastes and bitter chemicals behind in the waste water. Making fine brandy is an art that balances the requirement to remove the undesirable flavors with the necessity of preserving the character of the underlying fruit. Mass-produced brandies can be made out of anything as the intent of the people is to remove all of the flavors, both good or bad, and produce nothing but alcohol—taste is added later. Fine brandies are required to retain the concentrated flavor of the underlying fruit.

Fine brandy

1 The first step in making fine brandies is to allow the fruit juice (typically grape) to ferment. This usually means placing the juice, or must as it is known in the distilling trade, in a large vat at 68–77°F (20–25°C) and leaving it for five days. During this period, natural yeast present in the distillery environment will ferment the sugar present in the must into alcohol and carbon dioxide. The white wine grapes used for most fine brandy usually ferment to an alcohol content of around 10%.

2 Fine brandies are always made in small batches using pot stills. A pot still is sim-

ply a large pot, usually made out of copper, with a bulbous top.

3 The pot still is heated to the point where the fermented liquid reaches the boiling point of alcohol. The alcohol vapors, which contain a large amount of water vapor, rise in the still into the bulbous top.

4 The vapors are funneled from the pot still through a bent pipe to a condenser where the vapors are chilled, condensing the vapors back to a liquid with a much higher alcohol content. The purpose of the bulbous top and bent pipe is to allow undesirable compounds to condense and fall back into the still. Thus, these elements do not end up in the final product.

5 Most fine brandy makers double distill their brandy, meaning they concentrate the alcohol twice. It takes about 9 gal (34 l) of wine to make 1 gal (3.8 l) of brandy. After the first distillation, which takes about eight hours, 3,500 gal (13,249 l) of wine have been converted to about 1,200 gal (4,542 l) of concentrated liquid (not yet brandy) with an alcohol content of 26–32%. The French limit the second distillation (*la bonne chauffe*) to batches of 660 gal (2,498 l). The product of the second distillation has an alcohol content of around 72%. The higher the alcohol content the more neutral (tasteless) the brandy will be. The lower the alcohol content, the more of the underlying flavors will remain in the brandy, but there is a much greater chance that off flavors will also make their way into the final product.

6 The brandy is not yet ready to drink after the second distillation. It must first be placed in oak casks and allowed to age, an important step in the production process. Most brandy consumed today, even fine brandy, is less than six years old. However, some fine brandies are more than 50 years old. As the brandy ages, it absorbs flavors from the oak while its own structure softens, becoming less astringent. Through evaporation, brandy will lose about 1% of its alcohol per year for the first 50 years or so it is "on oak."

7 Fine brandy can be ready for bottling after two years, some after six years, and some not for decades. Some French cognacs are alleged to be from the time of Napoleon.

However, these claims are unlikely to be true. A ploy used by the cognac makers is to continually remove 90% of the cognac from an old barrel and then refill it with younger brandy. It does not take many repetitions of this tactic to dilute any trace of the Napoleonic-age brandy.

8 Fine brandies are usually blended from many different barrels over a number of vintages. Some cognacs can contain brandy from up to a 100 different barrels. Because most brandies have not spent 50 years in the barrel, which would naturally reduce their alcohol contents to the traditional 40%, the blends are diluted with distilled water until they reach the proper alcohol content. Sugar, to simulate age in young brandies, is added along with a little caramel to obtain a uniform color consistency across the entire production run. The resulting product can cost anywhere from $25 to $500 or even more for very rare brandy.

Mass-produced brandy

1 Mass-produced brandy, other than having the same alcohol content, has very little in common with fine brandy. Both start with wine, though the mass-produced brandies are likely to be made from table grape varieties like the Thompson Seedless rather than from fine wine grapes. Instead of the painstaking double distillation in small batches, mass-produced brandies are made via fractional distillation in column stills. Column stills are sometimes called continuous stills as raw material is continuously poured into the top while the final product and wastes continuously come out of the side and bottom.

2 A column still is about 30-ft (9-m) high and contains a series of horizontal, hollow baffles that are interconnected. Hot wine is poured into the top of the column while steam is run through the hollow baffles; the steam and wine do not mix directly. The alcohol and other low boiling point liquids in the wine evaporate. The vapors rise while the non-alcoholic liquids fall. As the still is cooler at the top, the rising vapors eventually get to a part of the still where they will condense, each type of vapor at a temperature just above its own boiling point.

The distillation of brandy.

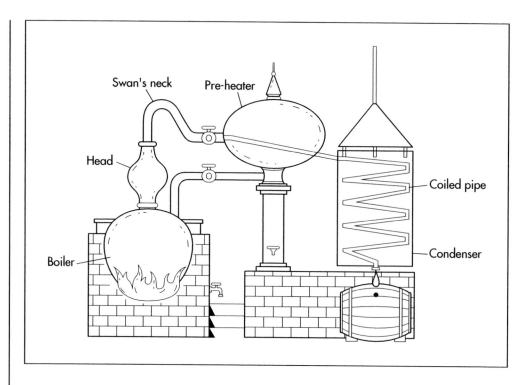

3 Once they have recondensed, the liquids begin to move downward in the still. As they fall, they boil again. This process of boiling and condensing, rising and falling, happens over and over again in the column. The various components of the wine fraction and collect in the column where the temperature is just below the boiling point of that component. This allows the ethyl alcohol condensate to be bled out of the column at the height where it collects. The resulting product is a pure spirit, colorless, odorless, and tasteless, with an alcohol content of about 96.5%. At 96.5% alcohol, it can be used to fuel automobiles. It can be diluted and called vodka or diluted and flavored with juniper berries and called gin.

4 Mass-produced brandies are also aged in oak casks and pick up some flavors from them. Like its fine counterpart, the brandies are blended, diluted to around 40% alcohol, and bottled.

Quality Control

The quality control process for fine brandies involves trained tasters with years of experience sampling brandy. A large cognac house might have 10,000 barrels of brandy in its cellars, each of which must be tasted annually. Hence, most of the brandy "tasting" involves only smelling, as tasting several hundred barrels of brandy in a day would result in alcohol poisoning. The tasters usually "taste" each of the barrels at least once a year to assess how it is aging and to evaluate it for its blending qualities. Brandies that pick up off-flavors during distillation are discarded.

As mass-produced brandies are manufactured to be odorless and tasteless, the only real quality control required is to check their alcohol content. Because alcohol is less dense than water, the alcohol content of brandy can be checked with a hydrometer. A hydrometer is a glass float with a rod sticking out the top of it. The rod is calibrated so that a line on the rod will be exactly at the liquid surface if the hydrometer is floating in water. As alcohol is less dense than water, the hydrometer will sink deeper in alcohol than it will in water. By calibrating the rod scale with different blends of known alcohol content, it can be used to determine the percentage of alcohol in a water/alcohol mixture.

Byproducts/Waste

The waste products from brandy production include the solids from the wine production and the liquids left over from the still. The solids from brandy production can be used for animal feed or be composted. The liquid wastes are usually allowed to evaporate in shallow ponds. This allows the residual alcohol in the waste to go into the atmosphere, but the United States Environmental Protection Agency does not consider this to be a major pollutant source.

The Future

For the foreseeable future, the vast bulk of all the brandies will be produced in column stills. However, there is an increasing interest in luxury goods throughout the world. Not just fine brandies, but Calvados (fine apple brandy) and slivovitz (fine plum brandy) are getting increasing amounts of attention from collectors and ordinary citizens.

Where to Learn More

Books

Faith, Nicholas. *Cognac.* Boston: David R. Godine Publisher, 1987.

Harper, William. *Origins and Rise of the British Distillery.* Lewiston, U.K.: The Edwin Mellen Press, 1999.

Periodicals

Kummer, Corby. "Don't Call It Cognac." *Atlantic Magazine* (December 1995).

Other

United States Environmental Protection Agency. *Emission Factor Documentation for AP-42, Section 9.12.2, Wines and Brandy.* (October 1995).

Jeff Raines

Breath Alcohol Tester

It is estimated that one person is killed every 32 minutes and another person is injured every 26 seconds in alcohol-related accidents.

Background

There is a serious need to ensure that alcohol-impaired drivers stay off the roads. It is estimated that one person is killed every 32 minutes and another person is injured every 26 seconds in alcohol-related accidents. Highway deaths increased slightly from 41,717 deaths in 1999 to 41,812 in 2000. Forty percent (16,725) involved alcohol, an increase from 38% the previous year.

Breath alcohol testers (BATs) depend on the blood to breath ratio. This ratio describes the relationship between the alcohol content of the breath and the alcohol content of blood at any given time. The accepted ratio of breath alcohol to blood alcohol is 2,100:1. This means 2,100 ml of a deep air lung sample contains the same amount of alcohol as 1 ml of blood. All breath testing instruments developed since 1939 use this ratio. The ratio is guided by Henry's Law, which states the quantity of gas that dissolves in a liquid at a standard temperature and pressure are directly proportional to the partial pressure of that gas in the gas phase.

In order to obtain an accurate reading of a person's alcohol content the device must test a person's deep lung air. The exchange of gasses, such as alcohol, between the blood and lung occurs in the alveoli. Each lung contains several million alveoli. If deep lung air is not exhaled into the device, the sample breath could be diluted with a lower alcohol concentration. Instruments usually require a person to blow for a minimum amount of time to ensure the air is captured from the deep lungs.

There are two main types of BATs used today. The first type uses infrared light to detect alcohol content. This device passes a breath sample through a narrow band of infrared light set to a frequency absorbed by alcohol. The amount of infrared light not absorbed by the alcohol tells the concentration of alcohol in the breath.

The second type of device uses fuel cells (which rely on chemical reactions) and is the most commonly used BAT. The alcohol in a person's breath is the energy for the fuel cell. The higher the concentration of breath alcohol, the more electricity that will be generated. The device measures the strength of the current to determine the breath/blood alcohol content (BAC).

The fuel cell itself varies only slightly from product to product and is a component that manufacturers purchase from outside venders. In a fuel cell, two electrodes are immersed in a liquid electrolyte, a substance that conducts electricity. An electrode is a solid electric conductor through which an electric current enters or leaves. The electrodes are coated with a platinum layer and have very fine pores. Between the electrodes is the thin electrolyte layer. The alcohol is pulled into the fuel cell by a pump and seeps through the electrodes, where it is then chemically converted. The fuel results in a flow of electricity between the electrodes.

The alcohol found in alcoholic beverages is ethyl alcohol, also known as ethanol. The molecular structure of ethanol has four major types of bonds: carbon to oxygen, carbon to carbon, carbon to hydrogen, and oxygen to hydrogen. In a fuel cell tester, the platinum material on the electrodes acts as a chemical catalyst and ionizes the hydrogen atoms by taking away their electrons. The hydrogen atoms are now positive. In this chemical conversion (which takes place at the top of the fuel cell) the hydrogen atoms then move lower in the fuel cell and combine with oxygen. Water is formed and one electron per positive hydrogen molecule is absorbed. Now there are more electrons at the top of the cell than at the bottom. The two surfaces are connected electrically through a wire. The electrons flow through the wire from the platinum electrode. The wire is connected to an electrical-current meter and to the platinum electrode on the other side. The result is a neutralizing current that flows through the fuel cell. The current indicates the amount of alcohol consumed by the fuel cell. The more fuel (alcohol) present, the higher the current.

History

Since the time of Hippocrates (c. 430 B.C.), physicians have known that human breath can provide clues to a medical diagnosis. Breath is one accurate way to measure a person's BAC because blood goes through the lungs, the site of gas exchange. The alcohol molecules are transferred from the blood to the lung air expelled in a breath.

In 1938, the first BAT was developed by Dr. R. N. Harger and called the Drunkometer. The year 1941 brought about the Intoximeter invented by Glenn Forrester and then the Alcometer developed by Professor Leon Greenberg. These machines calculated the blood alcohol to breath alcohol levels of deep lung air samples. The only way to determine the BAC before these instruments was through blood or urine tests. These methods were both time consuming and expensive. In 1954, the Breathalyzer was invented by Indiana State Policeman Robert Borkenstein. This was a portable, durable type of alcohol testing device that became the instrument of choice by police around the country.

Early breath alcohol testers required the person being tested to blow up a balloon. This ensured that a deep air lung sample would be taken. The balloon air was then released over photoelectric chemicals that changed color in the presence of alcohol; the deeper the color change, the higher the alcohol content. This device was often challenged in court because it could produce false results. For example, if a person used a mouthwash containing alcohol before taking the test, it could result in a higher BAC.

By the 1980s chemical breath tests were rarely used. Suspects were continuously challenging the results and courts were overturning them. Manufacturers focused on improving the accuracy, speed, and ease of fuel and infrared BAT use.

Raw Materials

Fuel cell breath alcohol testers are primarily composed of a fuel cell, pump, mouthpiece, printed circuit board (PCB), and a liquid crystal display (LCD) or light-emitting diode (LED) all contained within a plastic case generally made from low density polyethylene (LDPE), polypropylene (PP), or polystyrene (PS) plastic. The fuel cell is made from two platinum coated electrodes and a permeable electrolyte material. The pump is made of glass and nylon and used to pull the alcohol into the BAT. At the heart of the BAT is the PCB which controls the entire unit. The microprocessor contains the coding that the BAT uses to carry out the functions. The device also uses a LCD or LED to present instructions to the user, including the results or potential error messages.

Design

Both infrared and fuel-cell breath testing devices are used in three different types of instruments: evidential, screening, and passive breath alcohol testers. Evidential BATs collect samples from a person's breath. The results are accurate enough that the collected evidence can be used in a court of law. For the most part, these devices are large—the size of a desktop computer—and housed at the police department. Screening BATs are typically the most widely used BAC testers due to their accuracy and portability (they are about the size of a pocket calculator). These BAC testers require a person to blow into the device, and it reads either pass, fail, or provides a digital readout of the person's BAC. Passive BATs are also handheld devices, but require no action on the suspect's part. The device takes samples of the air around a person.

The Manufacturing Process

1 The first step in the production of a fuel cell BAT is to manufacture the case.

A fuel cell breath alcohol tester.

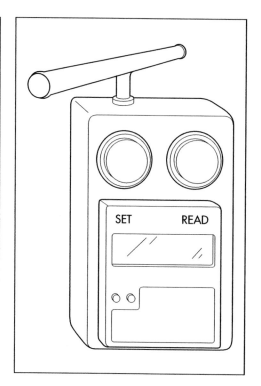

Plastic—LDPE, PP, or PS—pellets are fed into a hopper and heated until they are fully liquidized.

2 The hopper then releases the viscous liquid into a plastic injection mold. In this process, the plastic is poured into a die that is shaped like the desired case. After the liquid fills the die, it is closed and any excess material is drained. After the die cools, it opens and the case is ejected.

3 The PCB is made from a fiberglass epoxy resin with a copper coil bonded to one or both sides. These arrive at the plant assembled.

4 Two springs are soldered to the PCB that will connect to the batteries. When the case is closed, the battery compartment on the case lines up with the springs so the batteries are ready to fit in the compartment.

5 The pump is what draws a person's breath inside the testing device to the fuel cell. The manufacturer takes a three-volt motor, Pyrex cylinder, nylon piston, and a stainless steel lead screw. The pump turns the stainless steel screw and the screw moves the pistons back and forth, forcing the breath through the fuel cell. Pyrex is a heat-resistant, chemical-resistant glass. The nylon piston is a solid cylinder or disk that

fits snugly into the cylinder. An o-ring, a small ring made of rubber or plastic, is manually placed on the groove of the piston. The piston is then lubricated slightly then fit inside the Pyrex cylinder. Workers make sure the pieces are firmly held together. They then attach the motor and the lead screw to the pump with bolts.

6 After the pump is assembled it is tested on a test jig, a contraption that manufacturers build to test products. Every manufacturer has their own unique test jig but the goal is the same for all: to ensure the pump can properly cycle air in and out numerous times. Workers hook the pump to a machine that releases air through the pump about 200 times. They monitor the air current passing through the pump and make sure it is within specifications. Specifications require a set quantity of air moving through the device.

7 Once the pump is complete the rest of the pieces are assembled into the plastic case. The PCB is manually screwed into the case. Then the LCD is connected to the PCB. The LCD has a connector on one side. The connector is attached to a flexible strip consisting of wires and a flat piece of tape-like material. Workers push the connector into the PCB and it snaps into place.

8 Next, workers fasten the pre-assembled fuel cell to the rest of the BAT components. The fuel cell is attached to a porting block. The porting block is the plumbing of the device (a small plastic tube that the sample air moves in as it heads towards the fuel cell). Workers attach the fuel cell to the porting block with four screws. The porting block is then fixed to the microprocessor with two screws. The pump is attached to the porting block with piece of silicon tubing. The tubing ties the pump to the porting assembly.

9 Once the BAT is completely assembled manufacturers need to input measurements into the device. Manufacturers set the device by a process called calibration. Calibration is the process of setting-up the BAT to a known, standard alcohol level in order to accurately measure the alcohol concentration in the breath. The BAT will sample either a test gas or solution of a known alcohol concentration, called the alcohol standard. Once tested, the BAT will auto-

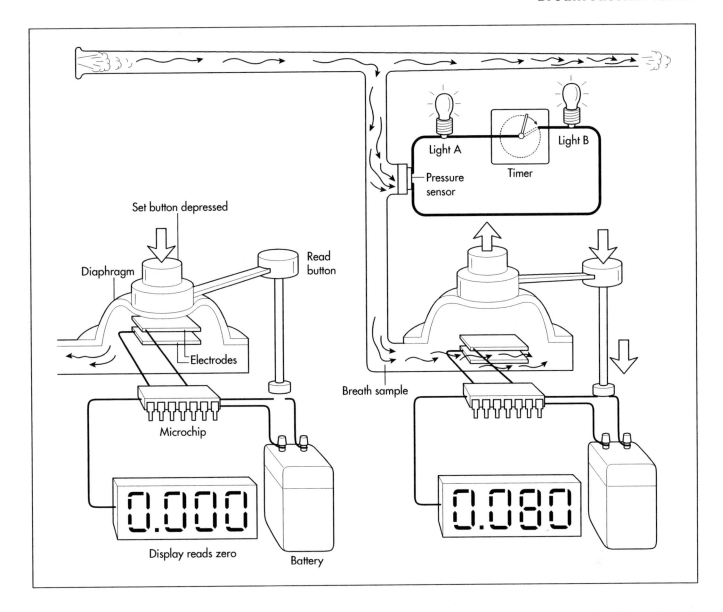

An example of how a fuel cell breath alcohol tester works.

matically adjust all internal parameters to provide accurate measurement.

10 After the BATs have been tested, they are packaged and shipped to the wholesaler or directly to individual customers.

Quality Control

Each BAT has a self-diagnosis test built into the PCB. Manufacturers test all the functions of the device. The units are cycled through simulators and tested repeatedly. Manufacturers can run a cycle of 50 tests at once. For example, the PCB is programmed to beep when the device has detected enough deep lung air entering it. It is also programmed to graph the air being blown.

Manufacturers will push a known amount of air through the machine to test multiple functions at once.

When manufacturers have completed their testing, they send their product to the United States Department of Transportation. The Department of Transportation must approve a device before placing it on their Conforming Products list. In terms of law enforcement, states are required to purchase only devices on the Conforming Products list. Government agencies will also only purchase devices from those approved by the Department. As these markets are the vast majority of alcohol testing devices sold in the United States, manufacturers regularly send their product to a Department of Transportation office for approval.

The Department of Transportation tests the device for performance. Laboratory workers use a wet or gas solution with a known concentration of alcohol. If using a wet bath, the device is hooked up to an instrument called a wet bath simulator. The solution is poured into the simulator and the instrument automatically heats the solution to a temperature of 93°F (34°C). The simulator has a gas outlet port. Workers connect a mouthpiece adapter to the port, then connect Bat's mouthpiece to the mouthpiece adapter. Vapor is pushed from the wet sample simulator into the BAT at a constant rate throughout the cycle. Workers then program the unit to 0.01. Testers can determine the exact concentration of alcohol that emerges from the solution by knowing the temperature. The simulator pushes vapor through the device and testers match up what the device reads with the known alcoholic concentration. BATs are required to meet specifications that are set by the Department. For example, among the requirements made by the Department of Transportation is that the device must distinguish alcohol from acetone when alcohol is at a 0.02 level. Laboratory workers repeat the test roughly 200 times, using different test conditions. They vary the known temperature from cold to warm, such as 50°F, 68°F, and 86°F (10°C, 20°C, and 30°C).

The Department of Transportation also ensures that enforcement personnel accurately use the alcohol-testing device. Law enforcement personnel must pass a certification that conduct breath alcohol testing by either their state or local government.

The Future

The push is to manufacture portable devices that can collect evidence on the spot. To accomplish this, there is an increasing move towards manufacturing alcohol fuel cell devices. As opposed to alcohol infrared devices, the fuel cells testers are much less expensive. Their accuracy and ease of use is comparable to infrared testers.

The desire for hand-held BATs made for the individual consumer market is also growing. People who want to know if they are over the legal alcohol limit can test themselves with these BATs. Establishments that sell alcohol also are a potential market for these devices. These products are about the size of a flashlight and far less expensive than the devices geared for professionals. They are not as precise as other testing devices, but have the benefit of ease and cost.

Law enforcement personnel are one of the main groups who commonly and regularly use breath testing devices to ensure drivers are not under the influence of alcohol. These devices are also used to ensure alcohol abuse does not occur in the workplace. The United States Department of Transportation requires alcohol testing for millions of safety-sensitive employees in the trucking, airline, rail, transit, and maritime industries. Testing of these employees is administered randomly. Non-regulated companies are increasingly administering alcohol tests on their employees.

In 2001, most states have a legal limit of a 0.1 BAC, with some states enforcing a 0.08. That translates to the BAC not exceeding 10 or 8 mg per 100 ml of blood.

Where to Learn More

Books

Considine, Douglas M., ed. *Van Nostrand's Scientific Encyclopedia.* 8th ed. New York: International Thomson Publishing Inc., 1995.

Periodicals

National Center for Statistics and Analysis. *Motor Vehicle Traffic Crash Fatality and Injury Estimates for 2000, 2000 Annual Assessment of Motor Vehicle Crashes.*

Other

"About Breath Tests." *Menssanna Research, Inc. Web Page.* December 2001. <http://www.menssanaresearch.com>.

Craig C. Freudenrich. "How Breathalyzers Work." *How Stuff Works Web Page.* December 2001. <http://www.howstuffworks.com/breathalyzer>.

Life Loc Technologies, Inc. Web Page. December 2001. <http://www.lifeloc.com>.

M. Rae Nelson

Bullet

Background

A bullet is a projectile, often a pointed metal cylinder, that is shot from a firearm. The bullet is usually part of an ammunition cartridge, the object that contains the bullet and that is inserted into the firearm. Cartridges are often called bullets, but this article will discuss only the projectiles fired from small or personal firearms (such as pistols, rifles, and shotguns).

History

Though there were cast lead bullets used with slings thousands of years ago, the history of the modern bullet starts with the history of firearms. Sometime after A.D. 1249, it was realized that gunpowder could be used to fire projectiles out of the open end of a tube. The earliest firearms were large cannons, but personal firearms appeared in the mid-fourteenth century. Early projectiles were stone or metal objects that could fit down the barrel of the firearm, though lead and lead alloys (mixtures of metals) were the preferred materials by 1550. As manufacturing techniques improved, firearms and lead bullets became more uniform in size and were produced in distinct calibers (the diameter of the bullet).

The Industrial Revolution produced further improvements. Firearms with rifled barrels (spiral grooves inside of the firearm barrel that impart stabilizing spinning motion to the bullet) led to the familiar conical bullet. More powerful smokeless powders replaced gunpowder (now called black powder) in the late nineteenth century, but they also required harsher firearm and bullet materials. Lead bullets left lead residue in the barrel; jacketed bullets (a harder metal layer sur-

rounds the softer lead core) were developed to stop this. The familiar metal ammunition cartridge (containing a bullet, a case, a primer, and a volume of propellant) was common by World War I.

Raw Materials

Bullets are made of a variety of materials. Lead or a lead alloy (typically containing antimony) is the traditional bullet core material. Traditional bullet jackets are made of copper or gilding metal, an alloy of copper and zinc. There are many other materials that are used in bullets today, including aluminum, bismuth, bronze, copper, plastics, rubber, steel, tin, and tungsten.

Bullet lubricants include waxes (traditionally carnauba wax made from the carnauba palm), oils, and molybdenum disulfide (moly). Modern wax and oil formulas are generally not made public. Moly is a recent innovation; this naturally occurring mineral sticks to metal on contact. The bullet making process can also use grease and oils to lubricate the bullet during machining and pressing steps. This lubrication prevents damage to the bullet or the machinery by allowing the bullet and machinery to move against each other without sticking. Solvents are used to remove grease and oil from the bullet afterward.

Design

There are several different uses for ammunition, such as military, law enforcement, hunting, marksmanship/target shooting, and self-defense, each requiring different bullet performance. There are also legal and public relations design considerations, such as lethality, threats to innocent bystanders, environmental impact, and appearance.

Public outcry in the United States has been greatest against so-called "cop-killer" bullets designed to penetrate body armor such as that used by police, and against expanding bullets such as the Black Talon, which has a tip that opens into six sharp "claws" on impact.

Bullet design is dependent on firearm design and vice versa. The bullet must fit into the barrel correctly. A bullet that is too small will not engage the rifling in the barrel, or it will bounce around in the barrel and not exit in a straight line. A bullet that is too large will jam in the barrel, possibly causing the firearm to explode from the pressure. The bullet weight must also match the amount of powder in the cartridge, so that it is fired at the correct speed.

Bullets are designed using calculations and data gathered from previous testing (firing) of bullets. This data can include variables such as accuracy (whether it hit the target), precision (whether more than one of the same bullet type produced similar results), speed of the bullet, effectiveness at a given range (distance to the target), penetration into the target, and damage to the target. Bullets are then tested against a target which resembles what they will be used against. There are several materials used to simulate the intended target, including bullet gelatin, a recently developed material used to simulate flesh.

Modern bullets can have many different features. Some of these features concern the shape of the bullet and others the materials of construction. Most bullets look like a cylinder with a pointed end. The cylindrical section to the rear of the bullet is the shank and the pointed section to the front of the bullet is the tip, though the tip may be flat instead of pointed. Bullets can be made of one or more materials.

Bullets made out of only soft material (such as lead) expand on impact causing more damage to the target. Bullets made out of only a harder material (such as steel) penetrate further into thicker targets, but do not expand much. A softer core can be enclosed or partially enclosed in a layer of harder metal called a jacket. This jacket can completely enclose the bullet or it can leave the softer tip exposed for expansion purposes. Varying the amount of jacketing alters the amount of penetration versus expansion.

The shank can have a flat base or a tapered base (boat tail). The flat base is heavier and provides greater penetration, but the boat tail provides greater accuracy over distance. The base of the shank can also have a base

plate of harder metal to prevent deformation of the bullet during firing. The base sometimes has a conical indentation (a gas check) that expands on firing to seal the base of the bullet against the firearm barrel and trap all of the energy from firing to propel the bullet forward. The shank may also have grooves used to contain lubricating grease that helps the bullet move freely in the firearm barrel. Sometimes a single groove, called a cannelure, is cut into the bullet to mark how far the bullet is to be inserted into the cartridge and to provide a feature to crimp the cartridge to the bullet.

The tip of the bullet is usually pointed. This point may be curved (called an ogive). Sharper tips provide greater penetration. Wadcutters are bullets with no point or a sharp shoulder behind the point used in target shooting to cut paper targets cleanly. Semiwadcutter bullets have a flat-tipped cone tip and can be used for target shooting, hunting, or self-defense. Target bullets are light and designed for speed and accuracy in a shooting range. They are usually not appropriate for other purposes.

Some tips are designed to expand on impact. This kind of bullet is banned from military use, but can be used for law enforcement, self-defense, and hunting. The tip or the entire bullet may be made of a soft material such as lead, but there are other design features that can aid bullet expansion. Hard material behind the softer tip provides more penetration and pushes the softer tip forward to expand more. The harder material can be the shank, a section of the tip, a partition of hard metal between the tip and the shank, or even a hard point on the tip that is driven backward on impact to expand the softer tip material.

Another feature that provides expansion is a hollow tip (or hollow point), an empty cone in the tip that points toward the rear of the bullet. When the bullet hits the target, the thin sides of the hollow tip expand outward. Even harder metals can expand, especially if they are scored (have grooves cut in them) to provide places to split apart.

Few bullets have separable parts. Some bullets have sabots, sleeves that surround the bullet while it is being fired but that fall off after leaving the firearm. Sabots allow smaller bullets to be fired from larger firearms at higher velocities than they would be fired from smaller firearms. Bullets can also contain multiple pellets or other particles that exit the bullet in a spray on impact or on leaving the target. This provides a higher chance of hitting something (from the many particles) or can cause many wounds in an easily damaged target.

Shotguns often fire shot (many small round pellets) or solid slugs (large, often soft bullets) out of an unrifled barrel, though some shotguns have rifled barrels. Air guns fire solid round or hourglass-shaped pellets.

Military bullets have special features, sometimes also used in law enforcement and self-defense. In order to get around the prohibition on expanding bullets, military bullets can be designed with heavier than normal back ends so that they tumble into the target on impact to create a larger wound. They can also be designed to break apart on impact with a similar effect. Some military bullets have incendiary (flammable) material in the base of the bullet that leaves a visible trail. This is known as a tracer bullet because it allows the shooter to track the bullet. Incendiary material can also be placed in the tip of the bullet so that it can start a fire on impact. Military bullets are usually made of harder materials or are fully jacketed. They are often designed for penetration. "Non-lethal" plastic or rubber bullets are sometimes used by the military and in law enforcement. These bullets are designed to temporarily incapacitate rioters and demonstrators, but they have the ability to kill.

Law enforcement and self-defense bullets should incapacitate the target. Many of these bullets are designed to expand or shatter after hitting the target, causing maximum damage. These bullets can be made of harder material that has greater penetration through materials such as heavy clothing and body armor. Police and self-defense bullets should not over penetrate (go through the target) and endanger bystanders.

Hunters have different requirements for different types of targets. Fast moving targets require faster, often lighter, bullets. Larger targets with heavy hides and large bones require bullets that can penetrate and inflict enough damage to drop the animal quickly. There are several different designs that ad-dress these conflicting demands. Many hunting bullets are designed to expand. Partitioned bullets and partially jacketed bullets are common for larger targets.

The Manufacturing Process

There are many types of bullet manufacturers, ranging from large companies and governments to smaller custom ammunition manufacturers to individuals who load and reload ammunition with a few simple tools. There are also many different bullet designs and a lack of consensus about which is most effective. Because of this, there is no uniform method of ammunition manufacture. Large ammunition manufacturers, including the United States government, automate some of the manufacturing steps. At appropriate points during the manufacturing process, special features may be added.

The solid bullet or bullet core

The two most common bullet-forming methods are casting and swaging. Hollow points can be formed by either method. Hard (harder than lead) solid bullets can be stamped (a metal punch cuts a bullet-shaped piece out of a bar or sheet of softer metal) and machined from metal stock. Machining includes any process where a machine is used to shape metal by cutting away portions. A typical machine used for bullets is a lathe. A lathe rotates the bullet metal against steel chisels to gradually cut away material.

CASTING A BULLET

1 Casting is pouring molten metal into a mold. This mold is hinged and when closed has a hollow space that is the shape of the bullet. The metal is melted in a crucible (a metal or ceramic pot that can hold molten metal safely) and then poured into the mold.

2 After the metal has cooled, the mold is opened and the bullet falls or is knocked out. Any imperfections are removed by cutting or filing. If the bullet is extremely deformed, it can be melted down and the process repeated.

3 To cast a bullet with multiple sections of different materials, the first material is poured into the mold to partially fill it. After

this material has cooled and partially or completely solidified, the second molten material is poured into the mold to fill it partially or completely. This can be done several times, but most often is done twice to create a bullet with a heavier section (for penetration) behind a softer section (for expansion).

SWAGING A BULLET

1 Swaging is a cold forming process, which means that it involves shaping metal without heating to soften or melt it. The appropriate amount of material to be swaged (measured in grains) is placed in a die. A die is a harder metal container with a cavity (an empty space) shaped like the bullet without the back end. The die is part of a larger stationary object or is held in place on a platform.

2 A metal punch that fits into the open end of the die is forced into the die to the appropriate depth. As the punch forces the bullet metal into the die cavity, the material takes the shape of the cavity. The pressure can come from a manual or hydraulic press, from repeated hammer blows, or from a threaded punch that is screwed on. Excess metal is squeezed out of bleed holes.

3 The punch is removed from the die and the bullet is pushed or pulled out of the cavity. Any imperfections are removed by cutting or filing.

4 Multiple swaging steps can be used to insert partitions, to create a bullet out of multiple materials, and to further define the shape of the bullet. Sometimes several steps are necessary to add features such as a hollow point.

The bullet jacket

Some bullets have jackets of harder metal surrounding a softer core.

5 A coin-shaped piece of jacket metal is punched out of a strip or a sheet. The punch is usually a round metal cylinder that is pushed through the jacket material into a depression in a table. Some punches are rounded so that the piece of metal is shaped like a cup. Sometimes, tubing is used instead of a coin or a cup of metal.

6 If the jacket material is too hard to be formed easily, it can be annealed. An-

nealing is heating the metal, often with a gas flame, to soften it and make it more workable.

7 The jacket material is then placed in a die or over a punch and the punch is forced into the die. There may be several different punches and dies used to form specific features in the jacket. One of usual steps is to make sure that jacket is of uniform thickness. The thickness is typically 0.03–0.07 in (0.08–0.17 cm). Some bullets have a thin jacket electroplated onto the core.

Bullet assembly

8 Jackets and multiple bullet parts can be joined by methods such as swaging them together, casting one section on top of another, soldering, gluing, or electrical welding. Soldering is a process of joining two pieces of metal together with solder, an alloy that is usually tin and lead. The solder is melted and sticks to both pieces of metal, gluing them together after it cools and solidifies. Glues for joining multi-part bullets are usually epoxies, plastics that are formed from two different fluids that harden when combined. The epoxy fluids are dispensed from tubes and mixed, then the pieces are joined together and held in place until the epoxy hardens. Electrical welding is the process of passing a strong electrical current through two metal parts that are in contact so that they soften and stick together. If the joining method is not strong enough, the bullet may fall apart prematurely.

9 Next, grooves may be cut or pressed into the shank of the bullet. The grooves can be pressed into a soft bullet by rotating the bullet against a ridge on a metal wheel, or they can be cut into the bullet on a lathe. Many cast bullets already have grooves.

10 The bullet is sometimes coated with a lubricant, usually wax, oil, or moly, which reduces bore fouling from soft bullets. Jacketed and hard bullets are not generally lubricated, though they can be, especially with moly. Bullets are often degreased (put in a solvent bath to remove grease from previous manufacturing steps) before the lubricant is applied.

11 Wax and oil lubricants can be applied by rubbing with a soft material such as a cloth wheel, spraying, pouring, or dipping.

Moly is applied by placing bullets in a container of moly powder and rotating the container so that the bullet and the moly particles tumble around until the bullet is coated.

12 The completed bullets are then manually removed and packaged.

Quality Control

Many firearm users want consistent performance from their ammunition. The larger ammunition manufacturers responded by instituting quality control programs in the 1980s and 1990s. These programs include statistical process control (SPC), total quality management (TQM), and random testing. SPC involves measuring a manufacturing process and determining statistically how to optimize it so that it produces correct and consistent results. TQM is the application of this kind of quality control to the whole business, not just the manufacturing part of the business.

Random testing involves periodically taking a manufactured part and testing it. Completed bullets are loaded into ammunition and fired to determine if they perform as expected. Unfinished bullets can be examined to determine if they are being produced correctly up to that point in the manufacturing process. Both finished and unfinished bullets can be weighed, measured for symmetry (bullets should be identical along every direction from an imaginary line drawn from the center of the tip to the center of the base), and cut apart to make sure that there are no air spaces and that internal features are correct (such as the thickness of a partition or a jacket). Commercial bullet sizes can vary by thousandths of an inch, but military and high quality bullets are more uniform.

Byproducts/Waste

Up to 24 toxic materials have been found in ammunition production. Solvents (often used to remove oil and grease) are dangerous to inhale and can be captured for disposal or purification and reuse, as can any oil. Scrap metal can be reused or disposed.

The most dangerous raw material is lead. Production workers and firearm users can be exposed to dangerous levels of lead from bullets, and firing ranges, including military ones, are being shut down because of high

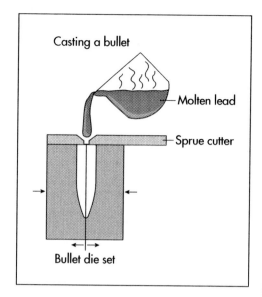

Casting a bullet
— Molten lead
— Sprue cutter
Bullet die set

The casting of a bullet.

lead levels. Lead can also leach into groundwater, further contaminating the environment. High levels of lead can lead to government intervention in the clean-up process, needing years of work to reach acceptable levels.

The Future

Companies continue to improve bullet performance to attract buyers, but social and political considerations are becoming more important. Health, safety, and environmental issues are leading to the replacement of toxic materials such as lead with materials such as tungsten, steel, bismuth, and plastic. Newer materials do not have the same performance characteristics as older materials, and this leads to newer ammunition designs.

There has been a legal struggle for decades over the lethality of police and self-defense weapons. Public outcry in the United States has been greatest against so-called "cop-killer" bullets designed to penetrate body armor such as that used by police, and against expanding bullets such as the Black Talon, which has a tip that opens into six sharp "claws" on impact.

Other innovations may be more radical. For example, tanks can fire shells with fins that pop out for stabilization at velocities that are too high for barrel rifling. This innovation could be scaled down for personal firearms. Self-propelled, finned rockets can also be shot out of pistol-sized launchers, though this type of projectile may no longer be called a bullet.

Where to Learn More

Books

Barnes, Frank C. *Cartridges of the World.* 9th ed. Ed. M. L. McPherson. Iola, WI: Krause Publications, 2000.

Grennell, Dean A. *The ABC's of Reloading.* 5th ed. Northbrook, IL: DBI Books, Inc., 1993.

Periodicals

"Brass Hats Led To Tungsten." *The Economist* 352, no. 8130 (31 July 1999): 68.

Petzal, David E. "Rifles: 2000 and After." *Field & Stream* 103, no. 5 (September 1998): 87.

Stolinksky, David C. "Stopping Power: Myth or Science?" *Handguns* 14, no. 4 (April 2000): 38.

Zutz, Don. "The Story Behind Winchester's Supreme Effort." *Shooting Industry* 34, no. 12 (December 1989): 90.

Other

Gunnery Network Web Page. December 2001. <http://www.gunnery.net>.

Hasenauer, Heike. "Bushels of Bullets." *Soldiers Magazine Online.* November 1998. December 2001. <http://www.dtic.mil/soldiers/nov1998/features/ioc3.html>.

Andrew Dawson

Candy Cane

Background

A candy cane is a hard candy usually peppermint flavored and decorated with stripes. The candy is long, thin, and bent at the top to resemble a walking cane. These candies are made using a batch process, which involves mixing and cooking the candy base, forming the stick shapes, and putting it in the appropriate packaging. First introduced in the seventeenth century, candy canes have been a favorite holiday candy for hundreds of years. Today, the candy cane makes up a significant amount of the $1.4 billion Christmas candy market. In fact, billions of candy canes are made and consumed each year.

Candy canes are traditionally a Christmas holiday candy. The classic candy cane is a white candy with red stripes infused with either peppermint or wintergreen flavors. They are usually about 6 in (15 cm) tall and about 0.25 in (6 mm) thick. Over the years, candy manufacturers have introduced various modifications on this classic look. Today, candy canes vary in size from about 2–12 in (5–30 cm) tall. Their widths can also vary ranging from about the width of a pencil to over 1 in (2.5 cm) thick. A variety of flavors are available such as apple, watermelon, cinnamon, strawberry, and even chocolate. The colors of these products are often modified to better reflect the candy's flavor. For example, a green apple candy cane might be colored green with red stripes. Some manufacturers have changed even the familiar "cane" shape. The result is a wide variety of candy cane type products that are sold throughout the year.

History

Candy was made as far back as 3000 B.C. In fact, the first candy maker was probably a caveman who discovered the pleasant taste of honey from beehives. Archeological evidence indicates that the ancient societies of Egypt, China, and Greece were all involved with candy production using honey mixed with fruits and nuts. During the Middle Ages, a method for refining sugar from sugar cane was developed in Persia. Over the years, this technology was improved and spread throughout Europe. Various hard candies and licorice were introduced and became a part of people's diets.

There have been numerous legends about the development of the candy cane. These typically suggest that candy canes were created as a religious symbol for early Christians. Stories say that it was developed to symbolize different aspects of the burgeoning Christian faith. For example, the red and white stripes are supposed to represent Christ's blood and purity. However, the historical evidence does not support these claims. In fact, the candy was clearly introduced well after the establishment of Christianity in Europe.

Candy canes were probably first introduced over 350 years ago. Professional candy makers had learned that sugar could be stretched and rolled into various shapes. This prompted them to produce straight, white sugar sticks that were easy to eat. During the 1600s, people began to decorate their homes at Christmas time. This typically involved a tree and various sweets like cookies, cakes, and stick candy. The historical evidence indicates that candy canes were first given the cane shape in 1670 by a German choirmaster at the Cologne Cathedral. He supposedly gave the children who sang in his choir sugar sticks that were bent like a shepherd's staff to keep them quiet during

Today, the candy cane makes up a significant amount of the $1.4 billion Christmas candy market. In fact, billions of candy canes are made and consumed each year.

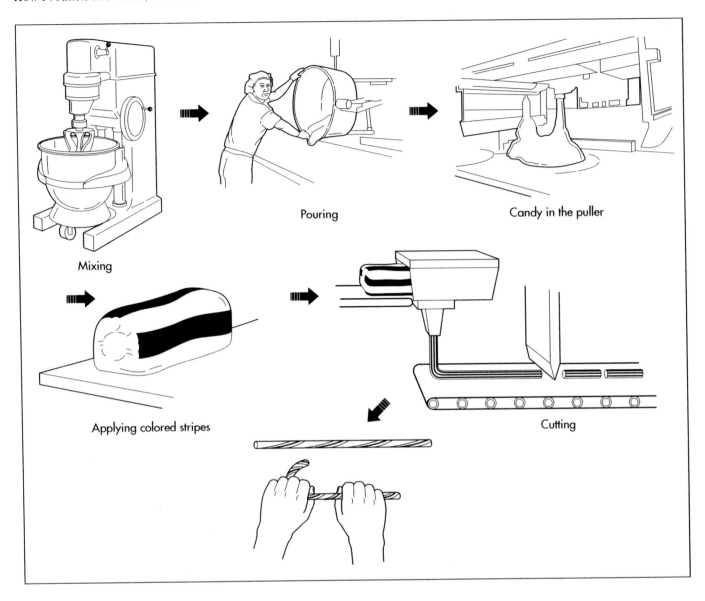

Mixing

Pouring

Candy in the puller

Applying colored stripes

Cutting

The candy cane making process.

long services. The tradition of handing out these candies during Christmas services spread throughout Europe.

The candy cane was first introduced to America in 1847 by a German-Swedish immigrant named August Imgard. He decorated a Christmas tree with candy canes in his Wooster, Ohio home. Evidently his creation had inspired others and a tradition was born.

While people had occasionally enhanced the appearance of the white candy cane with colored sugar prints, it was not until the early nineteenth century that candy canes got their stripes. It is not known exactly who gave candy canes this characteristic, but they have been produced that way ever since. This is also about the time when the

flavors of peppermint and wintergreen were added to make the product known today.

During the early part of the century, candy canes were made by hand. This process was extremely laborious. Candy canes were sold almost exclusively at a local level. In the 1950s a Catholic priest named Gregory Keller invented a machine that could make candy canes automatically. This sparked the mass production of the candy. Today, over 1.7 billion candy canes are sold each year.

Raw Materials

Confectioners have steadily refined candy cane recipes and production methods. By incorporating new information about the characteristics of ingredients and food production processes, they have been able to make candy cane manufacturing an efficient process. The raw materials used to make candy canes are specifically chosen to produce the appropriate texture, taste, and appearance. Sweeteners are the primary ingredients, but recipes also call for water, processing ingredients, colorants, and flavorings.

Sweeteners

Candy canes are primarily made up of sugar. When sugar (sucrose) is refined, it is typically provided as tiny grains or crystals. It is derived from beet and cane sugars. The sugar used in candy cane manufacture must be of high quality so that the proper texture and structure will be achieved. It is the unique physical and chemical characteristics of sugar that makes formation possible. When sugar is heated, it melts and becomes a workable syrup. The syrup can be manipulated, rolled, and fashioned. As it cools, the syrup becomes thicker and begins to hold its shape. When the candy is completely cooled, the sugar crystals remain together and form the solid candy cane.

Corn syrup is also used to produce candy canes. It is a modified form of starch, and like sugar it provides a sweet flavor. When it is mixed with the sugar, it inhibits the natural tendency of sugar to crystallize. Crystallization would result in a grainy appearance and a brittle structure. Corn syrup has the added effect of making the sugar concoction more opaque. Without the corn syrup and other ingredients, the candy would be transparent. The corn syrup also helps to control moisture retention and limits microbial spoilage. Beyond sugar and corn syrup, other sweeteners are sometimes incorporated into the candy cane recipe. These may include glucose syrups, molasses, or other crude sugars. Some low calorie candy cane recipes may incorporate artificial sweeteners like aspartame.

Processing ingredients

Certain ingredients are put in the candy cane recipe to aid in production. To dilute the sugar and make it workable, water is used. During the manufacturing process the water is steadily boiled off, and the end product has much less water than what it started with. Another processing ingredient is cream of tartar. This compound has the effect of producing air bubbles that help expand the sugar loaf and make it more stable. Salt also helps to adjust the chemical characteristics of the syrup. Typically, a small amount is used so that it is undetectable in the final product.

Colorants and flavorings

A variety of other ingredients may be incorporated into a candy cane recipe to produce various effects. To give the candy flavor and color, wintergreen or peppermint oils are added. Other natural flavors obtained from fruits, berries, honey, molasses, and maple sugar have also been used in candy cane production. Artificial flavors have also been added to improve taste. Additionally, fruit acids like citric acid and lactic acid can be added to provide flavor. Artificial colors such as certified Federal Food, Drug, and Cosmetic Act (FD&C) colorants are used to modify the color of the final product. In the United States, the federal government regulates these colors and qualifies each batch of colorant produced by the dye manufacturers. This ensures that no carcinogenic compounds are added to food products.

The Manufacturing Process

Making the batch

1 The first step of production involves blending the ingredients together in a large vessel. Typically, a stainless steel kettle is used that is equipped with automatic mixers. Ingredients can be poured or pumped into the batch by workers known as compounders. At this step, the water, sugar, corn syrup, and other processing ingredients are combined. They are then heated to over 300°F (141.5°C) and allowed to cook until they form an amber liquid.

Working the candy

2 While it is still hot, the sugar mixture is poured on water-cooled tables. The candy cools slightly and is sent to the working machines. These devices are equipped with arms that stretch the candy repeatedly until it looks silky white.

3 While the candy is being stretched, a line worker adds the proper amount of

A box of candy canes.

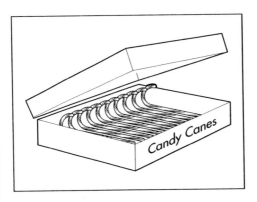

Candy Canes

flavoring. Also, coloring may be added at this point.

4 Another worker then takes a large portion (95 lb [43 kg]) of the warm candy and forms it into a loaf. Part of the loaf is put off to the side, dyed, and cut into strips. For the traditional candy cane, this portion is dyed red. It will become the red stripes in the final product. The 4 in-long (10 cm) red strips are then pressed at set intervals into the white loaf.

Extrusion and candy forming

5 The loaf can then be sent to the extruder machines to convert it into a candy cane. The loaf passes through the extruder and comes out the other side on a conveyor as a long strand of candy. The strand runs under cutters that slice it at set intervals to produce individual candies. They are then passed through a device that bends the candy. Since the candy is still slightly warm it can still be shaped as desired. Some extruders can handle over 2,000 lb (907 kg) of candy an hour.

6 After the candy cane is formed, it is put into its packaging. Some manufacturers wrap the candy cane in a clear plastic. This is done right as it is exiting the extruder. The plastic is then wrapped around the candy cane and sealed by a heat sealer.

7 In most instances, a set amount of candy canes are collected and boxed in secondary packaging. These boxes are passed through a shrink-wrap machine and sealed. This extra layer of packaging ensures that no moisture damages the product. The boxes are then put into shipping containers, put on pallets, loaded on trucks, and delivered to stores around the country.

Quality Control

Quality control is an integral part of all candy production. The first phase of control begins with tests on the incoming ingredients. Prior to use, lab technicians test ingredients to ensure they meet company specifications. Sensory evaluations are done on characteristics such as appearance, color, odor, and flavor. Other physical and chemical characteristics may also be tested such as liquid viscosity, solid particle size, and moisture content. Manufacturers depend on these tests to ensure that the ingredients used will produce a consistent batch of candy canes.

The next phase of quality control is done on the candy cane paste. This includes pH, viscosity, appearance, and taste testing. During production, quality control technicians check physical aspects of the extruded candy. A comparison method is typically used. In this method, the newly made product is compared to an established standard. For example, the flavor of a randomly sampled candy cane may be compared to a standard candy cane produced at an early time. Some manufacturers employ professional sensory panelists. These people are specially trained to notice small differences in tactile, taste, and appearance properties. Instrumental tests that have been developed by the confectionery industry over the years may also be used.

The Future

Modern candy cane production began in the 1950s. Since then manufacturers have steadily improved methods. In the future, improvements will be made to allow for even faster production with fewer workers. Improvements in the product may include lower calorie versions to appeal to more calorie conscious consumers. Manufacturers will undoubtedly create new flavors and colors to increase the number of candy canes sold in a year.

Where to Learn More

Books

Alikonis, Justin. *Candy Technology.* Westport: AVI Publishing Co., 1979.

Booth, R. Gordon. *Snack Food.* New York: Van Nostrand Reinhold, 1990.

Macrae, R., et al., eds. *Encyclopedia of Food Science, Food Technology and Nutrition*. San Diego: Academic Press, 1993.

Walburg, L. *The Legend of the Candy Cane*. New York: Zondervan Publishing House, 1997.

Other

National Confectioners Association. 24 September 2001. <http://www.candyusa.org>.

Perry Romanowski

Cash Register

Background

The cash register is an essential business tool that is often overlooked as one of the transforming mechanizations of the industrial age. A cash register records the amount of a sale, supplies a receipt to the customer, and keeps a permanent journal of daily transactions. Today, cash registers are highly automatic, and have many functions that help in the organized running of a store or restaurant. A more expensive and complex register system can be used to keep track of inventory and signal distant computers to re-order supplies. It can tally sales by department or by class of item, saving managers time and paperwork. This kind of machine is most often used by large chain retailers or restaurants and referred to as a point of sale (POS) terminal. The POS terminal may be a hodge-podge of components from different manufacturers. More conventional cash registers used by smaller establishments are generally one-piece machines with a built-in cash drawer, printer, and display. These are almost all manufactured in Asia but designed by distributors in the country where they will be used.

History

The cash register was apparently invented out of desperation. The creator was James Ritty, an Ohio restaurateur. Ritty ran a café in Dayton in the 1870s. The place was popular and always filled with customers. Nevertheless, the business continually lost money. Ritty blamed the dishonesty of his bartenders, who either kept money in their pockets or in an unlocked cash drawer, often nothing more than an old cigar box. This loose monetary system did not provide any way of keeping track of sales. If a customer returned to a shop after buying something, saying he had been overcharged or not given the correct change, there was no objective way to settle the dispute. The open box also meant that employees were always within reach of tempting cash. In Ritty's time, theft by clerks was a way of life, and shopkeepers had little defense against employee dishonesty. Ritty changed bartenders many times but continued to lose money until he was driven to a nervous breakdown.

To ease his mind, Ritty took a ship for Europe. On the ship he made friends with the ship's engineer, and spent hours in the engine room. There he observed the workings of an automatic device that recorded the revolutions of the ship's propellers. From this, Ritty imagined he could make a similar device that would record amounts of money passing through the cash drawer. He reputedly cut short his vacation to rush back and begin work on the prototype. Ritty assembled his first cash register in 1879, and patented a second, improved register later that year. Ritty went into business with "Ritty's Incorruptible Cashier" after perfecting a third model.

Ritty's early machines had two rows of keys running across the front, each key marking a money denomination from five cents through one dollar. Pressing the keys turned a shaft that moved an internal counter. This kept track of total sales for the day. The amount of each individual sale was shown to the customer on a dial similar to a clock face, with one hand for the cents and one for the dollars. Because the machine kept a daily total, any pilfering would be obvious. A later model kept the clock face and included a paper roll punched with pins to provide a more permanent record for the shopkeeper. However, Ritty was unable to ignite any excitement for his new device. Apparently he made only one sale, which was to John H. Patterson. Patterson ran a small coal business, but was so taken with the Incorruptible Cashier that he decided to buy Ritty's company.

Unfortunetely, Ritty had already sold his business to another party, Jacob Eckert. Eckert had made a vital addition to the machine, a bell that rang when a sale was made. Eckert ran the business as the National Manufacturing Company with several partners. John Patterson arrived in Dayton in 1884, eager to buy the small firm. After making a preliminary deal, he discovered that National Manufacturing was the laughingstock of Dayton. The company had not made any money, and no one believed that it could. Patterson tried to buy his way out of the contract, but was forced to complete the sale. Patterson changed the name of the firm to the National Cash Register Company.

The new company quickly improved the cash register. By 1890, the machines printed customer receipts as a standard feature. In 1906, the cash register was electrified. The company made a science of advertising and selling, becoming the role model for many other industries with its canned sales talks and innovative distribution of sales territories. By 1900, the company had sold over 200,000 registers and sent salesmen throughout Europe and South America. As early as 1896 it had sales in China, and by the end of World War I, National Cash Register was bringing in almost half its sales from overseas markets represented by at least 50 countries. The number of registers sold in 1922 alone was over two million. The company dominated the industry, buying up competitors when convenient. National Cash Register continued to develop its product line, coming out with new features to respond to customer demands. By 1944, the company had applied for 2,400 patents.

With the advent of micro processing technology in the 1970s, the cash register industry changed. Most of the manufacturing moved to factories in Asia, and eventually two basic types of cash register evolved. One type is the generally low-end, all-in-one machine usually referred to as an electronic cash register, or ECR. The other wing of the industry is the POS terminal, which is more than a cash register because of its superior data processing ability. Both are manufactured in similar ways, though the ECR may be shipped to the customer complete and ready to go, where the POS is made up of different components that may not meet up until the customer installs the terminal.

Raw Materials

Raw materials for cash registers are similar to materials used for other electronic products. The principal components for an ECR are an Acrylonitrile Butadiene Systrene (ABS) plastic casing, circuit board, metal printer, metal cash drawer, ABS plastic keyboard, and a liquid crystal display panel. ECRs are made at factories that also specialize in consumer goods such as televisions and VCRs. The materials and the construction process are virtually the same for all these products. Cash registers differ from other consumer goods, though, in the importance of the design process.

Design

Though cash registers are mostly made in Asia, they are used around the world. All except the lowest-end products need to be designed for the country and particular industry where it will be used. In the United States, most retailers or restaurants wanting a more specialized machine order their registers through a domestic distributor. The distributor works with the customer to understand the specific tasks the cash register needs to perform. Perhaps the register needs to be able to recall certain records of customer transactions. A cash register for a dim bar may need an easy-to-read display. The cash register in a restaurant may print one receipt for the customer, but print different information in the kitchen, telling the cooks what to prepare. The cash register distributor will design the software for these special functions or have it designed at a software company. The distributor then approaches the manufacturer with the list of needed features. In some cases, the new features can be made to fit in a pre-existing model or the manufacturer's engineers may have to redesign parts and processes.

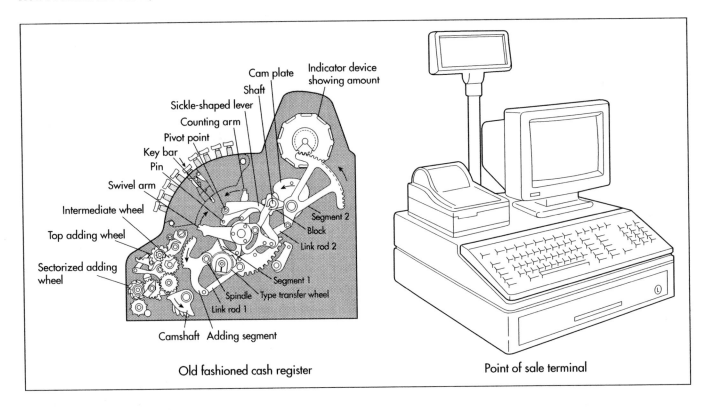

Cam plate
Indicator device showing amount
Shaft
Sickle-shaped lever
Counting arm
Pivot point
Key bar
Pin
Swivel arm
Intermediate wheel
Top adding wheel
Sectorized adding wheel
Segment 2
Block
Link rod 2
Segment 1
Spindle Type transfer wheel
Link rod 1
Camshaft Adding segment

Old fashioned cash register

Point of sale terminal

An early model cash register compared to a modern day POS terminal.

The Manufacturing Process

Cash registers are manufactured at large plants using a classic assembly line system. Twenty to 25 workers stand in front of a conveyor belt that may be 200 ft (61 m) long and move at 2–3 mi (3.2–4.8 km) per hour. Workers with screw guns and soldering irons attach parts as they come down the assembly line.

The cash drawer

1 Workers use lasers to cut the sheet metal to the customer-specified size. They then put the flat sheet of metal on the conveyor belt, which transports the sheet to a punch press. The hydraulic press has precut shapes that are clamped onto the sheet metal. The press is then closed and the shapes are cut out.

2 As the cut metal exits the punch press, workers weld the pieces together for the drawer and drawer case, these are then painted and dried.

3 The case and drawer move down the assembly line and workers attach the till and latch components, which are assembled out of the country. The cash drawer has a removable till that is opened by releasing a latch. The latch is activated by an electromagnetic device called a solenoid. When a current flows through the solenoid, it creates a magnetic field that moves a steel plunger, releasing the latch. Workers screw and solder the latch bracket sub-assembly and attach it to the back of the drawer.

On the assembly line

4 Some parts of the cash register, such as the drawer, may be made as sub-assemblies or bought from specialized suppliers. On the assembly line, workers begin by constructing the power supply.

5 Next, workers manually assemble the logic board by soldering, screwing, or snapping the circuits into pre-cut destinations.

6 Workers build the printer device, the display panel, and the keyboard, and attach these to the machine as it passes down the assembly line.

7 Finally the casing, which is usually made of plastic created through injection molding, is screwed onto the machine. A certain percentage of the cash registers may be taken off the assembly line at this point for quality control testing. The rest run

through a boxing machine, which packages them securely.

Shipping

8 The boxes are placed in cartons, and the cartons are labeled and sent for shipping. Distributors may warehouse them or send them directly to customers.

Installation

9 Because of the complexity of most cash registers, distributors help install them for customers and show them how to use the devices. In the case of a sophisticated POS the distributor makes sure that the terminal is properly integrated to work together as it should.

Quality Control

Quality control may be done both at the manufacturing plant and at the distributor's facility. The amount of quality control differs with the price of the product. A low-end ECR may have minimal quality checks. For a mid-grade machine or component, the manufacturer may check 10–15% of the devices as they come off the assembly line. On a top-quality machine, a higher percentage—up to 50%—may be checked. The distributor too generally gauges how much quality control to do according to the price and sophistication of the machine. Usually the distributor plugs in the machines to make sure the mechanical components are working. The distributor may leave the machines, or a sampling of the order, on all night to make sure they don't burn out. The distributor also runs diagnostics on the software. Other tests are also conducted depending on the order.

Byproducts/Waste

There are no unusual byproducts or waste associated with cash register manufacturing. Used cash registers can be reconfigured and upgraded. While some distributors concentrate on new machines, a subset of the industry specializes in rescuing used machines and updating them. Most cash registers used in the United States are 13–15 years old. Early cash registers are now highly valued as collectors' items.

The Future

Most advances in cash register technology come from the POS end of the industry, where large users such as giant chain retailers can take advantage of economies of scale and employ sophisticated new software or hardware. Much of this technology eventually trickles down into ECR manufacturing. Communications between registers was once an advanced feature, but it is becoming standard even on mid-level machines. Cash register software is constantly evolving in response to pressure from customers.

Where to Learn More

Books

Cortada, James W. *Before the Computer.* Princeton: Princeton University Press, 1993.

Crandall, Richard L., and Same Robins. *The Incorruptible Cashier.* Vestal, New York: The Vestal Press Ltd., 1988.

Marcosson, Isaac F. *Wherever Men Trade: The Romance of the Cash Register.* New York: Dodd, Mead & Company, 1945.

Periodicals

McCrory, Anne. "Jargon Judge: Point of Sale Device." *Computerworld* (20 July 1998): 51.

Angela Woodward

Catheter

The fastest growing segment of the catheter industry, the coronary catheter market, is expected to reach four billion dollars by 2003, growing at 11.2% annually.

Background

A catheter is a flexible tube made of latex, silicone, or Teflon that can be inserted into the body creating a channel for the passage of fluid or the entry of a medical device. For many years, the epidermal catheters used were plain tubes made of available industrial compounds, and design was largely based on current need. In the 1950s and early 1960s, a very common practice was to cut a suitable length of industrial polyvinyl chloride (PVC) or nylon tubing and have it sterilized with the other surgical equipment. Nowadays, there are many specialized catheter designs. For example, specific catheter designs allow catheters to be used in pulmonary, cardiac (vascular), neonatal, central nervous system, and epidural tissues. Catheters are designed to perform tissue ablation (tissue removal) and even serve as conduits for thermal, optics, and various medical devices.

The three major types of catheters are coronary, renal, and infusion. Coronary catheters are used for angiography (x-ray of blood vessels after injection of radiopaque substance), angioplasty (altering the structure of a vessel), and ultrasound procedures in the heart or in peripheral veins and arteries. The best-known renal catheters are Foley catheters, which have been commercially available since the 1930s. These catheters are equipped with an inflatable balloon at the tip and are used for urine incontinence, dying patients, and bladder drainage following surgery or an incapacitating injury or illness. The Foley catheter is relatively easy to use and used throughout the world in hospitals, nursing homes, and home-care settings.

History

The earliest precursor to the present day Foley catheter is documented in 3000 B.C. It is believed that Egyptians used metal pipes to perform bladder catheterizations. As early as 400 B.C., hollow reeds and pipes were used in cadavers to study the form and function of cardiac valves.

In 1844, Claude Bernard inserted a mercury thermometer into the carotid artery of a horse and advanced it through the aortic valve into the left ventricle to measure blood temperature. It is because of his work that the use of catheters became the method of standard for physiologists in the study of cardiovascular blood flow. Adolph Fick took another major step in the development of cardiac catheterization in 1870. His famous note on the calculation of blood flow is the basis for today's cardiac procedures.

Among the earliest published descriptions of human catheterization were done by Frizt Bleichroeder, E. Unger, and W. Loeb in 1912. They were among the first to insert catheters into the blood vessels without x-ray visualization. Interest in catheterization was also stimulated with the advent of chemotherapy. Early chemotherapy required the injection of drugs directly into the central circulation. Bleichroeder inserted catheters into dog arteries and assessed the effects after leaving them in place for several hours. He reported no complications or clots.

The Foley catheter came into existence in the 1930s. Frederick E. B. Foley began to experiment with different catheters of the time. He realized that urinary catheters would easily slip out of the bladder because there was no way to hold them in place. Foley experimented with different methods

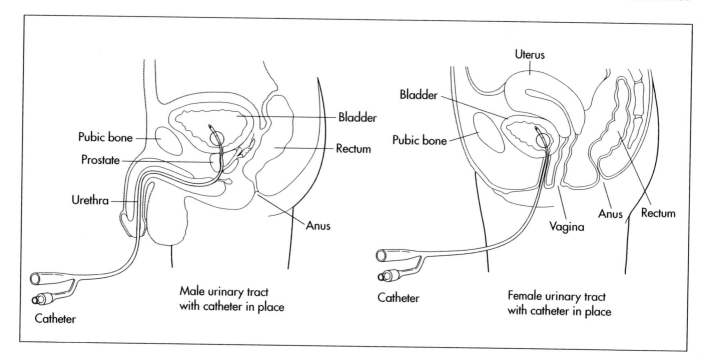

Male urinary tract with catheter in place

Female urinary tract with catheter in place

Catheter

of securing the catheter until he came up with the idea of attaching a balloon-like device to the end of the catheter. The device would then be able to be placed and then inflated from the outside. By 1934, Foley catheters were on the market. Other than in material, the Foley catheter remains relatively unchanged in design today.

Raw Materials

Foley catheters are made from either silicone or latex rubber, depending on the use.

Design

Foley catheters are made of latex or silicone rubber. Silicone rubber catheters are believed to be superior to latex catheters, as silicone is more biocompatible, causes less cell death, less likely to become encrusted, and more resistant to bacterial colonization. The catheter can either have two or three outlets. In a two-way catheter, one outlet acts a urine output and the other inflates the balloon. A three-way Foley catheter has the same function as a two-way catheter, but uses the third outlet for bladder irrigation.

Foley catheters vary in size from 12 fr to 30 fr (4 to 10 mm) in diameter, with the standard being 14 fr (4.6 mm). The balloon itself varies in size from 5 cc to 30 cc, depending on the needed use. The balloon can either be filled with sterile water or air. The catheter can also be attached to a drainage bag.

The Manufacturing Process

1 The first step in the manufacturing of a Foley catheter is the production of the long, thin tube that will be inserted into the bladder. The liquid rubber silicone is poured into a room temperature vulcanization (RTV) rubber mold. The mold is shaped like the desired catheter with either two or three outputs.

2 The silicone is then heat cured. This procedure can take anywhere from 0.5 to 40 hours. Once cooled, the tube is withdrawn from the mold.

3 A small opening is then punched in the distal end of the tube furthest away from the two outputs.

4 A thin band of cured latex is slipped over the tube by hand to form a sheath around the tube. It is positioned so that the latex covers the opening that has been punched in the tube.

5 To form the balloon, the entire length of the tube is dipped in latex, which creates an overcoat layer and bonds the band to the tube proximate to the distal and proximal ends of the band, forming the balloon. This

An example of how a Foley catheter is inserted into the male and female urinary tract.

A Foley catheter.

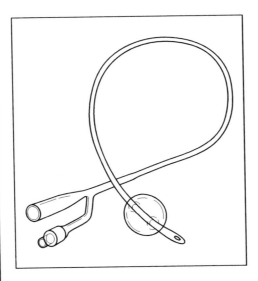

adds to the thickness of the balloon and is used to adjust the outer diameter of the tube to the desired size.

6 The catheter is then transported to the packaging center where it is put into a kit with a needleless syringe (to fill the balloon) and a drainage bag.

Quality Control

Quality control is built into each step of the manufacturing process. The machine operations check the final product of each stage in the process. The thermoplastic materials are immersed in liquid to ensure that defects are not present and that there will not be any leakage.

Byproducts/Waste

Any material deemed to be defective is either discarded or recycled depending on the severity of the damage. Since the product is directly related to human health, the materials must be of the highest quality.

The Future

A new use of the catheter is being tested in medical facilities for the purpose of dissolving clots or blockages in the coronary arteries. Once the catheter is positioned in the coronary artery, the tip of the catheter acts much like a showerhead, spraying six jets of saline around the clot. These saline streams break down the clot and the vacuum-like na-

tures of the pumps force the debris out of the artery. With the clot gone, doctors can proceed with balloon angioplasty to repair the fatty blockage, which caused the clot to lodge there in the first place. This method requires only mild intravenous sedation rather than the general anesthesia that would be required with bypass surgery. Such new technology lessens the physical and emotional strain on a patient.

In current ablation systems (catheter used for tissue destruction), the tip of a radio frequency ablation catheter can become quite hot. Blood can subsequently form coagulum on the catheter tip that prevents delivery of successful lesions. Another advance is active cooling of the catheter tip. This allows higher energy delivery at a cooler tip temperature without an increased risk of coagulum formation. The higher energy results in a better lesion.

Catheters are also being designed with safety features to prevent needlestick injuries along with Silver/Hydrogel-Coated Foley catheters to resist bacterial infection.

The fastest growing segment of the catheter industry, the coronary catheter market, is expected to reach four billion dollars by 2003, growing at 11.2% annually. The largest segment, however, is the renal market, which is comprised primarily of urinary catheters and dialysis catheters. Currently a four billion dollar segment, it is expected to reach 7.1 billion dollars in 2003.

Where to Learn More

Books

Topol, Eric J., ed. *Textbook of Interventional Cardiology.* Philadelphia: W. B. Saunders Co., 1993.

Periodicals

Mueller, R., and T. Sanborn. "The History of Interventional Cardiology." *American Heart Journal* 129 (1995):146–172.

Other

United States Patent Web Page. December 2001. <http://www.uspto.gov.com>.

Bonny P. McClain

Condensed Soup

Background

Condensed soup is a canned variety of soup prepared with a reduced proportion of water. The consumer then adds water or milk and the mixture is heated. Condensed soup was developed by John T. Dorrance, an employee of the Campbell's Soup Company, in 1899.

History

The advantage of boiling food in water as opposed to cooking it over an open flame is that it produces a denser food. Boiling several ingredients together causes the flavors to blend, thus creating a new taste. The practice of cooking meat in hot water dates to prehistoric times. Pots found from the Iron and Bronze Age excavation sites contain food residues. The Ancient Romans were known to eat a type of fish broth. Medieval cookbooks also included numerous recipes for soups. Robert May's *Accomplished Cook*, published in 1660, included recipes for "soops" with ingredients such as spinach, carrots, artichokes, potatoes, and parsnips.

For centuries, soups were poured over toasted bread. In fact, the word soup derives from the same source as "sop," a piece of bread soaked in liquid. In earliest times, soup was served in a communal pot and the broth was drunk directly from the pot. This changed with the invention of the spoon in the fourteenth century.

The practice of eating soup was not widespread in colonial America although there is some evidence that members of the upper and lower classes did enjoy it. In 1742 the first American cookbook was published in Williamsburg, Virginia by William Parks. It included recipes for "Soop Sante," "Pease Soop," "Craw Fish Soop," and "Brooth." German immigrants who settled in Pennsylvania introduced soups made from chicken, mutton, veal, beef, calf's head, rice, apples, and huckleberries. To the Pennsylvania Germans, soup making and eating was a ceremonious communal event.

The arrival of French immigrants during the French Revolution furthered the popularity of soup. One such immigrant, Jean Baptiste Gilbert Payplat dis Julien, opened a restaurant in Boston in 1794 and became known as the Prince of Soups. Turtle soup was a specialty of the house.

Soups soon began to appear in more American cookbooks. The *Virginia House-Wife* was published by Mary Randolph in 1824 and included 16 recipes for soup. It also included tips such as folding in butter and flour for a richer soup and using wine, tomatoes, cayenne, and curry powder for flavor.

However, at the end of the nineteenth century, soup was still not eaten in the United States to the extent that it was in Europe, as John T. Dorrance found after he returned from studying chemistry in Germany. Dorrance's uncle, Arthur Dorrance, was an executive with the Joseph Campbell Preserve Company. At the time, the company was a produce cannery. John Dorrance found that there were only two companies in the United States that were canning soup. Franco-American and Huckens sold ready-made soups in half-pint, pint, and quart containers. The soups' size and perishability made them difficult to ship and a not-very-profitable commodity.

Although compressing and concentrating soup into bouillon cubes was a common

In the United States, 2.5 billion bowls of condensed tomato, cream of mushroom, and chicken noodle soups are eaten annually.

practice in France, the resulting soup was often lacking in flavor. Using his knowledge of chemistry, Dorrance set about to create a better condensed soup. He reasoned that by reducing half of the soup's water, the weight would be considerably decreased. His challenge was to create a strong stock that would hold its flavor when water was added to reconstitute the soups. In effect, Dorrance was creating a sauce.

Within a year of embarking on his experiments, Dorrance introduced five varieties of condensed soup: tomato, consommé, vegetable, chicken, and oxtail. At first, Dorrance sold his soups door-to-door, working hard to convince American families that soup could be easily added to their daily diet. The ultimate success of Dorrance's innovation made the company profitable for the first time in its history, and in 1921 resulted in a corporate name change to the Campbell's Soup Company.

Dorrance closely guarded his recipes by dividing them into two separate parts. One was the list of ingredients with specific weights and measurements. The other part contained the directions for combining the ingredients. The production process was also divided between two separate departments of the company's plant.

Raw Materials

A wide variety of foods and seasonings are used to make condensed soup. Meat, such as beef and chicken, is used to create soup stock and as an ingredient. Vegetables can include broccoli, cabbage, corn, green beans, lima beans, okra, onions, peas, tomatoes, and white potatoes. Grains include noodles, rice, and barley. Spices and flavoring include allspice, bay leaves, celery seed, cloves, curry powder, parsley, pepper, salt, sugar, and thyme. Flour and mashed sweet potatoes are used as thickeners. Water is used in the cooking process, but generally cancelled out by the thickeners. Vegetable and soybean oil are used as a coating agent to prevent spillovers during the cooking process. Worcestershire sauce is also used in condensed soups. It is made from anchovy essence, clover, garlic, malt vinegar, meat extract, molasses, shallot, sugar, and tamarind. Preservatives, such as monosodium glutamate (MSG), may be added to pro-

long the soup's shelf life. MSG is a salt derived from seaweed and vegetable proteins.

The Manufacturing Process

The following process is for condensed tomato soup. Other types of condensed soups are made in a similar fashion.

Making the stock

1 Equal parts of lean beef and heavy bones are placed in 220-gal (833-l) iron kettles and covered with cold water. The meat is allowed to soak for 15–20 minutes before the heat is turned on. The water is slowly heated until it comes to a boil after approximately one hour. During this process, scum rises to the top. The scum is skimmed off regularly by mechanized long-handled skimmers and discarded.

2 The stock cooks for six to eight hours, until the meat is reduced to shreds. Water is added to keep the stock at a constant level. Spices are blended together and added to the stockpot. At the end of the cooking time, the stock is poured off. A screen in the bottom of the kettle traps the meat and bones.

Preparing the tomatoes

3 Tomatoes that pass inspection and receive a grading of number one are loaded into hanging baskets and sprayed with water to remove dirt. After this initial washing, the tomatoes are conveyed to trays where they are inspected and washed again.

4 The tomatoes are fed into a hammermill, a machine with two sets of interlocking fingers. The hammermill crushes the tomatoes to pulp. The pulp is piped into 600-gal (2,247-l) steam jacket copper breaker kettles. The pulp is simmered in the kettles and stirred constantly.

5 The kettles are equipped with spouts. The cooked pulp is emptied through the spout into cyclone machines. Spinning blades in the cyclones force the pulp through a fine wire mesh that separates out the seeds and skin. The remaining liquid is piped into stainless-steel tanks and transported to the soup-making area. Seeds are saved for future plantings; skins are discarded.

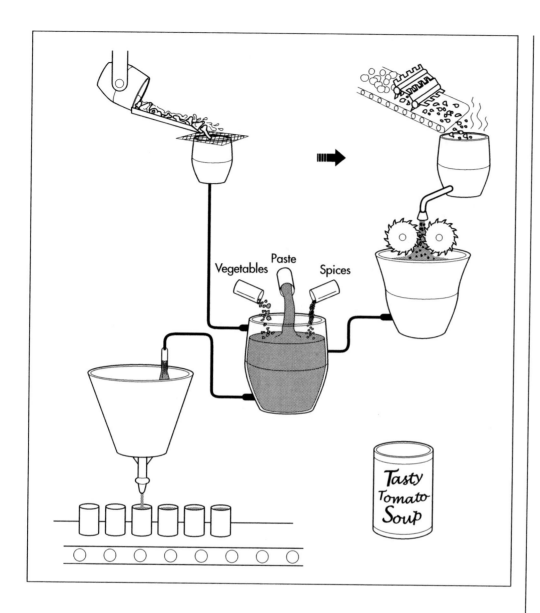

Seasoning the water

6 Proprietary amounts of celery seed, cloves, allspice, and bay leaves are poured into a kettle and covered with cold water. The water is quickly brought to a boil and allowed to boil for about 10 minutes. The seasoned water is drawn off through spigots and passed through several cheesecloth with varying levels of thickness.

Creating a paste

7 Flour and uncooked tomato pulp are combined in an industrial mixer until a paste is created.

Combining all parts

8 The cooked tomatoes are piped from the stainless steel tanks into 110-gal (416 l) kettles. Raw pulp is gradually stirred into the kettle until the kettle contains 95 gal (360 l) of tomato product. Salad oil is then piped into the kettle.

9 Garlic and onions, which have been mechanically chopped, are mixed together and added to the kettle. Then the tomato paste is added. The liquid spices are added along with pepper, salt, sugar, and Worcestershire sauce. The soup is mechanically stirred during this process.

10 The soup is boiled for two to three minutes until the flour thickens. It is then drawn off through a spigot and passed through a cyclone machine fitted with a #16 iron sieve and paddle brushes. The paddles rotate at about 250 revolutions per minute (rpm). From the cyclone, the soup is piped

through a brass wire screen. The wire screen is tightly woven, about 22 holes to an inch. As the soup passes through the screen, any black specks or other materials are sifted out and discarded.

Canning the soup

11 The soup is now more similar to a thick sauce due to the lack of water, thus making the soup condensed. The soup is next piped to the canning section where overhead valves pour it into cans. The valves are set to release the same amount of soup into each can. The filled cans are then mechanically sealed.

12 The cans are conveyed to a tall, cylindrical steam cooker called a retort. There the cans are subjected to blasts of heat for about 30 minutes for sanitation purposes.

13 The cans are mechanically lifted out of the retort and placed on a conveyer belt. Water is sprayed onto the cans to cool them.

14 The cans are conveyed to the labeling station and labels are affixed.

Quality Control

The the quality of ingredients in condensed soups is extremely important. Some companies, such as Campbell's, contract farmers to grow crops that are used in their products for quality assurance. Crops like tomatoes are processed as soon as possible after harvesting. The tomato pulp is processed at harvest and stored so that it may be used throughout the year. Certain companies also employ a staff of expert butchers to prepare the meat for stock and professional chefs who taste the soup throughout the manufacturing process. The United States Food and Drug Administration (FDA) enforces that many quality assurance steps are met. Any soups found to be harmful to the consumer are recalled back to the plant.

The Future

Although the production of ready-make soups has improved and increased since the early days of the Campbell Soup Company, condensed soups remain popular among consumers. In the United States, 2.5 billion bowls of condensed tomato, cream of mushroom, and chicken noodle soups are eaten annually. Internationally, Campbell's tailors its soups to various cultures, making watercress and duck gizzard soup in China and cream of chili poblano in Mexico. The company also promotes the use of condensed soups as sauces in many recipes. A soup company must remain competitive in order to keep up with the consumer's tastes and price range. Today, more than one million cans of soup are used every day, and Campbell's has 67 varieties of condensed soup. Trends in genetic research may help to grow a larger amount of above average-sized vegetables for soups. This would increase the volume that the soup companies would put out, and perhaps decrease the price to the consumer.

Where to Learn More

Books

Collins, Douglas. *America's Favorite Good: The Story of the Campbell Soup Company.* New York: Harry N. Abrams, 1994.

Other

"History of Campbell's Soup." *Campbell's Community Web Page.* December 2001. <http://www.campbellsoup.com/center/history>.

Smith, Andrew F. "History of Soup." *Chef Talk Web Page.* December 2001. <http://cheftalk.com/index.shtml>.

Mary McNulty

Corset

Background

The corset is an undergarment traditionally made of stiffened material laced tight to the body in order to slim a woman's waist. Evidence shows that some type of waist-cinching garment was worn by Cretan women between 3000 and 1500 B.C., but narrow waists became the fashion among women in Europe during the Middle Ages. Women from that period wore a forerunner of the corset, called a body or stay, or a pair of stays. The rigid, bust-to-hip corset became popular in the sixteenth century and persisted in various guises up through the middle of the twentieth century. It was considered beneficial to women's health by some doctors and writers, while others considered the constricting garment a virtual torture. Corset making was a specialized sub-sector of the garment industry. Tailors called staymakers were experts in the fitting and forming of corsets, which were sewn laboriously by hand. With the development of elastic textiles, corsets eventually became more yielding. Around the 1930s, women's fashions started emphasizing a more natural figure and the corset gradually became extinct. The closest thing to a modern corset is the all-in-one foundation undergarment.

History

Archaeological evidence shows that women wore surprisingly modern-looking undergarments as far back as 3000 B.C. in Babylonia. A Cretan figure dating from about 2000 B.C. was unearthed by British archaeologist Sir Arthur Evans in the late nineteenth century. It showed a bare-breasted woman with a tiny waist cinched tight by what looks like a ribbed belt. Ancient Greek writings refer to a women's undergarment made of linen or kid, cinching in the waist, and perhaps flattening the bust. Roman women also probably wore some sort of undergarments, but the general style was for long and loose clothing. This style persisted, for both men and women, through the Middle Ages. It was around 1150 that European women's clothing had a recognizable waistline. This was accomplished by lacing in an otherwise loose dress. A twelfth century British manuscript gives evidence of a tightly laced "shapemaker" worn as an outer garment.

The tailoring skills to make intricately cut and shaped clothing did not really develop in Europe until the middle of the fourteenth century. About this time, women began wearing an undergarment of stiffened linen, tightened by front or back laces. In the fifteenth century this item was known as a pair of stays or bodies in English and *corps* or *cors* in French. The English word corset presumably comes from a version of the French cors. At first corsets were made of two layers of linen, held together with a stiff paste. The resulting rigid material held in and formed the wearer's figure.

From the sixteenth century on, corset makers started using thin pieces of whalebone— shaped like quills or knitting needles—in between two layers of corset material. The whalebone corset was much more confining than the paste-stiffened one and often worn in conjunction with other undergarments that further exaggerated the female shape. In Queen Elizabeth's time, the fashion among the court classes was for a long, stiff corset reaching from the bust to below the natural waistline, paired with a huge, whale bone-stiffened hoop skirt called a farthingale. In the nineteenth century, women wore their corsets along with a cage-like hoop contrap-

Evidence shows that some type of waist-cinching garment was worn by Cretan women between 3000 and 1500 B.C., but narrow waists became the fashion among women in Europe during the Middle Ages.

tion—a crinoline—that held her skirts far out to the sides and back. The corset also accompanied the bustle, a padded device that emphasized the woman's backside. Corsets changed with fashion, becoming longer or shorter, supporting the bust or minimizing it, depending on the whim of the day.

Improvements in the manufacture of latex in the early 1930s led to workable elastic threads that could be woven or knitted into fabric suitable for undergarments. Soon the elastic corset became the norm. This was a much more flexible garment than the earlier rigid corset, and as the garment changed the name changed too. What had been called a corset became the roll-on, then came the step-in and the corselette. By 1940, women's underwear in Europe and the United States had evolved in favor of a two-piece arrangement; a brassiere for the bust and a roll-on or panty-girdle for the waist. The corset returned briefly after World War II in the guise of the waspie—a short, boned corset to wear with the tight-waisted dresses in high style at the time—but was never an everyday item again.

Health effects of the corset

European women throughout the Victorian era wore tightly laced corsets that were assuredly uncomfortable and in many cases actually injurious to health. Young girls were put in corsets to grow accustomed to the restrictiveness. Many illustrations and contemporary references from the turn of the century depict the painful process of tightening the corset. The corset wearer would lie on her stomach on the floor, while someone else put a foot on her back and pulled the laces. Women who perpetually wore tight corsets suffered from a variety of health problems, including deformed spines and ribcages, difficulty breathing, and compression of the internal organs. Around the turn of the century, several corset makers introduced new corsets designed by doctors. These aimed to support a woman's figure without undue compression.

In the early twentieth century, upper-class women had more access to physical activities such as sports and bicycling. With the tango craze just before World War I, women took to removing their corsets before a dance. Corset manufacturers introduced sports and dance corsets to accommodate these new activities. While some corsets were becoming looser and more comfortable, women were still admonished to wear them. Though some doctors spoke out about the danger to women's health of tight lacing, a conflicting and equally scientific-sounding opinion claimed that going without a corset was unnatural and unhealthy. Historical evidence—from the Cretan figurine to cave paintings—was used to uphold the idea that women had always needed figure support. One popular opinion was that evolution was more difficult for women than for men and the corset was essential to keep women upright. Thus only a small, radical minority actually advocated abandoning the corset.

Raw Materials

Corsets were made of a variety of materials, depending on the time period and the fineness of the article. The main fabric for the body of the corset might have been linen, stiffened with paste or starch. Lower-class women would have worn corsets of a cheap, sturdy cotton cloth. Corsets were also made of decorative fabrics like satin or silk.

The whalebone used to stiffen corsets was technically not bone at all but the teeth-like structures, called baleen, of a baleen whale. Baleen whales have hundreds of horny plates arranged in their upper jaws that serve to sieve tiny marine animals out of the water. Baleen is somewhat of an intermediary material between horn and hair, made up of many parallel hair fibers encased in hard enamel. Each baleen plate is about 10 in (25.4 cm) wide and 9–13 ft (2.74–3.96 m) long. Baleen can split along the parallel fibers and—when softened by steam—is easily shaped. Once dry, it holds its shape proving to be an enormously useful material for corset-making. Over-fishing led to the demise of baleen whale populations, and corset makers were driven to find substitute materials. They used cane or steel, and later plastic. The corset maker inserted thin slivers of whalebone into the corset to hold its shape. Whalebone was also used in some corsets for a front piece called the busk. The busk gave a smooth line to the front of the corset and was also sometimes made of wood, horn, or steel.

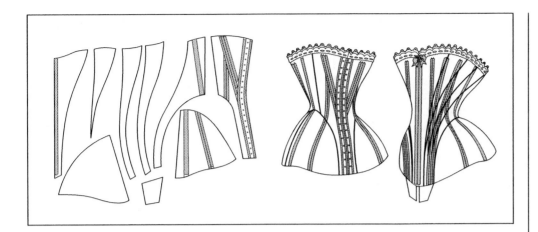

Metal eyelets for corset lacing were introduced in France in 1828. Elastic was used in corsets as early as the 1890s, but at first this material was suitable only for small shaped pieces called gussets. Around 1930, manufacturers learned to extrude latex into long fibers, making it possible to knit or weave a variety of elastic fabrics. Elastic became the norm in corsets and other undergarments in the 1930s.

Corsets were finished with a variety of decorative effects, including lace and ribbon. The thread used to stitch the corset together may have been strong silk or waxed cotton, depending on the garment.

Design

Corsets were designed to fit exactly to an individual wearer, otherwise the effect was lost or the garment would be even more uncomfortable. Though a corset maker might follow a standard design, each had to be modified for the individual customer's height, weight, and figure. For a fine corset, the wearer would be fitted twice. First, the corset maker made basic measurements of the customer's torso, then cut the material to measure. The garment was roughly sewn, using long stitches called tacking. The customer was then fitted again and any adjustments noted. The tacking was undone and the corset sewn back together, using fine, short stitches.

In terms of the fashion aspect of design, the corset changed along with the mode of dress. If dressmakers brought out a line of small-waisted gowns, then corset makers obliged them with tight corsets. The fashionable figure of the "Gibson Girl" in the early years of the twentieth century brought on a craze for the S-curve corset, which thrust the bust forward and the hips back. In the 1920s, the flapper style of dress needed no corset or only a straight-lined, non-constricting one. As noted above, several doctors designed what they considered healthful corsets, and corset makers also responded to cultural trends, such as the tango, by producing special use corsets.

The Manufacturing Process

Corsets were most often made by specialized corset makers. Elaborate corsets required great ingenuity in cutting and stitching and each had to be specially ordered and fitted, but simpler corsets for every day could be made at home. The following manufacturing process is for an eighteenth-century corset made by a professional corset maker.

1 The corset maker was usually a man and his assistants were usually women. He would start by taking measurements of the customer, either in her home or his shop. Then these measurements were used to make a pattern out of stiff paper.

2 The corset maker laid the paper pattern on a heavy material such as cotton drill or coarse linen. After tracing the pattern, it was cut out with scissors.

3 These cut pieces were laid on a different material (such as muslin) that would form the softer inner lining. The lining was also cut from the pattern.

4 Some corsets also had a third layer, an outer covering of some fine material

such as silk. These pieces would be cut in the same way.

5 The layers of the corset were then tacked together (sewn with long, light stitches). With a ruler, the corset maker made parallel lines 0.25 in (6.3 mm) apart, marking where the whalebone would go.

6 Then tight, straight stitches were sewn along the lines. This made cases between the two layers of cloth, to hold the bones.

7 Usually the corset maker had to cut the whalebone to size, but by the eighteenth century whalebone was available already split into strips. The corset maker cut the strips to size and rounded and filed the ends. Then the bones were pushed into the spaces in the corset pieces.

8 Next the eyelet holes were made. These would be punched with an awl and finished with a buttonhole stitch.

9 All the corset pieces were then tacked together. The corset maker steamed the whalebone into shape with a hot iron, and the corset was left to dry on a dressmaker's dummy.

10 Now that the corset was roughly put together, the customer was fitted again and any alterations were noted. Then the tacking was undone and the corset was stitched back together with strong thread and short stitches.

11 Once the corset was fitted to the customer, the maker added extra shaping bones and the busk. The busk was made of whalebone, horn, wood, or steel, and inserted through the center front of the corset. The corset maker shaped any additional whalebone with an iron and inserted these where needed, such as to hold in the waist or shape the bust.

12 Finally a layer of fine cloth was sewn on top if needed. Other finishing touches included sewing on loops to hold petticoats and stockings.

Quality Control

Corsets were generally very finely constructed articles made to order, so quality control was not an issue. In the 1930s, when corsets were waning in popularity, the corset industry made a concerted effort in the United States to train corset saleswomen in "scientific" fitting. Clerks in department stores specialized in corset fitting and generally spent a long time with customers, making sure each left with a suitable garment. Controlling the quality of the fit was very important and depended on a knowledgeable sales force.

Byproducts/Waste

The most notable byproduct of corset manufacturing was the whale. Though whales were also hunted for their oil, it is a fact that the craze for corsets and hoop skirts led to an over-fishing of baleen whales. By the end of the sixteenth century, the Atlantic Right whale was almost extinct in the popular Bay of Biscay fishing ground. When Biscay whales became hard to find, the whaling industry moved to waters off Greenland. This fishing ground was also seriously depleted by the late eighteenth century. After the 1840s, Bowhead whale were hunted for their whalebone, primarily caught by American fishermen in the Arctic. Whale oil was not used much after the discovery of petroleum in 1859, so whales hunted in the late nineteenth century were killed almost exclusively for their baleen. The Bowhead was almost completely extinct by the early twentieth century, just as the use of corsets was declining and new elastic materials made whalebone obsolete.

Where to Learn More

Books

Ewing, Elizabeth. *Dress and Undress.* London: B. T. Batsford, Ltd., 1978.

Shep, R. L. *Corsets: A Visual History.* Mendocino, CA: R. L. Shep, 1993.

Waugh, Norah. *Corsets and Crinolines.* London: B. T. Batsford Ltd., 1954.

Periodicals

Fields, Jill. "Fighting the Corsetless Evil: Shaping Corsets and Culture, 1900–1930." *Journal of Social History* (Winter 1999): 355 ff.

Angela Woodward

Cyclotron

Background

The modern cyclotron uses two hollow D-shaped electrodes held in a vacuum between poles of an electromagnet. A high frequency AC voltage is then applied to each electrode. In the space between the electrodes an ion source produces either positive or negative ions depending on the configuration. These ions are accelerated into one of the electrodes by an electrostatic attraction, and when the alternating current shifts from positive to negative, the ions accelerate into the other electrode. Because of the strong electromagnetic field, the ions travel in a circular path. Each time the ions move from one electrode to another they gain energy, their rotational radius increases, and they produce a spiral orbit. This acceleration continues until they escape from the electrode. The accelerated particles are extracted from the cyclotron when they reach the end of the spiral acceleration path. This beam of accelerated subatomic particles can be used to bombard a variety of target materials to produce radioactive isotopes.

Various isotopes are used in medicine as tracers that are injected into the body and in radiation treatments for certain types of cancers. Cyclotrons are also used for research purposes in academic and industrial settings, and for positron emission tomography (PET). Positron emission tomography (PET) is a technique for measuring the concentrations of positron-emitting radioisotopes within the tissue of living subjects. The usefulness of PET is that, within limits, it has the ability to assess biochemical changes in the body. Any region of the body that is experiencing abnormal biochemical changes can be seen through PET. PET has had a huge impact on the clinical applications of neurological diseases, including cerebral vascular disease, epilepsy, and cerebral tumors.

History

E. O. Lawrence and his graduate students at the University of California, Berkley tried many different configurations of the cyclotron before they met with success in 1929. The earliest cyclotron was very small, using electrodes, a radio frequency oscillator producing 10 watts, a vacuum, hydrogen ions, and a 4 in (10 cm) electromagnet. The accelerating chamber of the first cyclotron measured 5 in (12.7 cm) in diameter and boosted hydrogen ions to energy of 5–45 MeV depending on the settings. One mega electron volt (MeV) is 1.602×10^{13} J. (J stands for Joule, the standard unit for energy.) The design, construction, and operation of increasingly larger cyclotrons involved a growing number of physicists, engineers, and chemists. Lawrence was never certain as to whether his research should be classified as nuclear physics or nuclear chemistry.

Raw Materials

The magnets in the cyclotron are made from 25 tons of low carbon steel with two nickel plated poles. Physically, the cyclotron weighs 55 tons, and is located inside an inner vault with concrete walls and doors about 6.6 ft (2 m) thick to shield the surroundings from the nuclear radiation present when the machine runs. Fortunately, most of this radiation has a half-life of only seconds to minutes, so there are no long-term waste disposal problems. Actual dimensions are approximately $100 \times 100.5 \times 39$ ft ($30.5 \times 30.6 \times 11.9$ m). The coils are manufactured from annealed copper, insulated with fiber-

The earliest cyclotron was very small, using electrodes, a radio frequency oscillator producing 10 watts, a vacuum, hydrogen ions, and a 4 in (10 cm) electromagnet.

Ernest Orlando Lawrence.

E rnest Orlando Lawrence was born in South Dakota on August 8, 1901. He received his bachelor's degree in physics in 1922 from the University of South Dakota. Lawrence entered the University of Minnesota graduate school, completing his master's degree in one year. He received his Ph.D. at Yale in 1925, remaining there for three years as a fellow of the National Research Council, then as assistant professor. In 1928 he became associate professor at the University of California at Berkeley. Two years later Lawrence became the youngest full professor at Berkeley.

Lawrence conceived his most famous invention, the cyclotron, in 1929. He realized that to achieve particle energies of a few MeV (million electron volts) required for nuclear experiments, he could convert the particle's linear trajectory into a circular one by superimposing a magnetic field at right angles to the particle's path. Lawrence immediately proved that a particle's frequency of revolution depends only upon the strength of the magnetic field and the charge-mass ratio of the particle, not upon the radius of its orbit. This was the basic principle of the cyclotron, which Lawrence first reported in the fall of 1930.

In 1932, Lawrence married and had six children. He was elected to the National Academy of Sciences in 1934, awarded the Nobel Prize in physics in 1939, and received the Medal of Merit in 1946 and the Fermi Award in 1957. Lawrence remained at Berkeley until his death August 27, 1958 from an intestinal ulcer.

select different targets on each of the beamlines to be irradiated and are made primarily from aluminum, with a minimum of stainless-steel to minimize neutron activation.

Design

The design of the cyclotron varies according to the specifications of the purchaser. Ebco Technologies Inc. builds two different types of negative ion cyclotrons, one capable of accelerating protons to a maximum energy level of 19 MeV (TR19) and the other capable of accelerating protons to 32 MeV (TR32). The standard configuration of the TR19 cyclotron is with two external beamlines but there is a scaled down version with an option of one beamline. The TR19 standard target configuration is with two external beamlines and eight targets. There is a design option of two to four targets on one beamline, with the upgrade to up to eight targets at a later date. The TR19 is also available in a self-shielded or unshielded configuration. The self-shielded feature eliminates the need for a cyclotron vault or major upgrades to existing facilities. Additionally, the magnet gap in the TR19 is vertical to minimize space.

The radio frequency (RF) system consists of a RF amplifier, a coaxial transmission line from the RF amplifier to the cyclotron, a power supply, and instrumentation and readback devices, an oscilloscope, current/voltage, power gauges, and interfaces with the computerized control system. A mass flow controller, needle valve, and pneumatic valve regulate the gas pressure and flow.

A tungsten filament is placed inside the ion source and when heated will ionize the hydrogen gas. A plasma filter is placed on the ion source aperture to enhance conditions for negative ion production.

The negative ions generated will be injected into the cyclotron at its X-axis. The injection system is manufactured from a set of steering magnets to focus the negative ions onto the plane of acceleration by the tilted spiral inflector.

The Manufacturing Process

1 Project teams coordinate conduit, cable tray, floor duct, and related equipment

glass and covered with an epoxy resin. The aluminum vacuum tank is sealed by polyurethane o-rings. The ion source uses a tungsten filament to energize the hydrogen gas and borated polyethylene packing is used to reduce the build up of thermal neutrons around components of the cyclotron. The target changer allows the cyclotron operator to

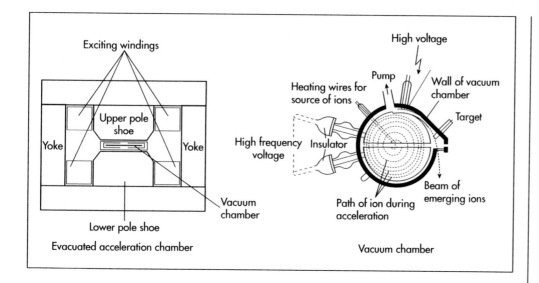

prior to the shipping, rigging, and installation of the cyclotron and its sub-systems.

2 The manufacturing process begins with the 25-ton steel magnet. It is machined from 10-in (25.4-cm) slabs and placed in-between the poles of a powerful electromagnet until the magnetic field area is precisely measured.

3 Two nickel plated magnetic poles are forged from low-carbon steel.

4 Two magnet coil assemblies are manufactured from annealed hollow copper and harden after being bent into shape. They are mounted in the yoke of the magnet, connected to water cooling headers, insulated with fiberglass, and coated in an epoxy resin.

5 The aluminum vacuum tank is placed between the nickel plated poles and bolted into place. The vacuum tank has cryopumps that are bolted externally to cool the tank close to -459°F (-273°C) in order to freeze out any gases that may be present.

6 The electrodes are machined from a single 0.06-in (1.6-mm) low resistively copper sheet (to optimize the energy transfer from the RF system to the accelerating hydrogen ions), cut out, and etched using boaring tools and drill bits.

7 Next, the tank is sealed with polyurethane o-rings after the copper electrodes are mounted inside. The electrodes are set, using nylon screws and spacers, into a round piece of industrial lisex nylon. A few holes are drilled in the nylon. Two are for the oscillator wiring. The third is meant for the vacuum pump; there is also a vacuum gauge attached to this port.

8 On top of the nylon and surrounding the electrodes is a ring of polyvinyl chloride (PVC) pipe. This has several holes drilled into it, the largest of which is the detector storage tube. Also located in this material are smaller holes sufficient for supplying a voltage source to the deflector plate, for the set screws required to control its position, and attachment holes for the solid brass hook that will be used to hang the complete apparatus on a set of Helmholtz coils.

9 Atop the PVC pipe is a piece of industrial strength clear plastic. This is both to allow people to see the inside workings of the mechanism, should anything go wrong, as well as increase the strength of the casing.

10 On either side of the PVC is silicon gel, in order to maintain a sufficient seal around the main chamber. This is so that the vacuum will be as efficient as possible. The vacuum is needed because the alpha particles are heavily influenced by particles of any kind, especially air. That is why alpha particles are considered so safe; by the time they contact a person through any medium, their energy has been so severely affected, they are not able to do damage.

11 The walls are guided in place by a thin cut in the face of both the top and bottom sheet and both electrodes are held together with the use of 2 in (5.1 cm) nylon

screws. No solder was used in these pieces so as to keep the inner chamber as clean and constant as possible. In one wall is cut a window, roughly 0.79 in (2 cm) long.

12 Pivoted on a nylon screw is a slightly smaller copper plate (the deflector) separated electrically from the rest of the component. Outlying set screws can control the deflector position and both it and each electrode have an electrical connection. This is to allow the oscillator to be supplied to the electrodes and a large negative charge to be put on the deflector plate.

13 The RF system is assembled inside a 19-in (48-cm) square, 6-ft (1.8-m) high metal chassis. Here, the resistors, transmitters, switches, tuning circuits, inductors, and capacitors are assembled by hand.

14 Power supply cabinets are purchased and assembled for the water-cooled targets and magnets, ion sources, cryopump, and the water circuitry.

15 The ion source will be injected after assembly of the cyclotron. A magnetic cylinder, 4 in (10 cm) in diameter and 4.7 in (12 cm) long comprises the ion source. Hydrogen gas will be injected through a capillary tube.

16 The tilted spiral inflector is enclosed by a grounded helical shaped electrode. The electrode is machined on a fixed axis milling machine.

17 Next, the target bodies are made of high purity silver, aluminum, and titanium and designed with helium-cooled thin foil windows. The two foil windows separate the target material from the high vacuum within the cyclotron.

18 A recirculating closed loop cooling system is placed in the target services metal cabinet to cool the foil windows with high speed streams of helium gas.

19 The tubing connections, solenoid valves, water-cooled beam stops, and electrically isolated collimators are assembled and attached to the target assembly.

20 The target assembly has a solid aluminum plug that is pierced by a 4 in (10 cm) hole that will act as the target collimator.

21 Grooves are machined onto the outside of the plug, and the o-ring is mounted to create the vacuum seal between the target body and the four position target changer.

22 A collimating disc is placed between the plug and the target body with a window on both sides.

23 Finally, the entire system is integrated with supervisory software to control and monitor the PLC hardware.

Quality Control

Each step of the manufacturing process must be monitored to ensure that the parts are of standard quality. If any of the components have a crack or leak, radiation may get into the environment. The steel used in the magnets of the cyclotron is carefully monitored to ensure it has the desired properties. Magnetic fields are constantly checked by Nuclear Magnetic Resonance (NMR).

Byproducts/Waste

The manufacturing process yields 2–3 tons of metal waste during production. This is recycled for future manufacturing processes. Due to the number of parts, the excess material from the manufacturing of the cyclotron is large. If any defective parts are found they are salvaged to the best of their ability, but the majority are scrapped.

The Future

The improvements in sealing the cyclotron unit are requiring that less concrete shielding be provided at the installation site and provide a safer and more compact cyclotron unit. More powerful cyclotron units are being designed for commercial isotope production. The latest series of cyclotrons are state of the art, compact, strong focusing, four sector negative ion cyclotrons, with external ions sources, cryopumps, high precision power and control systems, and superb manufactured quality. They are now modular in design and share a common technology irrespective of the size and type of cyclotron.

Where to Learn More

Books

Lawrence, Ernest O., and Irving Langmuir. *Molecular Films: The Cyclotron & The*

New Biology. New Brunswick: Rutgers University Press, 1942.

Periodicals

Burgerjon, J. J., and A. Strathdee, eds. *Cyclotrons—1972.* New York: American Institute of Physics, 1972.

Bonny P. McClain

Diving Bell

Diving bells were known as early as the fourth century B.C., when they were observed by the ancient Greek philosopher Aristotle.

Background

Commercial divers doing underwater construction or salvage often use a diving bell for transportation to the underwater site. Use of a diving bell (also known as a Personal Transfer Capsule, PTC) and a pressure chamber extends the amount of time a diver can safely stay underwater. Diving bells were known as early as the fourth century B.C., when they were observed by the ancient Greek philosopher Aristotle. More sophisticated diving bells were devised in the seventeenth century. Modern bells for commercial diving were developed after World War II, with the rise of the offshore oil industry.

Commercial diving (diving for pay) is divided into two main types, surface-oriented diving and saturation diving. In surface-orientated diving, divers in helmets work underwater, connected to a breathing apparatus on shore or on board a ship, barge, or platform. Typically divers work in pairs, one underwater and one at the surface tending the hoses and equipment. Surface-oriented divers can work safely at depths up to 300 ft (91.5 m), but divers can only spend a limited amount of time underwater. The effects of water pressure can lead to decompression sickness. Under pressure, nitrogen collects in the diver's body tissue, blocking the arteries and veins. If the diver rises too quickly, the nitrogen forms bubbles in the tissue, something like the way a soda bottle bubbles when uncapped. Gas bubbles in the tissue cause pain, paralysis, or death. After a deep dive, the diver needs to decompress gradually, returning very slowly to the surface pressure in order to avoid decompression sickness. Decompression time is related to the depth of the dive and the duration. With a deep dive of only one hour, decompression time can take days. Surface-oriented diving is only practical for small jobs.

The second type of commercial diving, saturation diving, is more useful for large-scale construction projects. In saturation diving, divers use a pressurized chamber, sometimes known as a Deep Diving System (DDS), attached to a diving bell. The chamber and bell begin on board a ship. A team of divers boards the chamber, which is then mechanically pressurized to simulate the environment at the depth of the planned dive. The chamber is a complete living environment—equipped with beds, shower, and furniture—and able accommodate a team of divers for weeks. When the divers are acclimated, they exit the chamber through a mating tunnel and enter the diving bell, which is also pressurized. A crane lifts the bell off the ship and drops it to the underwater site. Once at the site, one diver exits the bell in a diving suit and helmet and begins working. The other diver remains in the bell and tends the first diver's hoses and equipment. After an interval of perhaps two hours, they switch. Working from a bell, the divers may put in an eight-hour day underwater. Then they are ferried to the surface in the bell, enter the pressure chamber, and switch with the next shift of divers. When the entire job is completed, the team decompresses in the pressure chamber. Though they have submerged multiple times the team only needs to decompress once.

History

A bucket or barrel lowered straight into the water, open end down, will trap air inside it. Aristotle wrote of divers using air-filled cauldrons to breathe underwater. Alexander the Great was said to have gone to sea in a diving bell—reputed to be a barrel of white glass—in 332 B.C. He was said to have stayed deep underwater for days, though this is not plausible. There are several references to diving bells in the Middle Ages. In 1531 an Italian, Guglielmo de Lorena, made a workable diving bell that he used to recover sunken ancient Roman ships from the bottom of a lake. Other bells were invented and used in various places in Europe, mostly to salvage treasure. The forerunner of the modern diving bell was invented by Englishman Edmund Halley, who is best known for the comet bearing his name. In 1690 Halley built a diving bell that used leather tubes and lead-lined barrels to supply fresh air underwater. His bell was a wooden, open-ended cone, weighted with lead and fitted with a glass view port. Inside, Halley hung a platform for the diver to rest on, and a contraption of weighted barrels. The barrels were fixed so that when the diver pulled them into the bell, water pressure from below forced them to release fresh air into the bell. Helpers on the surface refilled the barrels with fresh air. Halley and a team of divers managed to stay underwater at a depth of around 60 ft (18.3 m) for as long as an hour and a half using his bell.

Others duplicated Halley's achievement, but the design was not significantly improved until 1788. In that year, a Scottish engineer, John Smeaton, made a diving bell that used a pump on its roof to force fresh air inside. Smeaton's bell was used by divers doing underwater bridge repair. A variety of diving equipment was invented in the nineteenth century, leading to workable diving helmets connected by hoses to an air supply on the surface. This equipment tended to be heavy and bulky, made with hundreds of pounds of metal to withstand deep water pressure. Workers on tunnels and bridges went down in huge cast iron bells or elevator-like chambers called caissons. As little was known about the hazards of pressure, many of these workers sickened and died of what was called caisson sickness, now know to be decompression sickness.

The groundwork for future commercial diving was laid after World War II. The Swiss diver Hannes Keller used a diving bell in 1962 to reach a depth of 984 ft (300 m). His bell was at a slightly higher pressure than his dive site. Keller breathed a mixture of heli-

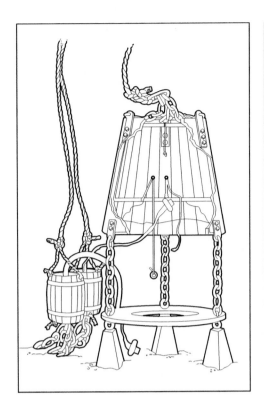

A Halley bell.

um and oxygen through hoses attached to a machine in the bell. He showed that the diving bell could be a valuable way-station for a deep diver, supplying not only breathable gas but also electricity, communication devices, and hot water to heat the diving suit.

Saturation diving was made possible by the work of Dr. George Bond, director of the United States Navy Submarine Medical Center in the mid 1950s. His experiments showed that a diver's tissue became saturated with nitrogen after a certain time of exposure. After the saturation point was reached, the duration of the dive was unimportant. A diver could remain under pressure for weeks or months. The time needed for decompression would be the same, whether the diver stayed at the saturation point for an hour or a week. Bond's experiments led to the development of Deep Diving Systems. These were used frequently by workers in the oil industry in the 1970s and 1980s, when deep offshore oil drilling platforms flourished.

The bathysphere and the bathyscaph

Two important modern diving bells were the bathysphere and the bathyscaph. These were deep sea diving vessels made for scientific observation. The bathysphere was built by

William Beebe, an American zoologist, and engineer Otis Barton in 1930. Beebe, fascinated with underwater life, conceived of the diving machine, and Barton was able to design it. Barton's idea was to make the chamber perfectly round to evenly distribute the water pressure. It was manufactured from cast steel a little over 1 in (2.5 cm) thick and 4.75 ft (1.5 m) in diameter. The bathysphere weighed an enormous 5,400 lb (2,449 kg), almost too heavy for the available crane to lift. Beebe and Barton made multiple dives off Bermuda in the bathysphere, reaching a depth of 3,000 ft (900 m) in 1932. Due to the great strength of the sphere the divers were protected from pressure, but the bathysphere proved unwieldy and potentially risky. It was abandoned in 1934.

A decade later, a Swiss father and son, Auguste and Jacques Piccard, designed a similar vessel called the bathyscaph. The bathyscaph resisted the effects of pressure, like the bathysphere, with a heavy steel spherical chamber. The chamber hung beneath a large, light, gasoline-filled container. Releasing air valves allowed the bathyscaph to lose buoyancy and sink to the ocean floor under its own power. To come up again, the operators released iron ballast, causing the vessel to slowly rise. The first bathyscaph was built in 1946, but irreparably damaged in 1948. An improved machine descended to 13,000 ft (4,000 m) in 1954. The Piccards built another bathyscaph, named the *Trieste*, in 1953. The United States Navy bought the *Trieste* in 1958. Jacques and Navy lieutenant Donald Walsh reached a record depth of 35,810 ft (10,916 m) in the Mariana Trench in the Pacific in 1960.

Raw Materials

Modern diving bells are made of high-strength, fine-grain steel. Windows are constructed from cast acrylic of a special grade designed for pressure vessels. The bell also needs an exterior girding made of thick aluminum to protect it from shocks. The bell is painted with a high-grade marine epoxy paint. Steel and aluminum specifications vary depending on the expected depth of the vessel.

Design

Diving bells are custom-built according to customer specifications. The customer ap-

proaches the manufacturer with an outline of what is needed. Depending the needs, the outline will specify bell shape, minimum number of occupants, number of windows, and any other special needs, such as racks to hold equipment. The manufacturer looks over the customer's plan, and then draws up a final design.

The manufacturing and design of diving bells is carried out under specific regulations provided by the American Society of Mechanical Engineers (ASME). ASME has a sub-section regulating what are generally called Pressure Vessels for Human Occupancy, or PVHOs. PVHOs include diving bells as well as submersible vessels, decompression chambers, recompression chambers, high altitude chambers, and others. ASME lays out strict standards for all aspects of diving bells, from the design through fabrication and testing. Manufacturers and their subcontractors must all follow the ASME guidelines step-by-step through the manufacturing process in order to receive an ASME stamp on the finished bell.

The Manufacturing Process

Making the bell

1 The body of the bell is formed from strong, fine-grained steel. Rolled steel plate is put on a conveyor belt and sent through an automated saw that cuts the plate into the top, bottom, and sides of the bell.

2 The sections are sent to a welding shop certified for this type of construction. Each section is manually welded together. The welds must be able to resist high pressure and be absolutely water tight. The welding shop follows the guidelines laid down by ASME.

3 Cast acrylic windows, either made by a sub-contractor or by the bell manufacturer, are fitted into place.

Inspection and testing

4 After the sections are welded together, the bell is inspected. It may undergo various tests, from visual inspection of the welds to ultrasonic scans. After these tests comes the "proof test." The bell is filled with water and pressurized for one hour at

one and a half times the pressure it was built to withstand. In other words, if the bell was designed to withstand the pressure found at a depth of 600 ft (183 m), 282 psi, the manufacturer subjects it to pressures found at 900 ft (274.3 m), or 415 psi. The bell should easily be able to withstand the proof test. It has been designed to withstand a pressure of four times its general use pressure, as a safety precaution.

Painting and finishing

5 Next the bell is painted. Mechanical sprayers coat the bell with a high-grade marine epoxy paint that is able to withstand the rough use the bell will endure underwater.

6 Then the interior of the bell is finished. The bell will hold a variety of devices such as a heater, instruments, lights, carbon dioxide remover, and fans. Brackets for these devices are bolted onto the inside of the bell. Piping and wiring cases are also bolted into place. The bell is not ready for use until all the equipment is in place.

Certification

7 If the bell passes all the tests and inspections, it is stamped with an ASME seal. This means that it has been built in accordance with ASME standards, and is presumed safe for human occupancy. The individual bell is also given a certificate recording where it was built, when, and by whom. Other records are also kept, such as the origin of the steel used for the body.

8 The manufacturer delivers the bell as a "raw" vessel. The customer then outfits it with all the needed machinery such as tracking devices, cameras, and radio transmitters.

Quality Control

Quality control is extremely important for a vessel used for inherently dangerous underwater work. Quality control is built into the diving bell manufacturing process, because manufacturers follow the standards laid down by ASME. Not only is the bell tested after construction, but even the preliminary design has been carried out in a way that satisfies ASME rules. The overall regulatory authority over diving, including

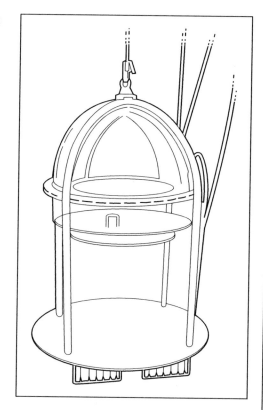

A modern Personnel Transfer Capsule (PTC).

commercial diving, in the United States is the Coast Guard.

The Future

The United States Navy also tests various diving equipment for its own use. It runs an Experimental Diving Unit that tests existing equipment and tries out cutting edge diving technology. The Experimental Diving Unit also employs doctors and researchers who investigate the physiological effects of diving. Some of this research may lead to regulations effecting commercial divers. This in turn may effect safety procedures and quality control tests for diving bells and other diving apparatus.

Commercial divers rely on diving bells every day for transportation between a pressurized chamber and a deep sea site. The development of saturation diving led to a much more efficient way of carrying out extensive underwater work, because divers only need to decompress once at the end of the job. Some current research, however, investigates ways to do without decompression altogether. Some researchers have investigated the possibility of equipping divers with artificial gills, allowing them to breathe oxygen directly from water. Another possible new technology is called liquid breathing. At deep pressure, if the lungs are filled with an oxygen-bearing liquid, they can theoretically continue to function. Hypothetically, a scuba diver might be able to breathe oxygenated liquid fluorocarbon from a portable tank. This would enable a diver to dive deeper without the use of a pressure chamber and diving bell. Another avenue of investigation is so-called biologic decompression. A special bacterium in the body could be used to metabolize the gases trapped in tissue that cause decompression sickness. This would eliminate the need for decompression in a chamber. If any of these technologies became viable for commercial divers, the existing system of pressure chamber and diving bell may alter.

Where to Learn More

Books

Beebe, William. *Half Mile Down.* New York: Dull, Sloan and Pearce, 1951.

Parker, Torrance R. *20,000 Jobs Under the Sea: A History of Diving and Underwater Engineering.* Palos Verdes Peninsula, CA: Sub-Sea Archives, 1997.

Piccard, Jacques, and Robert S. Dietz. *Seven Miles Down: The Story of the Bathyscaph Trieste.* New York: G. P. Putnam's Sons, 1961.

Periodicals

Bachrach, Arthur J. "The History of the Diving Bell." *Historical Diving Times* (Spring 1998).

Other

Diving Heritage Page. June 2001. <http://www.divingheritage.com>.

Angela Woodward

Drain Cleaner

Background

Drain cleaners, sometimes referred to as drain uncloggers, are solutions that are poured into sluggish or clogged drains in order to clear them. These solutions are devised to dissolve human hair, human waste, or food particles that stop up kitchen sinks or tub and shower drains. These drain cleaners may take a variety of forms. Some are powders, but most are liquids that may be poured directly into the drain. There are two types of drain cleaners available today. One is the conventional, chemical drain cleaner, the other is the bio-degradable, environment-friendly, and chemical-free drain cleaner.

The chemical drain cleaner is by far the most popular drain cleaner sold in the United States as it is far more effective than the bio-degradable drain cleaner. There are two types of chemical drain cleaners manufactured in the United States. One type is acid-based and includes chemicals such as sulphuric acid. This is the most effective drain cleaner of the two chemical types but is by far the most dangerous. Sulphuric acid immediately eats away at any organic material that it comes in contact with. This means the acid effectively eats away at human waste and hair that may be in a sink, quickly unclogging drains. However, it also means that the acid eats at organic material such as skin. So if the drain cleaner splashes onto the user, the user has little time to wash off the acid before it starts eating and burning at the skin. The other type of chemical drain cleaner is made of caustic sodium hypochloride. The well-known chemical drain cleaners made by large national firms primarily produce the caustic drain cleaners. Caustic drain cleaners are cheaper to make than acidic drain cleaners, not as effective, and a little safer to use

as sodium hypochloride does not immediately burn the skin and can be washed off before significant damage is done.

The chemical drain cleaners must be used with extreme caution. The manufacturers recommend that they be used with gloves and goggles because the chemicals have been known to splash into the face and eyes and cause burns to the eyes. More serious concerns revolve around a consumer's inadvertent mixing of drain cleaners with other household chemicals and generating extremely dangerous vapors that can harm the user. Finally, the chemicals should be kept far out of reach of young children.

The bio-degradable drain cleaners simply clean the drains and generally do not unclog them, as they are not particularly effective at eating away the organic material. Some plumbers recommend these safer products as they feel chemicals actually damage pipes and plumbing, as well as possibly wreaking damage to the water supply once dumped down the drains.

History

Drain cleaners do not have an extensive history as indoor drains hooked up to a municipal plumbing system were not common in middle-class homes until the early twentieth century. Prior to that time Americans poured water from pitchers into basins to wash up and then discarded the dirty water. Drains without significant mesh drain covers clog quickly. It seems likely that lye, a caustic substance made from hardwood ashes and water, was poured down the drains when they became sluggish. Less dangerous concoctions of vinegar and bak-

The chemical drain cleaner is by far the most popular drain cleaner sold in the United States.

ing soda were combined to make minor bubbling in the sink as well and was supposedly another way to clear a drain. Some plumbers physically snake out a drain using a long flexible, metal plumber's snake that physically pushes through the drain obstruction.

The large bleach manufacturers in this country realized by the mid-twentieth century that the sodium hypochloride they produced for bleach could also be used to make a liquid plumbing solution. Produced in somewhat greater strength than bleach, the caustic drain cleaner eats through the organic material that clogs the drains. Acidic drain cleaners, somewhat more expensive to make and hazardous to human contact, is produced by specialty or smaller firms and is often sold to plumbers or for commercial use.

Raw Materials

Drain cleaners require only the acquisition and mixing of a few chemicals. Acidic drain cleaners are made from sulphuric acid and a few additives (generally proprietary information). Sulphuric acid is an indirect byproduct of the purification process related to iron ore. Today, the majority comes from Canada where much of North America's iron ore is produce. Sulphuric acid is the chemical compound H_2SO_4, is colorless, odorless, oily, and extremely corrosive. It reacts immediately to organic material and is dangerous to human skin. It reacts with many metals, particularly zinc, and mixing tanks must be made from metals that do not react to it. Drain cleaner manufacturers use steel for mixing and filling tanks as steel does not react with the sulphuric acid.

Caustic drain cleaner is sodium hypochloride, an alkaline substance made from the mixing of chlorine, sodium hydroxide (sometimes called caustic soda), and water. Surfactants are added into the caustic drain cleaners. Surfactants, or surface-active agents, may be added to reduce surface tension, promote mixing of solutions, and reduce foaming.

The Manufacturing Process

Acid-based drain cleaner and alkaline drain cleaner are manufactured using very similar steps.

1 Huge tanks of sulphuric acid—brought up to the tanks via rail cars or very large tank trunks–is pump the chemical directly into large bulk tanks within the factory. Enclosures are often brought around the tankers to ensure that fumes do not escape from the hoses or tanks and dangerous fumes or chemicals are released into the atmosphere. Hoses connect the distribution tanks with the tanks within the building because the solution is very dangerous to human skin. Human operators never touch any of these materials directly in any part of the manufacture of chemical drain cleaners.

2 The sulphuric acid is then transferred into another tank known as a mixing tank. Other additives are added to the sulphuric acid in order to make the acid effective, but slightly more dilute (so that the mixture is not pure sulphuric acid). The combination or choice of additives is considered part of the manufacturers' trade secrets and is generally not revealed. This implies the kinds of additives and their relative strengths varies greatly from manufacturer to manufacturer.

3 These ingredients are carefully mixed so they are evenly distributed within the mixing tank.

4 When the mixing is complete, the solution (now drain cleaner) is ready for dispersal into containers. The drain cleaner leaves the mixing tank and is fed into a filler tank that is outfitted with a nozzle or nozzles. Empty containers enter a turntable in which a mechanical head lowers a nozzle into the container, fills it with a pre-determined amount of fluid, and the filled bottle is then moved to the side. The bottles are automatically sealed and capped. These individual containers are always heavy-duty and are generally of thick polyethylene plastic. The container must not be easy to shatter, leak, spill, or be etched by the acid as the drain cleaner is extremely dangerous to the touch. The acidic drain cleaner is now ready for distribution.

5 Caustic drain cleaner is manufactured using very similar manufacturing techniques. The primary difference is that the caustic drain cleaner must have more additives infused into it to make a viable drain cleaner. After mixing, just as in the production of acidic drain cleaner, the solutions are

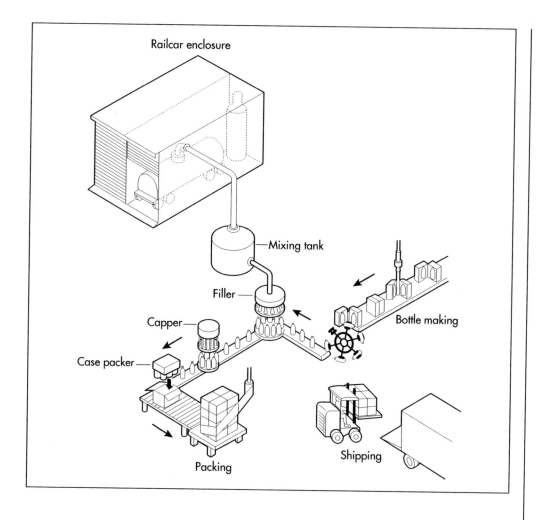

Railcar enclosure

Mixing tank

Filler

Capper

Case packer

Bottle making

Packing

Shipping

fed into the filler tank and dispersed into individual heavy-duty containers and sealed and capped.

Quality Control

Manufacturers heavily rely on suppliers who certify their products to be of the grade and strength of solution needed to produce effective drain cleaner that is still safe for ordinary household use. However, the solutions are periodically tested to ensure they are as presented to the manufacturer. The testing is carefully performed as the solutions are difficult to handle and can be dangerous to the tester. Hoses, fillers, and tanks must be carefully and vigilantly maintained to ensure that no dangerous solutions are leaked at any point during manufacture.

Byproducts/Waste

There is virtually no waste with the manufacture of drain cleaner. The solutions go from one tank directly into another without human

intervention. Unless batches are not mixed correctly, which very rarely occurs, all of the solution is placed into containers for sale.

The disposal of chemicals no longer wanted or needed by the consumer is a problem. Many suggest that the unused product must be taken to a hazardous household waste collection center available in larger municipalities. If not available, then it is suggested that the chemicals are poured down the sink with plenty of water. The chemicals must not be poured into a septic system. Similarly, the retail packaging once it is emptied by the consumer is somewhat of a problem. The container is considered contaminated with caustic or acidic chemicals and really should not be recycled despite recycle-compatible materials. They must be disposed of in the garbage, and then sent to landfills.

The Future

There is significant discussion about the continued use of acidic and alkaline drain cleaners. Many environmentalists are ardently opposed to their use as once they are poured down drains the acids (sulphuric acid) and alkalines (chlorine and sodium hydroxide) are released into the water supply. Some areas of the country have drains that have ties to lakes and ponds and these chemicals may be released into the environment. Environmentalists have worked hard to promote plumbing products that are deemed to be environmentally safe. These drain cleaners are varied in ingredients; the most successful of the products are bacteria or enzyme-based. These enzymes eat through sludge and are said to liquefy matter that clogs the drains such as grease, food particles, human waste and human hair. When released into the drain, and then into the water supply, these enzymes do not appear to upset the ecosystem nor harm the water supply for animal consumption.

While these environmentally-friendly products may be less harmful to the environment they are not particularly popular. They are not terribly effective and generally considered more of a cleaner rather than a drain unclogger. Some householders tend to prefer the less effective, environmentally-friendly drain cleaners simply because they are much safer to use. Sometimes the chemical solutions splash up into the eyes when poured into a clogged sink. Plumbers or homeowners may have to stick their hands in the water to check the drain and are then burned by the solutions they have poured into the drains. These chemicals can be difficult to use safely and are considered a hazard to have around children.

Where to Learn More

Books

Ierley, Merritt. *Open House: A Guided Tour of the American Home.* New York: Henry Holt and Company, 1999.

Ierley, Merritt. *The Comforts of Home.* New York: Clarkson Potter, 1999.

Rybczynski, Witold. *Home: A Short History of an Idea.* New York: Penguin Books, 1986.

Other

Clorox Web Page. December 2001. <http://www.clorox.com>.

Oral interview with Dennis West, Plant Manager of Rooto Corporation. Howell, MI. October 2001.

"Surfactants." *Encyclopedia Britannica CD Edition.* Encyclopedia Britannica Inc., 1994–1998.

Nancy EV Bryk

Dry Ice

Background

Dry ice is the name given to carbon dioxide when it is in a solid state. Carbon dioxide is found in the earth's atmosphere; it is a gas that humans exhale and plants use for photosynthesis. This chemical compound is colorless, odorless, tasteless, and about 1.5 times as dense as air. Carbon dioxide turns from gas to an opaque white solid while under pressure and at low temperatures, turning solid at -109°F (178.5°C). Dry ice is manufactured primarily in two forms, either as a block of dry ice which weighs over 50 lb (22.7 kg) or in small pieces that vary in size from the size of a grain of rice to a larger pellet. Dry ice does not melt, instead it sublimates, meaning the solid turns directly into a gas (bypassing the liquid state) as the temperature rises and the solid begins to dissipate. This unusual feature results in a smoking effect, and dry ice appears to be steaming as it sublimates. Thus, dry ice is often used to simulate fog or smoke.

Dry ice itself is not poisonous, but the surface of the solid is so cold that it should not be touched without gloves. Also, while the gas is stable and inert, it is heavier than air and can concentrate in low areas or enclosed spaces. When the concentration of carbon dioxide in the air exceeds 5%, the carbon dioxide becomes toxic. Thus, any area in which dry ice is used must be well ventilated.

It is relatively simple to turn carbon dioxide from a gas to a solid. Many dry ice manufacturers exist in the United States, and dry ice is shipped to all parts of the country for a variety of uses. It is an important refrigerant for keeping foods cold and preventing bacterial growth during shipment. Dry ice used for cooling or freezing foods must be very clean and considered "food grade" to ensure that food it may touch will not be contaminated. Because the solid sublimates rather than melts, large quantities of dry ice can be put into shipping containers without having to take into account volume of melting water that accumulates when ice is used as a refrigerant. Food-related uses are extensive and include quick freezing of foods for future use at food processing plants, retarding the growth of active yeast at bakeries, and keeping foods chilled for catering for the airline industry.

Other uses include: slowing the growth of flower buds at nurseries to keep plants fresh for consumers, flash freezing in the rubber industry during manufacture, absorbing ammonia refrigeration leaks, and creating smoke for theatrical productions. The most significant recent application of dry ice is dry ice blasting (or cleaning) in which dry ice pellets are hurled at a surface to be cleaned at high speed. The pellets strip the surface of the contaminants, sublimate into the atmosphere, and leave behind no toxic gases. The only residual is the dirt or paint left behind for disposal.

History

Dry ice was not invented, rather the properties of solid carbon dioxide were discovered in the early twentieth century. It was first produced commercially in the 1920s in the United States. A commercial venture trademarked the name *dry ice* in 1925 and solid carbon dioxide has been referred to as dry ice ever since. Until fairly recently, dry ice was often referred to as *hot ice*, a reference to the fact that when one touched the cold surface the hand felt burned.

Car body repair shops have discovered that applying dry ice to dents in the body can sometimes eliminate the disfiguration.

It appears that the Prest-Air Devices Company of Long Island, New York first successfully produced dry ice in 1925. Also in that year, Schrafft's of New York City first used the product to keep its famous ice cream from melting inside their parlor. Dry ice was far more extensively used for refrigeration and freezing of foods in the mid-twentieth century than it is today. Virtually every ice cream parlor in the world used dry ice for keeping ice cream frozen until well after the World War II, when electric refrigeration became affordable and efficient. The manufacturing of dry ice has not changed significantly in many decades and is a relatively simple process of pressurizing and cooling gaseous carbon dioxide. Uses for dry ice have diminished somewhat with the advent of better electric refrigeration. Some recent developments for its use include using the pellets in blasting or cleaning and its increasing use in transporting medical specimens, including hearts, limbs, and tissues, for reattachment and transplantation.

Raw Materials

The only raw material used in the manufacture of dry ice is carbon dioxide. This raw material is the byproduct of the refinement of gases emitted during the manufacture or refinement of other products. Most carbon dioxide used in the manufacture of dry ice in the United States is derived from refinement of gases given off during the refinement of petroleum and ammonia. The carbon dioxide emitted during these processes is sucked off and "scrubbed" to remove impurities for food grade carbon dioxide that will eventually become dry ice.

The Manufacturing Process

1 Carbon dioxide is liquefied by compressing and cooling, liquefying at a pressure of approximately 870 lb/in² (395 kg/cm²) at room temperature. Liquid carbon dioxide is pumped, via piping, into huge holding tanks so that dry ice manufacturers can remove the liquid required.

2 The liquid carbon dioxide is shipped in huge quantities, sometimes weighing many tons. Thus, most dry ice manufacturers choose to locate their factories close to the petroleum or ammonia refineries to keep transportation costs affordable. The pressurized, refrigerated liquid carbon dioxide is piped directly into a pressurized tank or rail car owned by the dry ice manufacturer and heads for the plant.

3 The tank trunk pulls up to the factory and dumps the liquid carbon dioxide into huge tanks on the premises. These tanks hold the liquid under pressure, keeping it refrigerated so that it remains in liquid state. These tanks are situated adjacent to the factory wall and, through piping, the liquid is brought directly inside when required for manufacturing.

4 The liquid carbon dioxide is released, again via piping, from the adjacent tanks through the factory wall and into the dry ice press. When the liquid moves from a highly-pressurized environment to atmospheric pressure, it expands and evaporates at high speeds, causing the liquid to cool to its freezing point which is -109°F (-78.3°C). A nozzle puts the liquid into the top block of a dry ice press, which stands approximately 16 ft (4.9 m) tall. This press includes a large block at the top that can exert extreme pressure on the product that is brought into it. When the liquid carbon dioxide hits the block of the dry ice press, it immediately solidifies since it is now at room temperature. The carbon dioxide now resembles snow.

5 This snow, now in the upper portion of the press, must be compressed into a block of dry ice. Thus, this top portion of the press goes up and down with extraordinary pressure (about 60 tons), squashing the snow into a solid block of dry ice. This is approximately a five minute process. When the block is solid, it is generally about 2 ft (61 cm) wide and 10 in (25 cm) high, weighing about 220 lb (100 kg).

6 This block of opaque white dry ice is pushed out of the press and onto a roller. A pneumatic saw cuts the block in half and the blocks are pushed to another saw that cuts the smaller blocks yet again. Thus, the single block made in the dry ice press is now in four pieces, each weighing about 55 lb (25 kg).

7 The smaller blocks are put into containers that keep the blocks cold so sublimation is kept to a minimum. If shipped as unwrapped pieces they must be tightly packed in a con-

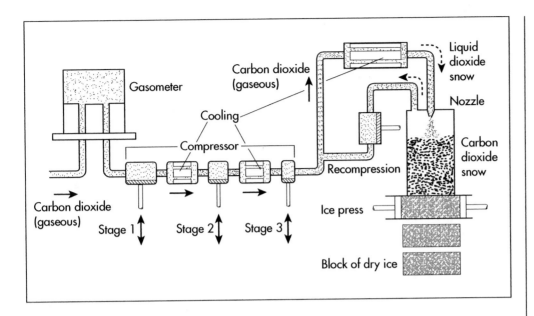

tainer, generally including four blocks, with very little air allowed inside to reduce sublimation. If a block is removed during shipping, the other blocks will quickly begin to dissipate. Many dry ice manufacturers wrap the blocks in paper using machines (it is wise not to touch the very cold surface) and send it distributors or wholesalers.

Quality Control

Quality control issues primarily revolve around the grade of the carbon dioxide used in the manufacture of food grade dry ice. Recently, the federal government set fairly stringent standards for the purity of carbon dioxide used in the manufacture of dry ice, causing manufacturers to certify and test the liquid carbon dioxide used, as well as the purity of the manufactured product.

Other quality control issues include ensuring that tanks and equipment are working precisely. If pressure is not properly maintained, the product cannot be produced. Moving the product quickly and efficiently from cutting to storage is very important, as dry ice quickly sublimates at room temperature, thus reducing the weight and price of the salable product. Shipping must pack the blocks densely to reduce sublimation in transit as well.

Byproducts/Waste

No significant chemicals are created in the production of dry ice. Since the product is made only from carbon dioxide, when the product sublimates the gases are emitted into the atmosphere. The only detectable waste is sublimation of the product in block form, which is kept to a minimum.

The Future

While the use of dry ice in refrigeration and food storage may be diminishing, its use in other areas holds some promise. As mentioned above, house cleaners and machinery operators are interested in the small dry ice pellets for their ability to bombard a house or machine at high pressure, remove dirt or other contaminants, and then dissipate into the atmosphere. Recently a telephone company used dry ice pellets to safely clean sensitive electronic testing equipment without using dangerous solvents. Car body repair shops have discovered that applying dry ice to dents in the body can sometimes eliminate the disfiguration. Also, tests on dry ice blocks advocate dropping it into gopher holes to eradicate the pests or putting it out in the backyard to attract mosquitoes in order to keep them away from humans.

Where to Learn More

Books

Russell, Allan S. *McGraw-Hill Encyclopedia of Science & Technology*. Vol. 5. New York: 1997.

The Handy Science Answer Book. Second ed. Detroit: 1997.

Other

"Dry Ice." *About.com Web Page.* August 2001. <http://inventors.about.com/library/inventors/bldryice.htm>.

dryiceInfo.com Home Page. July 2001. <http://www.dryiceinfo.com >.

Interview with Chuck Hines. Owner of Arctic Dry Ice Company. Baltimore. 31 July 2001.

Nancy EV Bryk

EEG Machine

Background

An electroencephalogram (EEG) machine is a device used to create a picture of the electrical activity of the brain. It has been used for both medical diagnosis and neurobiological research. The essential components of an EEG machine include electrodes, amplifiers, a computer control module, and a display device. Manufacturing typically involves separate production of the various components, assembly, and final packaging. First developed during the early twentieth century, the EEG machine continues to be improved. It is thought that this machine will lead to a wide range of important discoveries both in basic brain function and cures for various neurological diseases.

The function of an EEG machine depends on the fact that the nerve cells in the brain are constantly producing tiny electrical signals. Nerve cells, or neurons, transmit information throughout the body electrically. They create electrical impulses by the diffusion of calcium, sodium, and potassium ions across the cell membranes. When a person is thinking, reading, or watching television different parts of the brain are stimulated. This creates different electrical signals that can be monitored by an EEG.

The electrodes on the EEG machine are affixed to the scalp so they can pick up the small electrical brainwaves produced by the nerves. As the signals travel through the machine, they run through amplifiers that make them big enough to be displayed. The amplifiers work just as amplifiers in a home stereo system. One pair of electrodes makes up a channel. EEG machines have anywhere from eight to 40 channels. Depending on the design, the EEG machine then either prints out the wave activity on paper (by a galvanometer) or stores it on a computer hard drive for display on a monitor.

It has long been known that different mind states lead to different EEG displays. Four mind states—alertness, rest, sleep, and dreaming—have associated brain waves named alpha, beta, theta, and delta. Each of these brain wave patterns have different frequencies and amplitudes of waves.

EEG machines are used for a variety of purposes. In medicine, they are used to diagnose such things as seizure disorders, head injuries, and brain tumors. A trained technician in a specially designed room performs an EEG test. The patient lies on his or her back and 16–25 electrodes are applied on the scalp. The output from the electrodes are recorded on a computer screen or drawn on a moving piece of graph paper. The patient is sometimes asked to do certain tasks such as breathing deeply or looking at a bright flickering light. The data collected from this machine can be interpreted by a computer and provides a geometrical picture of the brain's activity. This can show doctors exactly where brain activity problems are.

History

The EEG machine was first introduced to the world by Hans Berger in 1929. Berger, who was a neuropsychiatrist from the University of Jena in Germany, used the German term *elektrenkephalogramm* to describe the graphical representation of the electric currents generated in the brain. He suggested that brain currents changed based on the functional status of the brain such as sleep, anesthesia, and epilepsy. These were revolu-

EEG machines are used for a variety of purposes. In medicine, they are used to diagnose such things as seizure disorders, head injuries, and brain tumors.

tionary ideas that helped create a new branch of medical science called neurophysiology.

For the most part, the scientific community of Berger's time did not believe his conclusions. It took another five years until his conclusions could be verified through experimentation by Edgar Douglas Adrian and B. C. H. Matthews. After these experiments, other scientists began studying the field. In 1936, W. Gray Walter demonstrated that this technology could be used to pinpoint a brain tumor. Walter used a large number of small electrodes that he pasted to the scalp and found that brain tumors caused areas of abnormal electrical activity.

Over the years the EEG electrodes, amplifiers, and output devices were improved. Scientists learned the best places to put the electrodes and how to diagnose conditions. They also discovered how to create electrical maps of the brain. In 1957, Walter developed a device called the toposcope. This machine used EEG activity to produce a map of the brain's surface. It had 22 cathode ray tubes that were connected to a pair of electrodes on the skull. The electrodes were arranged such that each tube could show the intensity of activity in different brain sections. By using this machine Walter demonstrated that the resting state brain waves were different than brain waves generated during a mental task that required concentration. While this device was useful, it never achieved commercial success because it was complex and expensive. Today, EEG machines have multiple channels, computer storage memories, and specialized software that can create an electrical map of the brain.

Raw Materials

Numerous raw materials are used in the construction of an EEG machine. The internal printed circuit boards are flat, resin-coated sheets. Connected to them are electronic components such as resistors, capacitors, and integrated circuits made from various types of metals, plastic, and silicon.

The electrodes are generally constructed from German silver. German silver is an alloy made up of copper, nickel, and zinc. It is particularly useful because it is soft enough to grind and polish easily. Stainless steel (which has a higher concentration of nickel) can also be used. It tends to be more corrosion resistant but is harder to drill and machine.

An adhesive tape is used to attach surface electrodes to the patient. Since the electric signals are weakly transmitted through the skin to the electrodes, an electrolyte paste or gel is typically needed. This material is applied directly to the skin. It may be composed of a cosmetic ingredient like lanolin and chloride ions that help form a conductive bridge between the skin and the electrode allowing better signal transmission. Polytetrafluoroethylene (Teflon) is used as a coating for the wires and various kinds of electrodes.

Design

The basic systems of an EEG machine include data collection, storage, and display. The components of these systems include electrodes, connecting wires, amplifiers, a computer control module, and a display device. In the United States, the FDA (Food and Drug Administration) has proposed production suggestions for manufacturers of EEG machines.

The electrodes, or leads, used in an EEG machine can be divided into two types including surface and needle electrodes. In general, needle electrodes provide greater signal clarity because they are injected directly into the body. This eliminates signal muffling caused by the skin. For surface electrodes, there are disposable models such as the tab, ring, and bar electrodes. There are also reusable disc and finger electrodes. The electrodes may also be combined into an electrode cap that is placed directly on the head.

The EEG amplifiers convert the weak signals from the brain into a more discernable signal for the output device. They are differential amplifiers that are useful when measuring relatively low-level signals. In some designs, the amplifiers are set up as follows. A pair of electrodes detects the electrical signal from the body. Wires connected to the electrodes transfer the signal to the first section of the amplifier, the buffer amplifier. Here the signal is electronically stabilized and amplified by a factor of five to 10. A differential pre-amplifier is next in line that filters and amplifies the signal by a factor of 10–100. After going through these amplifiers, the signals are multiplied by hundreds or thousands of times.

This section of the amplifiers, which receive direct signals from the patient, use optical isolators to separate the main power circuitry from the patient. The separation prevents the possibility of accidental electric shock. The primary amplifier is found in the main power circuitry. In this powered amplifier the analog signal is converted to a digital signal, which is more suitable for output.

Since the brain produces different signals at different points on the skull, multiple electrodes are used. The number of channels that an EEG machine has is related to the number of electrodes used. The more channels, the more detailed the brainwave picture. For each amplifier on the EEG machine two electrodes are attached. The amplifier is able to translate the different incoming signals and cancels ones that are identical. This means that the output from the machine is actually the difference in electrical activity picked up by the two electrodes. Therefore, the placement for each electrode is critical because the closer they are to each other, the less differences in the brainwaves that will be recorded.

A variety of output printers and monitors are available for EEG machines. One common device is a galvanometer or paper-strip recorder. This device prints a hard copy of the EEG signals over time. Other types of devices are also used including computer printers, optical discs, recordable compact discs (CDs), and magnetic tape units. Since the data collected is analog, it must be converted to a digital signal so electronic output devices can be used. Therefore, the primary circuitry of the EEG typically has a built-in analog to digital converter section. The software provided with some EEG machines can be used to create a map of the brain.

Various other accessories are used with an EEG machine. These include electrolytic pastes or gels, mounting clips, various sensors, and thermal papers. EEG machines used in sleep studies are equipped with snoring and respiration sensors. Other uses require sensory stimulation devices such as headphones and LED goggles. Still other EEG machines are equipped with electrical stimulators.

The Manufacturing Process

The different parts of an EEG machine are produced separately and then assembled by the primary manufacturer prior to packaging. These components, including the electrodes, the amplifier, and the storage and output devices, can be supplied by outside manufacturers or made in-house.

Electrodes

1 The EEG electrodes are typically received from outside suppliers and checked to see if they conform to set specifications. One type of electrode commonly used for the EEG machine is a needle electrode. These can be made from a bar of stainless steel. The bar is heated until it becomes soft and then extruded to form a seamless tube.

2 The tube is then drawn out to produce a fine hollow tube. These tubes are cut to the desired length, and then conically sharpened to produce a point.

3 To ensure easy insertion, the tube is passed through a bath of polytetrafluoroethylene (Teflon) to provide a slick, chemical resistant coating. As the tube exits the bath it is warmed to evaporate the solvent and allow the coating to adhere.

4 The tube is then mechanically placed in a plastic adapter piece that is made with an injection molding machine. This piece allows the disposable, individually packaged needles to hook up to the lead wire.

5 The shielded lead wire is fitted with an adapter that can be hooked up to the primary unit.

Internal electronics

6 The amplifiers and computer control module are assembled just like other electronic equipment. The electronic configurations are first printed on circuit boards. The boards can be fitted with chips, capacitors, diodes, fuses, and other electronic parts by hand or passed through an automated machine. This machine works like a labeling machine. It is loaded with numerous spools of electronic components and placing heads. A computer controls the motion of

A man undergoing an EEG, wearing a cap equipped with electrodes.

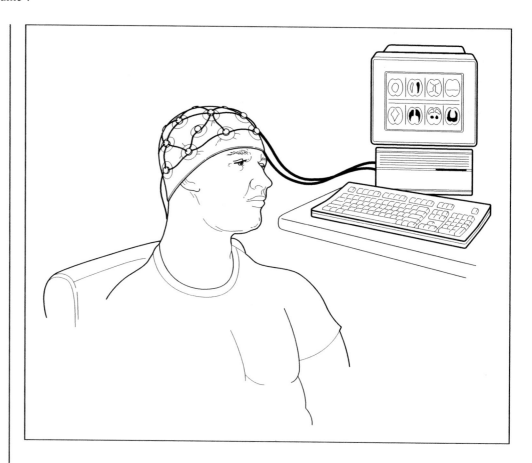

the board through the machine. When a board is moved under one of the component spools, a placing head stamps the electronic piece on the board in the appropriate positions. When completed the boards are sent to the next step for wave soldering.

7 In the next step, a wave-soldering machine affixes the electronic components to the board. As the boards enter this machine, they are washed with flux to remove contaminants that might cause short circuits.

8 Boards are then heated using infrared heat. The underside of the board is passed over a vat of molten solder. The solder fills into the needed areas through capillary action.

9 As the boards cool, the solder hardens and the electronics are held into place. Visual inspection is typically done at this point to ensure that defective boards get rejected.

Amplifier

10 The electronic boards for the amplifier are pieced together and affixed to a housing. This is typically done by line oper-

ators who physically place the pieces on pre-fabricated boards.

11 The housing is made of a sturdy plastic that is constructed through typical injection molding processes. In this process, a two-piece mold is created that has the inverse shape of the desired part. Molten plastic is injected into the mold and when it cools, the part is formed. For some EEG models, the amplifier is a separate box about the size of a textbook. The outer sides of the box have connectors where the electrodes and the computer connection lines are plugged in.

Computer control box

12 An EEG station consists of the amplifier and a computer control station. This control station typically has a desktop computer, a keyboard and mouse, a color printer, and a video monitor. These devices are all produced by outside manufacturers and assembled by the EEG manufacturer.

Final assembly

13 Each of the components of the EEG machine are brought together and

placed into an appropriate metal frame. This process is done by line operators working in extremely clean conditions. When the components are assembled they are typically put on a sturdy, steel cart to make the device portable.

14 The finished devices are then put into final packaging along with accessories such as electrodes, computer software, printout paper, and manuals.

Quality Control

At each step in the manufacturing process, visual and electrical inspections occur to ensure the quality of each EEG device being produced. Since circuit fabrication is sensitive to contamination, assembly work is done by line operators in air-flow controlled, clean rooms. Operators must also wear lint-free clothing to reduce the chance of contamination. The functional performance of each completed EEG device is also tested to make sure it works. This is done by powering up the device, turning it on, and running a series of standard tests. To simulate real-life use, these tests are done under different levels of heat and humidity.

In general, manufacturers set their own quality specifications for their EEG machines. However, in the United States the Food & Drug Administration (FDA) provides production recommendations that are usually adapted by the industry. Various other medical and governmental organizations also propose standards and performance suggestions. Some factors considered important are standardized input signal ranges, accuracy of calibration signal, frequency responses, and recording duration.

The Future

In the future, EEG machines will be improved in their manufacture and their applications. From a manufacturing standpoint, the components that makeup the internal electronics of the device will likely get smaller. This will allow for smaller, more portable machines. It will also make the devices less expensive. This will be important because some experts suggest that future ap-plications will make it desirable for individual consumers to have EEG machines.

While manufacturing improvements will come from research done in the general field of electronic manufacturing, specific research on EEG machines has focused on new uses and applications. For example, a device has recently been introduced that may make it possible to screen for Alzheimer's disease. This machine contains a cap that is fitted with electrodes. When worn it provides an electronic picture of a patient's brain activity. This picture is compared to the brain activity of healthy people and differences are noted.

A similar machine has been developed which can use information received from EEG electrodes to control computers. With this device the user wears an electrode-containing cap and looks at a computer screen. After a training session with the computer, users have been able to control the movement of a cursor on the screen just by using their thoughts. If fully developed, this technology could be a revolutionary development for paraplegics. Individual consumers may also benefit using such a device to control household lights, computers, and appliances just by thinking.

Where to Learn More

Books

Fisch, Bruce J. *Fisch and Spehlmann's EEG Primer*. Elsevier Science, 1999.

Othmer, Kirk. *Encyclopedia of Chemical Technology*. Vol. 22, 1992.

Webster, J. G. *Medical Instrumentation Application and Design*. 2nd ed. 1992.

Wong, Peter K. H. *Digital EEG in Clinical Practice*. Lippincott Williams & Wilkins, 1995.

Other

Sabbatini, Renato M.E. "Mapping the Brain." *Brain & Mind* 15 November 2001. <http://www.epub.org.br/cm/n03/tecnologia/eeg.htm#topography>.

Perry Romanowski

Electric Guitar

The earliest electric guitars were made in the 1920s and 1930s, but these were very primitive prototypes of the modern solid-body electrical guitar.

Background

Developed in the early part of the twentieth century, the electric guitar has become one of the most important instruments in popular music. Today's solid-body electric guitar derives from the acoustic guitar, an instrument first introduced in America as the Spanish-style guitar. Even though body designs of modern electric guitars often differ from their acoustic predecessors, all guitars are constructed with the same simple template. All guitars, acoustic or electric, are built with a bridge, body, and neck. The most significant difference is that acoustic guitars are hollow while electric guitars have a solid body.

For years, the acoustic guitar was limited to a supporting role in large musical ensembles because of its volume. Thus, the major motivation that drove the creation of the electric guitar was instrumentalists' desire for greater volume. Predecessors of the modern electric guitar were amplified acoustic guitars crudely modified by inventors who attached wires, magnets, and other "pickup" attachments. (Pickups are electromagnetic devices that increase volume.) However, as technology started advancing in the 1930s, newer versions became more complex, and the electric guitar became a solo instrument, a development that helped expand musical styles.

History

The earliest electric guitars were made in the 1920s and 1930s, but these were very primitive prototypes of the modern solid-body electrical guitar. The very first electrified guitar was said to have been invented by Paul H. Tutmarc. Inspired by the inner workings of the telephone, which employed magnetics to create vocal vibrations, Tutmarc experimented on the Hawaiian guitar, building a magnetic pickup out of horseshoe magnets and wire coils that amplified the vibration of the instrument's strings.

Around the same time, George Beauchamp and John Dopyera, two Los Angeles musicians, worked on creating even louder guitars. After experimenting with attaching amplifying horns to instruments, they, too, developed an electromagnetic pickup, this one comprised of two horseshoe magnets. Pleased with the effectiveness of the pickup, Beauchamp had a craftsman make a guitar designed with a wooden neck and body. Nicknamed the "frying pan" because of its shape, this became the first electric guitar. Beauchamp took the prototype to Adolph Rickenbacker. The two men formed a company and began manufacturing the first of the famous Rickenbacker line of electric guitars. Thus, Rickenbacker became the first manufacturer of electric guitars.

The first "Spanish-style" electric guitar was built and sold by Lloyd Loar, another early experimenter. His design was the direct predecessor of the modern electric guitar, and it inspired Orville Gibson, another guitar pioneer, to create the electric guitar model that revolutionized the instrument: the ES-150. Slide guitarist Alvino Rey developed the prototype of the ES-150, which has been called the first modern electric guitar. The final version was built by Gibson employee Walter Fuller. Though the guitar was an immediate success, it had some flaws. The vibrations from its hollow body were picked up and amplified, which created feedback and distortion. This led Les Paul, a guitarist and inventor, to develop the solid body electric guitar in 1940.

Paul's innovation, which was called "the Log" because of its solid body, involved mounting the strings and pickup on a solid block of pine to minimize body vibrations. The "Log" consisted of two basic magnetic pickups mounted on a 4 × 4 in (10.2 × 10.2 cm) piece of pine. To make it look more like a conventional guitar, Paul sawed an arch-top guitar in half and attached the pieces to his model. The solid body proved effective in eliminating the problems of the ES-150.

In 1946, Paul took his new guitar to Gibson, who was skeptical about the solid body. Leo Fender, however, understood the conception, and in 1949, he started selling the "Esquire," which became the first successful solid-body guitar. The guitar was later renamed the "Telecaster," one of the most famous guitar brand names. The Telecaster became extremely popular with country, blues, and rock and roll musicians. The Telecaster prompted Gibson to build his own solid-body model, which was named the "Les Paul."

In 1956, Rickenbacker introduced the student model Combo 400 guitar, with its so-called "butterfly-style" body. The guitar's unique construction featured a neck that extended from the patent head to the base of the body (known today as neck-through-body construction) and with the sides of the guitar body bolted or glued into place.

By the 1960s, the electric guitar was an established musical instrument. Innovations in design continued through the decade. In 1961, Gibson introduced "Humbucking" pickups into the Les Paul guitar that were designed to eliminate unwanted hum from the magnetic coils. (Humbucking pickups used two coils wrapped out of phase. This eliminated the common mode hum present in previous designs.) That same year, McCarty introduced the ES-335, a semi-hollow body guitar designed to incorporate the best of both the hollow body and solid body designs. Both Gibson and Fender had introduced futuristic looking designs. The Gibson SG and the Fender Stratocaster became familiar to audiences because they were frequently used by rock guitarists in the 1960s.

Raw Materials

Raw materials that go into the construction of the electric guitar include well-seasoned

Jimi Hendrix.

James Marshall Hendrix was born November 27, 1942 in Seattle. Hendrix taught himself to play guitar by listening to blues recordings; left-handed, he used a restrung right-handed guitar. Hendrix became known in the 1960s for playing the guitar behind his back, with his teeth, and setting it on fire. At times his stage pyromania overshadowed his musical pyrotechnics, but he is recognized as perhaps the most influential rock guitarist in history.

Hendrix began as a studio musician in the early 1960s, forming a band in 1965. The following year he created a new band, the Jimi Hendrix Experience, and started a new sound—acid rock—that employed intentional feedback and other deliberate distortions. His stage antics gained him notoriety at the 1967 Monterey Pop Festival, and the band had a Top 40 hit with their version of Bob Dylan's "All Along the Watchtower" in 1968. That year Hendrix directed his efforts to studio recordings, but appeared with his new group—Band of Gypsies—in 1969 at Woodstock, where he gave a memorable performance of "The Star-Spangled Banner."

Hendrix was named pop musician of the year by *Melody Maker*, 1967 and 1968; voted *Billboard* artist of the year, 1968; named performer of the year and honored for rock album of the year by *Rolling Stone*, 1968; presented with the key to Seattle, 1968; inducted into the Rock and Roll Hall of Fame, 1992; and received the Grammy Award for lifetime achievement, 1993. Hendrix died September 18, 1970 from asphyxiation resulting from a drug overdose.

hardwoods such as maple, walnut, ash, alder, and mahogany for the solid body. The denser the wood, the better sustain an instrument will have (sustain refers to how long a note can be held). Wood density can also have an effect on the tone. Some bodies are also constructed with plexiglass. Wood is also used in the construction of the neck, including maple, rosewood, and ebony. Other raw materials include glue to hold the pieces together, chrome for the hardware, and a nitrocellulose lacquer for finishing the body.

Design

The solid-body electrical guitar gets its volume from the magnetic pickup installed within its body. This pickup responds to the vibration of strings, transforming the energy into electrical impulses that are amplified by a loudspeaker system called an amplifier. For the best sound, the pickup needs to be stable and unaffected by vibrations from the body. Early electric guitar pioneers discovered that a pickup connected to a hollow-body acoustic guitar resulted in distortions and feedback. The need for stability is what led to the development of the solid body, the one feature that most characterizes the electric guitar. The solid body increases stability, and early electric guitar makers discovered, through experimentation, that guitar bodies made of high-density hardwood worked best.

In the late 1930s and 1940s, guitarists and inventors like Les Paul and Leo Fender developed the early designs of the solid-body electric guitar. Later, manufacturers moved away from traditional shapes and colors and came up with their own designs, many of which were quite fancy. More advanced models included the Fender Stratocaster and the Gibson Flying V.

The Manufacturing Process

The major components of the electric guitar include the bridge, the body and the neck. Secondary components include the fingerboard, strings, nut and tuning heads. A guitar manufacturing facility is, to a large extent, a woodworking facility, as wood selection and body design are large parts of the electric guitar construction process.

1 Wood is selected, inspected, and processed to be made into bodies, necks and fingerboards. Sometimes it must be cured first in a conventional or vacuum kiln to maximize its stability. Curing can take as long as a week, and it relieves stress and wetness. The wood that will be made into a body is loaded onto a scissors lift and transferred to a conveyor where it is planed on both sides. It then moves down to the cut-off saw worker, who cuts the wood to size. From there, the wood is sent to a machine called a KOMO, a computer-controlled router that drills weight relief holes to make the wood lighter. The machine also cuts a channel in the wood where wire will eventually be placed.

2 The wood then goes back into the rough mill, where it will have a maple top and mahogany back glued on in a glue mill under 900 lb (408 kg) of pressure. It is then placed on a glue wheel to dry for four hours. Up to this point, the wood is a square block. When dry, it is ready to be shaped. It is sent back to the KOMO, which is programmed to cut the periphery into the desired shape. The KOMO also routs the back electronic pockets.

3 The body then goes to the body line for its final shaping. First, a worker sands the body by hand with sandpaper, then it undergoes a process called "rabbeting." Rabbeting involves first making a machine cut that will accommodate the binding that the body needs. The worker maneuvers the body while the machine makes the cut. The body then moves down the line to the binding station. The worker takes the binding material, drenches it in glue by pulling it through a glue box, then wraps it around the rabbet cut made in the body. The worker then ties the body completely with rope to hold the glued binding material in place. Then the body is hung overnight to dry.

4 The next morning, the worker removes the rope, and the body moves to the next station, where it will be shaped by sanding into its finished contour. Using a rim sander, a worker sands off the excess glue and ensures that the binding and the wood are flush. The body then goes to the slack belt machine for smoothing. The worker, by hand, places it under a slack belt and pushes the body under the belt with varying pressure until all carved marks are smoothed out.

5 As the body of the guitar is built on the body line, the neck of the guitar is built on the neck line, where the neck is shaped and sanded by hand and the fingerboard and head veneer are applied.

6 Fingerboards are made of rosewood and ebony and are stabilized in kilns, shaped,

and slotted for frets. In shaping, the fingerboard first gets molded on a molder with a 12-in (30.5 cm) radius. From there, it moves into the rough board area, where location pin holes are drilled. Then it goes to the fret saw machine, where the fret slots are cut by a quick saw machine. A router then creates the inlay pockets on the fingerboard, and the inlays are added. The router is a powermatic tool that suctions the fingerboard down on a table and routs all of the pockets. The inlays themselves are placed in by hand at the inlay station. A worker places epoxy into the pockets, puts in the inlays, then places more epoxy on top of them. This eliminates any spaces. The fingerboard is then left to dry.

7 When dry, the fingerboard moves on to a surface grinder that cleans the dried epoxy off of the top. Now the frets are ready to be placed. A worker takes the fingerboard and puts glue into the fret slots and then, by hand, places the fret wire. Using a pneumatic snip, the worker first places the wire then cuts off the excess. From there, the fingerboard is put into a hydraulic press that presses the frets completely in place. The worker then hand-sands the frets to make them smooth. The fingerboard is then slotted to accommodate binding, then left to dry. When the frets are dry, the fingerboard is joined to the neck.

8 In the meantime, the neck has been built. This begins when the ten-quarter mahogany neck blanks are quarter-sawn for increased strength and straightness. Neck pattern templates are penciled, and then the neck blank is cut into the template shape with a bandsaw. The neck blank is then put on a rotary profile lathe. The lathe gives the neck its basic shape.

9 A worker then joins the fingerboard to the neck by tapping in the location pins on the fingerboard, applying the glue, putting the fingerboard and the neck together, placing the connected pieces into a glue press, and then allowing it to dry. The headstock veneer is also glued onto the neck blank. The neck is then sent down the line to be shaped and finished by machine rolling and hand sanding. Now the neck is ready to be fitted to the body.

In attaching the neck to the body, several methods are used by different manufactur-

ers. Some electric guitar necks are glued into place while others are bolted on. Many players prefer the glued-in neck, as they believe it gives a better joint that provides more sustain of notes. At Gibson, the necks on a Les Paul are always glued on.

10 On the body, the location of neck placement is then traced. A cavity is cut where the neck will be placed. The worker places the neck in the neck slot to see if the fingerboard, neck, and body are all flush. Neck fitting is all done by hand, with a worker using a chisel, a clamp, and glue. The neck is then placed in the joint until a seamless fit is made. The fit is glued, clamped, and left to dry for an hour. When dry, the worker sands off the excess glue. The pickup cavities and bridge holes are added by a computer-controlled router.

11 The guitar is now ready to for color preparation and finishing. Before ap-

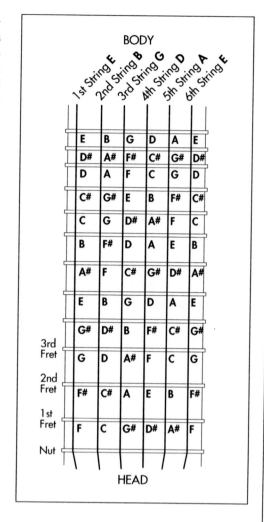

Note positions on an electric guitar.

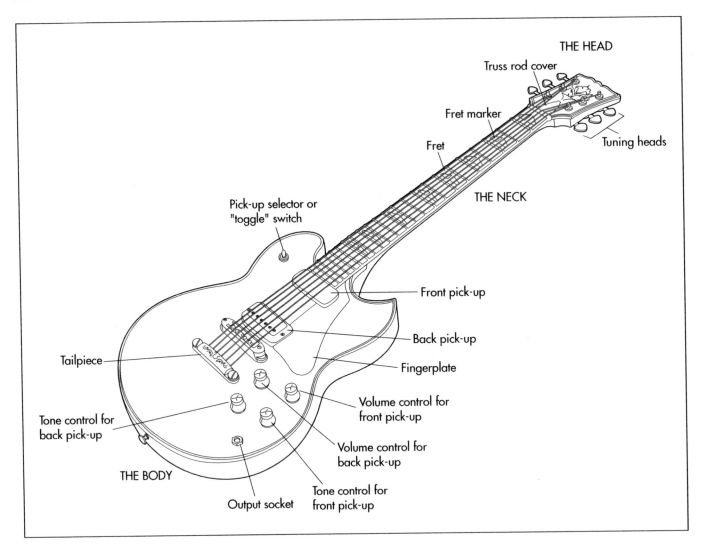

THE HEAD

Truss rod cover

Fret marker

Fret

Tuning heads

THE NECK

Pick-up selector or
"toggle" switch

Front pick-up

Back pick-up

Tailpiece

Fingerplate

Volume control for
front pick-up

Tone control for
back pick-up

Volume control for
back pick-up

THE BODY

Tone control for
front pick-up

Output socket

An electric guitar.

plying the finish, workers hand sand the guitar to smooth any sharp corners. Then a
wood filler and stain is applied to color the
wood and even out the grain pattern.

12 Before the body is sprayed with a finish, the body and neck are sealed to
ensure that paint will not be absorbed into
the wood. When the guitar dries, the finish
is applied by using automated electrostatic
methods that improve the consistency of the
finish. Afterward, the guitar is sent to the
scrapers, who remove any overspray with
metal tools.

13 After the guitar has dried and has been
sanded, it goes into the buffing department. Buffing is a three-step process. First
the guitar is buffed on a wheel. A jeweler's
rouge compound is used to remove any
rough spots in the finish. Two more buffings
are then done to achieve a brilliant gloss.

14 The guitar now awaits final assemblage, where all of its hardware and
electronics are installed. In general, at most
guitar manufacturing factories, the final assembly of an electric guitar involve the pickguard placement, vibrato installation, setting
the neck, tuner installation, installing strap
bottoms, fret dress, nut, bridge and vibrato set
up, string tree placement, and pick-up height.

15 Next, the hardware and electronics
are assembled and placed onto the
body and bridge. Hardware placed onto the
body include the pickguard, pickguard
shield, pickup compression spring, pickup
cover, pickup core assembly, lever knob,
pickup selector switch, volume knob, tone
knob, volume and tone potentiometers, ceramic capacitor, and output plug assembly.
Hardware placed on the bridge include base
plate, vibrato block, compression springs,
bridge bar, set screws, bridge cover, rear

cover plate, tension spring, tremelo tension spring holder, and lever assembly.

16 Builders install pickups, pots, tuning keys, jackplates and toggle switches. The adjusters notch the tailpiece and nut, string the guitar, check neck pitch and intonation, and adjust the bridge height. The cleaners remove smudges and dirt, install back plates, pickguards, truss rod covers and other hardware and then polish the chrome, nickel or gold hardware.

17 The guitar undergoes a final buff and polish and a final inspection.

Quality Control

During each stage of the process, the product is inspected. Even the smallest flaw in design such as a scratch or excess dried glue could send the guitar back down the line, or might even cause inspectors to scrap it. During final assembly, when hardware and wiring is installed, each component is tested separately to verify that it is working properly.

The Future

It is generally considered that the great part of the evolution of the electric guitar took place in between the late 1920s and the early 1960s, a period that saw the creation of the major innovations. However, guitar manufactures and inventors are still exploring ways to modify the instrument. These changes would include modification in design, materials, in pickups, or in finishes. Some guitar makers are looking to bodies made of plastic or graphite. Others are exploring designs that include hollow or semi-hollow bodies. For some time now, inventors have been trying to apply piezo to guitar pickup, or amplification. Piezo is a material with piezoelectric properties. If applied correctly to a musical instrument, it senses vibrations or changes in pressure. For a guitar, it could be applied in a contact microphone, or it could be placed on the guitar itself, where it would sense guitar vibration. Ultimately, it could enhance the sound of a guitar.

In the design area, a company has developed a mass 3D solid and surface modeling software that has attracted the attention of the Gibson, Warmoth, Suhr, and Tom Anderson Guitarworks guitar companies. The software would free designers from the limitations of two-dimensional planning and allow them to create complete three-dimensional designs before the manufacturing process began. In this way, they could be more experimental with designs. Potentially, the software would allow designers to create new designs in 3D without having to build prototypes or models. Designs could then be sent to a computerized woodworking station for a limited production run.

Books

Bacon, T., and P. Day. *The Ultimate Guitar Book*. New York: Alfred A. Knopf, 1992.

Denyer, R., I. Guillory, and A. M. Crawford. *The Guitar Handbook*. New York: Alfred A. Knopf, 1987.

Wheeler, Tom. *The Guitar Book*. New York: Harper and Row, 1998.

Other

Rickenbacker Web Page. December 2001. <http://www.rickenbacker.com/us/ehistory.htm>.

The Electric Guitar Web Page. December 2001. <http://www.si.edu/lemelson/guitars/noframes/00main.htm>.

Dan Harvey

Electric Tea Kettle

In its beginnings in China, tea was processed in blocks or cakes that had to be boiled after roasting and shredding; this required a tea kettle.

Background

The sole purpose of the tea kettle is to boil water. Water for coffee and for many cooking uses does not have to be boiled, but fresh, cool water that is brought fully to a boil is essential for tasty tea. Although it is associated with making tea, the kettle's contents make excellent instant drinks or soups that are also best with boiling water or to produce steam for remedying colds.

History

Tea itself has been processed into three different forms in its history, making specific utensils essential. In its beginnings in China, tea was processed in blocks or cakes that had to be boiled after roasting and shredding; this required a tea kettle. The Japanese method was more refined; powdered tea was whipped in porcelain bowls with bamboo whisks. Leaf-tea (the most common form in the Western World for about 200 years) consists of different methods of picking and processing tea leaves. This tea requires steeping in boiling water, so leaves are put in pots filled with boiling water from the tea kettle. Block, powdered, and leaf tea must all be steeped in boiling water.

The tea kettle evolved from the cooking kettle that was hung on a hook on an iron post in the cooking fire. The hook was turned to move the kettle over the fire, and a "tilter" helped to pour water from the kettle. Kettles were made of iron, one of the first metals to be mined and processed.

In Japan, the iron cooking kettle became a small, rounded bowl with two short arms or loops (one on either side of the bowl) for pulling it off the hearth and a lid. A classic example of a bowl-type iron kettle dates from 1517. As methods of casting iron became more sophisticated, the outsides of these kettles were decorated, and the two arms became a spout and better handle. Iron casters who made tea kettles were highly respected.

Beautifully decorated examples of Japanese iron tea kettles with the spouted tea kettle shape known today date from the late nineteenth century. Iron kettles could withstand cooking fires, but serving ware emerged from the porcelain industry. Kettles obviously existed before tea pots because pots copied the shape, spout, and handle.

In Russia, water is heated in a *samovar* (literally, self-boiler), which is not a tea pot but an elaborate tea kettle made of metal with a central chimney for containing fire and boiling water in the surrounding vessel. Russians learned about the samovar from Persians during border disputes and trade efforts. A strong concentrate of tea is kept in a tea pot and warmed constantly on top of the samovar. Concentrate is poured into tea cups, and boiling water from a spigot on the samovar fills the cups and dilutes the concentrate.

The English began making tea pots of unglazed earthenware in the mid-seventeenth century, but silver became a popular material in the early 1700s. The first known silver tea pot is dated 1670, but, by the turn of the century, all tea servingware was made of silver including kettles. Silver kettles are still made today, but they have been surpassed in importance by aluminum and stainless steel for both stove-top and electric types.

In both England and the United States, the tea kettle's development was closely linked to the evolution of the stove. When stoves replaced cooking fires, the kettle was pulled from the fire and given a place on the stove. Most kettles are shaped like modified globes with flat bottoms to sit on stove plates. Kettles became ornaments for kitchens when they were manufactured with different metals like copper and decorated with interesting handles and enamel.

Electrifying the kettle followed in the early twentieth century. Although the first kettles were seated on individual electric coils, heating elements were soon built in and more refined models appeared.

Raw Materials

Electric tea kettles are made predominately of steel, iron, silver, aluminum, or a combination of plastic and metal. High-grade steel is typically used for the housing of all-metal kettles. Heat resistant plastics, such as low density polyethylene (LDPE), polypropylene (PP), or polystyrene (PS), comprise the upper housings of the kettles in some models with metal enclosures on the bases that contain the heating elements.

Design

Depending on the quality of the kettle, the types of raw materials can vary greatly. For higher grade kettles, a higher grade of steel or even silver may be used. Lower quality kettles may be of simple aluminum. Ideas for electric kettles arise from several different sources. The Research and Development (R&D) Department in the manufacturer's firm develops its own concepts for entirely new designs and for freshening its existing models. Fashions for cookware and other kitchenware change along with many household conveniences and accessories. Fashion is not limited to color but includes up-to-date overall shapes, changes to parts such as bases or handles, and improved safety features.

The electrical components of a tea kettle include the heating element, a thermostat that turns off the kettle automatically if it is boiling dry, an on-off switch and its attachments, a connection linking the element to the cord, and the plug. These are received at the plant as assembled units. Their assembly is out-sourced (subcontracted) to other fac-

tories, often outside the United States. These manufacturers further subcontract the making of individual electrical parts.

The Manufacturing Process

1 Metal is received at the tea kettle factory in sheets or rolls appropriate for presses that will stamp out the parts. Stainless steel sheets are checked for thickness; a bill of lading specifies the alloy, but the receiving department checks the thickness of the steel with precision instruments. Polypropylene plastic that is usually heat reinforced is delivered as tiny pellets in the color the kettle manufacturer requests. Plastic pellets are randomly inspected for color and impurities. From large hoppers above the injection molding machines, the pellets flow into the machines by gravity. The electrical components are complete when they are shipped to the plant. One set per tea kettle is provided as a bundled or wrapped set; parts do not need to be sorted or assembled. The receiving department randomly inspects and tests some of the electrical sets.

2 In the factory, the sheet steel is mechanically placed in the die of a punch press. A hydraulic punch is released, creating a hollow shell in the shape of the kettle base. This piece is then removed. The second sheet of steel is placed in a punch press with a die shaped to hold the heating element. The steel shapes are called stampings. One piece forms the base of the tea kettle and the connector between the element and the cord. The second piece sits in the first like one bowl inside the other.

3 The metal stampings are attached to hangers on a conveyor system and carried through two different processes. The base stamping is conveyed to a paint booth where it is given a small electrical charge that attracts fine a paint mist to its surface. The base stamping is painted to match the upper plastic housing. The stamping that will house the electrical components is conveyed through an different coating process that treats the upper surface of the metal piece (the surface that will come into contact with water) so that mineral deposits will be easier to remove. These coating processes are called anodizing. This protects the steel parts from corrosion and gives the

A standard electric tea kettle.

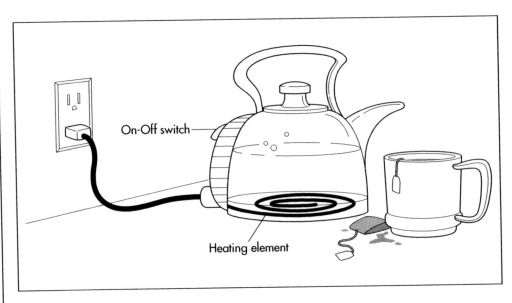

On-Off switch

Heating element

stampings a small electrical charge that attracts the coating.

4 Meanwhile, the upper part of the kettle is formed. The polypropylene plastic pellets are released through a hopper with a funnel-shaped bottom into an injection-molding machine that melts the pellets under high heat until the plastic is liquid. Under high pressure, the liquid plastic is forced into a steel box (called a die) with two halves. When the halves are locked together, the die contains a void that is an exact image of the plastic housing of the kettle. The two parts in the die form the outside and inside surfaces of the kettle. The cavity relief surface of the die shapes the outside of the kettle including the solid plastic handle, and the core relief surface is the inner surface. When the injected plastic has filled every tiny nook in the die, the die is opened and the plastic kettle housing, which is still hot, is ejected. The housings are allowed to cool until assemblers can handle them.

5 The painted and treated metal stampings are welded together to form the base of the kettle. The two major processes—injection molding and the punch press—allow large quantities of these to be made continuously. The kettle base is then placed on an assembly line conveyor belt, then equipped with power components at a workstation where the electrical set is removed from its packaging, and an assembly worker manually sets the connector between the heating element and the cord into the shaped opening

in the base. The kettle's rubber feet are also attached to the base.

6 The metal base (with its electrical components in place) and the plastic housing are sealed together to complete the tea kettle body.

7 At the next assembly station, metal flip-up covers are manually pressed onto the plastic spouts so the spouts will open and pour water; these wide spouts are also the openings for filling the kettles. Other models have metal lids with small plastic handles that are also injection molded. These models are filled through their opened lids.

8 The completed kettles are conveyed to the shipping department. Each kettle is wrapped in a plastic bag that also contains the instructions and boxed. The boxes are packed into shipping cartons.

Quality Control

Electric tea kettles are the subject of quality control in multiple locations in the factory. The steel and plastic materials are prepared, tested, and examined for any defects such wear, discoloration, or scratches. The sets of electrical components are also inspected in the subcontractors' factory where they are made. Underwriters' Laboratories must also test and approve electrical components before they become parts of larger products. At the manufacturer's plant, electric tea kettles are subjected to multi-tiered inspections that begin when all materials are received and continue through each manufacturing step. As a final quality check, each tea kettle is plugged in and operated before it is packed for the customer's use.

Byproducts/Waste

Metal and plastic wastes result from handling and processing both materials. Metal waste is steel only and trimmings are sold to iron or steel foundries. Plastic trimmings or flawed molded parts are reground and recycled. Only a small percentage of reground plastic can be used for other tea kettles in order to maintain the high quality of the plastic housings, but the remaining plastic can be used in other products.

The Future

The future of electric tea kettles is in the hands of consumers and retailers. Retailers provide sales data, consumer comments, and other input to the manufacturers. This information helps the producers to decide on new designs. Stove-top tea kettles have always had stronger sales figures than electric kettles, but electric models also have a devoted public.

The recent popularity of tea and flavored teas has awakened interest in electric tea kettles. Tea infusers and electric tea makers with water filtering systems are the newest ideas in this area of manufacture. The British have pioneered both of these developments that look something like coffee makers with glass tea pots above heating chambers. Unlike coffee makers, these appliances heat the water until it is boiling, which is a must for tea. A filtering system has a filter as a third layer or as a column on the back of the maker, and some can be connected directly to water faucets. These devices seem to point to the future of electric tea-making, but their stacked structure looks curiously like that Persian invention.

Where to Learn More

Books

Huxley, Gervas. *Talking of Tea: Here is the Whole Fascinating Story of Tea.* Ivyland, PA: John Wagner & Sons, Inc., 1956.

Kakuzo, Okakura. *The Book of Tea.* Rutland, VT: Charles E. Tuttle Company, 1972.

Papashvily, Helen, and George Papashvily. *Russian Cooking.* New York: Time-Life Books, 1969.

Pettigrew, Jane. *Tea & Infusions.* Carlton Books Limited, 1999.

Von Bremzen, Anya, and John Welchman. *Please to the Table: The Russian Cookbook.* New York: Workman Publishing Company, Inc., 1990.

Other

"About Tea Kettles." *Kyoto National Museum Web Page.* December 2001. <http://www.kyohaku.go.jp/mus_dict/hd11e.htm>.

Calphalon Corporation Web Page. December 2001. <http://www.calphalon.com>.

Nezu Bijutsukan Museum Web Page. "Tea Kettle of Shinnari Type." December 2001. <http://www.nezu-muse.or.jp/99-10-03/30187_e.html>.

TenRen.Com Web Page. December 2001. <http://www.tenren.com>.

Gillian S. Holmes

Epilation Device

The late 1700s saw the first manufactured razor, but it lacked safety capabilities and the search to develop a safety razor ensued.

Background

Epilation refers to removal of hairs from below the skin's surface. Examples of epilation devices include tweezers, waxes, electrolysis, and laser hair removal. These types of hair removal devices and methods are continually improved to diminish skin irritation, increase ease of use, increase the time between hair removal methods, and decrease pain. Epilation devices currently strive to remove the entire hair from the root using mechanical extraction or laser removal.

The original and most popular epilation device is the Epilady. Through tremendous and innovative marketing, Epilady quickly became a success in the hair removal arena. Despite legal issues surrounding the patent for this product, Epilady has remained a leader in the development and consumer satisfaction for electronic epilation devices. As a result of the consumer's familiarity with the product through advertising, the Epilady brand name has become generic, much like Kleenex, and may be used to refer to all electric epilation devices.

There are two commonly used designs for electronic epilation devices. One epilating head consists of a series of small tweezing mechanisms that grasp individual hairs and pluck them out. The second type of epilating head has a disc system. Several metal discs rotate in opposing directions to grip and pluck out the hair and its root. These methods do not destroy the actual follicle nor prevent regrowth permanently, but results may last several weeks.

History

Throughout history man has experimented with various means of removing hair. The reasons for hair removal range from social status to function to hygiene, and have resulted in diverse methods for performing this task. Many of the hair removal techniques used today are derived from these original methods, such as waxing, shaving, and using depilatory creams. The more advanced methods like electronic epilators, electrolysis, and laser hair removal, were developed only recently because the technology to create such devices was previously unavailable.

The late 1700s saw the first manufactured razor, but it lacked safety capabilities and the search to develop a safety razor ensued. It wasn't until the eighteenth century that the first instrument specifically designed as a safety razor appeared. Invented in 1762 by a French barber, Jean Jacques Perret, the safety razor had a metal guard placed along an edge of the blade to prevent the blade from accidentally cutting into the shaver's skin.

A traveling salesman-inventor named King Gillette launched the first shaving revolution almost single-handedly. In 1895, he conceptualized a razor with disposable blades. However, collaboration with William Nickerson, a professor at the Massachusetts Institute of Technology, was necessary before the product could successfully be brought to market in 1903. In 1915, Gillette introduced Milady Décolletée, a razor created especially for women.

In 1931, the year Gillette retired, Jacob Schick challenged the razor blade with the invention of the electric shaver. Schick's shaver was in turn challenged by Remington, who introduced the dual-headed shaver and the first electric shaver designed especially for women in 1940.

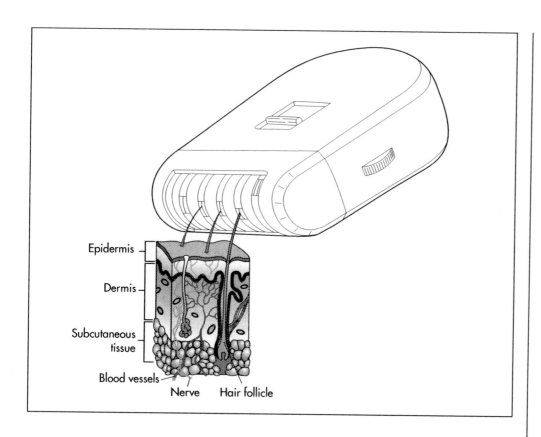

An epilation device that pulls the hair out at the root.

In 1986, an Israeli-based company began marketing Epilady, the first electronic hair removal device. This form of epilation used a tweezing method to extract hairs by the roots, which resulted in longer periods between hair removal. Despite its reputation for causing pain, it became an international success and revolutionized hair removal.

A newly designed Epilady arrived on the market in 1996 called "Discrette." This was the first epilator with a disc-operated head. It boasted a virtually pain-free experience because of the high velocity movement of the rotating discs.

Since then, several other versions of Epilady have been developed. Most of these specialize in specific areas like the face, underarms, and bikini area. In addition, other companies such as Phillips, Braun, and Emjoi have developed similar electronic epilators based on the original Epilady design. Due to enhancements that decrease pain, increase the area to be depilated, and provide longer lasting results, electronic epilation devices are reemerging as a popular means for hair removal.

Raw Materials

The housing for the epilation device can be made of a variety of insulative materials such as plastic, nylon, or any other nonconductive material. There is also a rubber-like gripping surface included on the housing to prevent slippage during use. Metal and electrical mechanisms such as wires and insulative components are manufactured for inclusion in electrical epilation devices.

Design

The two primary designs of an electric epilating device are a rotating disc epilator and a rotating tweezing barrel.

The original electronic epilators were designed using rotary cylinders with pivotal axes, which can adjust the zone of the closing action allowing the device to adapt to changing contour surfaces of the skin. Attached to the rotary cylinders are several small plucking components.

Disc-operated epilators are a newer design and usually cause less pain than the rotary tweezer head. These epilators use high velocity rotating discs that quickly grasp hairs and pluck them out at the root. Because the discs move so rapidly, they are able to catch shorter hairs and cover larger areas in less time.

In both cases, the hairs to be plucked-out are trapped between the projecting portions of the epilation blades or discs. These hairs are then gripped or pinched between the projecting zones of two successive tweezing parts. Finally, these hairs are ejected by centrifugal force during the stage of opening of the gripping zones as the rotating barrel or disc completes its cycle.

The Manufacturing Process

1 Motor-powered epilating devices are manufactured from a hollow cylinder enclosed by the housing that is used as a handgrip. A molding press is used to create the plastic housing. A pellet of encapsulating material, such as plastic, is placed in a die.

2 The encapsulating material is heated until it melts and fills the molding press. After allowing the encapsulating material to cool and solidify, the molding press is opened and the molded parts are separated.

3 The exterior of the housing is sprayed with a heat shrunk rubberized coating to provide a comfortable grip.

4 An epilator head and frame comprise the top of the device. The head frame is molded in a similar manner as the housing. It is usually formed by a molding press using a plastic-like material. The head frame has an opening through the top allowing exposure of the epilating head, and attaches to the main housing unit either by interlocking components or screws.

5 The epilator head is a roller-type component made of metal. Alternating arrangements of fixed and movable blades are mounted on the roller head. The moveable blades, which are the tweezing elements, and fixed blades are formed from a non-corrosive metal, which both heated and poured into a mold. Once the molded metal cools, the blades are sharpened with a filing saw. The various blades each have a polygonal shaped central opening to permit the insertion both of the rotary shaft and of the different bars and strips. Each piece is individually cut by a metal die cutter. The fixed and movable blades are held together with a series of small screws or by inserting a thin metal pin through holes at the base of each part. The pin itself is mounted into each end of the housing unit.

6 The rotary shaft of the epilation head is placed between two end plates that are soldered to the cover and carry the bearings in which the shaft-ends are mounted. A thin metal bar is positioned as an axis between the rotating head and the small electric motor that powers the device.

7 The axis is driven in rotation by the small electric motor, which is also housed within the casing. Power for the motor is derived from a dry cell battery positioned near the motor.

8 Against two opposite faces of the shaft, two flat metal strips are mounted that hold the epilation blades in position. Against each of the two other faces of the shaft are mounted two additional thin metal bars capable of sliding in a direction parallel to the axis of the shaft.

9 The sliding bars of each pair are machined with a series of transverse grooves. The teeth of the blades are engaged in the grooves of the two bars. Therefore, the epilation blades are coupled with each of the two sliding bars of one and the same pair in alternate sequence.

10 Once all of the metal components are fastened together and mounted to the motor, the remaining plastic housing is attached. This ensures that the unit is fully enclosed and interior components will be protected.

11 The completed unit moves on for final inspection and packaging prior to arriving in stores.

Quality Control

The major risk of motor driven devices is electrical shock, which can occur if the electrical components are not properly insulated. Therefore, manufacturers evaluate the insulating materials throughout the design process. Devices are left switched on in a protective container and effects of the accumulated heat are evaluated on the durability of the device and its safety.

Health risks associated with the hair removal appliances are typically infections resulting from unsanitary conditions. It is the consumer's responsibility to ensure that home devices are properly sanitized. The

United States Food and Drug Administration can take action against products that are found to cause damage. Laser hair removal devices are prescription devices and should only be used under a licensed practitioner's guidance. These devices have been approved by the USFDA for hair removal.

The company who manufactures the Epilady devices, Mepro, complies with ISO9002 regulations. These international standards help ensure quality control at all levels of the manufacturing process in order to reproduce parts and assemble them with efficient and repetitive methods.

Byproducts/Waste

The waste from the plastics and moldable parts that are incorporated for the housing are melted and reused. Electrical components are tested and either passed on for further inspection or dismantled and reassembled depending on the severity of the design.

The Future

Manufacturers of epilation devices have been making great strides to improve the effectiveness of their products, reduce the pain and irritation associated with use of epilators, and increase the amount of time between hair removal. Some of these enhancements include plating the tweezing components with 24-karat gold for better sanitation and hypo-allergenic qualities. The speed and number of tweezing components has also been increased so that a greater area can be covered in a faster amount of time. Quicker rotation of the plucking elements will also reduce pain.

Permanent depilatory devices are becoming more popular. They use high frequency electricity to destroy hair-producing papilla at the base of the hair shaft. Unfortunately, many of these devices involve large and expensive equipment requiring a visit to a salon. An improved tweezer mechanism using a compressible coil spring attached to a first and a second arm, where the hair is grasped between the coils of the spring and pulled from the body may be more cost effective and functional. The primary advantage is that the coil spring can grasp multiple hairs simultaneously.

Lasers have been used for many years for medical cosmetic procedures. Laser hair removal is non-invasive and can remove unwanted body hair without damaging the surrounding pores or structures of the skin. Optical energy is used that is of a wavelength sufficient to cause epilation in the cavity, therefore limiting damage to tissue outside of the hair root structure. A laser that may be operated in a pulse mode and contained in a hand piece having the optical system at the end generates the beam. During application of the laser energy, the root structure and the hair, if present, is progressively ablated and vaporized until the bottom of the hair bulb is reached at which time the beam is turned off.

Where to Learn More

Periodicals

Segal, Marian. "Hair Today, Gone Tomorrow." *FDA Consumer Magazine* (September 1996).

Other

"Electrolysis Manufacturers." *Hairfacts.com Web Page*. December 2001. <http://www.hairfacts.com/makers/epltrmfr.html>.

"Epilady Story." *Epilady Web Page*. December 2001. <http://www.epilady.com/profile.html>.

"The Shaving Historical Timeline." *The Quik Shave Body Razor Web Page*. December 2001. <http://www.quikshave.com/timeline.htm>.

United States Patent and Trade Office. December 2001. <http://patft.uspto.gov>.

Stacey L. Blachford
Bonny P. McClain

External Defibrillator

The idea of defibrillation was first conceived during the late 1800s, but it was not until 1947 that the first successful human test was done.

Background

An external defibrillator is a device that delivers an electric shock to the heart through the chest wall. This shock helps restore the heart to a regular, healthy rhythm. The device is generally sold as a kit that consists of a power control unit, paddle electrodes, and various accessories. The parts are made individually and pieced together via an integrated production process. Since then medical device manufacturers have introduced various defibrillators, internal and external, that have added years to patients' lives.

To understand how a defibrillator can restart a stalled heart, the physiology of the organ must be considered. The human heart has four chambers, which create two pumps. The right pump receives the oxygen-depleted blood returning from the body and pumps it to the lungs. The left pump receives the oxygenated blood from the lungs and pumps it to the rest of the body. Both pumps have a ventricle chamber and an atrium chamber and operate in a similar manner. The blood collects in the atrium and is then transferred to the ventricle. Upon contraction, the ventricle pumps the blood away from the heart.

The coordination of the pumping action is critical for the heart to function correctly. A pacemaker region, which is located in the heart's right atrium, is responsible for this control. In this region, a spontaneous electrical impulse is created by the diffusion of calcium ions, sodium ions, and potassium ions across the cell membranes. The impulse thus created is transferred to the atrium chambers causing them to contract, pushing blood into the ventricles. After about 150 milliseconds the impulse moves to the ventricles, which causes them to contract and pump blood out of the heart. As the impulse moves away from the chambers of the heart, these sections relax. In a normal heart, the process then repeats itself.

In some cases, the electrical control system of the heart malfunctions and results in an irregular heart beat such as ventricular fibrillation. Various conditions can cause ventricular fibrillation including blocked arteries, poor reaction to anesthesia, and electrical shock. Defibrillators are used to supply a strong electrical shock to the heart. Two electrodes are placed on the chest and a shock is given. A typical defibrillator device will deliver a shock for three to nine milliseconds. For reasons not quite understood, the shock essentially resets the natural ventricular rhythm and allows the heart to beat normally.

In practice, an external defibrillator can be operated at an emergency site or a hospital. The operator first turns on the machine and then applies a conductive gel to the paddle electrodes or patient's chest. The energy level is selected and the instrument is charged. The paddles are placed firmly on the patient's unclothed chest with a pressure of about 25 lb (11 kg). The buttons on the electrodes are pressed simultaneously and the electric shock is delivered. The patient is then monitored for a regular heartbeat. The process is repeated if necessary.

History

The discovery that a misfiring heart could be restarted using an electrical charge is one of the great developments of modern medicine. This idea was begun around 1888 when it was suggested by Mac William that ventricular fibrillation might be the cause of sudden death. Ventricular fibrillation is a condition in which the heart suddenly beats irregularly, preventing its blood-pumping ability that ultimately can lead to death. It can be caused by a coronary artery blockage, various anesthesia, and electric shock.

In 1899, Prevost and Batelli made the crucial discovery that large voltages applied across the heart could stop ventricular fibrillation in animals. Various other scientists studied further the effects of electricity on the heart during the early nineteenth century.

During the 1920s and 1930s, research in this field was supported by the power companies because electric shock induced ventricular fibrillation killed many power utility line workers. Hooker, William B. Kouwenhoven, and Orthello Langworthy produced one of the first successes of this research. In 1933, they published the results of an experiment, which demonstrated that an internally applied alternating current could be used to produce a counter shock that reversed ventricle fibrillation in dogs.

In 1947, Dr. Claude Beck reported the first successful human defibrillation. During a surgery, Beck saw his patient experiencing a ventricular fibrillation. He applied a 60 Hz alternating current and was able to stabilize the heartbeat. The patient lived and the defibrillator was born. In 1954, Kouwenhoven and William Milnor demonstrated the first closed chest defibrillation on a dog. This work involved the application of electrodes to the chest wall to deliver the necessary electric counter shock. In 1956, Paul Zoll used the ideas learned from Kouwenhoven and performed the first successful external defibrillation of a human.

In the 1960s, scientists discovered that direct current defibrillators had fewer adverse side effects and were more effective than alternating current defibrillators. In 1967, Pantridge and Geddes demonstrated that using a mobile, battery-powered DC defibrillator could save lives. The late sixties saw the introduction of an implantable defibrillator by Dr. Michael Mirowski. Both internal and external defibrillators were redesigned in the 1970s to automatically detect ventricular fibrillation. As improvements in electronics and computers became available these technologies were adapted to defibrillators.

William Kouwenhoven.

William Bennett Kouwenhoven was born January 13, 1886 in Brooklyn. Trained as an electrical engineer, his most enduring contributions to science came from the medical arena. Using his electrical engineering background, Kouwenhoven invented three different defibrillators and developed cardiopulmonary resuscitation (CPR) techniques.

In the 1920s, Kouwenhoven's interest crossed between electrical engineering and medicine. His engineering work focused on high tension wire transmission of electricity. Kouwenhoven became interested in electricity's possible role in reviving animals. He knew that when applied to the heart an electric current could start it again.

From 1928 through the mid-1950s, Kouwenhoven developed three defibrillators: the open-chest defibrillator, the Hopkins AC Defibrillator, and the Mine Safety Portable. These were intended for use within two minutes of the start of ventricular fibrillation, and at least one required direct contact with the heart. In 1956, Kouwenhoven began developing a non-invasive method. During an experiment on a dog, he realized the weight of the defibrillator's paddles raised the animal's blood pressure. Based on this Kouwenhoven developed CPR.

By the early 1960s, CPR was being used throughout the United States. Kouwenhoven's ground-breaking work was recognized by the medical community and the electrical engineering establishment. He was awarded the American Medical Association's (AMA) Ludwig Hekton Gold medal in 1961 and 1972, and the American Institute of Electrical Engineering's Edison Medal in 1962. Johns Hopkins bestowed Kouwenhoven with an honorary M.D. in 1969 (he is the only person to ever receive this honor). He won the Albert Lasker Clinical Research Award in 1973. Kouwenhoven died on November 10, 1975.

Today, defibrillation has become an integral part of the emergency response routine. In fact, the American Heart Association considers defibrillation a basic life support skill for paramedics and rescue workers.

Raw Materials

Biocompatible raw materials must be used in the construction of defibrillators because they interact with patients. The materials must also be pharmacologically inert, non-toxic, sterilizable, and functional in a variety of environmental conditions. The various parts of the defibrillator, including the control box casing, microelectronics, and the electrodes, are all made with biocompatible materials. Typically, the casing is made of a hard polystyrene plastic or lightweight metal alloy. The electrodes are made from titanium and silicone rubber. The microelectronics are made of modified silicon semiconductors. The primary materials used in battery construction can include numerous compounds such as lead acid, nickel-cadmium, zinc, lithium, sulfur dioxide, and manganese dioxide.

Design

The basic design of a external defibrillator includes a control box, a power source, delivery electrodes, cables, and connectors. While these devices are sometimes implanted in patients, this work focuses on portable units used in hospitals and emergency sites.

Controls

The control box is a small, lightweight, plastic case. It contains the power generating and storage circuits. In general, the charge that is delivered to the patient is generated by high voltage generation circuits from energy stored in a capacitor bank in the control box. The capacitor bank can hold up to 7 kV of electricity. The shock that can be delivered from this system can be anywhere from 30–400 joules. The control box also houses the control electronics and the operator input buttons. The typical controls on a defibrillator control box include a power control button, an energy select control, a charge button, and an energy discharge button. Certain defibrillators have special controls for internal paddles or disposable electrodes.

Electrodes

The electrodes are the components through which the defibrillator delivers energy to the patient's heart. Many types of electrodes are available including hand-held paddles, internal paddles, and self-adhesive, pre-gelled disposable electrodes. In general, disposable electrodes are preferred in emergency settings because they have advantages such as increasing the speed of shock and improving defibrillation technique. The paddle size affects the current flow. Larger paddles create a lower resistance and allow more current to reach the heart. Thus, larger paddles are more desirable. Most manufacturers offer adult paddles, which are between 3.1–5.1 in (8–13 cm) in diameter, and pediatric paddles, which are smaller.

Since skin is a poor conductor of electricity, a gel must be used between the electrode and the patient. Without this conductor, the level of the current reaching the heart would be reduced. Also, the skin may be burned. A variety of gels and pastes are available for this purpose. These are composed of cosmetic ingredients like lanolin or petrolatum. Chloride ions in the formula also help form a conductive bridge between the skin and the electrode allowing better charge transfer. Many of these materials are the same compounds used for other medical devices such as ECG scans.

Battery

Batteries are essentially containers of chemical reactions. In defibrillators, a variety of batteries are used. They are characterized by the chemical reactions contained in them and include lead-acid, lithium, and nickel-cadmium systems. These batteries can typically be recharged by an outside power source, and when not in use defibrillators are stored plugged in. Since extreme temperatures negatively affect the batteries, defibrillators are stored in controlled environments. Over time batteries wear out and are replaced. This is important because battery chemistries are inherently corrosive and potentially toxic.

Automated external defibrillators

In 1978, the automated external defibrillator was introduced. This device is equipped with sensors that are applied to the chest and determine whether ventricular fibrillation is actually occurring. If detected, the device calls out instructions to deliver an electrical shock. These automated devices greatly reduce the training required to use a defibrillator and have saved thousands of lives.

The Manufacturing Process

Defibrillators are sophisticated electronic devices. Typically, manufacturers rely heavily on suppliers to produce the component parts. These parts are then shipped to the manufacturer and pieced together to form the final product. The process is therefore not linear but an integrated one.

Making the batteries

1 One type of battery used in defibrillators is a lithium battery. The design of this type of battery involves the connection of multiple cells. For defibrillators, the cells are made up of lithium metal and sulfur dioxide gas. Working under oxygen free conditions, the lithium is shaped into a solid case and sulfur dioxide is added.

2 The cell is then hermetically sealed to prevent the sulfur dioxide gas from escaping and moisture from entering. In one battery design, four of these lithium cells are wired in series and packed in a solid housing. An 8-amp fuse is mounted to each cell for safety reasons. All of the cells have a vent that can be used to release pressure if it builds up too high.

Creating the casing

3 To make the casing and the outer housing for the electrodes, a process known as injection molding may be used. In this procedure plastic pellets are melted and forced into shape. The pellets of plastic are put into a holding bin attached to the injection-molding machine and melted.

4 The material is then passed through a hydraulically controlled screw. As the screw rotates, the plastic melts further. It is directed through a nozzle and injected into the mold. The mold is made up of two metal halves that form the shape of the part when brought together. When the plastic is in the mold, it is held under pressure for a moment and then cooled. As it cools, the plastic hardens and takes on the shape of the mold.

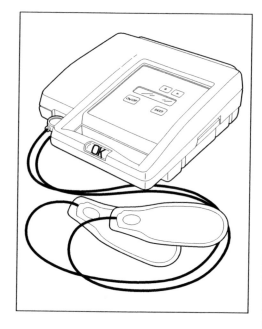

The mold pieces are separated and the plastic part falls out onto a conveyor. The mold then closes again and the process is repeated. After the plastic parts are ejected from the mold, they are manually inspected.

Making the electronics

5 The motherboard inside the casing contains all the electrical circuitry of the defibrillator including semiconductor chips, resistors, capacitors, and other devices. Using an intricate method known as hybridization, these components are combined to form a single complex circuit. Construction begins with a small board that has the electronic circuit configuration printout out on it.

6 The board is then moved through a computerized machine that places the appropriate components exactly where they are needed on the board. This action is accomplished by a placing head on the device. It holds the electronic component and presses it down on the board.

7 The electronic components are then affixed to the board by a soldering machine using a minimum number of welds. The circuitry is allowed to cool and is tested before being connected to the control box casing.

Assembly

8 The electronic boards are manually attached to the casing by line workers under extremely clean conditions. The boards are

The positioning of the defibrillator's paddles.

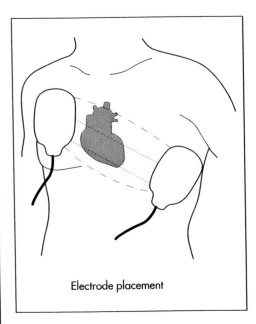

Electrode placement

attached with various screws and fasteners. The casings are fitted with control buttons and metal electrode adapters. The whole assembly is closed with screws and sent to an area for testing and final packaging.

Making paddle electrodes

9 In addition to the outer shell, the paddle electrodes are made up of a metal plate and a cable that connects them to the main machine. The metal plate is a conducting alloy such as tin. The plate is made using a machine called a "continuous caster." The caster converts molten tin to thin sheets by forcing it between large, water cooled rollers. A thin layer of stannous chloride is sprayed on the sheets, and they are cut into the appropriate size for the electrode.

10 The cable is produced by a drawing technique. In this step, metal is heated until it softens. It is then rolled out and drawn to produce a long wire.

11 The wire is then cut and bundled with other wires. The bundle of wires are coated with a relatively thick polymeric insulator and wrapped in an insulating sheath. One end of the wire is soldered to the metal plate.

12 The plate is then manually fitted onto the outer shell and the cable is fed through a hole in the back of the shell.

13 The end of the cable is fitted with an adapter that can be plugged into the control box.

Final assembly

14 When all of the components are completed, they are brought together in a final packaging. Line workers take the individual parts of the machine including the electrodes, battery, and control box and put them in a cushioned box. They also include cables, instruction manuals, and other information.

15 Before any product is shipped, it is tested to ensure that it delivers the proper charge.

Quality Control

Making visual and electrical inspections throughout the entire production process ensures the quality of each defibrillator. Electronic circuit fabrication is particularly sensitive to contamination so production is done in air-flow controlled, clean rooms. The clothes worn by line assembly workers must be lint-free to help reduce the chance of contamination. Since the batteries are critical and potentially dangerous, they are subjected to extensive performance, safety, and stability testing. The functional performance of each completed defibrillator is tested to make sure it works. This can be accomplished by charging the battery, discharging the device, and measuring the charge output. To simulate real-life use, these tests are done under differing environmental conditions. Quality testing is also done routinely after the defibrillators are purchased. Engineering personnel perform maintenance checks every three to six months depending on usage. This typically involves a charge-discharge test.

Each company that manufactures medical devices is required to register with the United States Food and Drug Administration (FDA). They must adhere to the FDA's quality standards known as "good manufacturing practices." This requires extensive record keeping procedures and also subjects the manufacturer to routine inspection of the facility for compliance.

The Future

In the future, defibrillators will be improved to become safer and more efficient. For example, designers are continually improving the electrode design to reduce the chances that the device operator will get shocked. A recent patent issued in the United States describes an electrode system that uses a Y shaped cable for just this purpose. Advances in the fabrication of integrated circuits will also make the devices easier to use and more lightweight.

Another important area of improvement will be found in battery technology. Scientists at the United States Department of Energy's Brookhaven National Laboratory have patented a new metal alloy that should greatly improve rechargeable battery performance. The alloy can be incorporated into a nickel/metal hydride battery to provide a significant increase in capacity for storing charge. In addition to these areas of advancement, improvements in defibrillator design such as the incorporation of more sensors to give vital information about a patient's condition will also be introduced.

Where to Learn More

Books

Carr, J. J. *Introduction to Biomedical Equipment Technology.* 2nd ed. Prentice Hall Career and Technology, 1993.

Fox, Stuart. *Human Physiology.* W. C. B. Publishers, 1990.

Oever, R. V. D. *Cardiac Pacing and Electrophysiology: A Bridge to the 21ˢᵗ Century.* Kluwer Acedemic Publishers, 1994.

Periodicals

Shakespeare, C. F., and A. J. Camm. "Electrophysiology, Pacing, and Arrhythmia." *Clinical Cardiology* 15 (1992): 601–606.

Other

Worthington, Janet Farrar. "The Engineer Who Could." *Hopkins Medical News.* 18 March 1998. 2 October 2001. <http://www.hopkinsmedicine.org/hmn>.

Perry Romanowski

Fabric Softener

In the early 1900s, preparations known as cotton softeners were developed to improve the feel of fibers after dyeing.

History

A fabric softener is a liquid composition added to washing machines during the rinse cycle to make clothes feel better to the touch. These products work by depositing lubricating chemicals on the fabric that make it feel softer, reduce static cling, and impart a fresh fragrance. The first fabric softeners were developed by the textile industry during the early twentieth century. At that time the process that was used to dye cotton fibers left them feeling harsh. In the early 1900s, preparations known as cotton softeners were developed to improve the feel of these fibers after dyeing. A typical cotton softener consisted of seven parts water, three parts soap, and one part olive, corn, or tallow oil. With advances in organic chemistry, new compounds were created that could soften fabric more effectively. These improved formulations soon found their way into the commercial market.

By the 1960s several major marketers, including Procter and Gamble, had begun selling liquid fabric softener compositions for home use. The popularity of these products dramatically increased over the next decade as manufacturers developed new formulations that provided improved softness and more appealing fragrances.

Despite their growing popularity, fabric softeners suffered from one major disadvantage: the softener chemicals are not compatible with detergents and therefore they can not be added to the washer until all the detergent has been removed in the rinse cycle. Initially, this restriction required the consumer to make an extra trip to the washing machine if they wanted to soften their clothes. In the late 1970s manufacturers found a way to deliver fabric softening benefits in a dryer sheet format. These sheets provide some of the benefits of fabric softeners but give the added convenience of being able to be added in the dryer instead of the washer rinse cycle. However, while dryer sheets are very popular today, liquid softeners are still widely used because they are more effective.

In the 1990s, environmentally minded manufacturers began test marketing ultra-concentrated formulations. These "ultra" formulations are designed such that only about one-quarter as much product has to be used and therefore they can be packaged in smaller containers. However the perceived value to the consumer is lower because there is less product and the price is higher. It remains to be seen if these ultra concentrates will succeed in today's marketplace.

By the end of the 1990s, annual sales of liquid fabric softeners in the United States reached approximately $700 million (in supermarkets, drug stores, and mass merchandisers). For the sake of comparison, about $400 million worth of dryer sheets are sold each year. The major manufacturers such as Procter and Gamble (Downy) and Lever Brothers (Snuggle), dominate about 90% of the market share while private label brands account for the remaining 10%.

Design

Product development chemists create fabric softeners that are designed to meet a series of specific marketing requirements. First, the formulations must deliver a variety of attributes desired by consumers such as superior softness, improved iron glide, reduced wrinkle formation during the wash cycle, improved wrinkle removal after washing, better color retention, and enhanced stain protection. In addition, the formulas must be safe to use, environmentally friendly, aesthetically pleasing, and cost effective. Chemists use technical evaluations in combination with consumer testing to design formulations that are both effective and affordable.

Raw Materials

Conditioning agents

Early fabric softener formulas were relatively simple dispersions of fatty materials that would deposit on the fabric fibers after washing. One of the most common ingredients used was dihydrogenated tallow dimethyl ammonium chloride (DHTDMAC), which belongs to a class of materials known as quaternary ammonium compounds, or quats. This kind of ingredient is useful because part of the molecule has a positive charge that attracts and binds it to negatively charged fabric fibers. This charge interaction also helps disperse the electrical forces that are responsible for static cling. The other part of the molecule is fatty in nature and it provides the slip and lubricity that makes the fabric feel soft.

While these quats do soften fabrics very effectively, they also can make them less absorbent. This is a problem for certain laundry items such as towels and diapers. To overcome this problem, modern formulations use quats in combination with other more effective ingredients. These newer compounds have somewhat lower substantivity to fabric which makes them less likely to interfere with water absorption.

One of the new classes of materials employed in fabric softener formulations today is polydimethylsiloxane (PDMS). Siloxane is a silicone based fluid that has the ability to lubricate fibers to give improved softening and ease of ironing. Other silicones used in softeners include amine-functional silicones, amide-functional silicones and silicone gums. These silicone derivatives are modified to be more substantive to fabric and can dramatically improve its feel.

Emulsifiers

The conditioning ingredients used in fabric softeners are not typically soluble in water because of their oily nature. Therefore, another type of chemical, known as an emulsifier, must be added to the formula to form a stable mixture. Without emulsifiers the softener liquid would separate into two phases, much like an oil and vinegar salad dressing does.

There are three types of emulsifiers used in fabric softener formulations: micro-emulsions, macro-emulsions, and emulsion polymers. Macro-emulsions are creamy dispersions of oil and water similar to hand lotions or hair conditioners. The emulsifier molecules surround the hydrophobic oil or silicone droplets and allow them to be dispersed in water. A micro-emulsion is chemically similar, but it creates oil particles that are so small that light will pass around them. Therefore, a micro-emulsion is characterized by its clarity and transparency as opposed to being milky white. Furthermore, one of the advantages of micro-emulsion is that the silicone particles are so tiny that they will actually penetrate into the fibers, while macro-emulsions only deposit on the fiber's surface. The third type, emulsion polymers, create dispersions that look similar to a macro-emulsion. This system does not use true emulsifiers to suspend and dissolve the oil phase. Instead, emulsion polymers create a stabilized web of molecules that suspend the tiny silicone droplets like fish caught in a net.

The emulsifying system used in softeners must be chosen carefully to ensure the appropriate level of deposition on the fabric. A blend of non-ionic emulsifiers (those that have no charge) and cationic emulsifiers (those that have a positive charge) are typically used. Anionic surfactants (which have a negative charge) are rarely used because the fabric conditioning agents have a positive charge which would tend to destabilize an anionic emulsion.

Other ingredients

In addition to conditioning agents and emulsifiers, fabric softeners contain other ingredients to improve their aesthetic appeal and to ensure the product will be shelf stable. For example, fragrance and color are added to make the product more pleasing to consumers. In addition, emulsion stabilizers and preservatives are used to ensure the product quality.

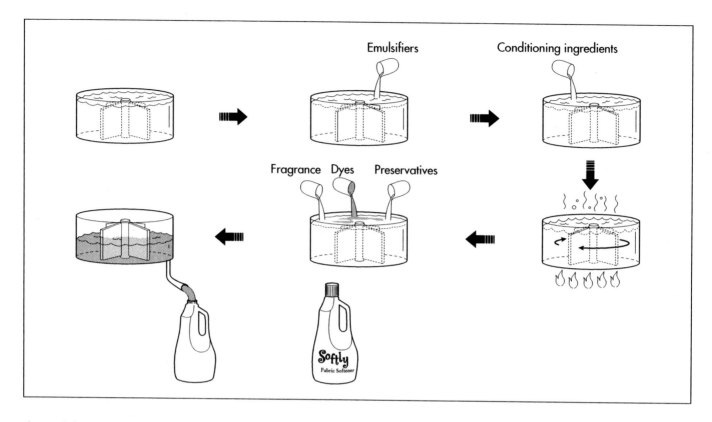

The emulsifiers and then conditioning ingredients are added to water. The batch is heated and mixed. Then the other ingredients are added.

The Manufacturing Process

1 The preferred method for manufacturing liquid softeners involves heating the ingredients together in one large mixing vessel. Mixing tanks should be constructed from high grade stainless steel to prevent attack from the corrosive agents in the formula. The tank is typically equipped with a jacketed shell that allows steam and cold water to be circulated, so the temperature of the batch can be easily controlled. In addition the tank is fitted with a propeller type mixer that is driven by a large electric motor. This kind of mixing blade provides the high shear that is needed to properly disperse the ingredients.

The first step in the manufacturing process is to fill the tank with the specified amount of water. Water is added first because it acts as a carrier for all the other ingredients. Deionized water is used because it is free from metal ions that can affect the performance of the batch. Conventional formulations can contain as much as 80–90% water.

2 Once the water has been added to the tank, heating and mixing is initiated. When the water has reached the appropriate temperature, the emulsifiers are added. Since these chemicals tend to be waxy solid materials they are added at relatively high temperatures (between 158–176°F [70–80°C]). While the order of addition depends on the specific formula, it usually more effective to disperse the emulsifiers prior to adding the less water-soluble materials. Emulsifiers are used between 1–10%, depending on the specific chemicals that are selected.

3 The conditioning ingredients used in softeners are not typically water soluble, so they are added to the water phase after the emulsifiers. For a typical strength formulation about 5% is used. For more concentrated formulations, levels of 10% are more common. When blends of quats and silicones are used, the silicones are used at levels as low as 0.5–1.5%.

4 When pre-emulsified silicones are used in the formula they are added late in the process when the temperature is lower and there is less mechanical agitation in the batch. If higher molecular weight silicones are used that have not been pre-emulsified they must be added to the batch at high temperatures with a high level of agitation to ensure the silicone oil droplets are evenly dispersed.

5 Heating and mixing continues until the batch is homogeneous. At this point cool water is circulated around the tank to lower the temperature. As the batch cools, the remaining ingredients, such as preservatives, dyes, and fragrance, are added. These ingredients are used at much lower concentrations, typically below no more than a few percent for fragrance and less than 1% for preservatives and dyes. When the batch is complete, a sample is sent to the analytical chemistry lab to ensure it meets quality control standards for solids, pH, and viscosity. The completed batch may be pumped to a filling line or stored in tanks until it is ready to be filled.

6 When the product is ready to be filled into the package, it is transferred to an automated filling line. Plastic bottles are fed onto a conveyor belt that carries them under a filling nozzle. At the filling head there is a large hopper that holds the formulation and discharges a controlled amount, usually set by volume, into the bottle. The filled package continues down the conveyor line to a capping machine that applies the closure and tightens it. Finally, the filled bottles are packed in cartons and stacked pallets for shipping.

Quality Control

The finished fabric softer formulations are tested using a number of different protocols. Simple laboratory tests are used to determine basic properties such as pH, viscosity, and percent solids. These tests can help confirm that the correct ingredients were added at the appropriate levels.

Other, more rigorous, tests are done to ensure the formulation is functioning correctly. One such evaluation is a water absorbency test, sometimes called the Drayes Wetting Test. This procedure involves dropping small pieces of treated fabric onto water and recording the length of time required for the fabric to sink. This measurement is taken 10 times to obtain an average result.

Anti-wrinkle properties can be evaluated by asking panelists to rate samples of fabric before they have been ironed. They are asked to numerically rate the amount of wrinkling between the test sample and the fabric softener treated sample. The test to measure ease of ironing is also done using trained panelists.

These tests are performed on swatches of identical fabrics with the only difference being that one fabric has been treated with softener and the other has been washed in detergent only. 100% cotton pillowcases are used for wrinkling and ironing tests while 100% cotton terry towels are used for evaluating softness and water absorbency. The swatches are dried in a controlled environment at 71.6°F (22°C) and 65% relative humidity for 24 hours before testing.

The Future

There are two formula related areas that will affect the future of fabric softeners. The first is the impact the ultra-concentrates will have on the market. At the time of this writing it is too soon to tell if they will be accepted by consumers. The second area is related to the role that multi-functionality will play in the future. As chemists develop new more efficacious ingredients there is more potential for additional consumer-perceivable benefits. At the turn of the millennium, multi-functional fabric softener formulations are the latest trend. These new products not only soften clothes but also improve the ease of ironing, reduce wrinkling in the dryer, and provide stain protection. Both Lever Brothers and Procter and Gamble have capitalized on this trend with new formulations that deliver multiple fabric care benefits.

Finally, manufacturers may turn to new delivery forms to make softeners easier to use. One new method introduced by P&G in the late 1990s is the "Downy Ball." This is a reusable plastic tennis ball sized sphere that is filled with liquid Downy and added to the washer at the beginning of the cycle. The ball stays sealed during washing but the spinning of the rinse cycle triggers it to open and release the softener. For consumer who do not have an automatic softener dispenser on their washing machines, the "Downy Ball" saves them from the trouble of adding the liquid in a separate step. Other innovative dispensing devices like this may become more common as manufacturers strive to differentiate their products from the competition.

Where to Learn More

Periodicals

Henault, Benoit. "A Fresh Look at Fabric Softeners." *Soap & Cosmetics* (June 2001).

Turcsik, Richard. "The Soft Sell: Super-markets are Hoping New Fabric Softener Formulations and Advertising Will Help Stem Market Share Erosion." *Supermarket News* (21 August 1995).

Randy Schueller

Feather Duster

Background

A feather duster is a cleaning device that uses bird feathers (certain feathers from a small number of species are preferred) to remove dust from objects. High quality dusters use feathers from the outer layers of an ostrich's feathers. Each has a quill (a hollow spine) near on edge of the feather and a fringe on the other side made up of barbs that lock together through a network of smaller barbs called barbules. These also interlock, making ostrich feathers highly desirable for dusters. The very fine, soft barbs will not scratch furniture, and when rubbed to build up static electricity they will capture and hold dust until shaken out. Down feathers from other birds like the turkey have smaller, simple shafts or quills down the centers of the feathers with very soft and loosely spaced barbs on either side of the quills.

Many housecleaning experts do not recommend feather dusters because they believe these dusters only scatter the dust or leave tiny fluffs of feathers. This is not true and shows that many people do not know how to use feather dusters properly. The buildup of static electricity is essential. It is the scientific property of the feathers that makes them trap dust, while the structural character of the feathers gives them enough tiny "fingers" to catch the dust. When the duster begins to leave dust or bits of feather, it should be taken outside and shaken to release the dirt. Like all handmade devices, the feather duster must be handled with care to prevent the feathers from falling apart. When feather dusters are used properly, they are more effective than most anti-static fabric dusters on the market.

History

Brushes, dusters, and other cleaning equipment were simple and crudely made until the middle of the nineteenth century. Housemaids used dusters routinely, and feather dusters in the late 1800s and early 1900s were symbols of social status. The feather duster is a patented invention. The patent was awarded in 1876 to Susan Hibbard, who had to fight her own husband, George Hibbard, in patent court because he claimed the invention as his own. Susan Hibbard's idea was prompted by seeing turkey feathers, which she thought could be useful, thrown away as waste. Other early feather dusters were made of goose feathers which were soft and sprang back to their shape. Feathers were also used to make small brushes for buttering loaves of bread to keep them soft; these were similar to modern basting brushes except that the use of feathers fell out of fashion. To make brushes for the hearth and the stove, wings still bearing their feathers from geese, chickens, and turkeys were popular "wing dusters," although they had to be kept from cats and dogs and were also meals for insects and moths. By 1900, dusters made of ostrich feathers were preferred for both light and demanding dusting because of their durability.

Raw Materials

South Africa is virtually the only place where ostriches are farmed for feathers and other products and where feather dusters are made. Feather dusters sold in the United States and other countries under the brand names of major manufacturers of home products are purchased directly from South African wholesalers.

Housemaids used dusters routinely, and feather dusters in the late 1800s and early 1900s were symbols of social status.

The plants in South Africa that make feather dusters both process the feathers and assemble the dusters. Ostrich feathers are plucked from live birds cared for on large farms of thousands of birds. These farms also produce leather and meat from separate stocks of ostriches. The birds have many layers of feathers for insulation that range in size from very short to extravagantly long. The processing and assembly plant receives the feathers in bulk shortly after they have been plucked from the farmed birds.

Handles for feather dusters are either wood or plastic. Wooden handles are also made in South Africa and are received at the feather duster plant as finished products. Specialized suppliers use hardwood from several species of South African trees. Cured wood is cut into blocks of suitable lengths and dimensions and turned on lathes to shape the handles. Holes are drilled in the tops of the handles so loops can be added for hanging the feather dusters. After the handles are turned and drilled, they are sanded in machines that sand in several stages using finer and finer grades of sandpaper. Finally, the handles are stained and varnished. After drying, they are packaged and shipped to the feather duster plant.

Plastic handles are made of Marlex plastic and are injection-molded at Asian plants. Other plastic parts include ferrules, the caps that fit over the feathers where they join the handles, and the hanging loops. Again, the feather duster maker receives these not as true raw materials but as supplied parts.

Contact glue is purchased in bulk from a specialized glue manufacturer. Contact glue is semi-sticky and is not designed to be the only fixative that holds feathers on the handles. Low carbon steel wire provides the strong binding. The sizes of wire are described in gauges that are measures of the diameter of the wire. Feather dusters are made with 16-gauge wire, a relatively fine size equaling 0.001 in (0.03 mm) in diameter. A wire manufacturer ships the wire to the feather duster factory on large reels or rolls.

Design

The design of feather dusters made with ostrich feathers is almost unchanged since about 1900. Probably the biggest design change in their history was the production of some lines of dusters with plastic handles when plastic design improved rapidly after World War II. Some plastic dusters are made in retractable designs with hollow handles that slip down over the feathers. Feather duster manufacturers will make design changes specifically for clients. These special orders are made for companies like large janitorial services or suppliers.

Design changes are more frequent for less expensive dusters made with turkey feathers. Usually, these feathers are dyed, and colors are changed occasionally depending on popular decorator colors. Handles for these dusters are modified in length and curvature to attempt to interest more purchasers. Foam grips on the handles are an example of a recent design change made to be both functional and eye-catching.

Interestingly, the care of ostriches is a "design change" that manufacturers are beginning to practice. Ostrich feather products have very little value compared to meat and leather from the birds; meat production generates 200 times more income than feather-related manufacture. Consequently, farmers and manufacturers who maintain flocks of ostriches are learning how to improve feather production. Ostrich care has changed little since ostrich farming began on a large scale, also about 100 years ago. Typically, it takes 12–14 months for an ostrich's feathers to grow enough so they can be harvested. If the birds are given quality feed and improved attention at the farms, this time can be reduced to two months and feather quality also improves. Workers in the feather duster plant are now also paid far more than South Africa's minimum wage, so they invest more care in the plucking and processing of the feathers. Designs of actual dusters made with ostrich feathers may be forced to change in the future if costs of materials skyrocket.

The Manufacturing Process

1 At the ostrich farm, individual birds are herded into corral-like pens where handlers pluck the birds to harvest the feathers. Feathers are plucked from the bodies of the birds where the ideal types of feathers (and those less likely to be damaged) grow but also from over the birds' bodies so no sin-

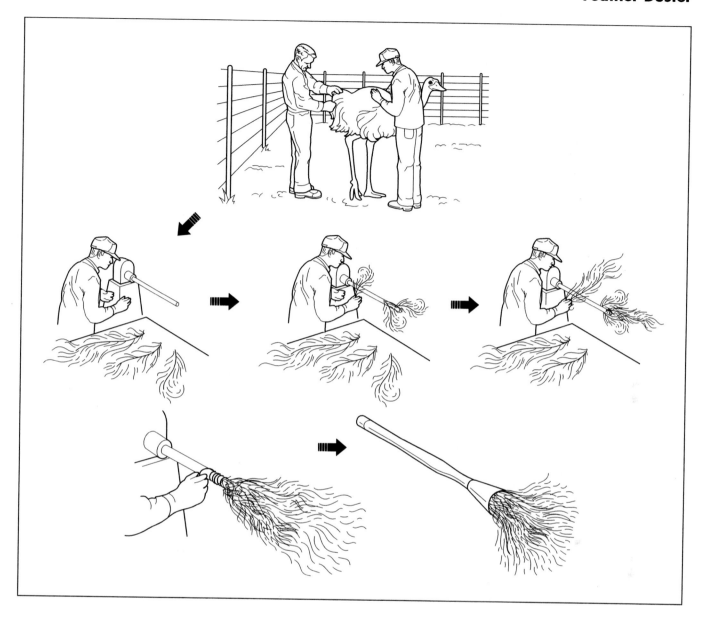

gle area is denuded. When feathers are plucked from slaughtered birds, damaged or heavily soiled feathers and types of feathers that are not useful are avoided. After the best feathers are selected, the remaining feathers are stripped off and destroyed before the hide is removed for tanning and the meat is processed.

2 Plucked feathers are delivered in bulk to the factory. They are washed, fumigated, graded by quality, and sorted by size. By the time the feathers are ready to be delivered to the duster assembly stations, they have been handled by at least 50 people.

3 At each assembly station, about 1 oz (25 g) of feathers have been sorted by size and are ready to be fixed to the handle. The handle is held in a small, lathe-like device that the worker can turn as feathers are fixed to the handle. Contact glue is spread on the bottom end of the handle. The contact glue is not very sticky and does not dry with a firm seal. It is only a temporary method of holding the feathers in place until wire can be wrapped around them.

4 After the first application of contact glue, the worker adds a small layer of short feathers. More glue is spread, and a layer of slightly longer feathers is put in place. The turning device keeps the handle rotating, more glue is applied, and a layer of even longer feathers is added. The process continues until all of the feathers previously

The feathers are plucked from the ostrich and then applied to the handle.

sorted by length have been fixed to the duster handle.

5 At each workstation, a roll of steel wire is mounted overhead above the turning device. As the layers of feathers are added, the worker wraps wire around the feathers already in place and ties and cuts the wire. Wire is not used with each layer, but it is applied several times during the process of building up the feather layers and after the last layer is added.

6 After the final feathers and wire are in place, a plastic cap (ferrule) is pushed over the top of the feathers and wire. Covering the wire makes the feather duster more attractive, but the cap also prevents any wire ends from hurting the person who will use the duster. A plastic cap is used with both wood and plastic handles. In fact, the supplier packs a plastic cap with each handle so they are ready for assembly. Loops for hanging the duster are tied through predrilled holes or formed in the handles.

7 Finished feather dusters are collected from the workstations and taken to the packaging department. Plastic sleeves are fitted over the feathers to cover them completely. The sleeves fit snugly but not too tightly to hold the feathers in place but not break them, and the sleeves are longer than the feathers so their tips do not get bent or broken. Other wrappings are put over the handles. Feather dusters are not packed and labeled for individual sale. Instead, they are packed in bulk in cardboard shipping boxes for distribution to consumers who use large numbers of feather dusters, like building maintenance firms and supply houses for janitors. They are also shipped to retailers who may change the packaging and add their own labels.

Quality Control

The handles, plastic ferrules, and hanging loops are inspected before they are distributed to the assembly stations. The process of controlling the quality of the feathers is continuous because the 50 people who handle the feathers during the cleaning, sorting, and sizing steps are responsible for rejecting poor quality feathers.

Inspectors observe all stages of feather processing and duster assembly. They intervene if they see inadequate materials or methods that the worker overlooks. The inspectors also perform final inspections of the finished feather dusters and look at each product before it is boxed.

Byproducts/Waste

Production of feather dusters does not generate any byproducts although feathers themselves are harvested during other processing of products from ostriches, such as meat. Waste from the manufacture of feather dusters consists mostly of feathers. Over 50% of those plucked from live birds or saved from meat-producing stock are wasted because of damage or poor quality. Feather waste is disposed by burning, which is legal in South Africa. No wood or plastic waste is generated because these components are made by outside suppliers and are not trimmed or modified at the feather duster factory. There are also no metal trimmings from the steel binding wire or wasted glue except for minor spillage.

Employees are also exposed to few hazards. Safety guards and emergency stops on the lathe-like turning devices prevent injury. The steel wire is enclosed in a plastic cover until it is near the handle that is being worked. The worker does not need to wear gloves to wind the wire on the handle although goggles are worn for eye protection because of the proximity of the wire. The glue is also inert and does not present hazards due to fumes or skin contact. Dust from feather processing is kept completely away from the duster floor, and, in the feather-processing area, it is swept up frequently and collected for burning with the other feather waste.

The Future

Feather dusters are a tried and true type of product and steadily becoming more popular. Companies that use the most feather dusters and also the largest sizes available, such as janitorial suppliers, know these products work and have no interest in abandoning them for electrostatic or chemical products that are easily wasted, costly to replace and stock, and less efficient. The popularity of feather dusters among individual homemakers is proven by the sales of feather dusters in mass retail stores and grocery and drug store chains. Products in these stores must demonstrate specified minimum unit sales or they will be moved to less prominent shelf space or will no longer be sold by that retailer. A new use for the feather duster is in interior decorating. A feather duster can be used to apply contrast or accent paint to interior walls because the feathers produce a unique effect.

Where to Learn More

Books

Aslett, Don. *Do I Dust or Vacuum First?* Cincinnati: C. J. Krehbiel Co. 1982.

Lantz, Louise K. *Old American Kitchenware: 1725-1925.* New York: Thomas Nelson Inc., 1970.

Norwak, Mary. *Kitchen Antiques.* New York: Praeger Publishers, 1975.

Patent, Dorothy Hinshaw. *Feathers.* New York: Dutton, Cobblehill Books, 1992.

Periodicals

Edwards, Mike. "Marco Polo, Part III: Journey Home." *National Geographic* (July 2001): 26–47.

Other

Feather Dusters. November 2001. <http://www.geocities.com/felicitax/Duster.htm>.

Ostriches On Line. November 2001. <http://www.ostrichesonline.com>.

PBS Online. "Special Feature: Forgotten Inventors." *PBS Online Web Page.* November 2001. <http://www.pbs.org/wgbh/amex/telephone/sfeature/index.html>.

Gillian S. Holmes

Felt

Some of the earliest felt remains were found in the frozen tombs of nomadic horsemen in the Siberian Tlai mountains and date to around 700 B.C.

Background

Most fabrics are woven, meaning they are constructed on a loom and have interlocking warp (the thread or fiber that is strung lengthwise on the loom) and weft (the thread that cuts across the warp fiber and interlocks with it) fibers that create a flat piece of fabric. Felt is a dense, non-woven fabric and without any warp or weft. Instead, felted fabric is made from matted and compressed fibers or fur with no apparent system of threads. Felt is produced as these fibers and/or fur are pressed together using heat, moisture, and pressure. Felt is generally composed of wool that is mixed with a synthetic in order to create sturdy, resilient felt for craft or industrial use. However, some felt is made wholly from synthetic fibers.

Felt may vary in width, length, color, or thickness depending on its intended application. This matted material is particularly useful for padding and lining as it is dense and can be very thick. Furthermore, since the fabric is not woven the edges may be cut without fear of threads becoming loose and the fiber unraveling. Felted fibers generally take dye well and craft felt is available in a multitude of colors while industrial-grade felt is generally left in its natural state. In fact, felt is used in a wide variety of applications both within the residential and industrial contexts. Felt is used in air fresheners, children's bulletin boards, craft kits, holiday costumes and decorations, stamp pads, within appliances, gaskets, as a clothing stiffener or liner, and it can be used as a cushion, to provide pads for polishing apparatus, or as a sealant in industrial machinery.

History

Felt may be the oldest fabric known to man, and there are many references to felt in ancient writings. Since felt is not woven and does not require a loom for its production, ancient man made it rather easily. Some of the earliest felt remains were found in the frozen tombs of nomadic horsemen in the Siberian Tlai mountains and date to around 700 B.C. These tribes made clothing, saddles, and tents from felt because it was strong and resistant to wet and snowy weather. Legend has it that during the Middle Ages St. Clement, who was to become the fourth bishop of Rome, was a wandering monk who happened upon the process of making felt by accident. It is said he stuffed his sandals with tow (short flax or linen fibers) in order to make them more comfortable. St. Clement discovered that the combination of moisture from perspiration and ground dampness coupled with pressure from his feet matted these tow fibers together and produced a cloth. After becoming bishop he set up groups of workers to develop felting operations. St. Clement became the patron saint for hatmakers, who extensively utilize felt to this day.

Today, hats are associated with felt, but it is generally presumed that all felt is made of wool. Originally, early hat-making felt was produced using animal fur (generally beaver fur). The fur was matted with other fibers—including wool—using heat, pressure, and moisture. The finest hats were of beaver, and men's fine hats were often referred to as *beavers*. Beaver felt hats were made in the late Middle Ages and were much coveted. However, by the end of the fourteenth century many hatmakers produced them in the Low Countries thus driving down the price.

The North American continent was home to many of the beaver skins used in European hatmakers' creations in the eighteenth and nineteenth centuries. North American Indians' second-hand skins, replete with perspiration, felted most successfully and were in extraordinary demand for hatmaking in both the New and Old Worlds. The beaver hat was surpassed in popularity in the second half of the nineteenth century by the black silk hat, sometimes finished to resemble beaver and referred to as beaver-finished silk.

The steps included in making felt have changed little over time. Felted fabric is produced using heat, moisture, and pressure to mat and interlock the fibers. In the Middle Ages the hatmaker separated the fur from the hide by hand and applied pressure and warm water to the fabric to shrink it manually. While machinery is used today to accomplish many of these tasks, the processing requirements remain unchanged. One exception is that until the late nineteenth century mercury was used in the processing of felt for hatmaking. Mercury was discovered to have debilitating effects on the hatter causing a type of poisoning that led to tremors, hallucinations, and other psychotic symptoms. The term *mad hatter* is associated with the hatmaker because of the psychosis that stemmed from the mercury poisoning. Hats of wool felt remain quite popular and are primarily worn in the winter months.

The use of felt has enlarged over the past century. Crafts enthusiasts use it for all types of projects. Many teachers find it to be an easy fabric for children to handle because once it is cut the edges do not unravel as do woven fabrics. Industrial applications for felt have burgeoned, and felt is found in cars as well as production machinery.

Raw Materials

Felt is produced from wool, which grips and mats easily, and a synthetic fiber that gives the felt some resilience and longevity. Typical fiber combinations for felt include wool and polyester or wool and nylon. Synthetics cannot be turned into felt by themselves but can be felted if they combine with wool.

Other raw materials used in the production of wool include steam, utilized during the stage in which the material is reduced in width and length and made thicker. Also, a weak sulfuric acid mixture is used in the thickening process. Soda ash (sodium chloride) is utilized to neutralize the sulfuric acid.

The Manufacturing Process

1 Since some felts use more than one type of fiber, the fibers must be mixed and blended together before any processing begins. To do this, the raw fibers are put into an opener with a big cylinder studded with steel nails that combine the fibers into a mass.

2 Next, these blended fibers must be carded. Carding machines are huge cylinders that mat the fibers into a web. Hopper-feeders allow a specific weight of fiber to pass into the cylinder in order to create a standardized web. The fibers in the web are pulled by the wires, or carded, so that they are parallel to one another.

3 Generally, at least two carding machines are used in the manufacturing process, each refining the web as it creates a new one. A transporter moves one web from the first carding machine to a second. The web is then fed into the second machine. This second carder generates a new web that is thicker and fully carded.

4 At the end of the second carding, a comb removes the carded web from the machine and rolls it up. There are two ways to remove the web from the machine: a cross-lapper may be used in which the web is perpendicularly rolled up, or across the direction of the fibers; or a vlamir may be utilized, in which the web is rolled parallel to the direction of the fibers.

5 Next, several different webs are combined to create one thick web. Four rolls of web are rolled up but are layered so that their fibers alternate in direction based on the way the webs were rolled, either cross-lapped or rolled using a vlamir. These four rolls are considered a standard single roll, sometimes referred to as a batt. This batt is considered a standard roll of material. Batts are layered in order to create different thicknesses of felt.

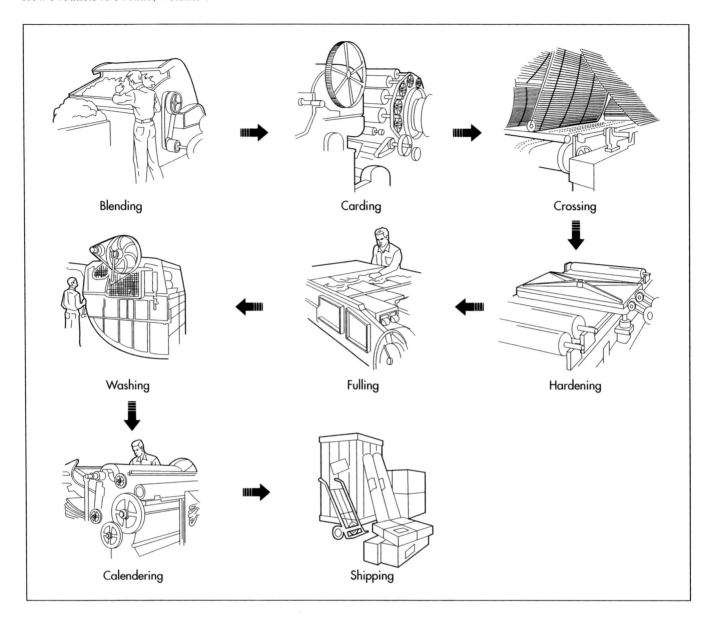

Blending

Carding

Crossing

Washing

Fulling

Hardening

Calendering

Shipping

The manufacturing process of felt.

6 The batts of felted material must be hardened or matted together in order to create thick, densely-felted material. The first step in this process is subjecting the batts to heat and moisture. In order to do so, the batts are passed through a steam table.

7 Now, the separate batts must be matted together and shrunk in length and width in order to create a dense felt. These batts must be subjected to heat, moisture, and pressure in order to be matted densely. First, the wetted batts are fed into a plate-hardener that shrinks the width of the fabric. The plate-hardener consists of a large, square flat bed with a large plate that drops down over the batts of wet, hot batts, exerting pressure on the material and compressing it.

At the same time, the plate-hardener oscillates from edge to edge, further matting the fiber to a specific width.

8 Next, the batts are fed into a fuller or fulling machine, which shrinks the length to a specific measurement. As it shrinks, the felt becomes more dense. The batts are fed through a set of upper and lower steel rollers that are covered with hard rubber or plastic and are molded with treads much like a car tire, enabling them to move across the batts. The felt is continuously wetted with a hot water and sulfuric-acid solution. The upper rollers remain stationary as the lower rollers are moved upwards to put pressure on the fabric and push it against the upper rollers. All of the rollers, both

upper and lower, move together forward and backward. The pressure, the acid, the hot water, and the movement causes the batts to shrink in length, making the felt even more dense. For example, a single piece of felt that is 38 yd (34.7 m) long may come out of the fuller at only 30 yd (27.4 m) in length.

9 The wet felt has sulfuric acid residue and must be neutralized. To do so, the felt is run through neutralizing tanks filled with a soda ash and warm water solution. This process is carefully timed so that specific yard lengths and widths are in for an exact amount of time.

10 The neutralized felt is then run through a refulling machine in which heavy rollers run over the surface of the fabric one last time to smooth out any irregularities.

11 If felts are to be dyed, the wet pieces are taken to a dye vat. Some industrial grades are not dyed but go directly to drying.

12 Some companies simply roll up the wet felt and send it to a centrifugal dryer that spins out the water. Others have huge dryers in which the felt is pinned in place on a dryer bed. Felt can also be open-air dried by either being hung or stretched out on a floor in a drying room.

13 Once dry, some companies press or iron the felt to ensure consistent thickness. Some manufacturers use this ironing to make dense felts even more dense as ironing can shrink it slightly.

14 The finishing step includes placing the felt on a gaging table in which the edges of the felt are neatly trimmed. The piece is now ready for packing, labeling, and shipping.

Quality Control

Quality control begins with the arrival of the materials. Materials are checked for quality and weight. Some companies purchase wool that has been scoured and baled; the purity of the bales is examined upon entry. Other important quality control checks include continuous monitoring of the carded webs, since the web sizes are important first steps in producing the desired length and width of the felt. Once the batts are shrunk in width and length, the company checks the weight, den-

sity, width, length, and evenness of the batts. When production is complete, visual checks may reveal that the surface of a batt is slightly uneven and additional pressing may occur to even out the surface. The acid baths are also very carefully monitored. The amount of time the fabric is in the acid bath is precisely calculated by weight and length of yard good, lest the piece is ruined. Finally, the company producing industrial felt has to check its goods against a governmental standard for the product. The government has determined that 16 lb (7.3 kg) density felt must be 1 in (2.5 cm) thick, 36 in (91.4 cm) wide, 36 in (91.4 cm) long, and weigh 16 lb (7.3 kg). If the felt weighs less than this, the fabric is not dense enough and does not meet government expectations for that grade of felt.

Byproducts/Waste

There is some waste generated in felt production. When the edges are trimmed, small pieces are cut off. These small pieces are often impregnated with oil and grease from the machinery and are unusable for other purposes. These materials are then sent to a landfill.

The Future

Due to its extreme versatility, the demand for felt is consistent. It is used in military applications for helmets, boots, small ammunitions, and rockets. The civilian uses of felt are too numerous to count. A unique use has been found for the excess white felt ground that is relatively clean and clear of oil and grease. It is ground up, colored, and put into an aerosol can. It is then sold as a spray to cover bald spots and has been somewhat successful in recent years.

Where to Learn More

Books

Gioello, Debbie Ann. *Profiling Fabrics.* New York: Fairchild Publications, 1981.

McDowell, Colin. *Hats: Status, Style and Glamour.* New York: Rizzoli, 1993.

Other

Design Arcade Web Page. November 2001. <http://www.designarcade.com/history/historyfelt.htm>.

Interview with Dick Pursell. Director of Sales, U. S. Felt. Sanford, ME. August 2001.

Sutherland Felt Company. *Manufacturing of Wool Felts Wet Process.* Troy, MI.

Nancy EV Bryk

Fluoride Treatment

Background

A fluoride treatment is a mineral solution applied to teeth in order to strengthen them and help prevent cavities. Fluoride containing products include commercially available toothpaste and mouth rinse, as well as more concentrated liquids and gels used professionally by dentists.

There are three primary factors that contribute to dental caries (tooth decay): a susceptible site on a tooth, an infective strain of bacteria (*Streptococcus mutans*), and sugars or other nutrients that stimulate the bacteria's growth. As these bacteria grow, they produce an acidic byproduct that can dissolve the minerals in the enamel and eventually destroy the tooth.

Dentists largely credit the use of fluoride treatments and fluoridated water with the drastic decline in tooth decay over the past several decades. One report indicates that for children ages 5–7 years old in the United States the average incidence of cavities has dropped from 7.1% in the 1970s to 2.5% in the 1990s. Similar improvements have been documented across almost all age groups. Still, tooth decay remains the most common infectious childhood disease and fluoride treatments remain an important tool in the fight against cavities.

Fluoride is actually a form of the element fluorine. In its elemental form fluorine is a toxic gas, but when it is chemically reacted with other compounds, such as tin, it takes on new cavity fighting uses. Once in the mouth, fluoride is diluted in saliva and deposited in bacterial plaque on the surface of the teeth. Here it works to protect the tooth in two ways. First, it directly inhibits bacter-ial growth so less acid is produced in the mouth. Second, the fluoride stored in plaque is released when the bacteria produce enough acid to lower the acid-base balance on the tooth. When this occurs fluoride diffuses into the tooth through tiny pores in the enamel. Fluoride ions replace the hydroxyl ions of the hydroxyapatite crystals, which are part of the tooth's enamel, and form a new compound called fluorapatite. This form of enamel is less soluble in the acids produced by oral bacteria and therefore helps protect teeth from decay.

History

Frederick S. McKay, a dentist practicing in Colorado Springs, Colorado in the early 1900s, was the first to discover that fluoride is an effective cavity fighter. McKay noticed that many of his patients had mottled enamel, or brown stains, on their teeth. By 1916, McKay and his researchers had found that the mottling was caused by something in the patients' drinking water. It took him 12 more years to understand how this effect was related to caries, and another three years to recognize the chemical mechanism causing this change. Finally, in 1931 in McKay verified that patients with mottled teeth were drinking water with unusually high levels of naturally occurring fluoride. This connection was studied in more detail throughout the 1930s and 1940s, culminating in the determination that one part per million was the ideal level of fluoride in drinking water, substantially reducing decay while not causing mottling.

This research led to the implementation of a fluoridation program by the federal government, and by the early 1950s most United States communities with a public water sys-

A report indicates that for children ages 5–7 years old in the United States the average incidence of cavities has dropped from 7.1% in the 1970s to 2.5% in the 1990s.

tem had adopted fluoride treated water. The idea of using fluoride in oral care products began in 1956 when Procter and Gamble launched "Crest with Fluoristan." Since the 1950s, scores of fluoride containing products have been introduced both for the general market and for dental professionals.

Even though fluoride has been used for decades, there are still concerns about its health effects today. Although the chemistry is not fully understood, researchers believe that high levels of fluoride can interrupt the natural formation of tooth enamel. They theorize that too much fluoride creates a hypomineralization, which leads to the chalky, cloudy, or opaque appearance that is characteristic of fluorosis. While dental proponents claim that fluoride is largely responsible for improved dental health, there are those who claim that it can cause a form of bone cancer. In the 1980s a study conducted by the National Toxicology Program found "equivocal evidence" of carcinogenicity based on testing done on rats. However, the panel eventually concluded that there was no solid data linking cancer, including osteosarcoma, directly to fluoridation.

Both United States and British Dental Associations continue to recommend that both adults and children brush twice daily with fluoride toothpaste, but they also recommend that to reduce the risk of fluorosis, children should not swallow the toothpaste. The subject remains a hot topic of political debate.

Raw Materials

There are a variety of fluoride compounds that are allowed by the Food and Drug Administration (FDA) for use in oral care products. Fluoride ingredients are listed according to the type of product in which they may be used and by the percentage that must be included in the formula.

Toothpaste. Sodium fluoride: 0.22%; sodium monofluorophosphate: 0.76%; stannous fluoride: 0.4%.

Treatment rinse. (The pH, or acid-base balance, of the formula can affect the functionality of the fluoride. The higher the pH of the product the more acid it contains. The lower the pH, the more basic it is.) Sodium fluoride acidulated with a mixture of sodi-

um phosphate, monobasic, and phosphoric acid to a level of 0.1 molar phosphate ion and a pH of 3.0–4.5, which yields an effective fluoride ion concentration of 0.02%; sodium fluoride acidulated with a mixture of sodium phosphate, diabasic, and phosphoric acid to a pH of 3.5, which yields an effective fluoride ion concentration of 0.01%; sodium fluoride 0.02% aqueous solution with a pH of approximately 7; sodium fluoride 0.05% aqueous solution with a pH of approximately 7; sodium fluoride concentrate containing adequate directions for mixing with water before using to result in a 0.02% or 0.05% aqueous solution with a pH of approximately 7; stannous fluoride concentrate marketed in a stable form and containing adequate directions for mixing with water immediately before using to result in a 0.1% aqueous solution.

Treatment gel. Stannous fluoride: 0.4%.

In addition to fluoride, these products contain a variety of other ingredients including solvents, thickeners, and pH control agents. Solvents include water or glycerine, which are used as a carrier for reasons of efficacy, safety, and cost. Deionized or demineralized water is used to prevent unwanted minerals from affecting the performance or stability of the product. The concentration of water in the formula may be 90% or more.

Thickening agents are added to control viscosity. These include xanthan, carrageenan, and various other gums and polymers used at concentrations between 0.1–2.0%.

Flavors and colors are added to make the products more appealing. Popular flavors include mint, bubble gum, and grape, and these are added at a few tenths of a percent. Dyes are used to impart color at very low levels (less than a hundredth of a percent). Since the product may be swallowed accidentally these dyes must be approved for use in food products.

Preservatives are added when necessary. Depending on the pH of the product, they may be required to prevent the growth of mold or bacteria in the product while it is stored the shelf. One or two tenths of a percent is a typical use level for a preservative.

Organic acids, such as phosphoric acid, may be added to control the product's pH. Some

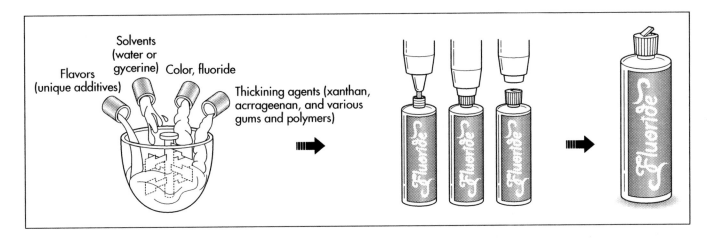

The fluoride ingredients are mixed and distributed into containers.

forms of fluoride require a low pH to be functional. These are added at a few tenths of a percent as well.

Design

Fluoride treatments are designed to provide an appropriate concentration of fluoride at a pH level that will help the correct amount of fluoride deposit on the teeth. If the fluoride level is too low the treatment will not be effective; if it is too high the patients might accidentally be poisoned. In the United States, laws have been created to ensure that these products are safe and efficacious. The FDA regulates them as either over-the-counter (OTC) drugs or as professional products for use by dentists. In addition, the FDA limits the size of commercially available products to reduce the possibility of accidental overdose. Finally, the organization determines labeling requirements for all commercial products and some aspects of professional ones. These requirements most be taken into consideration when designing fluoride treatments.

OTC fluoride-containing drugs include toothpaste and mouth rinse. Professional products are more concentrated and may be applied either as a gel, foam, or liquid. They may be designed to be applied using plastic trays that fit around the teeth. Depending on the type of product being formulated, the development chemist can choose from several types of FDA approved actives. Once again, these regulatory factors must be considered in product design.

Key examples of treatment formulations are: acidulated phosphate fluoride gel with 1.23% fluoride ion at pH 3.5, designed to flow easily during tray placement yet thickens during treatment so it does not drip down the patient's throat; sodium fluoride gel with 2% fluoride ion at pH 7.0, for use when etching of porcelain restorations is a concern; stannous fluoride liquid rinse with 0.63% fluoride ion, designed to prevent decay, reduce plaque accumulation, and help reduce gingival inflammation and bleeding; APF fluoride foam with 1.23% fluoride ion at pH 3–4.25, designed as an easy to use non-aerosol foam that reduces accidental ingestion by the patient.

The Manufacturing Process

Batching

1 The process of manufacturing a fluoride containing treatment is similar to the processes used to make other oral care products. Because most of the fluoride compounds used in treatments are water-soluble, these products are relatively easy to make. The manufacturing process involves simple mixing and does not require any special solvents or emulsification. Large batches can be made in stainless steel tanks as large as 3,000 gal (11,356 l). The first step is to charge the batch tank with water or glycerine, which makes up the largest percentage of the formula. The water is stirred with an electrically driven turbine mixer. The mixing speed is computer controlled to optimize the stirring conditions when other ingredients are added. While manufacturing procedures vary widely, it is common to add the color early in the batching process. The dyes that are used in these products are highly concentrated and if too much dye is added

by mistake it is easier and cheaper to dispose of the batch before the more expensive ingredients have been added.

2 The other ingredients may be added in succession. Depending on the solubility of the form of fluoride chosen, heating and cooling may be required to help dissolve the powders quickly. During this batching process samples may be taken periodically during the mixing process to check for clarity. Toward the end of the batching process the pH control agents are introduced. They ensure the batch has the proper acid base balance. Flavors are added at the end of the operation if they are heat sensitive.

Batch check

3 Once the batch is complete, it must be evaluated to ensure it is within specification. It is particularly important to ensure that the active ingredients are present at their designated concentrations. Tests such as pH analysis, weight percentage of solids, and fluoride concentration are used to maintain product specifications.

Filling operations

4 Once the batch is finished and approved, the filling operation can proceed. Depending on the nature of the process, the batch may be filled directly from the batch tank or it may be transferred to a secondary vessel or holding tank. High speed filling equipment is used but the filling speed depends on the product's viscosity. Thin, liquid products can be filled faster and more efficiently than thicker gels.

5 The filled package is fed down an assembly line where the closure is attached. While a bottle may only require a simple cap to seal it, products packed in tubes or foaming dispensers may involve more complicated sealing mechanisms. After the package has been filled and closure has been attached, the unit is ready for final packing. Multiple units may be shrink wrapped together or placed in cartons for shipping.

Quality Control

In addition to the chemical tests conducted during the manufacturing process, fluoride treatments are subject to special testing considerations to establish product performance. Historically these tests have involved expensive human clinical studies, but since 1988 the Dental Panel has allowed the use of new laboratory tests to determine the efficiency of fluoride treatments.

The Future

While the political future of fluoride treatments may be uncertain, they continue to be an important tool in the fight against cavities. There are new technological advances that may some day lead to fluoride free cavity fighters. British researchers have discovered a new kind of anti-caries agent that stops tooth decay for up to three months. Their new ingredient is a protein fragment, called peptide p1025 that works by attaching itself to the tooth surfaces where cavity-causing bacteria normally bind. The protein blocks the bacteria from attaching to teeth so they are easily washed away. Breakthroughs like this could someday provide fluoride-free ways to prevent tooth decay.

Where to Learn More

Books

Wolinsky, L. E. "Caries and Cariology." In *Oral Microbiology and Immunology.* 2nd ed. Ed. R. J. Nisengard and M. G. Newman. Philadelphia: W. B. Saunders Company, 1994.

Periodicals

Brady, Robert P., and Abbe Goldstein. "Keeping Faith in Fluoride." *Chemist & Druggist* (24 May 1997): 24.

"Mouthwash Cancels Cavities." *Popular Mechanics* (February 2000): 15.

"Postmenopausal Osteoporosis Treatment with Fluoride." *American Family Physician* (January 1996): 302.

Sheikh, Aamir, and Alice M. Horowitz. "Benefits of Fluoride Toothpaste." *Journal of School Health* (October 1999): 299.

Other

Connelly. "Caries Treatment with Fluoride." United States Patent 5738113, 1998.

Randy Schueller

Fountain Pen

Background

Humans have used various instruments to convey thoughts and feelings. Man's first writing instrument was his finger, using it to form symbols in the dirt. Later, pieces of metal or bone were used. Ancient Greeks used a stylus to mark on wax-coated writing tablets, while early Egyptians used hollow reeds as writing tools and papyrus as their writing surface.

During the Middle Ages, quill pens, made from bird feathers whose ends had been split and sharpened, became the writing tool of choice until the development of the steel dipping pen in the early 1800s. Steel pens, which used steel tips called nibs, did not require the frequent sharpening that quill pens did. However, they still needed to be dipped in an inkbottle because they did not contain their own ink.

Even as the steel pen was gaining in popularity, attempts to design a more practical writing instrument were being made. These efforts eventually resulted in one of the most popular writing tools still used, the fountain pen. Lewis Edson Waterman, a New York Insurance Agent, produced the first practical fountain pen in 1884. While both the quill and steel pens had to be dipped in ink, the fountain pen was the first to hold its own ink within a self-contained reservoir. Because of its practicality and durability, the fountain pen became the most popular writing instrument and remained so until the development of the ballpoint pen in 1938. The fountain pen remains popular for its elegance and prestige, both as a writing instrument and a valuable collector's item.

History

Attempts to develop a self-feeding pen that did not require sharpening were made as far back as the beginning of the tenth century. Numerous ideas were developed, but it was not until 1884 that success was finally achieved. Waterman's pen worked as flawlessly as a dipping pen but without the need for an external inkbottle. Waterman started producing these pens at a rate of 36 per week and selling them at his New York City cigar kiosk. However, the demand quickly soared, prompting Waterman to open a six-story production facility on Broadway, which he expanded even more in later years. The fountain pen dominated the writing instrument market for the next 60 years.

Capitalizing on Waterman's success, other companies joined the writing instrument manufacturing business. In 2001, the Writing Instrument Manufacturers Association, an organization comprised of companies that produce fountain pens and other writing tools, had approximately 25 members. The larger companies now use an automated process to produce fountain pens, while some smaller companies and individuals still produce them by hand, just as Waterman did back in 1884.

Although fountain pens are available in a variety of styles offering unique features, each is comprised of the same basic components: the nib, or point; the barrel, which holds the ink reservoir; and the cap, which fits over the nib of the pen to protect it from damage. Ink flows from the reservoir to the nib at a balanced rate of flow by means of a force called capillary attraction. This is the same force that causes a blotter to absorb ink or kerosene to flow up the wick of a flame.

During the Middle Ages, quill pens, made from bird feathers whose ends had been split and sharpened, became the writing tool of choice until the development of the steel dipping pen in the early 1800s.

The first nibs were made of gold alloys, often dipped in a hard metal called iridium for strength and resistance to corrosion. However, when gold alloy nibs became too expensive to mass-produce, steel was adopted as the material of choice. Solid gold, ranging from 18–22 karat, is still used for the nibs on some pens. Each nib has a slit at its tip that controls the flow of ink.

The first barrels were made of black hard rubber, chosen because it is ink-resistant and easily machinable. Postwar pens are more commonly made from durable plastic. However, barrels can be made from gold, silver, brass, wood, bone, or even crushed velvet.

The first fountain pens were filled with medicine droppers, which were later replaced with rubber sacs. First used in 1890, these sacs had a short life because the rubber material they were made from was not able to withstand the chemical action of the ink. Rubber compounds were later improved, and a long-lasting rubber sac was introduced in the late 1920s. This sac was later replaced by an even better semi-transparent, plasticized vinal resin sac containing no rubber. Various forms of sac depression mechanisms have been used throughout the years. The first sacless pen was introduced by the Parker Pen Company in 1932.

Raw Materials

Fountain pen barrels can be made from a wide variety of materials. Finer, more expensive pens are made from materials such as brass, silver, or gold. Modern pen manufacturers generally use less expensive materials for pen barrels, including: acrylic resin, also known as Lucite or Perspex, which is used for Parker 51 models; cellulose acetate; and various other injection-moldable polymers. Handmade pens can be created from wood or almost any other material that is solid, stable, and can be worked with standard woodworking tools. Examples include plywood, crushed velvet, bone, leather, and even antlers. Stainless steel is generally used to make the nibs, although gold or sterling silver may also be used. The clips and other fittings are usually made from a gold alloy that has been electroplated, or they may be gold or gold filled on finer pens.

Design

Fountain pens are available in a variety of designs and styles. Some are mass produced while others are custom-designed. With custom-designed pens, the creator must decide ahead of time what special features the pen will have and choose the appropriate tools and process to use based on those features. Some possible variations on design include laminating strips to produce intriguing patterns and color combinations, changing the style of the clip to give the pen a different look and feel, carving or burning a unique design into the surface of the pen, or inlaying gemstones or other materials into the pen surface. All of these design variations require some extra preparation and materials, but help make the pen unique, and sometimes, more valuable.

Refill mechanisms

A variety of mechanisms can be used to fill fountain pens. These include levers, buttons, pistons, and squeeze bulbs. Lever-fillers have a tiny lever built into the side of the pen. Lifting the lever causes the ink sac to compress. Then, after the nib is dipped in ink, closing the lever causes the sac to reinflate. Button-fillers have a button on the end of the pen. The button works similarly to the lever; pressing the button causes the sac to deflate, and releasing the button causes the sac to reinflate after the pen has been dipped in ink. Piston-fillers use a screw mechanism to move a piston inside the barrel, taking in and expelling ink, while squeeze bulb fillers are filled by repeatedly squeezing the bulb. Each one of these mechanisms are installed on the pen during final assembly.

The Manufacturing Process

The larger pen manufacturers use automated processes to produce fountain pens. However, some smaller companies and individuals continue to create pens manually. The materials and processes used by different companies and individuals vary. Here is a common process used to create handmade pens from wood. This same process, or similar processes, can be used to create pens from other types of materials as well.

Manufacturing individual fountain pens from wood.

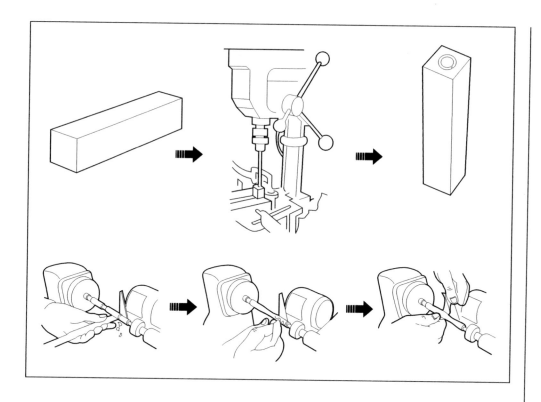

Preparing the blank

1 The first step in creating the pen is to prepare a blank, a rectangular piece of wood that will be filed and shaped into the two halves of the pen barrel. Virtually any type of wood can be used. However, it must be dry when the process begins. Colorful woods with interesting patterns are preferable because they make for a more attractive finished product. To prepare the blank, the wood is first cut into a rectangle approximately 0.75 in^2 × 5 in long (1.9 cm^2 × 12.7 cm).

2 Then, the blank is marked with a line (called a match mark) extending about 1 in (2.5 cm) beyond where a cut will be made separating the blank into two halves, an upper and a lower. The blanks are then crosscut into lengths that are slightly longer than the brass tube that will eventually be inserted into them. One way for the crosscut to be made is with a bandsaw equipped with a miter gage and an auxiliary slip fence, which is an adjustable fence especially made for cutting different lengths of pen blanks.

3 The blanks are then held in place using a jaw clamp or some other similar method, and a hole is drilled down the center of each half of the blank in preparation for the inser-

tion of the brass tube. Each blank section is held in place with the match marks facing up.

Inserting the brass tube

4 Once the holes are drilled, polyurethane glue is applied to both the outside of the brass tube and the inside of each drilled hole. Then, the tube is inserted into the blank using a rotating motion to distribute the glue evenly. After approximately 10 minutes, the glue is usually dry and the tube stays securely in place.

5 With the pen secured, the tube is trimmed to fit precisely in the blank using an adapter sleeve and a barrel trimmer until a brass curl is produced.

Mounting and turning the blank

6 Once the tube is inserted, the upper and lower blanks are ready for mounting on a mandrel in preparation for turning. A split mandrel system, consisting of two short steel mandrels, is generally used. Each blank is mounted separately on the split mandrel along with the appropriate bushings.

7 As each blank spins, the wood that does not belong on the pen is carefully cut away using a gouge or similar tool. Finishing cuts are made using a skew chisel.

8 A notch is carved around the upper blank, completely through to the tube, using a thin parting tool. This notch is where a metal trim ring will later be inserted.

Sanding and finishing

9 Now the blanks have turned into pen barrels, ready for sanding and finishing. Sandpaper, starting with 120-grit and progressing to 180-grit, 220-grit, and finally 320- or 400-grit, is used to ensure a smooth finish. If Dymondwood is used, a grit as fine as 600 may be necessary for a high gloss finish.

10 The pen barrels are coated with a wax or plastic polish, applied with a soft cloth while the barrels are still spinning.

Preparing for assembly

11 Prior to assembly, the sharp corners on each end of the finished barrels are sanded slightly.

12 If a clip is to be used, a notch is cut in the upper barrel in preparation for its insertion. To avoid chipping the wood, a grinding wheel or a file is used to cut this notch.

Clips, fittings, and bands

13 The clips and metal fittings used on fountain pens are stamped, just as when the fountain pen was first produced. Bands and overlays are rolled from sheet metal and hard soldered.

Final assembly

14 Now the ring, clip, and other metal trimmings are attached to the finished barrels to make a completed pen. A drill press and press jig may be used to press the pen parts together. The drill press is used as a vice, while the jig (a long, flat piece of wood with metal posts sticking up) is used to keep the parts in line during assembly. First, a nib coupler is pressed into the large end of the lower barrel. Then the lower barrel is flipped over, and an end cap is pressed into the smaller end. The nib is then inserted into the nib coupler.

15 Next, the trim ring and clip assembly are inserted into the previously created notches in the upper barrel. Then a closing cap is screwed in place using a cap adjuster tool or Phillips screwdriver to complete the assembly of the upper barrel.

16 Finally, the two barrels are screwed together and the fit is tested and adjusted with the cap adjuster or screwdriver if necessary.

Automated process

The process used to make fountain pens at larger companies is generally more automated. These companies use specially tooled machines to mold the pen barrels, usually from molten plastic. Machines also do the stamping and crimping of the metal parts, assemble the final product, and even take care of the polishing and cleaning.

Quality Control

Although there are no official guidelines governing the manufacture of fountain pens, most companies have a series of set inspections to ensure quality. They test for defects in the surface of the pen, the quality of the ink flow, the fit of the cap, and so on. Some pens even come with certifications to attest to the quality of the finished product. For example, the Parker 75, one of the Parker Pen Company's more famous models, undergoes 792 inspections and comes with a certificate of quality signed by the final inspector.

The Future

Although ballpoint pens have replaced the fountain pen as the universal writing tool, fountain pens continue to be popular with collectors as well as those who desire a more elegant and sophisticated writing tool. According to sources from Parker, the fastest growing markets for fountain pens in 2001 are in the Far and Middle East and in Europe.

One trend in the fountain pen market is the growing number of individuals who have begun manufacturing and selling pens on their own. This is made possible by the availability of pen kits containing all the materials and instructions needed, and by the popularity of the Internet as a means of selling handcrafted pens to a larger market.

Where to Learn More
Books

Christensen, Kip, and Rex Burningham. *Turning Pens and Pencils*. United Kingdom: Guild of Master Craftsman, 1999.

World Book Encyclopedia. Illinois: Field Enterprises Educational Corporation, 1963.

Other

Development of the Fountain Pen. 10 September 2001. <http://barnyard.syr.edu/~vefatica/fountain.txt>.

Nishimura, David. "Filling Instructions." *Vintage Pens Web Page.* 10 September. 2001. <http://www.vintagepens.com/fill. htm>.

"Parker Plant Tour. July/August 1998." *Pen World International Magazine Web Page.* 10 September 2001. <http://www.penworld. com/Issues98/julyaug98/parkernib.htm>.

WoodenPen.Com Web Page. 10 September 2001. <http://www.woodenpen.com/how. htm>.

Writing Instrument Manufacturers Association Web Page. 10 September 2001. <http://www.wima.org>.

Kathy Saporito

Furnace

One of the earliest forms of a furnace was invented by the Romans and called a hypocaust. It was a form of underfloor heating using a fire in one corner of a basement with the exhaust vented through flues in the walls to chimneys.

Background

A furnace is a device that produces heat. Not only are furnaces used in the home for warmth, they are used in industry for a variety of purposes such as making steel and heat treating of materials to change their molecular structure.

History

Central heating with a furnace is an idea that is centuries old. One of the earliest forms of this idea was invented by the Romans and called a hypocaust. It was a form of underfloor heating using a fire in one corner of a basement with the exhaust vented through flues in the walls to chimneys. This form of heating could only be used in stone or brick homes. It was also very dangerous because of the possibility of fire and suffocation.

Furnaces generate heat by burning fuel, but early furnaces burned wood. In the seventeenth century, coal began to replace wood as a primary fuel. Coal was used until the early 1940s when gas became the primary fuel. In the 1970s, electric furnaces started to replace gas furnaces because of the energy crisis. Today, the gas furnace is still the most popular form of home heating equipment.

Wood and coal burning furnaces required constant feeding to maintain warmth in the home. From early morning to late at night, usually three to five times a day, fuel needed to be put in the furnace. In addition, the waste from the ashes from the burnt wood or coal must be removed and disposed.

Raw Materials

Today's modern furnace uses stainless steel, aluminized steel, aluminum, brass, copper, and fiberglass. Stainless steel is used in the heat exchangers for corrosion resistance. Aluminized steel is used to construct the frame, blowers, and burners. Brass is used for valves, and copper in the electrical wiring. Fiberglass is used insulate the cabinet.

Design

The original gas furnace consisted of a heat exchanger, burner, gas control valve, and an external thermostat, and there was no blower. Natural convection or forced air flow was used to circulate the air through large heating ducts and cold air returns to and from each room. This system was very inefficient—allowing over half of the heated air to escape up the chimney.

Today's gas furnace consists of a heat exchanger, secondary heat exchanger (depending on efficiency rating), air circulation blower, flue draft blower, gas control valve, burners, pilot light or spark ignition, electronic control circuitry, and an external thermostat. The modern furnace is highly efficient—80–90%, allowing only 10–20% of the heated air to escape up the chimney.

When heat is requested from the thermostat, the burners light and throw heat into the primary heat exchanger. The heated air then flows through the secondary heat exchanger (90% efficient furnace only) to the exhaust flue and chimney. The average furnace has three heat exchangers each producing 25,000 BTUs for a total of 75,000 BTUs. A flue draft blower is placed in the exhaust flue to supercharge the burners and increase efficiency. The heat exchangers perform two functions: transfer heated air from the burners to the home and allow dangerous exhaust

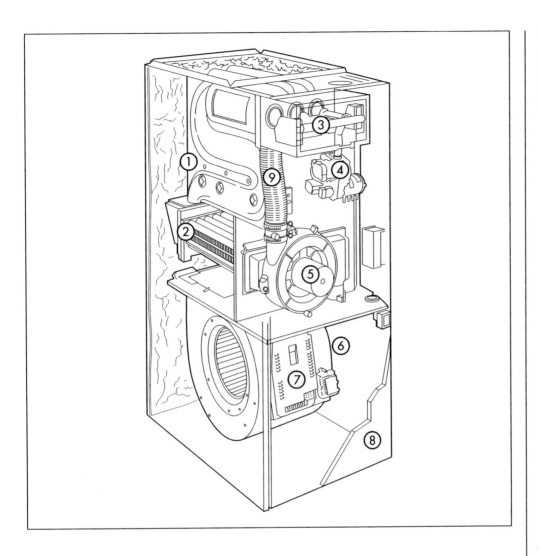

1. An "S"-curve primary heat exchanger. 2. Secondary heat exchanger. 3. In-shot burners. 4. Gas valve. 5. Draft motor. 6. Fan motor. 7. Ignition system. 8. Steel cabinet. 9. Exhaust pipe.

gases to escape up the chimney. The transferred heated air is circulated throughout the home by a large blower and heating ducts.

The Manufacturing Process

1 The primary heat exchanger is formed from two separate pieces of 409 stainless steel sheet. Each half is formed into shape by a 400 ton hydraulic press. The two halves are then fused together by a 25 ton hydraulic press.

2 The secondary heat exchanger is formed from 29-4C stainless steel tubing and fins. The fins are welded to the tubing to form a radiator type configuration.

3 The primary heat exchanger is crimped to the secondary heat exchanger through a transition box. The flue draft blower is attached to the secondary heat exchanger.

4 The burners are constructed of aluminized steel and arrive at the plant preformed. They are then attached to a plate on the input side of the primary heat exchanger. There is one burner for each heat exchanger in the furnace.

5 The vendor supplied gas control valve is mounted to the heat exchanger and burner assembly. It is connected to the burner through a pipe.

6 The air circulation blower housing is formed through the same hydraulic press formation as the primary heat exchanger. The vendor supplied motor and squirrel cage rotor are connected and attached to the blower housing with brackets.

7 A plate is then attached for mounting the blower assembly to the heat exchanger assembly. Another mounting plate containing the vendor supplied furnace control cir-

cuitry and transformer are attached to the blower housing.

8 The air circulation blower assembly is then mounted to the heat exchanger assembly with screws and nuts.

9 The cabinet consists of two doors and the cabinet housing. The cabinet housing is supplied as a flat pre-painted sheet of steel and placed in a hydraulic press to form a three sided configuration. Sheets of fiberglass insulation are glued to the sides of the cabinet.

10 The cabinet is installed around the furnace assembly and secured with screws and nuts. The doors are installed on the front of the cabinet assembly. The completed assembly is boxed and prepared for shipment.

Quality Control

Each completed furnace undergoes an extensive series of tests. Checks for proper operation of the flue draft and air circulation blowers are performed. The gas valve is checked for proper operation. The heat output of the furnace in BTUs is measured. A dielectric test is performed for shorts.

Byproducts/Waste

Scrap metal from cutting and forming operations are collected and sent to recycling plants for reclamation. Any excess piping is either reused or discarded. Defective steel sheets can be sent bac tot he manufacturer and reformed, depending on the extent of the damage. The majority of the components of the furnace are able to be recycled.

The Future

Furnaces have come a long way in the past 30 years. A primary focus by manufacturers is zone control of every room in the home. Each room will have a thermostat that will regulate heat flow to that individual room. As technology advances, these thermostats will be able to process voice commands or commands placed through a cell phone or computer.

Where to Learn More

Other

"Hypocaust." *The Romans in Britain Web Page.* December 2001. <http://romans-in-britain.org.uk>.

Armstrong Air Conditioning Web Page. December 2001. <http://www.aac-inc.com>.

Ernst S. Sibberson

Gas Lantern

Background

A gas lantern is a lightweight, portable device that supplies bright, efficient light while protecting its contents from wind and rain. Rural dwellers and outdoorsmen alike have relied on variations of the modern gas lantern for roughly 100 years, allowing access to barns, cabins, campgrounds, and wooded paths beyond the daylight hours.

This style of lantern is more practical than its ancestors because it operates on the principle of incandescence—rather, it relies on light produced by heat. The heated mantles in a gas lantern emit far more light than the flame of an oil lamp, therefore providing better visibility in a larger area. Mantles are chemically saturated fabric shells that, when heated by the lantern's flame, become a powerful source of white light—up to 300 candlepower, or the rough equivalent of a 300-watt bulb.

History

For an untold number of years, the open flame was humankind's only source of controlled light. Early ceramic lamps dating back to the Roman era were little more than earthenware pots with tubes to supply vegetable oil to a wick and spout. Centuries of development attempted to master the potential of lamplight, employing variations of fuel and wick materials to boost efficiency, but it was not until the nineteenth century that scientists and inventors began to make vast improvements in light quality.

By the 1830s, a portable lamp had been developed using a pressure mechanism to force fuel oil to the burner. This concept, paired with the arrival of the first durable working mantle in 1885, led to the modern styles of portable lanterns used during the last hundred years.

Austrian chemist Carl Auer von Welsbach is credited with the invention of the modern thorium mantle. Through his work with rare earth metals, Auer von Welsbach discovered that certain oxides would give off incandescent light when heated. Original Welsbach mantles came in the form of loosely woven silk fabric impregnated with magnesium and lanthanum oxides. Six years later, he had settled on a mix consisting of 99% thorium, a silvery-white metal with a melting point of almost 6,000°F (3,300°C). This ability to withstand immense heat allowed it to emit higher levels of brilliant white light. Historians note that Auer von Welsbach's progress in this area was partially driven by a sense of developmental urgency; his work was in direct competition with that of the incandescent electric light.

It would be decades before reliable electric service would reach outside of urban communities, and the demand for usable light swelled in rural homes and workplaces. A forerunner to the modern lantern was known as the Efficient Lamp, manufactured by the Connecticut-based Edward Miller Company. The portable Efficient Lamp used a pressure system to vaporize gasoline, mix it with air, and ignite it in a burner to heat the mantles. In 1900, part-time typewriter salesman W. C. Coleman happened across an Efficient Lamp in the window of an Alabama drugstore. Fascinated by the lamp's intensity, Coleman sought out its owners and immediately began selling the product himself. Two years later, he bought the rights to the design, made some improvements, and renamed it the Coleman Arc Lamp. Over the next decade, variations of pressure mantle

By the 1830s, a portable lamp had been developed using a pressure mechanism to force fuel oil to the burner.

147

lamps emerged from Coleman and several competitors, including the Western Lighting Company (now Aladdin), whose founder was similarly inspired by a German kerosene mantle burner called the "Practicus."

The Coleman Arc Lantern, introduced in 1914, was the first in a long succession of portable gas lantern models. Able to illuminate a circle 100 ft (30 m) in diameter, the Arc Lantern featured a protective metal hood to ward off wind, rain, and curious insects. Its bail (handle) and stout shape allowed the Arc Lantern to be easily carried, hung from a branch or rafter, or set to the ground.

Revisions of the Arc lantern were to remain in production for the next 53 years. Initial improvements in the 1920s introduced the "Instant-Lite Lanterns," which eliminated the need to preheat the generator. In earlier models, the generator would have to be manually heated before it would vaporize fuel; this involved holding a match or a burning piece of felt (usually fuel-soaked) up against it. Later innovations brought multi-fuel lanterns that would burn kerosene, gasoline, benzine, petrol, or paraffin. Advances in metallurgy following World War II led to non-corrosive steel founts, or fuel tanks. The development and use of heat-resistant glass also solved a key design issue: that a hot glass globe tended to shatter when hit with cold rain.

Enhancements over the years have made the traditional lantern brighter, lighter to carry, and simpler to use. Newer electric-start models no longer require a match. Propane bottle fuel now eliminates the need for building pressure manually. However, even considering these alterations, the straightforward design of a portable lantern has remained essentially unchanged since the early decades of this century.

Raw Materials

High-grade steels comprise the majority of a lantern's components. The ventilator hood and fount are usually draw-quality, meaning that the steel is flexible and will not crack under the pressure of a deep press. Various brass alloys are used to make parts of the fuel delivery system; the grade used for each part depends on how much heat that particular piece needs to withstand. Other steel alloys are used for smaller parts such as the

bail, collar, and pressure and ignition systems. The standing base and control knobs in more recent models have been made of molded plastic or rubber.

While some globes are made of a metal mesh, heat-resistant borosilicate glass is still the most prevalent material used in globe production. Often sold under the brand name Pyrex, the glass is formed from a combination of silica sand and boric oxide.

Mantles consist of a silk or rayon mesh saturated with various chemicals. Thorium is still commonly used but often criticized— applications for the slightly radioactive thorium include the manufacture of nuclear weapons. In response to safety concerns, makers in the United States now substitute the pricier but non-radioactive element yttrium, which gives off a more yellowish tone.

Design

Modern designs are tailored to different needs. Though the standard, durable lantern of past decades still enjoys a devoted market, design engineers now consider convenience, utility, and even cosmetic concerns in the development of new models. For serious campers and climbers, a class of small, lightweight lanterns is available; the light output is minimal, but consumer concern in this case is for portability. For standard uses, however, design competitors experiment with higher grades of steel, better fuel efficiency, and a hardier shell. Features like metal cages around the globe, self-gauging pressure pumps, electric ignitions, and non-slip rubber bases are becoming aspects of a new production standard for gas lanterns. The mantles themselves have also been subject to improvements in shape, material, and size.

The Manufacturing Process

Making steel components

1 To form molten steel, iron ore is melted with coke, a carbon-rich substance that results when coal is heated in a vacuum. Depending on the alloy, other metals such as aluminum, manganese, titanium, and zirconium may also be introduced. After the steel cools, it is formed into sheets between high-pressure rollers and distributed to the manufacturing plant.

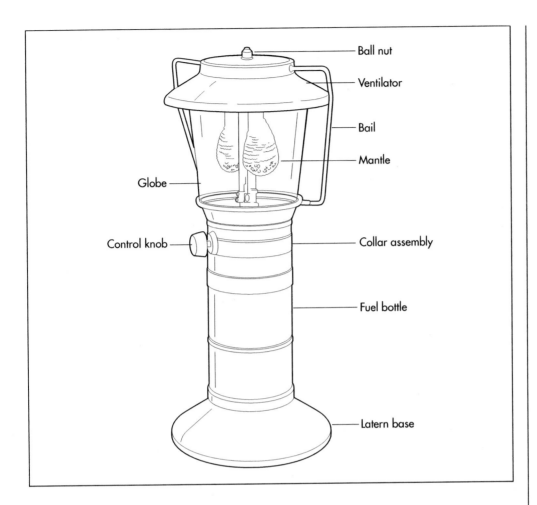

Ball nut
Ventilator
Bail
Mantle
Globe
Control knob
Collar assembly
Fuel bottle
Latern base

A double mantle propane lantern.

2 There, metal presses shape the steel into the appropriate parts. This process is not entirely mechanized, however; multi-step manual operations are required to move the steel from press to press.

Enameling the steel

3 This is usually done via "e-dip," a water-based process used to give lanterns their signature colors. The steel components are cleaned and manually set on a large conveyor. These parts then receive an electric charge, which determines the thickness of paint when dipped and ensures an even coating.

4 After primer, paint, and topcoat dips, the parts are then bake-dried. However, because e-dipping is expensive, smaller components are often enameled by an automated paint sprayer. This is a process in which static electricity attracts paint to the object, minimizing overspray and airborne toxins. A significant amount of hand labor is involved in this method,

which requires parts to be hung on hooks before enameling.

Making the plastic parts

5 Small plastic pieces such as knobs and buttons are often made by outside vendors. To form these objects, plastic pellets are added to the hopper of an injection molding machine. The plastic is melted, and a hydraulic screw pushes the substance through a nozzle, where it is injected into a pre-shaped mold, held under pressure, and cooled. Factory staff transport the finished parts, but the process is otherwise fully automated.

Making the globes

6 Globe production involves a multi-cavity horizontal wheel, usually with six molds. Hot borosilicate glass is pushed in tube form off a feeder nozzle and onto the wheel. A layer of compressed air is then blown against the molds and the wheel is spun, forming the globe shapes. The glass edges are fired automatically, and the glass is left to cool.

Making the mantles

7 Silk or synthetic string is shipped to the factory by vendors, with the rest of production done in-house. The delicacy of mantles requires that the "sock" be made by hand with the help of sewing machinery, with some automated conveyor systems employed to move the process along more efficiently.

8 Factory personnel then hang the unfinished mantles in preparation for an automated chemical dip. Chemical impregnation processes vary and are usually considered trade secrets by mantle manufacturers.

Assembly

9 Before the lantern is fully assembled on the main conveyor lines, a process called sub-assembly gathers the smaller parts and connects them into larger systems. Main assembly involves a "square line," a four-sided conveyor manned by three or four personnel. Pre-assembled parts, such as the fuel and pressure systems, are screwed to the fount. Down the line, workers use nuts and screws to complete the final assembly phase, which involves mounting the collar and attaching the globe, ventilator, and bail.

Quality Control

The feature that buyers consistently look for in a lantern is durability. These products are expected to last, trouble-free, for decades. Because of these standards, visual and mechanical inspection is necessary at every step. During the design process, in-house quality assurance teams brainstorm and troubleshoot in an effort to form individual specifications for each product. This includes the required grade levels of materials, inspection protocol, and machinery pressure and temperature management. Manufacturers must also adhere to governmental regulation; these standards include those related to occupational safety, emissions, and the transport and packaging of products containing potentially volatile fuels.

Byproducts/Waste

No byproducts result from the manufacture of gas lanterns. Waste is minimal due to the fact that most of the production materials can be reused. The yttrium used in mantles, since it is fairly expensive, is conserved and recycled for the purpose of efficiency. Metal alloys are recycled as much as possible, but scraps do comprise one example of industrial leftovers. The only examples of hazardous waste are called VOCs (volatile organic compounds), which are formed in the enameling process. However, the technologies used in this stage are designed to keep VOC levels to a minimum and as far below government limits as possible.

The Future

When new technologies become available, research and development teams present these options to the engineering and design staff, who then decide whether to incorporate them into a product. Gas lanterns, however, are less susceptible to drastic change because of their simple design. Although lanterns using alternative light sources are widely sold, employing battery, electric, and solar power, the rustic and utilitarian appeal of a gas lantern will likely keep the product from undergoing any major system overhauls. Nevertheless, new possibilities for materials and ease of operation are always a significant priority.

Where to Learn More

Books

Hobson, Anthony. *Lanterns that Lit Our World, Book Two.* New York: Golden Hill Press, 1997.

Other

"A Brief History of the Incandescent Mantle Pressure Lamp." *Pressure Lamps Unlimited Web Page.* 1998. December 2001. <http://ourworld.compuserve.com/home pages/awm/history.htm>.

Coleman Company, Inc. *A Brief History of the Use of Coleman Lamps and Lanterns.* Pamphlet, 1980.

"Dr. Carl Auer von Welsbach: Portrait." *Auer-von-Welsbach Museum Web Page.* December 2001. <http://www.althofen.at/welsbach.htm>.

"For a Better Lantern—Borax." *Corning Museum of Glass Web Page.* December 2001. <http://www.cmog.org>.

Oral interview with Richard Long, Senior Engineer at Coleman Company, Wichita, KS. December 2001.

Kate Kretschmann

Glucometer Test Kit

Background

Diabetes mellitus effects an estimated 16 million people in the United States. An additional five million people have the disease and do not realize it. Diabetes is a chronic metabolic disease that affects the pancreas's ability to produce or respond to insulin. The two major forms of diabetes are type I and type II. Both types of diabetes can have elevated blood sugar levels due to insufficiencies of insulin, a hormone produced by the pancreas. Insulin is a key regulator of the body's metabolism. After meals, food is digested in the stomach and intestines. Carbohydrates are broken down into sugar molecules—of which glucose is one—and proteins are broken down into amino acids. Glucose and amino acids are absorbed directly into the bloodstream, and blood glucose levels rise. Normally, the rise in blood glucose levels signals important cells in the pancreas—called beta-cells—to secrete insulin, which pours into the bloodstream. Insulin then enables glucose and amino acids to enter cells in the body where, along with other hormones, it directs whether these nutrients will be burned for energy or stored for future use. As blood sugar falls to pre-meal levels, the pancreas reduces the production of insulin, and the body uses its stored energy until the next meal provides additional nutrients.

In type I diabetes, the beta-cells in the pancreas that produce insulin are gradually destroyed; eventually insulin deficiency is absolute. Without insulin to move glucose into cells, blood sugar levels become excessively high, a condition known as hyperglycemia. Because the body cannot utilize the sugar, it spills over into the urine and is lost. Weakness, weight loss, and excessive hunger and thirst are among several indicators of this disease. Patients become dependent on administered insulin for survival.

Type II diabetes is by far the more common diabetes. Most type II diabetics appear to produce variable amounts of insulin, but have abnormalities in liver and muscle cells that resist its action. Insulin attaches to the receptors of cells, but glucose does not get inside a condition known as insulin resistance. While many patients can control type II diabetes with diet or with medications that stimulate the pancreas to release insulin, commonly the condition worsens and may require insulin administration.

Blood glucose levels that remain high (above 150 mg/DL) can lead to health complications such as blindness, heart disease, kidney disease, and nerve damage. One way that diabetics monitor blood glucose concentration is by testing blood samples several times throughout the day and injecting the appropriate dose of insulin. Upon doctors' recommendations and using such products, patients typically measure blood glucose level several (three to five) times a day. Generally these blood samples are taken from the finger, but can be taken from other places. A finger-stick comprised of a lancet is used to prick the finger and withdraw a small amount of blood that is placed on a test strip. The test strip is placed in a monitoring kit typically based on the electroenzymatic oxidation of glucose. While there is no known cure for diabetes, studies show that patients who regularly monitor their blood glucose levels and work closely with their healthcare providers have fewer complications in relation to the disease.

Using a typical glucometer and lancing device, the sampling and measurement process

Blood glucose levels that remain high (above 150 mg/DL) can lead to health complications such as blindness, heart disease, kidney disease, and nerve damage.

is generally as follows. First, the user prepares the meter for use by removing a test strip from a protective wrapper or vial and inserting it in to the meter. The glucometer may confirm the proper placement of the test strip and indicate that it is prepared for a sample. Some glucometers also may require a calibration or reference step at this time. The user prepares the lancing device by removing a cover from the lancing device, placing a disposable lancet in the lancing device, replacing the cover, and setting a spring-like mechanism in the lancing device that provides the force to drive the lancet into the skin. These steps may happen simultaneously (e.g., typical lancing devices set their spring mechanisms when one installs the lancet). The user then places the lancing device on the finger. After positioning the lancing device on the finger, the user presses a button or switch on the device to release the lancet. The spring drives the lancet forward, creating a small wound.

After lancing, a small droplet of blood appears at the lancing site. If adequate, the user places the sample on a test strip according to manufacturer's instructions. The meter then measures the blood glucose concentration (typically by chemical reaction of glucose with reagents on the test strip).

History

In 2001, Dr. Helen Free was inducted into the National Inventor's Hall of Fame in Akron, Ohio. In the 1940s, Dr. Free developed the first self-testing kits allowing diabetics to monitor their blood sugar by checking their urine at home. In the past, diabetics had to go to a doctor's office to get their blood-sugar level checked. Early indicators for home analysis were based on urine testing. Dr. Free was involved in over seven patents that led to improvements in design and function for home testing of glucose. In the late 1950s and early 1960s blood glucose levels were analyzed to detect more accurate levels for monitoring and treatment.

For years the solution for diabetics was one of several urinalysis kits that provided imprecise measurements of glucose in the blood. Later, reagent strips for urine testing were developed. Testing urine for glucose, however, is limited in accuracy particularly since the renal threshold for glucose spillage

into the urine is different for each individual. Moreover, sugar (glucose) in urine is a sign that the glucose was too high several hours prior to the test because of the time delay in glucose reaching the urine. Readings taken from the urine, therefore, are indicative of the glucose level in the blood several hours before the urine is tested.

More accurate readings are possible by taking readings directly from blood to determine current glucose levels. The advent of home blood tests is considered by some to be the most significant advance in the care of diabetics since the discovery of insulin in 1921. Home blood glucose testing was made available with the development of reagent strips for whole blood testing. The reagent strip includes a reactant system comprising an enzyme, such as glucose oxidase, capable of catalyzing the oxidation reaction of glucose to gluconic acid and hydrogen peroxide; an indicator or oxidizable dye, such as o-tolidine; and a substance having peroxidative activity capable of catalyzing the oxidation of the indicator. The dye or indicator turns a visually different shade of color depending upon the extent of oxidation, which is dependent upon the concentration of glucose in the blood sample.

Raw Materials

There are many raw materials used to produce a glucose monitoring kit. The test strips consist of a porous fabric or material such as polyamide, polyolefin, polysulfone, or cellulose. There is also a water-based hydroxyl elastomer with silica and ground titanium dioxide. Water, tramethylbenzidine, horseradish peroxidase, glucose oxidase, carboxymethyl-cellulose, and dialyzed carboxylated vinyl acetate ethyl copolymer latex are also used.

The meter itself is composed of a plastic case that houses the printed circuit board and sensors. There is a liquid crystal display (LCD) that will show the readings of the blood glucose.

The lancet is composed of a stainless steel needle encased in a plastic housing.

Design

There are many different forms of glucose test kits. Some glucometers have needles already installed. The user only presses the release button and the meter ejects the needle prick and withdraws a sample. Others require a separate lancet and test strips. These are the most commonly used forms of glucose kits.

The meter itself typically has a LCD display at the top of the machine. In the middle towards the bottom is a horseshoe-shaped slot in which to fit the test strip. Underneath this slot is a sensor that transmits the readout from the blood sample. The device runs off of batteries and usually has a short term memory built in to remember past glucose readings. Some devices can be hooked up to computer programs to track these readings and printout charts and diagrams depicting drastic shifts.

The Manufacturing Process

Test strips

1 The test strip is preferably a porous membrane in the form of a non-woven, a woven fabric, a stretched sheet, or prepared from a material such as polyester, polyamide, polyolefin, polysulfone, or cellulose.

2 A test strip is manufactured by mixing 40 g of an anionically stabilized (3.8 parts by weight sodium lauryl sulfate and 0.8 parts by weight dodecyl benzene sulfonic acid) water-based hydroxyl elastomer, containing about 5% by weight colloidal silica and 5 g of finely ground titanium dioxide. Then 1 g of tetramethylbenzidine, 5,000 units horseradish peroxidase, 5,000 units glucose oxidase, 0.12 g tris, and 10 g of water (hydroxymethyl) aminomethane (buffer) are mixed into the batch.

3 After mixing to ensure a homogeneous blend, the batch is cast onto a polyethylene terephthalate sheet for added structural integrity in a carrier matrix, and dried at 122°F (50°C) for 20 minutes.

4 Next, 100 mg of 3-dimethyl amino benzoic acid, 13 mg of 3-methyl-2-benzothiazolinone hydrazone, 100 mg of citric acid monohydrate-sodium citrate dihydrate, and 50 mg of Loval are added in dry form to a 50 ml tube.

5 These dry materials are mixed with a spatula, then 1.5 g of 10% water solution of carboxymethylcellulose is added and mixed thoroughly with the above solids.

6 Next, 2.1 g of dialyzed carboxylated vinyl acetate ethyl copolymer latex is added and thoroughly mixed.

The latex copolymer had been dialyzed (separation of larger particles from smaller particles) by placing about 100 g of carboxylated vinyl acetate/ethylene copolymer emulsion into a membrane tubing. The filled membrane was soaked in a water (distilled) bath at 68°F (20°C) for 60 hours to allow low molecular weight particles, unreacted monomer, catalyst, surfactant, etc. to pass through the membrane. During the 60 hours the water was continuously changed using an overflow system. The remaining dialyzed emulsion was then used in preparing the reagent layer.

7 Then 0.18 ml of glucose oxidase is pipeted to the tube as a liquid. Next, peroxidase is pipeted as a liquid to the tube and tartrazine is pipeted to the tube. The resulting mixture is mixed thoroughly. This mixture is allowed to stand for approximately 15 minutes.

8 A polished-matte vinyl support prior to being coated with the above solution was cut to form cell rows and then wiped clean with methanol. The mixture is pulled into a 10 ml syringe and approximately 10, 6 mm drops are placed on each cell row. The coated cell row is heated in an oven at 98.6°F (37°C) for 30 minutes followed by 113°F (45°C) for two hours. This process of coating and spreading the mixture is repeated for each cell row. The cell rows were then cut into strips of the desired size.

9 These strips were packaged with absorbent packs of silica gel and dried overnight at approximately 86°F (30°C) and 25 mm/Hg vacuum.

The glucometer

1 A molding press is loaded within the mold cavities, and a pellet of encapsulating material (thermoplastic resins used in the injection molding such as phenol resin, epoxy resin, silicone resin, unsaturated polyester resin, and other thermosetting resins) is placed in a receiving chamber.

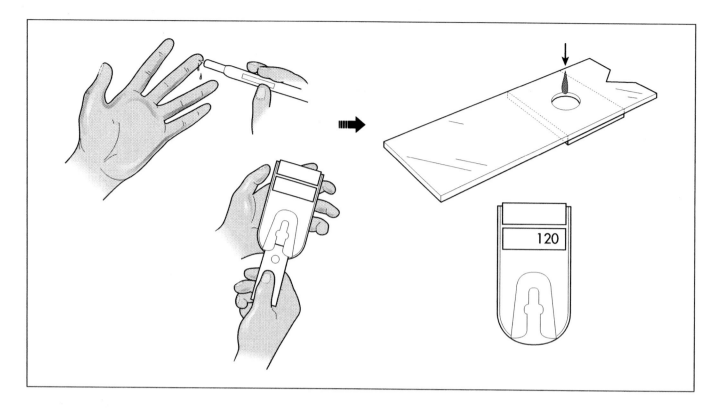

The patient pricks his finger and applies a sample to the test strip. The test strip is then inserted into the glucometer. After a period of about 10–15 seconds, the blood glucose reading appears.

2 Encapsulation of the integrated circuits (of the glucose detector) is achieved by heating the encapsulating material pellet and pressing it within the chamber using a transfer plunger, which causes the pellet to liquefy and flow into the mold cavities through small passages between the chamber and the mold cavities.

3 After allowing the encapsulating material to solidify again, the molding press is opened and the mold parts are separated.

4 After removal of the encapsulated integrated circuits, the open molding press is ready to receive new inserts and encapsulating material pellet to repeat the encapsulating process.

The lancet

1 Blood lancets today are generally manufactured using either an injection-molding process or an assembly process. In the injection-molding process, the wire is held in place by the adherence of the wire to the surrounding finger grip material.

2 The finger grips are generally made of plastic material such as polyethylene. The sharp point of the wire is embedded in a

point cover with a narrow neck attaching the point cover to the finger grips.

3 The point cover maintains the wire point clean until use. When the lancet is to be used, the point cover is twisted off at the neck, exposing the wire point for use.

4 The assembly process involves attaching the wire to the finger grips with an adhesive such as thermal epoxy, two-part epoxy, or ultra violet adhesive.

5 A cap is then placed over the wire point for protection and sterility. When the lancet is to be used, the cap is twisted off at the neck, thus exposing the wire point for use.

Byproducts/Waste

Plastics and various polymers used for the housings can be recycled in order to be melted and poured into molds. Chemicals used as reagents on the test strips are disposed of as lab waste. The majority of the parts are able to be recycled, therefore there is little waste.

The Future

Research on implantable sensors is progressing well. Several systems have been developed and may soon enter clinical testing. These will be very tiny needles implanted under the skin. Chemicals in the tip of the needle react with the glucose in the tissue and generate an electrical signal. The process is similar to the process used in most glucometers. The electrical signal can then be telemetered to a wristwatch-size receiver that can interpret the signal as a glucose value to read on the watch.

An alternate system being developed uses a tiny laser beam to drill a microscopic hole in the skin through which a tiny drop of tissue fluid is drawn. The device can then measure the glucose in the fluid in a manner similar to the previously described device. The developers of this device hope to combine in the same wristwatch-size receiver a mechanism to infuse insulin through the skin using a process called reverse iontophoresis. This process uses an electrical current to make insulin pass through the skin without a needle stick. Both these devices are still several years away from general use.

Bloodless meters that measure blood glucose without pricking the finger are an ultimate dream. At Kansas State University, a similar technology was developed for the food industry using a laser beam to measure the sugar content of fruit and other foods without breaking the skin of the food. Unfortunately, this technology is more difficult for use in humans. Skin thickness varies from person to person, and temperature varies the accuracy. This technology, dubbed "The Dream Beam," is still possible, but it is still some time away before it is cheap enough or accurate enough to be of practical use in the future.

Synthetic Blood International (SYBD) has developed an implantable glucose biosensor to monitor blood glucose eliminating the need for finger sticks. The glucose biosensor uses an enzyme specific for glucose. Once implanted in subcutaneous tissue, the biosensor, which is about the size of a cardiac pacemaker, provides continuous, accurate monitoring of blood glucose. The latest technology is still several years away from being able to offer a closed loop system where insulin will be delivered based on the digital readings on the biosensor. Ultimately, the glucose biosensor will be linked to an implanted insulin pump, creating a closed-loop mechanical pancreas.

Where to Learn More

Other

Abbott Laboratories Web Page. December 2001. <http://www.abbott.com>.

American Diabetes Association Web Page. December 2001. <http://www.diabetes.org>.

Synthetic Blood International Web Page. December 2001. <http://www.sybd.com/Synthetic.html>.

United States Food and Drug Administration Web Page. December 2001. <http://www.fda.gov>.

Bonny P. McClain

Goalie Mask

In 1930, Clint Benedict of the Montreal Maroons was the first goalie to wear any type of facial protection in a hockey game.

Background

As a formal game, hockey began to be played in North America in the 1870s in Halifax, Nova Scotia. The first organized hockey league began with four teams in Kingston, Ontario in 1885. Hockey has been played for over 100 years in North America and over 500 years in Europe, but amazingly the goalie mask is a relatively new invention. In the early beginnings of hockey, goaltenders had never thought to wear any facial or head protection. Shots were low and were not as strong and fast as they are now.

Clint Benedict of the Montreal Maroons was the first goalie to wear any type of facial protection in a hockey game. In 1930, he wore a modified leather mask that covered his broken nose and cheekbone, the result of getting hit by a puck during a game. There are two theories as to the origin of the facemask that he wore. One theory is that it was a modified football faceguard, and the other is that it was a boxer's sparring mask. Benedict ended up wearing the mask for only two games because it blocked his vision. This, however, was not recorded as the first goalie mask in history.

In 1934, Roy Mosgrove, who wore eye glasses all the time, first put on a wire cage to protect his glasses. The wire cage was originally worn by baseball catchers. It took decades for the goalies to incorporate the wire cages into the fiberglass masks of today.

As the game of hockey progressed, the players got stronger and faster, and players were shooting the puck harder and higher. The slapshot was beginning to hit players in the face, resulted in broken facial bones and skin lacerations. Goaltenders began experi-

menting with facial protection such as wire cages and clear shatter-proof shields in the 1950s, but they were only used during practice. These forms of facial protection were not satisfactory since they fogged up, there was a glare from the lights in the rink, and they had many blindspots, making them unsuitable for game use. Goalies also did not wear any form of facial protection during games because they worried that other players and fans would lose respect for them, thinking them to be weak. Goaltenders at that time wore their injuries like badges.

History

Clint Benedict in 1930 is thought to have worn the first mask-like protection in a game but his mask was not formally recorded. The first recorded goalie wearing a mask came on November 1, 1959, when Jacques Plante of the Montreal Canadians was hit with a rising shot in the nose and was knocked unconscious by New York Ranger Andy Bathgate. Plante was forced to leave the game to be stitched up, but later returned wearing a flesh tone mask constructed of fiberglass with cutouts for the eyes. By this time, Plante had broken his jaw, both cheekbones, and his nose, his skull had suffered a hair-line fracture, and he had gotten over 200 stitches from past injuries received during games.

Plante's mask was the product of a Canadian company called Fiberglass Canada. Bill Burchmore, a sales and promotional manager for the company, envisioned the mask. One evening he had gone to watch a game in which Plante was goaltending. He witnessed Plante getting hit in the forehead with a puck, resulting in a 45 minute delay of game while he was being stitched up. While at work the next day, Burchmore was looking at a fiberglass mannequin head when he realized the he could design a contoured, lightweight fiberglass mask that would fit the goalie's face like a protective second skin. Burchmore gave Plante his idea, and Plante was persuaded by his trainers to give it a try. A mold was taken of Plante's face. He had to put a woman's stocking over his head, cover his face with Vaseline, and breath through a straws stuck in both nostrils while his head was covered with plaster. Burchmore layered sheets of fiberglass cloth saturated with polyester resin on top of the mold. The result was the flesh toned 0.125 in (52 mm) thick mask that weighed only 14 oz (397 g).

By January of 1960, Burchmore had come up with a second goalie mask design. His new mask, called the pretzel design, used fiberglass cloth and was made up of fiberglass bars contoured to the goalie's face. The new mask weighed 10.3 oz (292 g) and allowed more air to circulate around the goalie's face. In 1970, Plante added hard ridges on the forehead and down the middle of his mask to deflect pucks, and protruded the mask over the ears. This new mask was made with epoxy resins which were able to absorb impact better. This mask withstood a test in which pucks were fired out of an air cannon at speeds of 120 miles per hour (193 km/h). The only thing that broke were the hockey pucks. More and more goaltenders began to wear masks not only in practice but also in games. The year 1973 was the last year that a goaltender went without a mask in the National Hockey League (NHL). Now it is a rule in the NHL and other hockey leagues that all goaltenders must wear masks for protection. A hockey puck weighs 6 oz (170 g) and can get up to speeds in excess of 100 mph (161 km/h), the speed at which many players in the NHL can shoot the puck at the goaltender.

Raw Materials

Masks are still made of fiberglass and epoxy resins, but now have added materials such as Kevlar, carbon fiber, and capron nylon resin. Fiberglass is still used because it is a light material, has a high tolerance to damage, and is easy to handle and mold. It also comes in

Manon Rheaume.

Manon Rheaume was born in 1972 and raised in Lac Beauport, a suburb of Quebec City in Canada. Learning to skate by age four, Rheaume spent hours practicing hockey with her brothers. When she was five, her father was short a goalie for a local tournament. Rheaume volunteered, and the minute she tried her skills in real competition she was hooked.

Rheaume's ability landed her on boys' teams all through school and the youth leagues. After high school she made Canada's Junior B league and even played briefly on the Junior A level—the level just below the NHL. With women's teams she was nothing less than a star. As the goaltender for the Canadian national women's team she helped to win a gold medal at the 1992 women's world championships in Finland. Rheaume gave up just two goals in three games in the world championship tournament.

Given a tryout for the Tampa Bay Lightning in 1992, Rheaume composed herself bravely in the Tampa Bay tryout, giving up two goals and making seven saves in the first period of an exhibition contest. Tampa Bay general manager Phil Esposito signed her to a three-year contract and sent her to the minors in Atlanta. Rheaume became the first woman ever to play in an NHL game.

different styles and weights. Kevlar is the material used in bullet proof vests. It adds strength to the mask, but at the same time is very light weight. Carbon fiber is similar to fiberglass, but it has higher strength and stiffness. It is also more expensive than fiberglass, therefore it used in limited amounts in goalie masks. Carbon fiber is also used in snowboards, mountain bikes, and race car bodies. Rubber and foam are used as padding inside the mask. The caging in masks is made of stainless steel rods or titanium. Cages began to be used more often after goaltender Bernie Parents was hit in the eye by a stick, resulting in an injury that ended his career.

An example a mold being taken and the mask being formed.

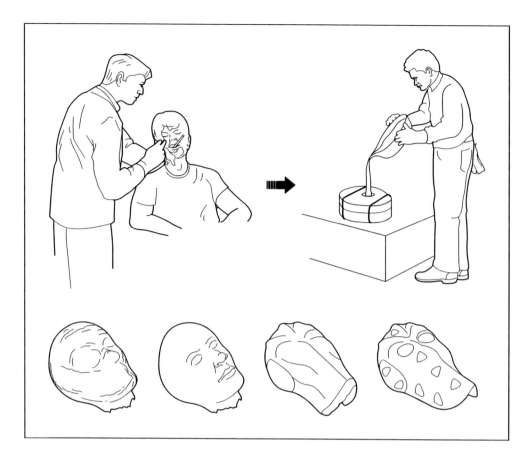

The mold taken of the goaltender's head is typically made from alginate, the same gel-like material used in dental molds.

Design

Goalie masks vary in size and color, depending on the goalie's preferences. The first goaltender to decorate his mask was Gerry Cheevers, who painted stitches on his mask where he had been hit with a puck or stick. The masks of goaltenders, especially in the NHL, are often the canvas of an artist. The first artistic mask in the NHL was owned by Glenn "Chico" Resch of the New York Islanders. In 1976, Linda Spineela, a friend of the trainer and an art student, was allowed to paint Resch's plain white mask. Masks may be decorated by a combination of painting and airbrushing in various ways, such as with team colors, images that reflect the team name, or where the team is from. For example, the San Jose sharks goaltender has a shark painted on his mask. For decorative painting Epoxy primers, basecoats, automotive paints, and urethane clearcoats are used. To ensure

that the paint will not chip it is clear coated, sanded, polished, and then baked.

The Manufacturing Process

Custom made masks ensure a comfortable fit to each goalie's head. Custom made masks are still manufactured in a similar fashion as in the 1960s, and are still made by hand without the aid of automated processes.

1 The first step is to make an impression of the goaltender's entire head. A bald cap is placed over the goalie's hair and straws are placed in both nostrils. A gel-like substance called alginate is spread over the player's face. The substance adheres to the skin, but will not stick when removed. After about four minutes, the gelatin-like mold is removed.

2 After the alginate impression has completely air dried, plaster is pored into the mold and a bust is made.

3 Once the bust is dry the molding process of the mask begins. The beak area—the area over the nose and mouth—and other

extrusions, such as around the ears, must be sculpted onto the plaster bust using clay or paper maché.

4 Once the sculpting is finished, the laminating process begins. Multiple sheets of fiberglass coated with epoxy are laid on the bust until the desired thickness is achieved. The mask may also be reinforced with Kevlar or carbon graphite composite. Epoxy resins are used to form a bond between the layers. It is important for the manufacturer to smooth out any air pockets when laying the layers of fiberglass. If there are air pockets in the mask it will result in a weak spot that may break when hit by a puck.

5 Finally, once the resin has dried, the mask is broken from the mold. It is then properly fitted to the goaltender and the facial opening is cut out. The mask is primed, and a top coat of enamel paint is applied.

6 Cages are now placed in the facial opening and are attached with stainless steel fasteners. Cages are made out of steel, and occasionally titanium. They protect the eyes from being penetrated with a stick or a puck.

7 The interior of the mask is lined with high impact rubber padding for comfort and extra protection. Chin straps are added to hold the mask onto the goaltenders head. The result is a strong mask that is comfortable and lightweight. Energy from a puck is transferred more evenly because of the custom, even fit.

Byproducts/Waste

A goaltender's mask will endure many hits, but parts that are not made out of fiberglass (such as the padding and cage) will deteriorate after time. Many mask manufacturers offer reconditioning of masks when they reach this stage of wear. Therefore, instead of throwing the mask away and spending hundreds or thousands of dollars for a new one, the goaltender can have it reconditioned and continue to use it. Reconditioning includes putting in new interior padding, sweat bands, and chin cup. A new cage may be added if the old one is dented. The cage could also be chrome or gold painted. If there are chips in the mask they can be repaired with fiberglass layers.

The Future

As NHL players become stronger and the puck becomes faster, injuries will continue to grow. New technology in the manufacturing of hockey pucks, sticks, and skates, as well as rink developments, will result in new safety regulations. With the development of new materials, masks continue to vary in design and offer more protection.

Where to Learn More

Books

Hunter, Doug. *A Breed Apart: An Illustrated History of Goaltending.* Triumph Books, 1998

Other

Classic Mask Web Page. December 2001. <http://www.classicmask.com/evolution.html>.

Dillon's Custom Goalie Mask, LTD Web Page. December 2001. <http://www.dillonmask.com>.

NHL.com Web Page. December 2001. <http://www.nhl.com>.

Pro-Masque Web Page. December 2001. <http://www.promasque.com>.

Sportmask Web Page. December 2001. <http://www.sportmask.com>.

The Science of Hockey Web Page. December 2001. <http://www.exploratorium.edu/hockey>.

Carrie Lystila

Grenade

Grenades were first used in the late fifteenth or early sixteenth century.

Background

Grenades come in a variety of sizes and shapes depending on their function, but all have two things in common. First, they are hollow to allow filling with explosive or chemical filler. Second, they contain a threaded hole into which a fuse can be inserted.

A grenade is essentially a small bomb, but works like a simple firecracker. A firecracker is made up of a paper body filled with gun powder containing a small fuse. When the fuse is lit, it burns down to the powder and causes the gun powder to explode. A grenade works exactly the same way, except that a mechanical device instead of a match lights the grenade's fuse.

The grenade is held in the throwing hand with the thumb placed over the safety lever while the safety pin is pulled. When the grenade is thrown (safety lever released), a spring throws off the safety lever and rotates the striker into the primer. The primer contains material similar to the head of a match. When struck, it ignites and sets fire to the fuse. The fuse burns at a controlled rate, providing a time delay (about four to five seconds). When the flame of the fuse reaches the detonator (a small blasting cap), it causes the grenade to explode.

History

The inventor of the grenade is not documented, but grenades were first used in the late fifteenth or early sixteenth century. Originally they were just hollow metal balls filled with gunpowder. These early models were ignited with a slow burning fuse and thrown distances over 100 ft (30.5

m). As the range and accuracy of firearms increased during the eighteenth century, grenades were virtually abandoned. They were re-introduced to the infantryman's equipment arsenal during the trench warfare of World War I. During World War II over 50 million fragmentation grenades were manufactured in the United States alone. Grenades have remained a part of the modern arsenal, and are delivered by a variety of methods: throwing, single shot launcher, or rapid-fire cannon. Modern grenades are ignited by either a timed fuse or impact fuse.

Raw Materials

There are many polymers that can be used for the fragmentation grenade casing depending on the strength and processing requirements (e.g., polypropylene, polyamides, polyacetals, polycarbonates, polyesters, polyethers, aldehyde/phenolic condensates, melamine resins, and urea resins). A variety of metals and shapes are used for the fragments and the casing. The molding method can also be varied.

The steel fragments are obtained by cutting wire with a gauge of about 0.09 in (2.2 mm) to form pieces, each having a length of about 0.09 in (2.2 mm) with subsequent round-hammering of the wire pieces and hardening of the hammered particles.

The explosive component of a grenade consists of: a high-explosive main charge (e.g., Ammonium Nitrate Fuel Oil [ANFO]), a primer or booster charge (e.g., Pentolite or Cyclotol), and a primary initiator (e.g., a blasting cap, electronic detonator or Low Energy Detonating Cord [LEDC]).

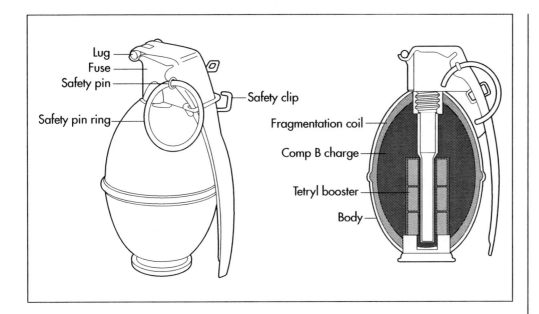

Design

Five basic types of hand grenades were used in Korea by United States forces: offensive grenades, fragmentation grenades, chemical grenades, practice grenades, and training grenades.

Offensive grenades contain explosive charge filler of flaked TNT in a body with sheet metal ends and pressed fiber sides. They are designed for demolition effect and to stun the enemy in enclosed places.

Fragmentation grenades contain an explosive charge in a metal body, designed to break into fragments upon the charge exploding. They typically weigh about 21 oz (595 g), and are constructed of cast iron.

Chemical grenades are designed to produce a toxic or irritating effect, a screening or signal smoke, an incendiary action, or some combination. Some of these grenades—as with the TH M14 thermite—come with metal straps that prevent rolling and an M200A1 igniting fuse with only about two seconds of delay after safety lever release. Baseball-type tear gas grenades are special issue for riot control.

Practice grenades contain a reduced charge for safe use in training.

Training grenades contain no explosive charge or chemical, and are used solely for throwing practice.

The Manufacturing Process

1 The casting mold may contain a wide variety of features to manipulate the shape of the cast segment (hatch marks, identifying material, or a manufacturers logo).

2 The components (polymer powder and metal fragments) can be introduced into the mold either separately or as non-compressed mixture through the threaded mold opening that can be sealed by the threaded pin.

3 To achieve a substantially uniform thickness of the polymer layer, the mold is generally rotated in a tumbling manner, that is, about at least two different rotational axes, for example about the vertical axis and one horizontal axis of the mold.

4 The tumbling, rotating shell mold is heated externally by a fluid (such as hot air) to temperatures that lead to coalescing and melting of the polymer particles, or to thermal cross-linking (activation of cross-linking catalyst). The use of catalytically cross-linking polyalkanes or polyalkylenes is preferred, the term "catalytically" describes cross-linking triggered by radicals (e.g., as formed upon decomposition of a peroxy compound). An advantage of cross-linking of the polymer resides in an increased thermal and mechanical strength of the casings of the grenade. Another advantage of cross-linking is that the increasing heat transfer from the fluid into and

through the mold wall does not lead to an overly fluid melt, as cross-linking tends to increase viscosity. By a cross-linking that proceeds gradually as the temperature rises, the polymer layer can be kept sufficiently viscous for achieving and maintaining uniform layer thickness even at very low rates of rotation.

Shell mold

1 A shell mold is provided with a thermally conducting mold wall and has an interior surface shaped to correspond with the outer shape of the casing described above.

2 Mold parts are pressed together with a clamp.

3 Next, the assembled mold, filled with polyethylene powder and steel fragments is closed by a threaded plug, mounted in a conventional tumbling frame, and placed in a hot air oven.

4 After 30 minutes under these conditions the mold is cooled and at 122°F (50°C) the rotation is discontinued and the casing is discharged from the mold.

5 The mold halves are separated by releasing the clamp.

6 The fragmentation grenade casing is supplied with a charge of high explosive and closed by threading in a conventional time fuse device. Upon detonation the casing disintegrates and forms a multiplicity of high-speed fragmentation projectiles by the steel fragments.

Quality Control

Due to the grenade's desired effect, the quality is extremely important. The ram compresses the explosive composition into the detonator housings. If the ram pressure exceeds the maximum designed pressure due to excessive conductive explosive volume within the housing or misalignments of the parts, the grenade provides a maximum pressure switch for turning off power to the ram. This switch also triggers a reject mechanism by which the housing is removed and

another is positioned for compression. If insufficient conductive explosive composition is present, a member connected to the ram causes a volume acceptance switch to open and thereby open the power circuit to the ram and at the same time activate a recycle mechanism or the rejection mechanism. This recycle mechanism provides for the loading of additional conductive explosive composition in the insufficiently filled detonator housing.

For additional testing purposes, multiple casings are obtained and cut in a vertical plane for investigation. The density and tensile strength (determined according go ASTM Standards at 2 in [5.1 cm] per min) of 2,600 tons per square inch is randomly tested. A toluene boiling test quantifies the degree of cross-linking to determine the stability of the casings at elevated temperatures and the absence of embrittlement at low temperatures in the testing range for explosives.

The Future

Grenade manufacturers continue to design safety features to prevent injury to the person launching the grenade. Minimizing the smoke trails emitted as the grenade travels towards its target prevents the location of the launcher from being detected. Manufacturing quality control measures continue to remove duds and unsafe devices from final production further increasing the safety of military personnel.

Where to Learn More

Books

Hogg, Ian V. *Ammunition: Small Arms, Grenades and Projected Munitions.* Greenhill Books, 1998.

Other

United States Patent Office Web Page. December 2001. <http://www.USPTO.gov>.

Bonny P. McClain

Guillotine

Background

The guillotine evokes images of horrifying and bloody public executions during the French Revolution in the eighteenth century. Many historians consider this device the first execution method that lessened the victim's pain and the first step in raising public awareness of the morality of the death penalty. It is difficult, however, to think of the guillotine as humane when descriptions of blood flowing in the streets of Paris paint such a gruesome picture.

The guillotine was used for a single purpose, decapitation. The device releases a blade that falls about 89 in (226 cm). With the combined weight of the blade and the mouton (a metal weight), the guillotine can cut through the neck in 0.005 seconds. Expert craftsmen, such as carpenters, metal workers, and blacksmiths, made parts of the guillotine separately and then others assembled the parts at the site of the execution. The guillotine was never mass-produced.

History

Although history links the guillotine to the French Revolution, an earlier version of a similar instrument was used as early as 1307 in Ireland. In Italy and Southern France, another guillotine-like device called the *mannaia* was used in the sixteenth century, but only to execute nobility.

Dr. Joseph Ignace Guillotin was a physician and a deputy of the National Assembly of France, an early stage of the Revolutionary government. He recognized and promoted the guillotine's use in 1789. Dr. Guillotin believed this swift method of execution would reform capital punishment in keeping with human rights. Other Assembly members rejected his championing of the guillotine with laughter.

In 1792, a public executioner named Charles-Henri Sanson recommended reconsideration of the guillotine and Dr. Antoine Louis (the secretary of the Academy of Surgeons) supported him. In April 1792, Tobias Schmidt (a German piano maker) built the first working model in less than a week. On April 17, 1792, the executioner tested the prototype by decapitating sheep, calves, and corpses from the local poorhouse. On April 25, Nicolas Pelletier (a thief who viciously assaulted his victims) entered the history books as the first criminal beheaded by the guillotine.

In its earliest days, the guillotine was called the "louison" or "louisette" after Dr. Louis who had pressed it into service. Later, the name changed to commemorate Dr. Guillotin, who—although he had never constructed a single instrument—came to resent this association. Most commonly, it was simply called "the machine."

The most famous victims of the guillotine were King Louis XVI and his queen, Marie-Antoinette. The King was convicted by the Revolutionary government in 1793 for treason. He was decapitated on January 21, 1793. His wife, Marie-Antoinette, was imprisoned for nine months after the King's death until she was also executed by the machine's blade. Charles-Henri Sanson executed the King and his son, Henri, dispatched the Queen.

Estimates of the number of lives taken by the guillotine during the French Revolution range from 17,000 to 40,000 citizens. It is thought that three-quarters of the executed

With the combined weight of the blade and the mouton (a metal weight), the guillotine can cut through the neck in 0.005 seconds.

King Louis XVI.

Louis-Auguste (Duke of Berry) was born August 23, 1754. He was the third son of Louis the *dauphin,* heir to the throne of Louis XV. After the death of his brothers and father, in 1765 Louis became the sole heir. In 1770 he married Marie Antoinette, and in 1774 Louis XVI became king of France.

Louis restored the powers of the Parliament, but he was indecisive, easily influenced, and lacked the strength to support reformation against opposition whose positions were threatened by change. By 1788, France was on the verge of bankruptcy. Pressure mounted to invoke the Estates General to handle the fiscal crisis. In May 1789, the Estates General met at Versailles, opening the French Revolution. A Parisian crowd forced the court to move from Versailles to Paris, where it could be controlled more easily. In June 1791, Louis sought to escape from Paris to eastern France. However, at Varennes the royal party was recognized and forced to return to Paris, where Revolutionaries had lost all confidence in the monarchy.

In September 1791, the National Assembly adjourned and was succeeded by the Legislative Assembly. On April 20, 1792, France declared war on Austria, which was soon joined by Prussia. France was incensed by the manifesto of the Prussian commander, the Duke of Brunswick, threatening punishment on Paris if the royal family were harmed. On August 10, 1792, the crowd forced the Legislative Assembly to suspend Louis, who—with the royal family—became prisoner of the Commune of Paris. The National Convention, which succeeded the Legislative Assembly, abolished the monarchy and tried "Citizen Capet," as Louis was now called, for treason. He was found guilty, sentenced to death, and on January 21, 1793, guillotined.

were innocent. In its "glory" days, the guillotine took 3,000 lives in one month. Paris was responsible for only 16% of executions; in cities with many counter revolutionaries,

like Lyons, many more faced the blade. The locations of public executions were moved frequently. After beheadings, blood continued to pump out of the bodies, overtopping the gutters, and running down the streets. In France, the guillotine remained the official execution device until the last use of the "national razor" in 1977. French President François Mitterand abolished the death penalty in 1981.

Raw Materials

The platform, posts, déclic for the rope, crossbar, the bascule (bench supporting the body), and the lunette (the device holding the head) were made of hard wood. The mouton was the metal weight to which the blade was attached. The extra weight ensured a swift, clean cut. The blade itself was made of steel, and the heavy-duty rope was cotton. Leather straps restrained the victim's body around the arms and to the bench around the back and legs. A leather bag or basket was also used to catch the falling head.

Design

Very few design changes occurred during the history of the guillotine. The primary modification was the adaptation of the size and weight of the machine to a horse-drawn cart when portability was needed to increase the efficiency of the machine. These moveable guillotines were mounted on horse-drawn carts that were also made of wood with wooden wheels strapped with iron. Wood braces were attached to the wheels when the guillotine was used to keep it motionless.

The Manufacturing Process

Guillotines were hand crafted locally and were relatively simple to make because they were without ornamentation or refined finishes. The craftsmen were very experienced with wood construction and the honing (shaping and sharpening) of the steel for the blade.

1 Construction of the guillotine began with the platform or scaffold. A skilled carpenter cut the lumber for the major pieces including post supports, interconnecting beams, the floorboards, and the steps for the stairway underneath the platform. The stairs

bottomed at one open end of the scaffold (on the front side of the guillotine) and opened in an entry or hatch near the other end of the platform at the back of the guillotine. The platform also had an open railing around three sides of the scaffold; the side without the railing was toward the front of the machine and the bottom of the stairs.

2 The supports and beams were all nailed together to form a base. The floor was either built as a separate unit with an underside of wood sheets, much like modern rough-grade plywood, and a top face of long, thin floorboards. The two layers reduced weathering and other damage. The unit could then be lifted in place and nailed to the edges and cross beams of the scaffold.

If the guillotine was constructed at the execution site, construction of the platform continued by adding the side rails. The stairway was built while the platform was being constructed by making a four-sided base with interior braces for strength. One side was the front face of the first stair, the back extended from the ground up to form the back of the top stair, and the two identical sides had bottom and back edges forming a 90° angle. Both sides were cut to hold the tops and backs of the set of stairs.

3 While the platform was being constructed, work began on the steel blade and mouton. The width between the posts and the maximum thickness of the blade were provided to the forger or blacksmith. This specialist made a mold for the blade. The cutting edge angled up from one side of the blade (in an oblique angle) to the opposite post. The angle allowed the blade to cut more quickly and cleanly; a blade with an even edge (parallel with the upper cross beam) would have encountered more friction as it tried to cut through the wider back of the neck. Molten steel was poured into the mold. The craftsman sharpened the cutting tip by repeated filing, hammering, and reheating. Worn blades were also resharpened this way. The steel blade generally weighed about 15 lb (7 kg).

4 The mouton was manufactured the same way. The craftsman would melt the metal down and pour it into a mold. After the mold cooled, it would be taken out. The mouton typically weighed 66 lb (30 kg).

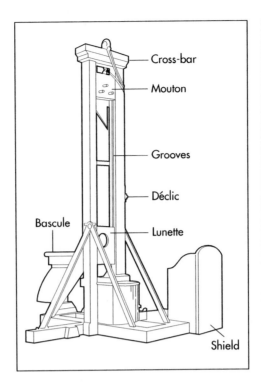

A guillotine.

Cross-bar
Mouton
Grooves
Déclic
Bascule
Lunette
Shield

5 Workers would then screw the blade to the mouton with three bolts, two in the bottom corners and one in the middle. The bolts would then be welded into place.

6 When the platform was complete or if other carpenters were available, construction of the machine frame began. A small-diameter tree for each post was cut to create a four-sided post, then a groove was cut out on the inside of each post and chiseled so the falling blade would drop smoothly. At the base of the machine, the posts were mounted in a wide crossbar. The blade and mouton were fitted in the post grooves, and a crossbar at the top that was exactly the width between the side posts was fitted in place. The upper crossbar also had a hole in the top for the rope and a groove along the top and side to guide the rope. Metal rings were fixed to the outside of the top crossbar and one or two points down the post to guide the rope. Wood braces were fitted to the outsides of the posts and extended down at angles to the base crossbar for added strength.

7 On the back side, where the victim and the executioner stood, another crossbar was mounted to hold the lunette, which consisted of two separate pieces of relatively thin wood with a hole big enough for the victim's neck. Half of the hole was in the bottom section of the lunette, and the match-

ing half-moon was in the top portion. The upper half was hinged on the post so it could be raised for the prisoner's head. The machine as a separate piece was complete and could be hauled on a cart to the site.

8 The bascule was carved out of wood by a carpenter and transported to the site of the execution. The end of the bascule nearest the blade had leather straps to restrain the victim's arms, and straps crossing the bench kept the back and legs tied down.

9 The déclic was a wooden handle that opened the grooves in the posts. It was attached to the outside of one of the vertical posts so that the executioner could easily release the blade.

10 The rope was is made from natural fibers and twisted into yarn. The yarn is then woven and twisted rope. The rope is tied securely to the top of the mouton, through the hole in the upper crossbar, through the rings, and wrapped around the déclic. In the early days of the guillotine, the executioner cut the rope with a sword to drop the blade, but it became too time-consuming to readjust the rope so they changed the design to incorporate the déclic.

Quality Control

The executioner usually owned the guillotine and accessories. Executioners in major cities owned several guillotines and cycled them in and out of use for repair. Quality control of construction and maintenance were entirely the executioner's responsibility.

The executioner also maintained a fleet of eight to 10 tumbrels for transporting the victims from the prison to the guillotine. A coach maker constructed and repaired the tumbrels and carts for hauling the guillotine's pieces, but the executioner had to approve the work.

With this particular product, quality control was also required for the execution process. Five to eight assistants helped the executioner lead the victim to the machine, remove any clothing around the neck, and cut the victim's hair. They strapped the victim down, placed the victim's head across the lunette, and lowered the top of the lunette around the victim's neck in a series of smooth motions. The executioner released

the déclic, the head and body were separated in a split second by the weight of the blade and mouton, and the head fell into a leather bag or lined basket. An assistant raised the head for the crowd's approval, and several other assistants took the head and body back down the stairs where they were thrown into carts for disposal. Heads of well-known victims had the added distinction of being impaled on poles.

The Future

The guillotine has been relegated to history and lore and is no longer used for executions. In isolated cases, craftsmen make guillotines for entertainment (films and television), but these are built with sophisticated safety systems and often as models. There are books and kits available to make models of the guillotine.

The guillotine has since been replaced by other so-called humane ways of executing criminals, such as lethal injection, hanging, gas chambers, a firing squad, and the electric chair. Thirty-eight of the United States apply the death penalty, but Texas leads the number of executed criminals with a total of 253 as of January of 2001.

Where to Learn More

Books

Banfield, Susan. *The Rights of Man, The Reign of Terror: The Story of the French Revolution*. New York: J. B. Lippincott, 1989.

Doyle, William. *The Oxford History of the French Revolution*. Oxford: Clarendon Press, 1989.

Guillon, Edmund Vincent. *Build Your Own Guillotine: Make A Model That Actually Works*. New York: Putnam, 1982.

Schama, Simon. *Citizens: A Chronicle of the French Revolution*. New York: Alfred A. Knopf, 1989.

Vallois, Thirza. *Around and About Paris*. Vol. 1. London: Iliad Books, 1999.

Periodicals

"Dr. Guillotin's Killing Machine." *Maclean's* 102, no. 20 (May 1989): 34.

Lawday, David. "The Heirs of Madame Guillotine: The Descendants of France's Dynasty of Executioners Today Ponder the Paradoxes of the Revolution." *U.S. News & World Report* 107, no. 3 (17 July 1989): 46–49.

Other

"The Guillotine." *Mutimedia World History* December 2001. <http://www.historywiz.com>.

Gillian S. Holmes

Hair Dryer

The first handheld hairdryer appeared on the market in 1925 and weighed over 2 lb (1 kg).

Background

A hair dryer, also known as a blow dryer, is an electrical device used to dry and style hair. It uses an electric fan to blow air across a heating coil; as the air passes through the dryer it heats up. When the warm air reaches wet hair it helps evaporate the water. Hair dryers may be used with a variety of brushes and combs to achieve different hair styles.

History

The first handheld hairdryer appeared on the market in 1925. It produced only 100 watts of heat and therefore did not have sufficient power to dry hair quickly. It weighed over 2 lb (1 kg) because it was made of heavy steel and zinc. Over the next 20 years engineers improved the design and managed to triple the heat output, raising it to 300 watts. By the 1960s, further improvements in electrical technology allowed the production of hairdryers with up to 500 watts of power.

In the late 1970s, manufacturers began to focus on improving the safety of dryers. Early hairdryers were dangerous because if they accidentally came in contact with water they would short circuit and cause an electrical shock. There are hundreds of recorded cases of accidental electrocutions because a hairdryer was dropped into a bathtub or sink full of water while it was being used. In the late 1970s the Consumer Products Safety Commission (CPSC) recommended guidelines for hair dryer manufacturers to follow that would create safer products.

The power of hair dryers was limited by the electric motors available. As smaller, more efficient motors were developed, greater airflow and greater heat output could be achieved. By the 1990s portable hairdryers could produce over 1500 watts of heat. Improvements in plastic technology and the discovery of new insulating materials made possible a new generation of lightweight hairdryers. Modern hair dryers can produce up 2000 watts of heat and can dry hair faster than ever before.

Design

One of the key factors to consider when designing a portable hairdryer is the amount of heating power it can produce. Since warm air is capable of absorbing more moisture than cold air, the temperature of the airflow is critical. By calculating the specific heat of the air and understanding the maximum temperature that can be used without burning the skin, engineers can calculate the amount of power required for the heating element. This ensures that the device will generate enough heat to dry the hair quickly. However, it is not enough to simply raise the temperature of the hair; the air must also pass rapidly through the hair for efficient moisture removal. Therefore, the efficiency of the fan is also a critical consideration.

Another key design criteria is the safety of the device. First, the hair dryer must not become so hot that it burns the user during operation. The plastic housing must remain at a comfortable temperature and cannot overheat or else it could melt or catch fire. To solve this problem, engineers have developed temperature cutoff switches that prevent the heating coil from getting too hot. This cut off switch also turns off the heating coil if the blower motor on the fan stops functioning. Second, the hair dryer must not cause electric shock. A special shock safe-

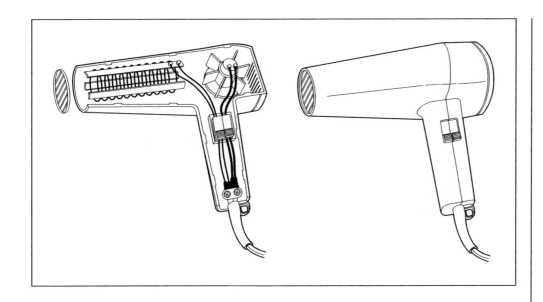

guard, a Ground Fault Circuit Interrupter (GFCI), is used in modern hair dryers to prevent accidental electrocution.

Other factors to consider include the weight of the unit and a user-friendly design. For the sake of convenience, hair dryers are designed to weigh only about 1 lb (500 grams), and they are made to be easy to handle during operation.

Raw Materials

Hairdryers are assembled from a series of components including the electrical motor, the fan blade, copper wiring, switching mechanisms, and various other electrical components. The plastics used to make the outside shell of the hair dryer must be durable, yet light-weight.

The Manufacturing Process

1 The electric motor and fan blades arrive at the manufacturing plant pre-assembled.

2 Hairdryers produce heat in the same fashion that a toaster does: by passing electric current through a wire. The wire has a high level of electrical resistance that causes it to generate heat as the current passes through it. Hair dryers use a metal heating element that is made of nichrome, an alloy of nickel and chromium. Unlike other electrical wires made of copper, nichrome will not rust at high temperatures. This wire looks like a coiled spring and may be up to

12 in (30 cm) long. It is wrapped around an insulating board so that the entire heating element is only a few inches long.

The insulating board is usually made of mica, a mineral that can stand high levels of heat. Two flat pieces of mica several inches long are connected to form what looks like a three-dimensional "x." Notches are cut in the edges of this board such that when the nichrome wire is wrapped around the board it fits snugly into these slots. At the end of the wire there is a connection to the circuit that controls the power supply. Depending on how this circuit is wired, current can be fed to part of, or all of, the heating coil. More heat is produced as current is fed to a greater portion of the wire. The heating element used in modern hair dryers can produce up to 2000 watts of heat energy.

3 The body of the hairdryer consists of a gun shaped plastic shell. This shell is divided into two sections to allow for easy assembly. The plastic parts are created by injection molding, a process that involves injecting hot, molten plastic (such as polypropylene) into a stainless steel die. After the mold is filled with hot plastic, cold water is circulated around the die to cool the plastic and make it harden. When the die is opened, the plastic parts are removed. One side of the plastic shell has a series of holes molded into it while the matching half has a series of short pins. These help align the two halves during manufacture and keep them firmly attached.

The shell is molded with multiple compartments to hold the various electrical components. The handle of the gun contains the switch apparatus and the controls to operate the dryer. The electric motor and fan are found in the central part of the drier located just above the handle. The long barrel of the device contains the heating element. Next to the motor is an air intake air inlet. This inlet is covered with a fine mesh metal screen to prevent objects from accidentally getting caught in the fan blades. Some newer models even had a removable lint screen over the air inlet that can be taken off and easily cleaned. At the end of the barrel is another protective screen that prevents anyone from sticking their fingers or other long objects into the heating element.

4 Other components of the hairdryer are designed to ensure its safe operation. Dryers contain a safety cut off switch that prevents the temperature of the drier from exceeding 140°F (60°C). This switch is a bimetallic strip, which is made from a sheet composed of two metals. These metals expand at different rates when they are heated. As the temperature inside the drier increases, the strip will bend one way or another as the metal strips expand at different rates. When a certain temperature is reached the strip bends to appoint where it trips a safety switch that cuts off the power to the drier.

Another safety device is a thermal fuse built into the electrical circuitry. This fuse has a small metal strip that melts if the temperature of the circuit exceeds a certain amount. This breaks the circuit and instantly cuts power to the drier. Both these safety features are designed to prevent overheating and stop a fire from occurring.

A third type of safety control is the Ground Fault Circuit Interrupter (GCFI) that is built into dryers to prevent electrocution, The GCFI senses how much current is flowing through the circuit and can shut it off if it detects a leak or a short-circuit.

5 The dryer components can are put in place on an assembly line using a combination of automated equipment and manual labor. First, the electrical components are fitted into the bottom half of the plastic shell. Once this step has been completed, the top half of the shell is locked into place.

These pins and holes are lined up when the shell hands are assembled.

6 Screws and other fasteners are used to anchor the plastic parts together and hold them in place. Early hair dryers used dozens of screws to lock the shell hands to place. Because of more efficient designs, modern models only require a few key screw components. This helps control cost and reduce assembly time.

7 After assembly warning labels showing that the hair dryer should be kept away from water must be attached. These labels are attached to the cord as required by the Consumer Products Safety Commission guidelines. Once the dryers have been fully assembled they are boxed along with an instruction booklet and additional safety warning materials and are packaged for shipping.

Quality Control

All electrical appliances may be dangerous if misused. Hair dryers are particularly dangerous because they may accidentally dropped into a sink or bathtub full of water. Therefore, special quality control precautions must be taken to reduce the chance of electrocution. Beginning in 1980, manufacturers were required to include a warning picture on hair dryers to show they should not be used near water. This warning label must be permanently attached to the drier cord. In 1985, manufactures began adding a polarized electrical plug that would help ground the appliance and prevent accidental shock. In 1991, products were required to have design feature that prevents the possibility of a short-circuit whether or not the device is turned on.

Modern hair dryers use GFCI to prevent any power flowing into the device when a short-circuit is detected. By the year 2000, recorded deaths due to electrocution by hairdryer had already dropped to less than four a year, and it is anticipated that this additional safety feature will completely prevent accidental electrocutions once all the older hair dryer models have disappeared from the market.

The Future

While the quality of portable hair dryers has improved over the last 70 years there has been little change to their fundamental design. The improvements that have been made have greatly enhanced the safety of the devices and have increased their power nearly 20 fold. It is unlikely that future dryers will be made much more powerful because they already produce close to the theoretical maximum amount of heat that the user can safely be exposed to without danger of burning their hair or skin.

Other future design improvements seem more likely. Dryers can be made to run more quietly or to be easier to hold and operate. Some manufacturers are focussing on improving the appearance of their products and are attempting to please younger uses by making dryers in designer colors or with clear plastic.

It is also possible that technological breakthroughs may make new types of hair dryers possible. In 2000, one manufacturer introduced a new ionic hair dryer which is claimed to use ions, electrically conductive chemical species, to evaporate water faster with less damage to the hair. At the time of this writing it is not yet known is this new type of hair dryer will be successful in the market place.

Perhaps one of them most interesting areas for future development of hair dryers really has nothing to do at all with hair. Imaginative consumers have identified that hair dryers can be used for alternate functions such as removing chewing gum from hard surfaces and helping to dry nail polish. It remains to be seen if manufacturers will respond by marketing hair dryers modified for these special uses.

Where to Learn More

Books

Dalton, John W. *The Professional Cosmetologist.* New York: West Publishing Company, 1985.

Randy Schueller

Hairspray

The first commercial hairsprays were marketed in the late 1940s. These early products used shellac, a natural resin, to hold hair in place.

Background

Hairsprays belong to a class of personal care products that help hair to hold a desired style. These products contain film forming ingredients that are applied as a fine mist. When dry, these chemicals form tiny glue-like spots that hold the hair shafts together. Hairsprays are formulated as aerosols that are powered by pressurized gasses or non-aerosols that are dispensed by manually depressing a pump.

History

Women have used natural compounds (such as clays and gums) to hold their hair in place since antiquity. The modern hairspray was not born until aerosol spray containers were developed by the United States Army during World War II. The army created this technology to enable them to spray insecticides over large areas. Over the next decade or so, aerosol technology spread to other industries such as paints and coatings and personal care. The first commercial hairsprays were marketed in the late 1940s. These early products used shellac, a natural resin, to hold hair in place. In the last 50 years researchers have improved both the quality of hairspray ingredients and the aerosol packaging used to deliver them. Hairsprays have since become an important part of the hair care market.

Raw Materials

Holding agents

Hairsprays work by coating the hair with polymers, which are long chain chemical compounds. Polymers are sometimes called resins because natural resin materials (like shellac) were used in the first hair sprays. The key property that makes polymers useful as hair holding agents is their ability to form films upon drying. Once hair spray is applied to the hair, the liquid drops run down the hair shaft until they reach the intersection of two hair fibers. When the drops dry at this fiber intersection, they create an invisible film that bonds hairs together.

Chemists may choose from many different polymers when developing hair styling products. For example, polyvinylpyrrolidone (also known as PVP) is a common polymer used in hair styling products. However, it does not provide a strong hold and it tends to pick up moisture from the air. To remedy this situation chemists combine PVP with another polymer, vinyl acetate. The resulting mixture, PVPVA, is a copolymer that has improved humidity resistance and will therefore hold curls better. However, if not properly formulated PVPVA copolymers can be so waterproof that they become difficult to wash out of the hair. Another common copolymer is made from vinyl acetate and crotonic acid. This ingredient is popular because it provides the proper balance of hardness, solubility, and moisture susceptibility.

Solvents

Solvents make up the largest portion by weight of an aerosol hairspray. They are used as a carrier for the active ingredients in the formulation and are selected based on their compatibility with the chemical actives and the propellant. Water is a popular solvent due to its low cost. Unfortunately, formulations that contain water take longer to dry and are less soluble in many propellant systems. Water also increases the chance of corrosion inside the can. Ethanol, although somewhat more expensive, is another popular solvent. However, ethanol belongs to the class of volatile organic compounds (VOCs) whose use in aerosols have been restricted because they contribute to air pollution. To date, no acceptable replacement solvents have been approved.

Additives

Hairsprays contain a number of chemical additives in addition to polymers and solvents. For example, plasticizers are added to modify the effects of polymers. These include chemicals such as isopropyl myristate, diethyl phthalate, and silicones that can make hairspray films more flexible and less brittle. Neutralizing and anti-corrosion agents, like aminomethyl propanol (AMP), ammonium hydroxide, morpholine, cyclohexylamine, and borate esters are added to control resin solubility and help prevent the inside of the can from rusting.

Propellants

Propellants, as the name implies, are responsible for propelling the hairspray out of the can. These are gasses that can be stored under low pressure in the can. Originally chlorofluorocarbon gasses (CFCs) were used, but they have been banned due to their suspected complicity in the depletion of the ozone layer in the upper atmosphere. Hydrocarbon propellants like butane and propane were used as replacements for CFCs. These gasses are mixtures of butane and propane designed to deliver a certain amount of pressure in the can. For example, the propellant Butane 40 is a mixture of butane and propane that has a vapor pressure of 40 lb/in^2 (18 kg/cm^2). Hydrocarbon propellants were extremely popular until the 1980s when California and a few other American states began to legislate how much of these gasses could be used in hairsprays because they were shown to contribute to air pollution. This legislation led to decreased use of hydrocarbons.

Scientists working at Dupont developed a new class of propellants. These are known as hydrofluorocarbons (HFCs), and they have many of the properties of CFCs but are not as polluting. Popular HFCs are 1,1,-difluoroethane (Propellant 152A) and 1, 1, 1, 2,-tetrafluoromethane (Propellant 134A). While they are relatively expensive, they are used to formulate fast drying hairsprays.

Packaging

Aerosol hairsprays have traditionally been packaged in containers made from tin plated steel or aluminum. The package is fitted with a valve that both seals the can and dispenses the contents. The valve is connected to a plastic dip tube that carries the liquid product up through the can to the top. The upper portion of the valve is fitted with a button, or actuator, that is pressed to open the valve and release the product.

The valve itself is a very complicated piece of equipment. It consists of a body section with a tail piece that attaches to the diptube. The central part of the valve body serves as a mixing chamber where the propellant and liquid concentrate blend together. This mixing process is very turbulent and helps break the hairspray into very fine mist. At the top of the valve body is a stem that feeds product from the mixing chamber into the dispensing button. The entire valve assembly is housed in a metallic cup which is a ring shaped piece of either aluminum or tinplate steel.

Design

Chemists must consider several key parameters when designing hairspray formulations including efficacy, safety, cost, consumer appeal, and regulatory considerations. The functionality of the product is a key consideration. Formulating chemists have hundreds, if not thousands, of ingredients to choose from. They begin by selecting a resin that will give the desired hold characteristics and combine it with plasticizers and other ingredients that give the proper feel on the hair.

For a polymer to make a good styling resin, it must have certain characteristics. It must deposit a film that is substantive on the hair yet that can still be easily washed away; it must hold hair with flexibility so the hair can move without breaking the film; it must be transparent so it does not reduce the hair's natural gloss; it should not flake when the hair is brushed; and it must not absorb moisture from the atmosphere and become sticky.

Because the ingredients used in hairsprays may be in contact with the skin for an extended period of time, they must be designed to be non-irritating and non-sensitizing. To make sure they are not hazardous when in-

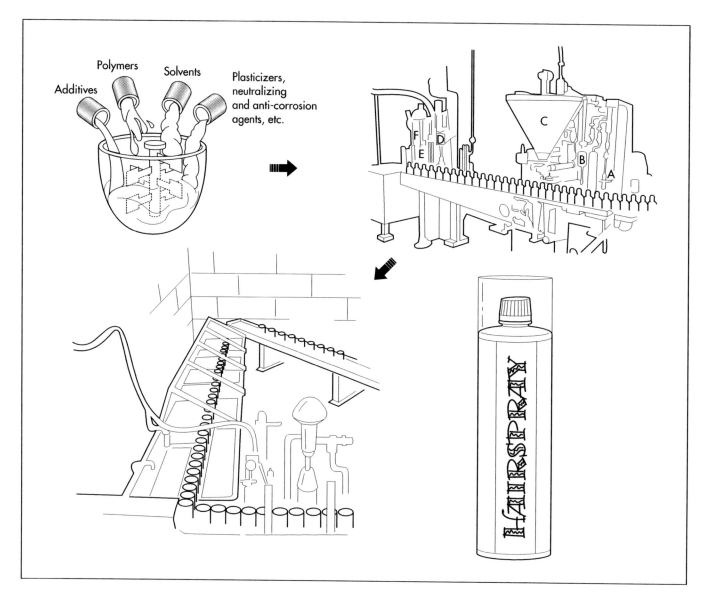

Additives
Polymers
Solvents
Plasticizers, neutralizing and anti-corrosion agents, etc.

A. Compressed air blower. B. Pressure tester. C. Product filler. D. Air evacuator. E. Rotary capper. F. Propellant pressure filler valve.

haled, their medical safety can be determined by animal inhalation testing, although an increasing number of non-animal alternative tests are becoming available. This is less of an issue for non-aerosol products because the particles produced by the pump dispenser are much larger than those produced by pressurized gasses and are therefore less likely to travel into the lungs. For this same reason, non-aerosols can be wetter and stickier than their aerosol counterparts.

Chemists design hairsprays to be more consumer appealing by employing pleasant fragrances and easy to use packaging. Some fragrances are designed to simply cover the odor of the chemicals in the formulation; others are designed to be highly fragrant. Packaging design options also help increase the product's aesthetic appeal. These include easy to use dispensers like toggle valves and large actuator buttons.

Typically hairsprays are designed to deliver 8 oz (237 ml) of product but both smaller and larger sizes are available. The pressure in the can is generally between 40 and 80 lb/in^2 (18 and 36 kg/cm^2) depending on the type of propellant that is used. It is also important to note that hairsprays are designed to have a minimum shelf life of three years but have been known to last five years and longer.

The Manufacturing Process

The manufacture of aerosol hairspray requires highly automated equipment. Because of the cost involved operating this equipment, most hairspray marketers choose to have their products produced by specialized vendors known as contract manufacturers.

Batching

1 During the manufacturing process the formula is prepared as a liquid concentrate in large batch tanks composed of stainless steel, coated aluminum, or fiber glass reinforced polyester. These batches may be as large as 2,000 gal (7,570 l). The tanks are equipped with a large turbine mixer with blades that are several feet in diameter. The solvent is charged into the tank first and then followed by the other ingredients. The solvent makes up the largest proportion of the formula and may be present at 80% by weight or greater. Other ingredients range in concentration from a few tenths of a percent for some of the pH control agents, to a few percent for fragrance, to approximately 10% for the resin. Depending on the solubility of the ingredients in given formula, this mixing step may take as little as 30 minutes or as long as several hours. Since some of the ingredients are in the form of powders, the mixture must be carefully monitored during the batching process to ensure they dissolve properly. After mixing is complete the concentrate is tested to ensure it conforms to specifications and is then transferred to a holding tank prior to filling.

Filling

2 The packaging components are staged on the filling machinery. As the empty cans move down the conveyor belt, a jet of compressed air removes any dirt and dust.

3 At the next stage of the filling line, there are a series of nozzles, known as filling heads, that are connected to tubes that transfer the liquid concentrate from the tank where it is stored. A piston mechanism injects a precise amount of liquid into the can.

4 The cans proceed down the line to the next station where two actions occur at once. The gaseous propellant is shot into the cans and the valve cup is immediately crimped into place. The metal cup is crimped onto the rim on the opening of the can. This tight seal prevents the gas and liquid from leaking out.

5 After the gassing operation the cans are fed through a long trough filled with hot water. As the cans slowly move underwater they are checked visually for escaping bubbles, which would indicate a bad valve seal or a leaky can. Leaking cans are removed during this stage of the operation.

6 After exiting the water-bath, cans are dried by compressed air jets. A cap is placed over the valve at the end of the filling line; this prevents the aerosol from being accidentally activated during shipping.

7 Finally the finished units are packed into boxes and stacked on pallets for shipping.

Byproducts/Waste

The most obvious byproducts of aerosol hairspray are the environmentally unfriendly VOCs emitted into the atmosphere. In addition to depleting the ozone layer, these chemical byproducts are also used in inhalant abuse. Although companies have developed alternatives (such as manual pump sprays), many consumers prefer and continue to use the aerosol can.

Quality Control

During the manufacturing process, samples are pulled from the conveyor line and checked to ensure they meet all specifications. Key properties that are monitored include the fill weight, the concentration of active ingredients, and the pressure of the can. Spray characteristics such as spray rate (the amount of product delivered per second) and the spray pattern (the physical size and shape of the spray) are also monitored. Long-term stability testing is conducted to ensure that the cans do not clog when sprayed and that they remain free of internal corrosion.

The Future

The future of hairsprays depends not only market considerations but also on the regulatory actions of state and federal agencies. Since the 1970s, the industry has struggled with the severe limitations imposed by the government. To circumvent these limitations, researchers continue to experiment with polymers that can be incorporated into non-polluting, water-based formulas. In addition, they are evaluating several non-traditional aerosol delivery systems that release fewer contaminants into the atmosphere. Pump sprays are the most well-known alternative to aerosols, and they operate by the physical force generated when a spring inside the pump is compressed. Other aerosol alternatives include the bag-in-the-can system that uses the physical force of a stretched rubber bladder to spray the contents. Hairspray marketers will have to continue to find cutting edge technology like this to keep pace with an ever changing regulatory environment.

Where to Learn More

Books

Dallal, Joseph, and Colleen Rocafort. *Hair Styling/Fixative Products, in Hair and Hair Care.* Marcel Dekker, 1997.

Schueller, Randy, and Perry Romanowski. *Beginning Cosmetic Chemistry.* Allured Publishing, 1999.

Randy Schueller

Handcuffs

Background

Handcuffs are standard law enforcement and security industry tools used for restraining and controlling dangerous or unreasonable people. Police officers routinely use handcuffs in their work. When using handcuffs, police officers need to employ great care and good judgment. Officers can be civilly or criminally liable for the improper use of handcuffs, especially when injury results.

Handcuffs restrain an individual with handcuffs when the open bracelet is placed upon a body part, usually the wrist, and the ratchet is locked in place. When open, the ratchet pivots freely on a pivot. When closed, the teeth of the ratchet engage the teeth of the spring-loaded pawl located inside the bracelet. The pawl is forced against the ratchet, which locks the two sets of teeth together. To open the handcuff, the pawl must be disengaged from the ratchet teeth. This is accomplished by the handcuff key that unlocks the primary lock.

History

Handcuffs have been used as a means of restraint for several centuries. However, before 1862, they were essentially a "one size fits all" device. These early cuffs, which were simply metal rings that locked in place, created discomfort for people with thick wrists and were ineffective when used on people with thin wrists. That changed in 1862, when W. V. Adams revolutionized the device with the invention of adjustable ratchets that could bind wrists tightly or loosely. An Adams cuff consisted of a square bow with notches on the outside that engaged with a lock mechanism shaped like a teardrop. Several years later, Orson C.

Phelps patented a version of the ratchet handcuff that placed the ratchet notches on the inside of the square bow.

In 1865, an entrepreneur named John Tower used the Adams and Phelps patents to start his own handcuff company. The Tower Company manufactured handcuffs until World War II, and though it always employed the Adams patent, its products set the standards for precision, craftsmanship and effectiveness. The first Tower cuff was based on the Phelps design, with notches on the inside and a three-link chain connecting the two cuffs. The second cuff was more similar to the Adams handcuff. The keyhole location was moved from the side of the lock case to the bottom, and it featured a round bow. Also, it featured three round rings between the cuffs, instead of a chain. The outer two rings were perfectly round while the middle ring was bent, just like the rings on the Adams cuff. In addition, the lock case of this second cuff was smaller. Tower applied for his first handcuff patent in 1871. His design innovation was a round bow meant as an improvement to the square bow of the Phelps and Adams cuffs. The patent was finally issued in 1874.

With the adjustable-fit handcuffs, a design flaw was evident: the spring loaded mechanism that allowed the cuffs to be adjustable also allowed the cuffs to be tampered with so that the restrained person would be able to spring the lock and escape. Tower solved that problem with the introduction of the double lock handcuff, which he patented in 1879. The lock had two settings: a single-lock mode and double-lock mode. The first mode acted just like a single-lock handcuff. The user could turn the key clockwise to set the cuff into double-lock mode, which froze that

Handcuffs have been used as a means of restraint for several centuries. However, before 1862, they were essentially a "one size fits all" device.

177

A standard handcuff set.

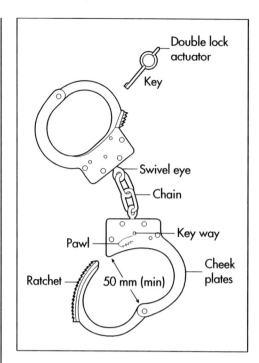

catch or bolt, effectively preventing any escaping. The double lock model set the standard both for effectiveness and engineering.

Throughout the nineteenth and twentieth centuries, the handcuff industry continued to change and handcuff design innovations were implemented. One inventor (E. D. Bean in 1882) created a release button to keep the locking mechanism from engaging until the officer released a button and then the cuffs were locked. This mechanism's purpose was to stop cuffs from closing and locking (during a struggle, for example) before the officer could get the cuffs around the offender's wrists. This problem was more effectively solved by another inventor's patent for a swing-through ratchet. This design ensured that the cuffs would not lock unless placed on a wrist. The resulting Peerless cuffs (patented in 1912) set a new standard for the handcuff industry, and became the model for modern cuffs.

Raw Materials

Raw materials include chrome steel, nickel plating for the handcuff plating, and metal springs.

Design

Handcuffs typically do not vary much from the standard. A standard pair of modern hand-

cuffs weighs no more than 15 oz (425 g). The minimum opening of the bracelet is 2 in (5 cm). The minimum inside perimeter of the bracelet is 7.9 in (20 cm) when the ratchet is engaged at the first notch. The maximum perimeter is 6.5 in (16 cm) when the ratchet is engaged at the last notch. The maximum overall length of the handcuffs is 9.4 in (24 cm).

The Manufacturing Process

1 Handcuff manufacturing starts with the construction of each shackle bracelet. The bracelet consists of three separate parts: the cheek plates, the ratchet (a cheek bar with inclined teeth designed to engage the pawl), and the pawl. The bracelet is made from molten chrome steel poured into a mold and cooled. Once the steel has cooled, it is taken out of the mold.

2 A spring-loaded pivot bar is then constructed inside each bracelet, also made out of steel.

3 The ratchets are constructed in the same format as the cheek plate. The rotatable shackle bracelet, including ratchet teeth, is attached to the shackle base at a protruding end of a shackle base. The ratchet is designed to advance in only one direction.

4 The shackle base is also molded from molten steel and includes a flat keyhole.

5 The shackle bracelet is jointed to the protruding shackle end at a pivot. This permits the shackle bracelet to rotate about the pivot relative to the shackle base.

6 Next, the shackle bracelet end is shaped and dimensioned.

7 The two ring-shaped shackle bracelets are then connected to a short chain. The chain is welded at either end to the shackle base of each handcuff.

8 At the conclusion of the manufacturing process, the handcuffs are marked with the manufacturer's name or trademark, model number, and serial number.

Quality Control

According to National Institute of Justice Standards, a finished pair of handcuffs must be free of 16 defects to be deemed acceptable for use. Defects include corrosion, broken or loose parts, or cracked or incomplete welding.

Finished handcuffs are put through a variety of tests to ensure practicality. They are blasted with salt for 12 hours. After that time the product should not have severely corroded or discolored, and they should function as normal. Handcuff standards also dictate that a handcuffs cannot be opened when a tensile force of 495 lbf is applied for a minimum of 30 seconds.

Parts cannot be missing, broken, malformed, loose, or not in proper alignment. Rivets and pins must be secure. The rivets and pins must be free of any burrs, slivers, sharp edges, dents, or tool marks, and the metal must not be split or cracked. The end of the pin must be set below the exposed surface of the plate. Welding must be complete and free of cracks.

Manufacture markings must be present, visible, legible, correct, and permanent. The key must be able to unlock the handcuff. The handcuff must be able to be double-locked. Also, it should require no force to remove the handcuff. The openings and closings of the handcuffs must function properly.

Byproducts/Waste

In the manufacturing of handcuffs, there is not much waste. Any defective steel can be either recycled or melted down and remolded. Waste from the salt testing is minimal. Most of the salt is reused.

The Future

As the technology of metals and steels grow, so will the evolution of handcuffs. The durability of handcuffs will increase as will the locks. Criminals will no longer be able to pick the locks or maneuver out of them. New device are being used, but it is doubtful that they will make handcuffs obsolete. Police are using a variation of plastic ties to secure criminals. The tie is made of tough, high quality plastic and securely wraps around the person's wrist. To remove this tie, it is simply cut off. Also, handcuffs are considered a more humane way to detain criminals, as opposed to rubber bullets and sprays.

Where to Learn More

Books

Harris, James. *A Study of Handcuff Improvements*. 1989.

Peters, John G. *Tactical Handcuffing for Chain and Hinged Style Handcuffs*. Ventura, CA: Reliapon Police Products, Incorporated, 1989.

Other

Brave, M. A., and J. G. Peters Jr. *Liability Assessments and Awareness International, Inc. Web Page*. December 2001. <http://www.laaw.com>.

Lauher, Joseph W. *Handcuffs.org Web Page*. December 2001. <http://www.handcuffs.org>.

Dan Harvey

Headstone

By 2015, cremation is expected to be the preferred method over burials in Canada.

Background

Headstones are known by many different names, such as memorial stones, grave markers, gravestones, and tombstones. All of which apply to the function of headstones; the memorialization and remembrance of the deceased. Headstones were originally made from fieldstones or pieces of wood. In some localities, stones (referred to as "wolf stones") were placed over the body to prevent scavenging animals from uncovering a shallow grave.

History

Archeologists have found Neanderthal graves that date back 20,000–75,000 years. The bodies have been discovered in caves with large rock piles or boulders covering the openings. It is thought that these grave sites were accidental. The wounded or dying had probably been left behind to recover, and the rocks or boulders were pushed in front of the cave for protection from wild animals. The Sharindar Cave in Iraq was home to the remains of a person (c. 50,000 B.C.) with flowers strewn about the body.

Various other methods of burial have developed as time moved on. The Chinese were the first to use coffins to contain their dead some time around 30,000 B.C. Mummification and embalming were used about 3200 B.C. to preserve the bodies of the Egyptian pharaohs for the afterlife. The pharaohs would be placed in a sarcophagus and entombed with statues representing their servants and trusted advisors, as well as gold and luxuries to ensure their acceptance in the world beyond. Some kings required that their actual servants and advisors accompany them in death, and the servants and advi-

sors were killed and placed in the tomb. Cremation, which started about the same time as mummification, was also a popular method of disposing of the dead. Today it accounts for 26% of disposal methods in the United States and 45% in Canada.

As religions developed, cremation came to be looked down upon. Many religions even banned cremation, claiming it was reminiscent of pagan rituals. Burial was the preferred method, and sometimes the dead were laid out for days in the home so people could pay their respects. In 1348, the Plague hit Europe and forced people to bury the dead as soon as possible and away from the cities. These death and burial rituals continued until cemeteries were overflowing and, due to the numerous shallow graves, continuing to spread disease. In 1665, the English Parliament ruled in favor of having only small funerals and the legal depth of graves was made to stand at 6 ft (1.8 m). This decreased the spread of disease, but many cemeteries continued to be overpopulated.

The first cemetery similar to those seen today, was established in Paris in 1804 and called a "garden" cemetery. The Père-Lachaise is home to many famous names such as Oscar Wilde, Frederick Chopin, and Jim Morrison. It was in these garden cemeteries that the headstone and memorials became elaborate works. One's social status determined the size and artistry of the memorial. Early memorials depicted horrible scenes with skeletons and demons to instill fear of the afterlife in the living. Later in the nineteenth century, headstones evolved in favor of peaceful scenes, such as cherubs and angels leading the deceased upward. The United States established its own rural cemetery, The Mount Auburn Cemetery in Cambridge, Massachusetts, in 1831.

Raw Materials

Early headstones were made out of slate, which was available locally in early New England. The next material to become popular was marble, but after time the marble would erode and the names and particulars of the deceased were indecipherable. By 1850, granite become the preferred headstone material due to its resilience and accessibility. In modern memorials granite is the main raw material used.

Granite is an igneous rock composed primarily of quartz, feldspar, and plagioclase feldspar with other small bits of minerals mixed in. Granite can be white, pink, light gray, or dark gray. This rock is made from magma (molten material) that is slowly cooled. The cooled magma is unearthed through shifts in the earth's crust and erosion of soil.

Design

There are countless ways to personalize a headstone. Epitaphs range from scripture quotes to obscure and humorous statements. Accompanying statuettes can be carved into, placed on top of, or beside the stone. Size and shape of headstones also varies. Generally, all stones are machine polished and carved, then finely detailed by hand.

The Manufacturing Process

1 The first step is to choose the type (typically marble or granite) and color of the stone. The granite block is then cut from the bedrock. There are three ways of doing this. The first method is drilling. This method uses a pneumatic drill that bores vertical holes 1 in (2.54 cm) apart and 20 ft (6.1 m) deep into the granite. The quarrymen then use 4 in (10.1 cm) long steel bits that have steel teeth to cut away at the core of the rock.

Jet piercing is much faster than drilling, about seven times so. In this method, 16 ft (4.9 m) can be quarried in one hour. The process uses a rocket motor with a hollow steel shaft to expel a blend of pressurized hydrocarbon fuel and air in the form of a 2,800°F (1,537.8°C) flame. This flame is five times the speed of sound and cuts 4 in (10.2 cm) into the granite.

The third way is the most efficient method, quieter, and produces almost no waste. Water jet piercing employs water pressure to cut the granite. There are two systems of water jet piercing, low pressure and high pressure. Both emit two streams of water, but the low pressure system streams are under 1,400–1,800 psi, and the high pressure streams are under 40,000 psi. The water from the jets is reused, and the method minimizes the mistakes and wasted material.

2 The next step is to remove the block from the quarry bed. Workers take large pneumatic drills tipped with 1.5–1.88 in (3.81–4.78 cm) steel bits tipped with carbide and drill horizontally into the block of granite. They then place paper-wrapped blasting charges into the holes. Once the charges are set, the block makes a clean break from the rest of the rock.

3 Granite blocks are usually about 3 ft (0.9 m) wide, 3 ft (0.9 m) high, and 10 ft (3 m) long, weighing about 20,250 lb (9,185 kg). Workers either loop a cable around the block or drill hooks into either end and attach the cable to the hooks. In both ways the cable is attached to a large derrick that lifts the granite block up and onto a flatbed truck that transports it to the headstone manufacturer. The quarries tend to be independently owned and sell the granite to manufacturers, but there are some larger companies that own quarries.

4 After arriving at the manufacturing house, the granite slabs are unloaded onto a conveyor belt where they are cut into smaller slabs. The slabs are generally 6, 8, 10, or 12 in (15.2, 20.3, 25, and 30.4 cm, respectively) thick. This step is done with a rotary diamond saw. The saw is equipped with a 5 ft (1.5 m) or 11.6 ft (3.54 m) solid steel diamond blade. The blade usually has about 140–160 industrial diamond segments and has the ability to cut an average of 23–25 ft² (2.1–2.3 m²) an hour.

5 The cut slabs are passed under a varying number of rotating heads (usually eight to 13) with differing levels of grit arranged

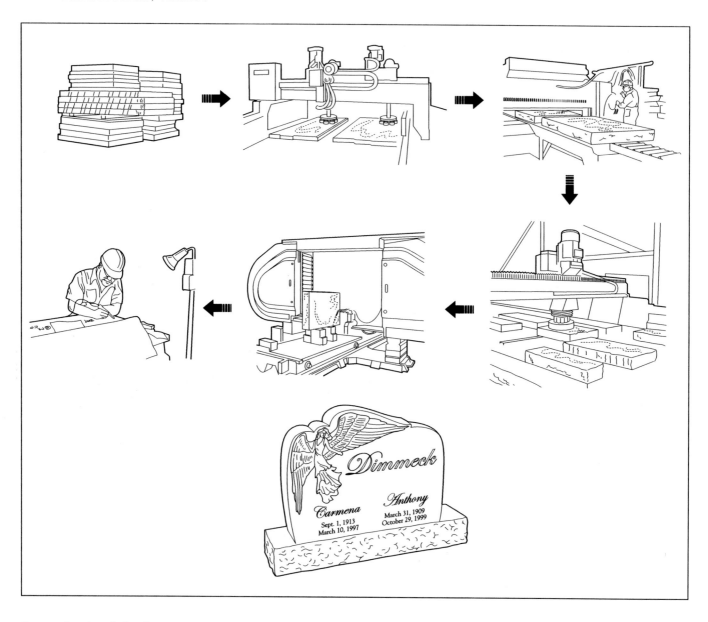

The manufacturing of a headstone.

from the most abrasive to the least. The first few heads have a harsh diamond grit, the middle heads are for honing, and the last few heads are equipped with felt buffer pads. These pads have water and aluminum or tin oxide powder on them to polish the stone to a smooth, glossy finish.

6 The polished slab is then moved along the conveyor belt to the hydraulic breaker. The breaker is equipped with carbide teeth that exert close to 5,000 psi of hydraulic pressure on the granite slab, making a vertical cut through the stone.

7 The cut stone is then fashioned into the appropriate shape. This is either done by hand with a chisel and hammer, or more

precisely with a multi-blade diamond saw. This machine can be set to hold up to 30 blades, but usually is only loaded with eight or nine. Equipped with nine blades, this multi-bladed diamond saw can cut 27 ft^2 (2.5 m^2) an hour.

8 The surfaces of the stone are then polished again. In a highly automated process, 64 pieces can be polished at a time.

9 The vertical edges are polished by an automated polishing machine, similar to the surface polisher. This machine chooses the harshest grit head and works it across the vertical edges of the stone. The machine then works its way through the other grits until the edges are smooth.

10 The radial edges are ground and polished at the same time using two diamond grinding drums. One has a harsh grit diamond, and the second has a finer grit. The stone's radial edges are then polished.

11 If intricate stone shapes are needed, the polished stone is moved to the diamond wire saw. The operator adjusts the saw and starts the process, which uses computer software to etch the shapes into the headstone. Any fine etching or detailing is finished by hand.

12 The headstone is then ready for finishing. Rock Pitching entails chiseling the outer edges of the stone by hand, giving a more defined, personal shape.

13 Now that the headstone is polished and shaped, it is time for the engraving. Sandblasting is generally used. A liquid glue is applied to the headstone. A rubber stencil is applied over the glue and then covered with a carbon-backed layout of the design. The carbon transfers the design prepared by the draftsman, onto the rubber stencil. The worker then cuts out the letters and design features that are wanted on the stone, exposing them to the sandblasting. The sandblasting is either manually done or automated. Either method is done in an enclosed area due to the dangers of the process. The worker is entirely covered to be protected from the grains reflected off the stone. The course cutting abrasive is exerted at a force of 100 psi. Dust collectors collect and save the dust for reuse.

14 The stone is then sprayed with high pressure steam to get rid of any leftover stencil or glue. It is again polished and closely inspected, then packaged in cellophane or heavy paper to protect the finish. The package is placed in crates and shipped to the customer or funeral director.

Quality Control

Quality control is strongly enforced throughout the manufacturing process. Each slab of rough granite is checked for color consistency. After each polishing step, the head stone is examined for flaws. At the first sign of a chip or scratch, the stone is taken off the line.

Byproducts/Waste

Depending on the cutting process used at the quarry, waste varies. Drilling is the least precise method of quarrying, thus producing the most waste. The water jet method produces the least amount of noise pollution and dust. It is also more fuel efficient than the other processes, and enables the water to be recycled. In sandblasting there is little waste also since the sand particles are collected and reused too. Any defective granite stones from the manufacture are generally sold off to other manufacturing companies or exported overseas. Other substandard stones are discarded.

The Future

There are many new techniques that use innovative software to etch designs on headstones. Laser etching is an upcoming development that allows pictures and more intricate designs to be put on the headstone using a laser beam. The heat from the laser pops the crystals on the surface of the granite, resulting in a elevated, light-colored etching.

The depletion of granite is not foreseeable in the near future. As quarries are mined, new resources develop. There are many regulations that limit the amount of granite that can be exported at a time. Alternative methods of the disposal of the dead are also factors that may limit the production of headstones. In 2015, cremation is expected to be the preferred method over burials in Canada.

Where to Learn More

Other

Cold Spring Granite Brochure. 17 October 2001. <http://www.coldspringgranite.com>.

Elberton Granite Association, Inc. *Elberton Granite: The Quarrying and Manufacturing Process.* 19 October 2001. <http://www.egaonline.com/index.htm>.

Monumnet Builders of North America. *The Monument Industry Certification Manual.* July 1993. <http://www.monumentbuilders.org>.

Rock of Ages Web Page. 17 October 2001. <http://www.rockofages.com>.

Deirdre S. Blanchfield

Ice Resurfacing Machine

Ice resurfacing machines are widely known by the brand name Zamboni.

Background

An ice resurfacing machine shaves the ice, removes the shavings, washes and squeegees the ice, and has enough carrying capacity to clear the ice surface in one run, making it completely smooth. Ice resurfacing machines are widely known by the brand name Zamboni.

The main component of an ice resurfacing machine is the sled (also known as the conditioner). The sled houses the blade, which shaves the ice, and the towel, which acts as a squeegee. It also distributes the wash water and the fresh water onto the ice. When the sled unit is lowered to the ice the machine is in operative condition. As the machine is driven forward, the blade takes a light shaving cut off of the ice to clean the surface. The shaved ice is moved toward the center of the sled unit by the worm screw conveyor, picked up by the chain conveyor, and dumped into the snow box. The spreader receives water from the tank. Surplus water is picked up from the ends of the spreader and discharged into the snow box. These ice machines can only get up to a speed of 9 mph (14.5 km/h).

History

Before Frank J. Zamboni invented the original ice resurfacing machine, he and his brother started up an ice block business. They manufactured ice for storage and transport. Building on his knowledge of refrigeration, in 1934 Zamboni opened a 20,000 ft^2 (1,858 m^2) open-air ice skating rink in Paramount, California. Soon the weather and use began to wear away at the ice, so Zamboni installed a dome to cover the rink. Still the ice would deteriorate quickly. The ice-resurfacing process at the time required three to five workers to follow behind a tractor and shave the ice surface, shovel the shavings, wash down the ice, push the dirty water off the surface by hand using large rubber squeegees, and then wait for the ice to dry. The entire process took about an hour.

Some time around 1942, Zamboni began modifications on the tractor. He experimented with different blades that would scrape the ice and suction it up at the same time. Soon Zamboni's tractor was able to scrape the ice, remove the shavings, and leave a trail of water that would quickly freeze, repairing any damage to the original ice surface. By 1949, the Zamboni ice resurfacing machine was in widespread use.

Original models of ice resurfacing machines were built on the top of four-wheel drive vehicles. As more space was needed for water and snow, just the chassis (the frame) was used. Twenty to 30 chassis (built to order) were lined up, the roof portions cut off, and all of the new ice-resurfacing parts added by hand. In 1964, the dump tank was invented to simplify the removal of the snow from the machine. The dump tank stores the excess ice scrappings in a tank similar to a dump truck. When it is full, the ice resurfacing machine leaves the ice rink and dumps the shavings outside.

Raw Materials

The main three raw materials used in the production of an ice resurfacing machine are plastic for the housing, rubber for the hoses and towel, and steel for the conveyors, tanks, and other miscellaneous parts.

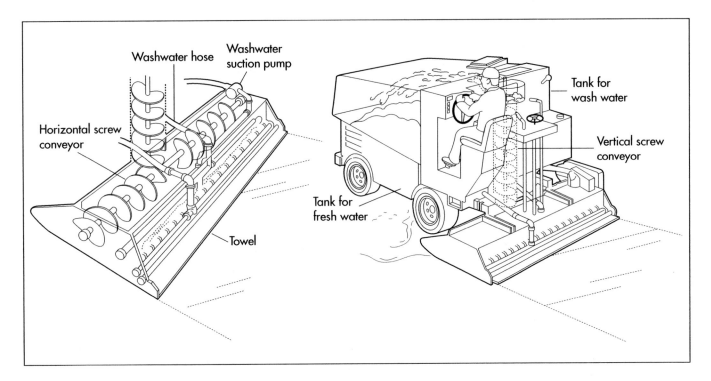

The main components of an ice resurfacing machine are mostly contained in the sled: a horizontal screw conveyor, wash water hose, wash water suction pump, the blade, and the towel. Other components are directly connected to the sled (also knows as the conditioner), such as the horizontal screw conveyor and the separate tanks which hold the wash water and the fresh water. The rest of the machine has similar components to an automotive vehicle, such as tires, metal, and a range of plastic materials. The frame of the device is welded steel with metals and rubber components.

Design

There are several different ice resurfacing models that vary in size, with smaller machines designed for small ponds, and the largest machines designed for speed-skating rinks. The engine of the machine also differs. There are ice-resurfacing machines that are electric and several models that are primarily powered by propane.

Many of the components of the machine are patented and each can vary its specified design. The standard machine used on professional hockey rinks uses horizontal and vertical screw conveyors that are 10 in (25.4 cm) in diameter. It can hold a total of 264 gal (1,000 l) of water. The standard machines are usually about 13 ft (4.1 m) long and 7.1 ft (2.16 m) high. The shaving blade is $0.5 \times 5 \times 77$ in ($1.27 \times 12.7 \times 195.6$ cm). A full machine weighs 8,780 lb (3,983 kg). Some of the towels are able to get to varying levels of proximity to the ice. Some ice resurfacers come equipped with a dump tank, others simply push the ice shavings off the rink to a designated spot outside. The machine operates off of a 2.5 liter engine.

The Manufacturing Process

1 The first step is the manufacturing of the engine and chassis. Typically, both are manufactured from an outside automobile plant and shipped to the ice resufacer plant. A metal frame is then welded to the chassis. The frame will hold the water tanks.

2 The water tanks are made out of injection molded plastic. Plastic pellets are melted in a hopper. The liquid plastic is injected into the die, which is shaped like the water tank. After the die cools, the molded plastic is ejected.

3 The tanks are then fit into the metal frame.

4 The tanks and fenders are bolted onto the chassis. Tanks stored underneath the bin

and in front of the driver store water for conditioning and cleaning the ice.

5 The sled is made from two transversely extending angle irons joined to end plates. The bottom edges of the end plates are bent inwardly to form runners on the ice.

6 Then, mounted on the rear of the unit, perpendicular to the direction of travel is the conditioner with a blade bolted to the underside of an inverted T-shaped beam. The blade shaves a thin layer of ice as screw conveyors remove the shavings.

7 The worm screw conveyor has two oppositely wound helix blades mounted on the inside face of the end plates. A sprocket wheel is secured to the shaft, and paddles pick up shaved ice pushed in to the center by the worm screw conveyor. At the top of the housing is another sprocket mounted on a shaft and the conveyor chain passes over the last sprocket and leaves the structure. Side wings keep the snow from falling away from the paddles until the snow passes over the snow box and drops into the snow box.

8 The spreader is connected to the sled unit by two slings that hang from the ends of rear extending arms. The arms are fixed to opposite ends of a rock shaft which is held in place by bearings mounted on the back of the unit.

9 The towel is attached to the sled. The towel is a thick rubber that arrives at the plant in sheets. The sheets are placed on a conveyor belt and passed under an automatic punch press. The hydraulic press has the desired shape imbedded in steel. The press then punches down and cuts through the rubber.

10 The towel is fit into the sled and bolted into place. It sprays the water from holes in a discharge pipe at 180°F (82°C). The large bin in front of the ice resurfacing machine catches the ice shavings scraped by the blade in the conditioner.

11 The final step is to manufacture the housing of the ice resurfacing machine. The housing is also injection molded. Once it has cooled sufficiently, it is attached to the machine and bolted into place.

Quality Control

During the manufacturing, each part is tested separately for damages. The tanks are filled and checked for leaks, and the blade is checked for sharpness. After the product has been assembled, it is completely filled and also checked for any leakage. The machine is then tested on ice to see if a sufficient amount is shaven and enough water is being distributed.

Byproducts/Waste

When defects are found in the plastics or rubbers, they are either sent back to the manufacturer or melted down and remolded. This can only happen a limited amount of times before the quality of the product will be effected. Any defects found in the engine or chassis are sent back to the manufacturer to be corrected.

If any defects are found after the machine has been sold, it is returned to the plant. When the defects are too severe to be fixed, undamaged parts are salvaged and the rest of the machine is scrapped.

The Future

Ice resurfacing machines continue to evolve additional features to fulfill the demands of maintaining ice surfaces. Portable models are available for smaller rinks and ponds. Automatic machines are in the works that will not require a driver, but will be equipped with sensors that can detect how close the machine is to the boards. It is doubtful that this machine will be widely available for many years.

Where to Learn More

Other

The Zamboni Web Page. December 2001. <http://www.zamboni.com>.

United States Patent and Trademark Office Web Page. December 2001. <http://www. uspto.gov/patft/index.html>.

Zamboni, Richard F. "Ice-Resurfacing Machines." *Scientific American Web Page.* December 2001. <http://www.sciam.com>.

Bonny P. McClain

Insulin

Background

Insulin is a hormone that regulates the amount of glucose (sugar) in the blood and is required for the body to function normally. Insulin is produced by cells in the pancreas, called the islets of Langerhans. These cells continuously release a small amount of insulin into the body, but they release surges of the hormone in response to a rise in the blood glucose level.

Certain cells in the body change the food ingested into energy, or blood glucose, that cells can use. Every time a person eats, the blood glucose rises. Raised blood glucose triggers the cells in the islets of Langerhans to release the necessary amount of insulin. Insulin allows the blood glucose to be transported from the blood into the cells. Cells have an outer wall, called a membrane, that controls what enters and exits the cell. Researchers do not yet know exactly how insulin works, but they do know insulin binds to receptors on the cell's membrane. This activates a set of transport molecules so that glucose and proteins can enter the cell. The cells can then use the glucose as energy to carry out its functions. Once transported into the cell, the blood glucose level is returned to normal within hours.

Without insulin, the blood glucose builds up in the blood and the cells are starved of their energy source. Some of the symptoms that may occur include fatigue, constant infections, blurred eye sight, numbness, tingling in the hands or legs, increased thirst, and slowed healing of bruises or cuts. The cells will begin to use fat, the energy source stored for emergencies. When this happens for too long a time the body produces ketones, chemicals produced by the liver. Ketones can poison and kill cells if they build up in the body over an extended period of time. This can lead to serious illness and coma.

People who do not produce the necessary amount of insulin have diabetes. There are two general types of diabetes. The most severe type, known as Type I or juvenile-onset diabetes, is when the body does not produce any insulin. Type I diabetics usually inject themselves with different types of insulin three to four times daily. Dosage is taken based on the person's blood glucose reading, taken from a glucose meter. Type II diabetics produce some insulin, but it is either not enough or their cells do not respond normally to insulin. This usually occurs in obese or middle aged and older people. Type II diabetics do not necessarily need to take insulin, but they may inject insulin once or twice a day.

There are four main types of insulin manufactured based upon how soon the insulin starts working, when it peaks, and how long it lasts in the body. According to the American Diabetes Association, rapid-acting insulin reaches the blood within 15 minutes, peaks at 30–90 minutes, and may last five hours. Short-acting insulin reaches the blood within 30 minutes, it peaks about two to four hours later and stays in the blood for four to eight hours. Intermediate-acting insulin reaches the blood two to six hours after injection, peaks four to 14 hours later, and can last in the blood for 14–20 hours. And long-acting insulin takes six to 14 hours to start working, it has a small peak soon after, and stays in the blood for 20–24 hours. Diabetics each have different responses to and needs for insulin so there is no one type that works best for everyone. Some insulin is

In the 1980s, researchers used genetic engineering to manufacture a human insulin. In 1982, the Eli Lilly Corporation produced a human insulin that became the first approved genetically engineered pharmaceutical product.

Frederick Banting.

In 1891, Frederick Banting was born in Alliston, Ontario. He graduated in 1916 from the University of Toronto medical school. After Medical Corps service in World War I, Banting became interested in diabetes and studied the disease at the University of Western Ontario.

In 1919, Moses Barron, a researcher at the University of Minnesota, showed blockage of the duct connecting the two major parts of the pancreas caused shriveling of a second cell type, the acinar. Banting believed that by tying off the pancreatic duct to destroy the acinar cells, he could preserve the hormone and extract it from islet cells. Banting proposed this to the head of the University of Toronto's Physiology Department, John Macleod. Macleod rejected Banting's proposal, but supplied laboratory space, 10 dogs, and a medical student, Charles Best.

Begining in May 1921, Banting and Best tied off pancreatic ducts in dogs so the acinar cells would atrophy, then removed the pancreases to extract fluid from islet cells. Meanwhile, they removed pancreases from other dogs to cause diabetes, then injected the islet cell fluid. In January 1922, 14 year-old Leonard Thompson became the first human to be successfully treated for diabetes using insulin.

Best received his medical degree in 1925. Banting insisted Best also be credited, and almost turned down his Nobel Prize because Best was not included. Best became head of the University of Toronto's physiology department in 1929 and director of the university's Banting and Best Department of Medical Research after Banting's death in 1941.

sold with two of the types mixed together in one bottle.

History

If the body does not produce any or enough insulin, people need to take a manufactured version of it. The major use of producing insulin is for diabetics who do not make enough or any insulin naturally.

Before researchers discovered how to produce insulin, people who suffered from Type I diabetes had no chance for a healthy life. Then in 1921, Canadian scientists Frederick G. Banting and Charles H. Best successfully purified insulin from a dog's pancreas. Over the years scientists made continual improvements in producing insulin. In 1936, researchers found a way to make insulin with a slower release in the blood. They added a protein found in fish sperm, protamine, which the body breaks down slowly. One injection lasted 36 hours. Another breakthrough came in 1950 when researchers produced a type of insulin that acted slightly faster and does not remain in the bloodstream as long. In the 1970s, researchers began to try and produce an insulin that more mimicked how the body's natural insulin worked: releasing a small amount of insulin all day with surges occurring at mealtimes.

Researchers continued to improve insulin but the basic production method remained the same for decades. Insulin was extracted from the pancreas of cattle and pigs and purified. The chemical structure of insulin in these animals is only slightly different than human insulin, which is why it functions so well in the human body. (Although some people had negative immune system or allergic reactions.) Then in the early 1980s biotechnology revolutionized insulin synthesis. Researchers had already decoded the chemical structure of insulin in the mid-1950s. They soon determined the exact location of the insulin gene at the top of chromosome 11. By 1977, a research team had spliced a rat insulin gene into a bacterium that then produced insulin.

In the 1980s, researchers used genetic engineering to manufacture a human insulin. In 1982, the Eli Lilly Corporation produced a human insulin that became the first approved genetically engineered pharmaceutical product. Without needing to depend on animals, researchers could produce genetically engineered insulin in unlimited supplies. It also did not contain any of the animal contaminants. Using human insulin also took away any concerns about transferring any potential animal diseases into the insulin. While companies still sell a small amount of insulin produced from animals—mostly porcine—from the 1980s onwards, insulin users increasingly moved to a form of human insulin created through recombinant DNA technology. According to the Eli Lilly Corporation, in 2001 95% of insulin users in most parts of the world take some form of human insulin. Some companies have stopped producing animal insulin completely. Companies are focusing on synthesizing human insulin and insulin analogs, a modification of the insulin molecule in some way.

Raw Materials

Human insulin is grown in the lab inside common bacteria. *Escherichia coli* is by far the most widely used type of bacterium, but yeast is also used.

Researchers need the human protein that produces insulin. Manufacturers get this through an amino-acid sequencing machine that synthesizes the DNA. Manufacturers know the exact order of insulin's amino acids (the nitrogen-based molecules that line up to make up proteins). There are 20 common amino acids. Manufacturers input insulin's amino acids, and the sequencing machine connects the amino acids together. Also necessary to synthesize insulin are large tanks to grow the bacteria, and nutrients are needed for the bacteria to grow. Several instruments are necessary to separate and purify the DNA such as a centrifuge, along with various chromatography and x-ray crystallography instruments.

The Manufacturing Process

Synthesizing human insulin is a multi-step biochemical process that depends on basic recombinant DNA techniques and an understanding of the insulin gene. DNA carries the instructions for how the body works and one small segment of the DNA, the insulin gene, codes for the protein insulin. Manufacturers manipulate the biological precursor to insulin so that it grows inside simple bacteria. While manufacturers each have their own variations, there are two basic methods to manufacture human insulin.

Working with human insulin

1 The insulin gene is a protein consisting of two separate chains of amino acids, an A above a B chain, that are held together with bonds. Amino acids are the basic units that build all proteins. The insulin A chain consists of 21 amino acids and the B chain has 30.

2 Before becoming an active insulin protein, insulin is first produced as preproinsulin. This is one single long protein chain with the A and B chains not yet separated, a section in the middle linking the chains together and a signal sequence at one end telling the protein when to start secreting outside the cell. After preproinsulin, the chain evolves into proinsulin, still a single chain but without the signaling sequence. Then comes the active protein insulin, the protein without the section linking the A and B chains. At each step, the protein needs specific enzymes (proteins that carry out chemical reactions) to produce the next form of insulin.

STARTING WITH A AND B

3 One method of manufacturing insulin is to grow the two insulin chains separately. This will avoid manufacturing each of the specific enzymes needed. Manufacturers need the two mini-genes: one that produces the A chain and one for the B chain. Since the exact DNA sequence of each chain is known, they synthesize each mini-gene's DNA in an amino acid sequencing machine.

4 These two DNA molecules are then inserted into plasmids, small circular pieces of DNA that are more readily taken up by the host's DNA.

5 Manufacturers first insert the plasmids into a non-harmful type of the bacterium *E. coli*. They insert it next to the *lacZ* gene. LacZ encodes for ß-galactosidase, a gene widely used in recombinant DNA procedures because it is easy to find and cut, allowing the insulin to be readily removed so that it

does not get lost in the bacterium's DNA. Next to this gene is the amino acid methionine, which starts the protein formation.

6 The recombinant, newly formed, plasmids are mixed up with the bacterial cells. Plasmids enter the bacteria in a process called transfection. Manufacturers can add to the cells DNA ligase, an enzyme that acts like glue to help the plasmid stick to the bacterium's DNA.

7 The bacteria synthesizing the insulin then undergo a fermentation process. They are grown at optimal temperatures in large tanks in manufacturing plants. The millions of bacteria replicate roughly every 20 minutes through cell mitosis, and each expresses the insulin gene.

8 After multiplying, the cells are taken out of the tanks and broken open to extract the DNA. One common way this is done is by first adding a mixture of lysozome that digest the outer layer of the cell wall, then adding a detergent mixture that separates the fatty cell wall membrane. The bacterium's DNA is then treated with cyanogen bromide, a reagent that splits protein chains at the methionine residues. This separates the insulin chains from the rest of the DNA.

9 The two chains are then mixed together and joined by disulfide bonds through the reduction-reoxidation reaction. An oxidizing agent (a material that causes oxidization or the transfer of an electron) is added. The batch is then placed in a centrifuge, a mechanical device that spins quickly to separate cell components by size and density.

10 The DNA mixture is then purified so that only the insulin chains remain. Manufacturers can purify the mixture through several chromatography, or separation, techniques that exploit differences in the molecule's charge, size, and affinity to water. Procedures used include an ion-exchange column, reverse-phase high performance liquid chromatography, and a gel filtration chromatography column. Manufacturers can test insulin batches to ensure none of the bacteria's *E. coli* proteins are mixed in with the insulin. They use a marker protein that lets them detect *E. coli* DNA. They can then determine that the purification process removes the *E. coli* bacteria.

PROINSULIN PROCESS

11 Starting in 1986, manufacturers began to use another method to synthesize human insulin. They started with the direct precursor to the insulin gene, proinsulin. Many of the steps are the same as when producing insulin with the A and B chains, except in this method the amino acid machine synthesizes the proinsulin gene.

12 The sequence that codes for proinsulin is inserted into the non-pathogenic *E. coli* bacteria. The bacteria go through the fermentation process where it reproduces and produces proinsulin. Then the connecting sequence between the A and B chains is spliced away with an enzyme and the resulting insulin is purified.

13 At the end of the manufacturing process ingredients are added to insulin to prevent bacteria and help maintain a neutral balance between acids and bases. Ingredients are also added to intermediate and long-acting insulin to produce the desired duration type of insulin. This is the traditional method of producing longer-acting insulin. Manufacturers add ingredients to the purified insulin that prolong their actions, such as zinc oxide. These additives delay absorption in the body. Additives vary among different brands of the same type of insulin.

Analog insulin

In the mid 1990s, researchers began to improve the way human insulin works in the body by changing its amino acid sequence and creating an analog, a chemical substance that mimics another substance well enough that it fools the cell. Analog insulin clumps less and disperses more readily into the blood, allowing the insulin to start working in the body minutes after an injection. There are several different analog insulin. Humulin insulin does not have strong bonds with other insulin and thus, is absorbed quickly. Another insulin analog, called Glargine, changes the chemical structure of the protein to make it have a relatively constant release over 24 hours with no pronounced peaks.

Instead of synthesizing the exact DNA sequence for insulin, manufacturers synthesize an insulin gene where the sequence is slightly altered. The change causes the resulting

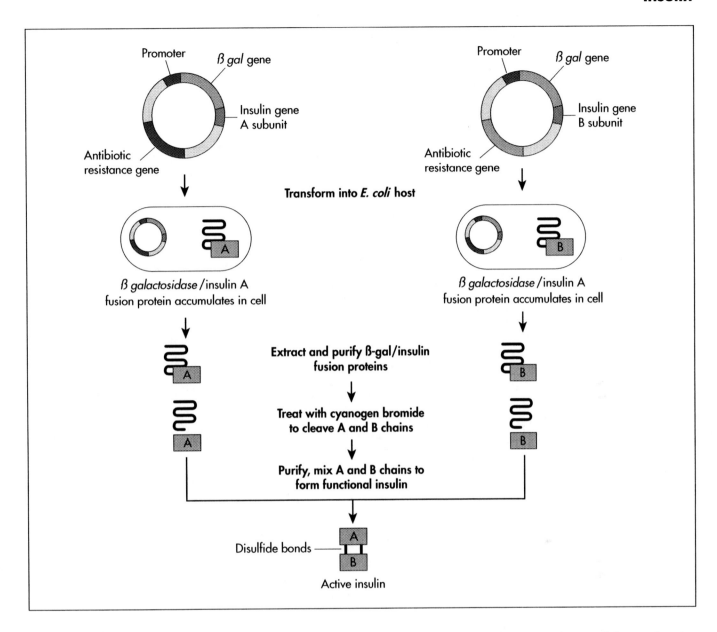

A diagram of the manufacturing steps for insulin.

proteins to repel each other, which causes less clumping. Using this changed DNA sequence, the manufacturing process is similar to the recombinant DNA process described.

Quality Control

After synthesizing the human insulin, the structure and purity of the insulin batches are tested through several different methods. High performance liquid chromatography is used to determine if there are any impurities in the insulin. Other separation techniques, such as X-ray crystallography, gel filtration, and amino acid sequencing, are also performed. Manufacturers also test the vial's packaging to ensure it is sealed properly.

Manufacturing for human insulin must comply with National Institutes of Health procedures for large-scale operations. The United States Food and Drug Administration must approve all manufactured insulin.

The Future

The future of insulin holds many possibilities. Since insulin was first synthesized, diabetics needed to regularly inject the liquid insulin with a syringe directly into their bloodstream. This allows the insulin to enter the blood immediately. For many years it was the only way known to move the intact insulin protein into the body. In the 1990s, researchers began to make inroads in synthesizing various devices and forms of insulin that diabetics can use in an alternate drug delivery system.

Manufacturers are currently producing several relatively new drug delivery devices. Insulin pens look like a writing pen. A cartridge holds the insulin and the tip is the needle. The user set a dose, inserts the needle into the skin, and presses a button to inject the insulin. With pens there is no need to use a vial of insulin. However, pens require inserting separate tips before each injection. Another downside is that the pen does not allow users to mix insulin types, and not all insulin is available.

For people who hate needles an alternate to the pen is the jet-injector. Looking similar to the pens, jet injectors use pressure to propel a tiny stream of insulin through the skin. These devices are not as widely used as the pen, and they can cause bruising at the input point.

The insulin pump allows a controlled release in the body. This is a computerized pump, about the size of a beeper, that diabetics can wear on their belt or in their pocket. The pump has a small flexible tube that is inserted just under the surface of the diabetic's skin. The diabetic sets the pump to deliver a steady, measured dose of insulin throughout the day, increasing the amount right before eating. This mimics the body's normal release of insulin. Manufacturers have produced insulin pumps since the 1980s but advances in the late 1990s and early twenty-first century have made them increasingly easier to use and more popular. Researchers are exploring the possibility of implantable insulin pumps. Diabetics would control these devices through an external remote control.

Researchers are exploring other drug-delivery options. Ingesting insulin through pills is one possibility. The challenge with edible insulin is that the stomach's high acidic environment destroys the protein before it can move into the blood. Researchers are working on coating insulin with plastic the width of a few human hairs. The coverings would protect the drugs from the stomach's acid.

In 2001 promising tests are occurring on inhaled insulin devices and manufacturers could begin producing the products within the next few years. Since insulin is a relatively large protein, it does not permeate into the lungs. Researchers of inhaled insulin are working to create insulin particles that are small enough to reach the deep lung. The particles can then pass into the bloodstream. Researchers are testing several inhalation devices much like that of an asthma inhaler.

Another form of aerosol device undergoing tests will administer insulin to the inner cheek. Known as buccal (cheek) insulin, diabetics will spray the insulin onto the inside of their cheek. It is then absorbed through the inner cheek wall.

Insulin patches are another drug delivery system in development. Patches would release insulin continuously into the bloodstream. Users would pull a tab on the patch to release more insulin before meals. The challenge is finding a way to have insulin pass through the skin. Ultrasound is one method researchers are investigating. These low frequency sound waves could change the skin's permeability and allow insulin to pass.

Other research has the potential to discontinue the need for manufacturers to synthesize insulin. Researchers are working on creating the cells that produce insulin in the laboratory. The thought is that physicians can someday replace the non-working pancreas cells with insulin-producing cells. Another hope for diabetics is gene therapy. Scientists are working on correcting the insulin gene's mutation so that diabetics would be able to produce insulin on their own.

Where to Learn More

Books

Clark, David P, and Lonnie D. Russell. *Molecular Biology Made Simple and Fun.* 2nd ed. Vienna, IL: Cache River Press, 2000.

Considine, Douglas M., ed. *Van Nostrand's Scientific Encyclopedia.* 8th ed. New York: International Thomson Publishing Inc., 1995.

Periodicals

Dinsmoor, Robert S. "Insulin: A Never-ending Evolution." *Countdown* (Spring 2001).

Other

Diabetes Digest Web Page. 15 November 2001. <http://www.diabetesdigest.com>.

Discovery of Insulin Web Page. 16 November 2001. <http://web.idirect.com/~discover>.

Eli Lilly Corporation. *Humulin and Humalog Development.* CD-ROM, 2001.

Eli Lilly Diabetes Web Page. 16 November 2001. <http://www.lillydiabetes.com>.

Novo Nordisk Diabetes Web Page. 15 November 2001. <http://www.novonet.co.nz>.

M. Rae Nelson

Ironing Board

Clothing and linens were pressed on table tops or large pieces of board that were covered with padding, pillowcases, or ironing blankets until nearly 1900.

Background

An ironing board is generally a large, flat piece of board or metal that is covered with a heat-safe padding on which clothing or linens may be ironed safely. Modern ironing boards take a surprising number of forms. The standard, inexpensive American ironing board has two primary parts and includes a flat bed for ironing and collapsible legs that are hinged or slip into the top and folded down for easy storage. The top of the standard American board is generally a flat pan that may have some holes to accommodate paint run-off during manufacture or to disseminate the heat from the steam iron that runs over it. Legs are generally lightweight and tubular with padded feet so that they do not mar floors. More expensive domestic ironing boards may be mesh tables with expanding metal tops to accommodate larger goods to be ironed. No matter what the configuration, every ironing board made of metal must have a pad and cover so that the metal bed of the ironing board does not become too hot from the iron. Most ironing boards are sold with a foam pad and decorative cover.

Many of the ironing boards sold in American stores are made overseas where labor is far cheaper than in the United States. In 2001, there were only two manufacturers of ironing boards in the United States. Some expensive designer-based ironing boards are available in this country and come from European home design firms; these boards may be five times more expensive than the ordinary collapsible ironing board made in the United States.

History

Clothing and linens were pressed on table tops or large pieces of board that were covered with padding, pillowcases, or ironing blankets until nearly 1900. In fact, it is more logical to refer to the modern ironing surface as an ironing table but the device is referred to as an ironing board because the earliest devices were composed of wooden boards. Some housekeeping advisors of the nineteenth century urged women to use large boards that could be placed between a table and a chair back that they could pad and iron upon. One advisor named Catherine Beecher described in 1841 what appears to be the shaped ironing board known today. She recommended that this wooden form be cut wide on one side and narrower on the other and referred to this type of ironing board as a skirt board. Of course, this was the era before electric irons and sad irons were used with these ironing boards. These heavy cast irons were heated at the hearth or on wood or coal burning stoves and the heavy, hot iron pressed out stubborn wrinkles.

Manufacturers quickly caught on to the notion of offering skirt boards ready-made by the late nineteenth century. By about 1898, the skirt board came equipped with legs that could be taken down and enabled the board to be set up anywhere. These early manufactured ironing boards had a leg in each corner that could be unfolded but were difficult to manipulate. By 1914, an inventor named Springer devised a table with three support points that was easier to set up. These early manufactured tables were of wood that was supposed to resist warping although they still warped. Early manufacturers made metal-top boards but they often rusted despite painting. Some buckled under the heat of the iron. The J. R. Clark Company of Minneapolis began making metal tops of mesh which permitted steam to escape and

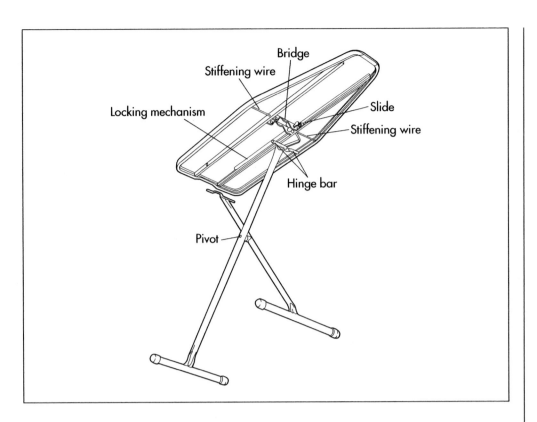

prevent buckling and rusting. By 1940, a few manufacturers were producing all-metal collapsible ironing boards. Soon thereafter all were made entirely of metal. The design of the ordinary, collapsible metal ironing board has changed little over the ensuing decades.

Raw Materials

The conventional, inexpensive American-made ironing board is made of few raw materials. The board uses many different widths of cold rolled steel, which are long rolls of rolled steel that begin the forming process as a cold metal. The widths vary according to the part under production and include widths as small as 0.75 in (1.9 cm) to over 28 in (71 cm) in width. Typically the rolls may be several hundred feet in length and may weigh between 500–9,000 lb (227–20,000 kg) depending on width. These rolls are then flattened and rolled or pressed into different parts.

Other raw materials used in construction of ironing boards include metal rivets, plastic or rubber tips on the feet of the standing portions of the ironing board, oil-based paint (either powdered or in liquid form) and degreasing agents sprayed upon metal parts prior to painting that remove oily effluvia

deposited on the metal portions that could affect paint adherence. Some companies sew a cotton pad and cotton cover for the ironing board. A company might purchase such a product from another manufacturer and slip it onto the completed ironing board before packaging.

The Manufacturing Process

1 The factory is supplied with various widths of strip steel for manufacture of the various parts of the ironing board. These widths vary between 4–28 in (10.2–71 cm) in width depending on intended use. These tightly-coiled rolls must be straightened. They are fed into a reel or de-coiler which flattens them.

2 While the rolls are very long, sometimes hundreds of feet in length, they are of finite length. The ends must be seamed or joined together so that the process of cutting out parts does not stop and start when one roll ends and another begins. So, the ends of the rolls are butt-joined (meaning the ends are not overlapped but butt up against one another) and are butt-welded together. An operator-controlled welder comes across this seam and welds it together. The un-

coiled steel now awaits being formed into the legs and the top of the board.

3 Tubular legs for the ironing board are next formed. The cold steel is fed into the receiving end of a tube mill that is used to form flat steel into tubing. The tube mill brings the ends of the steel together and forms into a tube. The tube goes through an electrode which shoots electricity through the piece, melting the ends of the metal that join it into a tube. This electrode is, essentially, a high-frequency welder that sears the ends together.

4 The formed tubing is pulled through the machine to the next station. Here, a sharp carbide tool is scraped over the rough edge of the weld so there are no metal burrs at the juncture.

5 Next, the tubing must be cooled down from the welding process that has melted those ends together. The long tubing (not yet cut apart into separate legs)is pulled to the next station and is flooded with coolant that cools it down somewhat. The metal is still warm at this point, however.

6 The warm tubing is fed into a sizing mill that pushes in on the warm tubing to ensure it is of the right size. The warm metal is configured to the dimensions of the sizing mill.

7 After leaving the sizing mill the tubing reaches the cut off station, in which the long length of tubing is cut off to the requisite length needed for the legs. The feet on the legs are then attached to the legs by being riveted in place. The legs are now complete and await the painting process.

8 The top is the now formed. The most common type of American-made ironing boards have a pan top. This top is formed of approximately 16-in (40.1-cm) width steel that is pressed and stamped into configuration. After de-coiling, the metal strip is placed into a press that is begun by an operator. The press pulls in the metal and curls the edges on the side, punches a hole on the flat pan so that paint may later drain from the flat surface, and knocks the piece off the punch in order for another piece to be drawn onto the press.

9 The pan top must be attached to a set of ribs and cross-members that will be at-tached underneath in order to keep this flat pan rigid. So, ribs and cross members are roll formed and stamped. The lock assembly (the lock which secures the collapsible ironing board in place)is composed of a spring and handle. This locking assembly is attached to the ribs and cross members at this point.

10 The pan top and the rib assembly (with locking mechanism attached) meet up on the assembly line at this point. The two components are placed together by hand. The pan and assembly are then fed into a top welder, which is a huge machine that welds the ribs, cross members, and locking mechanism to the flat pan top. The ironing board top comes out of the welder and is inspected to ensure the pressure points are securely welded. The top of the ironing board is now complete and awaits painting.

11 Before the legs and top are connected, both components must be painted. First, the components are hung on a paint line. The parts are sprayed with a degreaser. Then the parts are sent to a drying room until the degreaser evaporates. Next, the parts are sent through a paint sprayer that shoots either powdered or liquid paint. The parts are then sent to the bake area in which the paint is baked on the surface.

12 Now the top of the board must now be attached to the legs. The legs are fed onto a roller conveyor and meet up with the top. An operator slips the legs in slots on the underside of the pan top and the operator clinches the tabs in place in order to secure the legs in place. An operator then puts rubber or plastic tips on the feet of the ironing board. If the board is to receive a pad and top, they are slipped on the pan top at this point as well. The board is now complete and ready for packaging.

Quality Control

Quality control is evident in all aspects of the production of the ironing board. First, the incoming raw materials undergo thorough inspection upon receipt. The cold rolled steel deliveries are assessed to ensure they comply with manufacturer's specifications. The metal undergoes thickness and hardness tests. The surface of the metal is physically inspected for signs of rust; if rust is found the batch is deemed unacceptable.

All cold rolled steel is expected to be delivered with a fine coating of oil which inhibits rust. Finally, the metal is assessed to ensure the steel is free of camber, which is a term for metal twisting as such twisting weakens the metal.

Roving inspectors on each shift assess all aspects of the operation. Operators on each shift are empowered to fix any problems that occur and understand that they are accountable for quality and problem-solving. Machines integral to the processes are maintained scrupulously to ensure there are no breakdowns in the process or that tolerances in manufacturing are unacceptable. As noted under the manufacturing process above, key operations performed by machines, such as the welding of the ribs and locking mechanism to the top of the ironing board, are double-checked by an operator by hand. The operator puts pressure on the key welds to ensure that the welds are strong. Finally, after the products are completed the quality inspectors at the end of the line inspect completed products to make sure that the overall product is acceptable.

Byproducts/Waste

Excess metal that is the result of cutting pieces from the cold rolled steel is gathered up into a scrap hopper and when the hopper is full the metal is sold back to steel mills that supply the metal so that it may be re-used. The de-greasing agent that is sprayed on the metal parts prior to painting runs off and forms a sludge that must be treated. When the sludge becomes sizable, a company that specializes in dealing with such compounds pumps it out, fires it and burns the sludge. Other volatile organic compounds (referred to as VOCs) present in the oil-based paints used on the steel components are used according to specifications set down by the federal government; the company must apply for permits for their use.

Where to Learn More

Books

Beecher, Catherine. *A Treatise on Domestic Economy of 1841*. New York: Schocken Books, 1977.

Litshey, Earl. *The Housewares Story*. Chicago: The National Housewares Manufacturers Association, 1973.

Ierley, Merritt. *The Comforts of Home*. New York: Clarkson Potter, 1999.

Other

Oral interview with Joseph Deppen, Vice President of Manufacturing, Home Products International. Chicago, Illinois and Seymour, Indiana. October 2001.

Nancy EV Bryk

Juice Box

The juice box was introduced to the United States in 1980, and by 1986, it made up approximately 20% of the United States juice market.

Background

For centuries, people all over the world have been drinking fruit juice. Today, it is available in both frozen concentrate and liquid form and packaged in a variety of ways, including bottles, cans, and—most recently—boxes. A juice box is an individual-sized container that usually holds 4–32 oz (118–946 ml) of juice and generally comes with an attached straw that can be removed and inserted for drinking. A juice box is considered an aseptic container, meaning it is manufactured and filled under sterile conditions and requires no refrigeration or preservatives to remain germ-free. Along with its portability and convenience, the juice box has gained widespread popularity due to the brick-shaped container's composition of unbreakable materials and tight seal.

History

The aseptic container was invented in the 1960s by a Swedish man named Ruben Rausing. In 1963, Rausing was trying to figure out a more efficient method to get milk to the market. He needed a container that was smaller and less cumbersome than the metal canisters being used. Rausing developed a precursor to the juice box: a brick-shaped box he named the Tetra Brik. Because of their rectangular shape, Tetra Briks, when stacked on top of each other, took up half the space of the old containers. Five years later, Rausing made an even bigger breakthrough when he figured out how to fill the Tetra Briks under completely sterile, or aseptic, conditions.

Once the juice box was introduced to the United States in 1980, competitors began entering the market at a rapid rate. These companies began implementing all sorts of ideas to gain larger shares of the market, including filling the juice boxes with a variety of different flavors, adding vitamins and other nutrients, and making packaging changes to widen the juice box's appeal. By 1986, juice boxes made up approximately 20% of the United States juice market.

When juice boxes first entered the market, they were often filled with diluted juice drinks rather than real fruit juice. However, realizing that Americans were becoming more health conscious, the juice box industry responded by filling the boxes with healthier beverages. A number of companies added vitamins, such as A and C. In the early 1990s, Minute Maid became the first company to add calcium to its juice boxes. Other companies soon followed.

Environmental concerns

Despite their growing popularity, not everyone had positive things to say about juice boxes. Environmental groups were worried about the effect that juice boxes and other aseptic containers could have on the environment. Specifically, these groups were afraid that aseptic containers would fill the nation's landfills because they are not as easy to recycle as other types of packages. The state of Maine even went so far as to ban the sale aseptic containers. This ban was later repealed, but other states have considered adopting similar legislation.

In response to this opposition, the Aseptic Packaging Council (APC), a trade association that represents the major United States manufacturers of aseptic packages, was formed in 1989. Their primary mission was to inform the American public about the product benefits and environmental attributes of aseptic packaging. Since its inception, the APC as been working closely with communities nationwide to encourage the inclusion of juice boxes in recycling programs. These efforts have already proven successful in some communities. In addition to recycling efforts, juice box manufacturers argue that aseptic containers are actually friendlier to the environment than other types of containers. For one, they take up less room on trucks when being transported from factory to store, thus conserving energy by requiring fewer trips and using less fuel. The aseptic filling process itself also requires less energy than traditional canning and bottling methods. The manufacturers also point out that packaging makes up only 4% of the weight of a filled aseptic container in contrast to filled glass bottles, which are typically 30–40% packaging. This leaves less packaging to dispose of when dealing with an aseptic container.

In the late 1990s, attitudes about the environmental friendliness of the aseptic package began to change. In 1996, the aseptic carton won the Presidential Award for Sustainable Development, and the aseptic packaging industry was recognized for demonstrating environmental responsibility throughout the product life cycle. In 2001, this increased acceptance by environmental activists, combined with the industry's efforts toward incorporating new and innovative marketing ideas, continues to make the juice box the driving force behind the juice industry.

Raw Materials

Juice boxes are typically made up of six layers of paper (24%), polyethylene (70%), and aluminum foil (6%). The paper provides stiffness and strength and gives the package its brick shape. Polyethylene serves two purposes. On the inner most layer, it forms the seal that makes the package liquid tight. On the exterior, it provides a protective coating that keeps the package dry and provides a printing surface for nutritional and marketing information. The aluminum foil forms a barrier against light and oxygen, eliminating the need for refrigeration or preservatives to prevent spoilage. The straws are made of plastic and wrapped in cellophane. Multipacks contain six or more juice boxes, and are often wrapped in a cardboard sleeve that displays the name of the product and other specifications, then shrink-wrapped in plastic.

Design

Although they are available in a variety of sizes, virtually all juice boxes have the same basic design features. Each of these features was designed to serve a specific purpose. First, the rectangular, brick-shaped design was chosen for its convenience, particularly during transport. Second, the materials from which juice boxes are made were selected to keep the beverages inside safe and fresh.

The third basic design feature is the drinking mechanism. This can be either a straw affixed to the side of the package that can be removed and inserted into a preformed hole in the top, or a pull tab incorporated into the top of the package that may or may not be resealable. The type of drinking mechanism used depends on the size of the juice box and/or who will be using it. For example, juice boxes designed for small children often use a straw, while boxes with more adult appeal may use a pull tab. Boxes that contain more than one serving would typically use a resealable tab.

The Manufacturing Process

The aseptic packaging process is considered a major breakthrough in the beverage industry. During the process, the juice is sterilized outside the package using an extremely high temperature (195–285°F [91–141°C]) and then cooled before being poured into the specially designed pre-sterilized juice box. This sterilization process is called flash heating and cooling because it is accomplished within a very short amount of time, usually three to 15 seconds, substantially reducing energy use and nutrient loss associated with conventional sterilization. This process is so revolutionary that it has won an award for innovation from the Institute of Food Technologies.

Creating the carton blanks

1 The juice box itself is made up of six layers of paper, polyethylene, and aluminum foil. First the raw paper, which is rolled on a

A. Filler operation and control. B. Sleeve extraction and forming. C. Base folding. D. Transfer station. E. Sterile air treatment. F. H_2O_2 injection. G. Drying zone. H. Filling station. I. Ultrasonic top sealing. J. Paddle wheel ejector. K. Discharge conveyor.

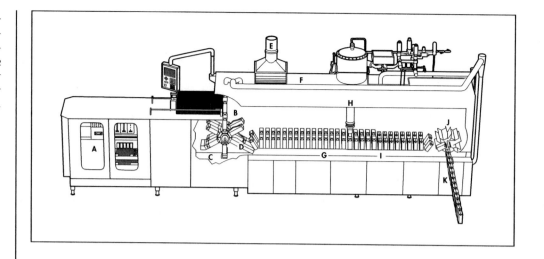

mother roll, is printed with the appropriate marketing and nutritional information. Then the layers of polyethylene and aluminum are added. The layers are bonded together using special extrusion-lamination equipment.

2 Next, automated, high-speed machines crease and cut out several carton blanks, or sleeves, from the roll. The sleeves are now ready to be formed into cartons, sterilized, and filled in specially designed filling machines. This filling is generally performed at another location, separate from the plant that created the sleeves.

Sterilizing and filling the juice boxes

3 A programmable logic controller (PLC) monitors and controls the filling machine that is run by an operator. First, the pre-formed sleeves are loaded into a magazine directly from the shipping box. The sleeves are then extracted individually by suction, shaped into a rectangle, and slid onto a mandrel.

4 The inner layer of the sleeve, which is made of polyethylene, is thermally activated by convection heating.

5 As the mandrel wheel transports the sleeves to a pressing station, the sleeves are folded for bottom sealing. Then the sleeve bottoms are formed and sealed, creating a carton base with the top still open. The cartons are then transferred from the mandrel to the pocket chain where the tops are pre-folded.

6 Once the tops are pre-folded, the cartons enter the aseptic zone, where fresh air is

sterilized by filters. Once in the aseptic zone, the cartons are sterilized with hydrogen peroxide vapor. Using compressed air, liquid hydrogen peroxide is forced through a nozzle into a heater where it is vaporized before being injected into the cartons. Sterile air is heated and blown into the cartons repeatedly to dry out the hydrogen peroxide, while a fan extracts vapors from the aseptic zone.

7 Once the cartons are sterilized, they are filled with the pre-sterilized product. Any foam that is produced during filling is extracted from cartons as necessary.

8 The tops of the filled cartons are folded and sealed ultrasonically above the product level. The ears are convection heated and folded down against the side panels. The finished cartons are then discharged from the filling machine on a conveyor belt.

Finishing touches

9 The next step is to add the drinking mechanism. The most common mechanism is a straw, although some companies offer alternative methods such as pull tabs. If a straw is used, it is wrapped in plastic and glued to the side of the box with a temporary adhesive that will allow the straw to later be pulled off the box, unwrapped, and inserted into a hole punched in the top of the box. The straw hole is created via laser cutting. If a pull tab is used it is added to the top of the box, also using a laser cutting process for the opening. This completes the creation of an individual juice box.

10 Often, several juice boxes are packaged together to form multipacks. The

individual boxes are wrapped in a cardboard sleeve with nutritional and other information printed on it, then shrink-wrapped in plastic for shipping.

Quality Control

To ensure quality and safety standards are met, a PLC monitors and controls the operation of the filling machine during the sterilization of the liquid and the filling of the juice boxes. This controller is run by an operator from a console that complies with all United States Food and Drug Administration (FDA) reporting requirements. Hundreds of manual and automatic quality checks are preformed before, during, and after the sterilization and filling processes to ensure that the temperature of the liquid and speed of the process remain in the proper range; that the sterility, nutritional content, and flavor of the beverage is never compromised; that the boxes themselves remain intact with no leaks; and that the drinking mechanisms are properly attached.

Byproducts/Waste

Despite early skepticism from environmentalists, juice boxes and the aseptic packaging process used to fill them have proven to be highly environmentally friendly, resulting in much less waste and energy use than traditional beverage packaging methods.

Also, because of their light weight and unique brick shape design, juice boxes help save energy by taking up less space during transport than bottles or cans. In addition, aseptic packages do not require refrigeration during transport or storage, which also reduces energy use. During the aseptic filling process, the time and temperature are carefully monitored to ensure maximum energy efficiency without compromising the integrity of the product.

Recycling of used juice boxes helps reduce waste as well. In the 1990s there was an increase in the number of communities including juice boxes as part of their curbside recycling programs. Through a process called hydrapulping, the paper is separated from the polyethylene and ground into pulp to be used to produce other paper products.

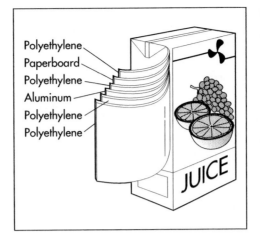

A juice box has several layers of polyethylene, paper, and aluminum foil.

The Future

In the 1990s, sales of single-serve beverages such as juice boxes experienced record growth, and experts expected such growth to continue into the twenty-first century. Factors contributing to this growth include expansion of international operations by industry leaders such as Coca-Cola, Tropicana, and Pepsi-Cola; continued implementation of new flavors and marketing ideas to appeal to wider market segments; and new venues for selling the product such as health clubs and bike shops.

One factor that may hurt the juice industry is the fact that in 2001 the American Academy of Pediatrics recommended a reduction in juice consumption by children, explaining that drinking too much juice can lead to obesity and other health problems. It remains to be seen whether this recommendation will offset the many benefits that parents have experienced with juice boxes so far and the innovative ideas yet to be implemented by the juice box industry.

Where to Learn More

Periodicals

Kelly, Kristine Porney. "Bountiful Growth for Juices, Juice Drinks." *Beverage Industry* 86, no. 9 (September 1995): 10.

Kulma, Linda. "Junking the Juice Box Habit." *U.S. News and World Report* 130, no. 20 (21 May 2001): 71.

Skenazy, Lenore. "Juice Boxes, Practical, Convenient, Fun." *Knight-Ridder/Tribune News Service* (3 November 1998): K7326.

Other

Aseptic Packaging Council Web Page. <http://www.aseptic.org>.

Combibloc, Inc. Web Page. <http://www. combi-blocks.com>.

Kathy Saporito

Kerosene

Background

Kerosene is an oil distillate commonly used as a fuel or solvent. It is a thin, clear liquid consisting of a mixture of hydrocarbons that boil between 302°F and 527°F (150°C and 275°C). While kerosene can be extracted from coal, oil shale, and wood, it is primarily derived from refined petroleum. Before electric lights became popular, kerosene was widely used in oil lamps and was one of the most important refinery products. Today kerosene is primarily used as a heating oil, as fuel in jet engines, and as a solvent for insecticide sprays.

History

Petroleum byproducts have been used since ancient times as adhesives and water proofing agents. Over 2,000 years ago, Arabian scientists explored ways to distill petroleum into individual components that could be used for specialized purposes. As new uses were discovered, demand for petroleum increased. Kerosene was discovered in 1853 by Abraham Gesner. A British physician, Gesner developed a process to extract the inflammable liquid from asphalt, a waxy petroleum mixture. The term kerosene is, in fact, derived from the Greek word for wax. Sometimes spelled kerosine or kerosiene, it is also called coal oil because of its asphalt origins.

Kerosene was an important commodity in the days before electric lighting and it was the first material to be chemically extracted on a large commercial scale. Mass refinement of kerosene and other petroleum products actually began in 1859 when oil was discovered in the United States. An entire industry evolved to develop oil drilling and purification techniques. Kerosene continued

to be the most important refinery product throughout the late 1890s and early 1900s. It was surpassed by gasoline in the 1920s with the increasing popularity of the internal combustion engine. Other uses were found for kerosene after the demise of oil lamps, and today it is primarily used in residential heating and as a fuel additive. In the late 1990s, annual production of kerosene had grown to approximately 1 billion gal (3.8 billion l) in the United States alone.

Raw Materials

Kerosene is extracted from a mixture of petroleum chemicals found deep within the earth. This mixture consists of oil, rocks, water, and other contaminates in subterranean reservoirs made of porous layers of sandstone and carbonate rock. The oil itself is derived from decayed organisms that were buried along with the sediments of early geological eras. Over tens of millions of years, this organic residue was converted to petroleum by a pair of complex chemical processes known as diagenesis and catagenesis. Diagenesis, which occurs below 122°F (50°C), involves both microbial activity and chemical reactions such as dehydration, condensation, cyclization, and polymerization. Catagenesis occurs between 122°F and 392°F (50°C and 200°C) and involves thermocatalytic cracking, decarboxylation, and hydrogen disproportionation. The combination of these complex reactions creates the hydrocarbon mixture known as petroleum.

The Manufacturing Process

Crude oil recovery

1 The first step in the manufacture of kerosene is to collect the crude oil. Most

Mass refinement of kerosene and other petroleum products began in 1859 when oil was discovered in the United States.

oil supplies are buried deep beneath the earth and there are three primary types of drilling operations used to bring it to the surface. One method, Cable-Tooled Drilling, involves using a jackhammer chisel to dislodge rock and dirt to create a tunnel to reach oil deposits that reside just below the earth's surface. A second process, Rotary Drilling, is used to reach oil reservoirs that are much deeper underground. This process requires sinking a drill pipe with a rotating steel bit into the ground. This rotary drill spins rapidly to pulverize earth and rock. The third drilling process is Off Shore Drilling and it uses a large ocean borne platform to lower a shaft to the ocean floor.

2 When any of these drilling processes break into an underground reservoir, a geyser erupts as dissolved hydrocarbon gases push the crude oil to the surface. These gases will force about 20% of the oil out of the well. Water is then pumped into the well to flush more of the oil out. This flushing process will recover about 50% of the buried oil. By adding a surfactant to the water even more oil can be recovered. However, even with the most rigorous flushing it is still impossible to remove 100% of the oil trapped underground. The crude oil recovered is pumped into large storage tanks and transported to a refining site.

3 After the oil is collected, gross contaminants such as gases, water, and dirt are removed. Desalting is one cleansing operation that can be performed both in the oilfield and at the refinery site. After the oil has been washed, the water is separated from the oil. The properties of the crude oil are evaluated to determine which petroleum products can best be extracted from it. The key properties of interest include density, sulfur content, and other physical properties of the oil related to its carbon chain distribution. Since crude oil is a combination of many different hydrocarbon materials that are miscible in one another, it must be separated into its components before it can be turned into kerosene.

Separation

4 Distillation is one type of separation process involves heating the crude oil to separate its components. In this process the stream of oil is pumped into the bottom of a distillation column where it is heated. The lighter hydrocarbon components in the mixture rise to the top of the column and most of the high boiling-point fractions are left at the bottom. At the top of the column these lighter vapors reach the condenser which cools them and returns them to a liquid state. The columns used to separate lighter oils are proportionally tall and thin (up to 116 ft [35 m] tall) because they only require atmospheric pressure. Tall distillation columns can more efficiently separate hydrocarbon mixtures because they allow more time for the high boiling compounds to condense before they reach the top of the column.

To separate some of the heavier fractions of oil, distillations columns must be operated at approximately one tenth of atmospheric pressure (75 mm Hg). These vacuum columns are structured to be very wide and short to help control pressure fluctuations. They can be over 40 ft (12 m) in diameter.

5 The condensed liquid fractions can be collected separately. The fraction that is collected between 302°F and 482°F (150°C and 250°C) is kerosene. By comparison, gasoline is distilled between 86°F and 410°F (30°C and 210°C). By recycling the distilled kerosene through the column multiple times its purity can be increased. This recycling process is known as refluxing.

Purification

6 Once the oil has been distilled into its fractions, further processing in a series of chemical reactors is necessary to create kerosene. Catalytic reforming, akylkation, catalytic cracking, and hydroprocessing are four of the major processing techniques used in the conversion of kerosene. These reactions are used to control the carbon chain distribution by adding or removing carbon atoms from the hydrocarbon backbone. These reaction processes involve transferring the crude oil fraction into a separate vessel where it is chemically converted to kerosene.

7 Once the kerosene has been reacted, additional extraction is required to remove secondary contaminants that can affect the oil's burning properties. Aromatic compounds, which are carbon ring structures such as benzene, are one class of contaminant that must be removed. Most extraction processes are conducted in large towers that

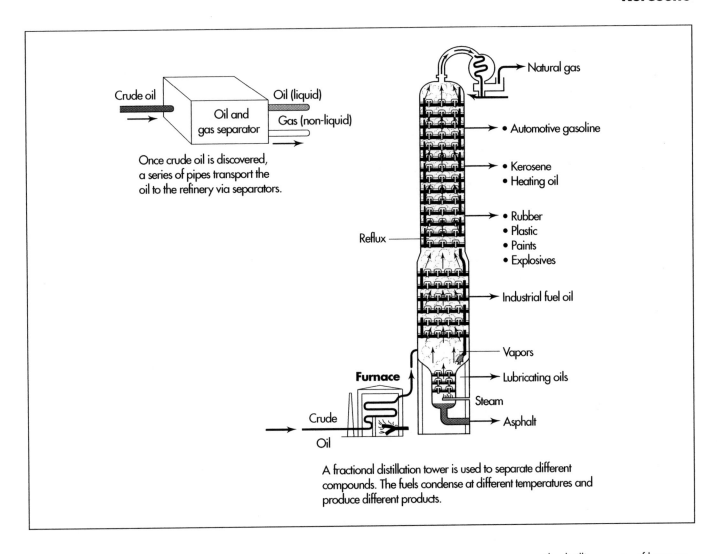

Crude oil

Oil (liquid)

Oil and gas separator

Gas (non-liquid)

Once crude oil is discovered, a series of pipes transport the oil to the refinery via separators.

Natural gas

• Automotive gasoline

• Kerosene
• Heating oil

• Rubber
• Plastic
• Paints
• Explosives

Reflux

Industrial fuel oil

Vapors

Furnace

Lubricating oils

Steam

Crude Oil

Asphalt

A fractional distillation tower is used to separate different compounds. The fuels condense at different temperatures and produce different products.

The distilling process of kerosene.

maximize the contact time between the kerosene and the extraction solvent. Solvents are chosen based on the solubility of the impurities. In other words, the chemical impurities are more soluble in the solvent than they are the kerosene. Therefore, as the kerosene flows through the tower, the impurities will tend to be drawn into the solvent phase. Once the contaminants have been pulled out of the kerosene, the solvent is removed leaving the kerosene in a more purified state. The following extraction techniques are used to purify kerosene.

The Udex extraction process became popular in the United States during the 1970s. It uses a class of chemicals known as glycols as solvents. Both diethylene glycol and tetraethylene glycol are used because they have a high affinity for aromatic compounds.

The Sulfolane process was created by the Shell company in 1962 and is still used in

many extraction units 40 years later. The solvent used in this process is called sulfolane, and it is a strong polar compound that is more efficient than the glycol systems used in the Udex process. It has a greater heat capacity and greater chemical stability. This process uses a piece of equipment known as a rotating disk contractor to help purify the kerosene.

The Lurgi Arosolvan Process uses N-methyl-2-pyrrolidinone mixed with water or glycol which increases of selectivity of the solvent for contaminants. This process involves a multiple stage extracting towers up to 20 ft (6 m) in diameter and 116 ft (35 m) high.

The dimethyl sulfoxide process involves two separate extraction steps that increase the selectivity of the solvent for the aromatic contaminants. This allows extraction of these contaminants at lower temperatures. In addition, chemicals used in this process are non-toxic and relatively inexpensive. It uses a specialized column, known as a Kuhni column, that is up to 10 ft (3 m) in diameter.

The Union Carbide process uses the solvent tetraethylene glycol and adds a second extraction step. It is somewhat more cumbersome than other glycol processes.

The Formex process uses N-formyl morpholine and a small percentage of water as the solvent and is flexible enough to extract aromatics from a variety of hydrocarbon materials.

The Redox process (Recycle Extract Dual Extraction) is used for kerosene destined for use in diesel fuel. It improves the octane number of fuels by selectively removing aromatic contaminants. The low aromatic kerosene produced by these process is in high demand for aviation fuel and other military uses.

Final processing

8 After extraction is complete, the refined kerosene is stored in tanks for shipping. It is delivered by tank trucks to facilities where the kerosene is packaged for commercial use. Industrial kerosene is stored in large metal tanks, but it may be packaged in small quantities for commercial use. Metal containers may be used because kerosene is not a gas and does not require pressurized storage vessels. However, its flammability dictates that it must be handled as a hazardous substance.

Quality Control

The distillation and extraction processes are not completely efficient and some processing steps may have to be repeated to maximize the kerosene production. For example, some of the unconverted hydrocarbons may by separated by further distillation and recycled for another pass into the converter. By recycling the petroleum waste through the reaction sequence several times, the quality of kerosene production can be optimized.

Byproducts/Waste

Some portion of the remaining petroleum fractions that can not be converted to kerosene may be used in other applications such as lubricating oil. In addition, some of the contaminants extracted during the purification process can be used commercially. These include certain aromatic compounds such as paraffin. The specifications for kerosene and these other petroleum byproducts are set by the American Society for Testing and Materials (ASTM) and the American Petroleum Institute (API).

The Future

The future of kerosene depends on the discovery of new applications as well as the development of new methods of production. New uses include increasing military demand for high grade kerosene to replace much of its diesel fuel with JP-8, which is a kerosene based jet fuel. The diesel fuel industry is also exploring a new process that involves adding kerosene to low sulfur diesel fuel to prevent it from gelling in cold weather. Commercial aviation may benefit by reducing the risk of jet fuel explosion by creating a new low-misting kerosene. In the residential sector, new and improved kerosene heaters that provide better protection from fire are anticipated to increase demand.

As demand for kerosene and its byproducts increases, new methods of refining and extracting kerosene will become even more important. One new method, developed by ExxonMobil, is a low-cost way to extract high purity normal paraffin from kerosene. This process uses ammonia that very efficiently absorbs the contaminants. This method uses vapor phase fixed-bed adsorption technology and yields a high level of paraffin that are greater than 90% pure.

Where to Learn More

Books

Kirk Othmer Encyclopedia of Chemical Technology. Vol. 18. John Wiley and Sons, 1996.

Periodicals

Kovski, Alan. "New Kerosene Laws Get off to Bumpy Start." *The Oil Daily* 48 (1998).

"Paraffins, Normal." *Hydrocarbon Processing* 80 (2001): 116.

Randy Schueller

Laser Pointer

Background

The laser pointer is a low cost portable laser that can be carried in the hand. It is designed for use during presentations to point out areas of the slide or picture being presented, replacing a hand held wooden stick or extendable metal pointer. It is superior over older pointers because it can be used from several hundred feet away in a darkened area and because it produces a bright spot of light precisely where the user desires. It has also caught on as an all-purpose pointing tool and has become so commonplace that laws have been passed to restrict its use.

History

Technically, the word *laser* is an acronym that stands for "light amplification by stimulated emission of radiation," but the term has become so commonly used that it is no longer capitalized. The radiation is the light that is emitted from the laser; this light can be visible or invisible to the human eye. Technically, only some lasers use light amplification, but the name laser is still used for a device that produces monochromatic (all one color or wavelength), coherent (the light waves are similar enough to move in one direction) radiation.

All lasers have a lasing medium, a source of energy, and a resonator. The lasing medium is a material that can be pumped (energized) by an energy source (such as light or electricity) to a higher energy state. After being pumped, the lasing medium can release that energy as monochromatic radiation. The resonator is an area that allows the released energy to build up before being released. A basic resonator is a pair of mirrors at either end of the lasing medium. One mirror is completely reflective so that all light striking it reflects back into the lasing medium; the other is partially reflective so that some of the light striking it reflects back into the lasing medium and some of the light passes through it to exit the laser. The pair of mirrors causes the light to reflect back-and-forth through the lasing medium and align itself in one direction, which produces the coherency of the light.

The theory used to produce lasers was published in 1958 by researchers at Bell Labs. The first laser, built in 1960 at Hughes Aircraft, used a piece of ruby for a lasing medium, light for an energy source, and mirrors to produce a resonator. The semiconductor laser was invented in 1962. It used a semiconductor material, similar to the materials used in transistors and integrated circuits for a lasing medium. It also used direct current (DC) electricity, the current produced by batteries, for an energy source. It still used resonator mirrors. The first semiconductor lasers produced non-visible infrared radiation. Current semiconductor lasers can also produce visible light, with red being the least expensive type of semiconductor laser and green, blue, and violet being increasingly more expensive. Semiconductor lasers used in laser pointers are also known as diode lasers, because they are a type of semiconductor diode. A diode passes electricity easily in one direction; light emitting diodes and laser diodes produce light when electricity passes through them. Semiconductor electronics have become less expensive to produce since the late 1950s. They have also become smaller and require less energy. They became inexpensive enough to be used in consumer electronic devices such as laser pointers in the 1980s.

The word laser *is an acronym that stands for "light amplification by stimulated emission of radiation," but the term has become so commonly used that it is no longer capitalized.*

Current laser diodes are the size of a blood cell. They produce light that is less collimated (moving all in one direction) than most lasers because the shortness of the resonator space. Because of this, they need some sort of external optics (lenses) to focus the light into a tighter beam. Laser diodes, like many semiconductor devices, are delicate and need to be protected from the environment and from power surges. Power control circuitry, which usually includes a photodiode (a diode that produces electricity when light strikes it) to monitor the output of the laser diode, prevents the diode from receiving too much or too little power. The diode is protected from the environment by a plastic case so that is resembles most other semiconductor devices that are used on circuit boards.

The first laser pointers cost hundreds of dollars, but the demand and improved methods of fabrication have resulted in a price below five dollars for the most inexpensive types. There are also several items which incorporate laser pointers, or at least the components, such as laser sights for guns and projectors with built-in laser pointers.

Raw Materials

A laser diode is less complicated than many types of consumer electronic equipment. It consists of a laser diode, a circuit board, a case, optics, and a case. Some of the electrical components on the circuit board and the laser diode are made of semiconductor materials, metals, and ceramics. The semiconductor materials include compounds (materials made of more than one pure element) made of aluminum, gallium, arsenic, phosphorus, indium, and similar elements. These compounds are used in a variety of semiconductor products. Semiconductors also contain metals such as aluminum, gold, and tantalum.

The circuit board is typically made of a resin (plastic) such as epoxy with glass fibers in it to strengthen it. Electricity is conducted to the various components on the circuit board with lines of metal such as aluminum and copper. Individual components placed on the circuit board include diodes, the laser diode, capacitors, and resistors. Semiconductor parts, such as the diodes are encapsulated in plastic with metal leads that are con-

nected to metal pads on the circuit board with solder (a metal alloy traditionally made of tin and lead, but now containing less lead and other metals as substitutes). Non-semiconductor parts, such as resistors and capacitors, are made of a variety of metals, plastics, and ceramics (including glass).

The collimating optics can be glass, but less expensive acrylic plastics are used in most laser pointers. The case can be made of any material, such as metal, plastic, or even wood. It contains metal (usually brass) contacts for the batteries.

Design

The design of the laser pointer depends on the electrical requirements of the laser diode, the desired lifetime of the power supply, and the drive to produce smaller consumer products. The smallest laser pointers are less than two inches in length, but some laser pointers are designed to look like pens. The longer laser pointers can hold AAA or AA batteries, which provide a longer lasting power supply than the watch batteries used in the shorter laser pointers. Most laser pointers use two or three batteries.

The Manufacturing Process

The red laser pointer is the most common laser pointer. Other laser pointers use different laser diode assemblies, but are produced in a similar fashion, so the red laser pointer manufacturing process and diagram are used in this article.

The laser diode

1 The laser diode is produced in a semiconductor fab (a factory where semiconductor materials are produced in very clean and carefully controlled conditions). The substrate is the base material on which other materials will be deposited. A wafer of the substrate is produced, cleaned, and prepared. Then it goes through several steps where layers of material are deposited on it. Some of these layers are only several atoms thick. These layers can be conductive (metals such as aluminum and gold) or semiconductors (as described above). These layers can also be altered by exposure to other chemicals. After all materials are added to the wafer, it

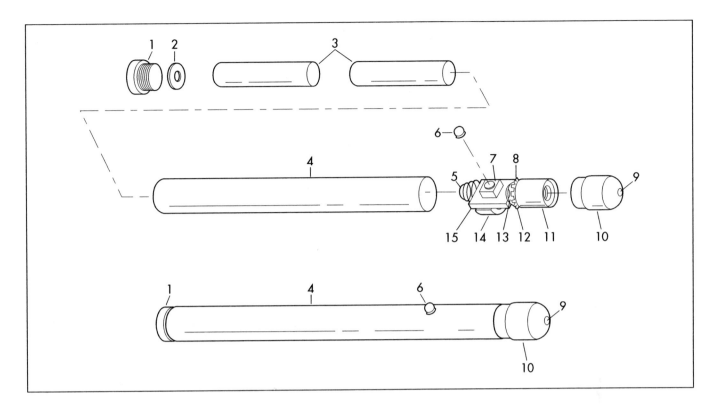

is diced (cut apart, usually into rectangular sections) into individual diodes. The diodes are tested either on the wafer or after separation, and nonfunctioning ones are scrapped (thrown away). Working laser diodes are then packaged in a plastic container with metal leads for electrical connection.

The circuit board

2 The circuit board contains the circuitry that makes the laser pointer function. It contains the switch, the laser diode, and the components of the control circuitry, typically a photodiode, diodes, resistors, and capacitors. These parts are placed on the circuit board, sometimes with an adhesive, and then are soldered in place. Soldering is a process where two metal objects are placed in contact and solder is melted around them so that when it cools, it surrounds both of them and holds them together. Solder is used instead of glue because it sticks to metal and because it conducts heat and electricity.

The collimating optics

3 The collimating optics in a laser pointer consist of a single lens that focuses the cone of light exiting the laser diode into a narrower beam that produces a narrower

spot over a longer distance. Plastic lenses are injection molded, a process wherein molten plastic is forced into a mold. The plastic cools and solidifies, then the mold is pulled apart and the lens is removed. It is ground and polished to a smooth surface so that the light from the laser diode will not bounce off of imperfections on the surface.

The laser diode assembly

4 The laser diode and the collimating optics are put together with a plastic holder to form the laser diode assembly. Most laser diode assemblies have a metal spring attached at the back. This spring makes contact with the batteries in the laser diode and is part of the circuit that draws electricity from the batteries.

Case construction and final assembly

5 The case is a tube with space for the laser diode assembly and the batteries. The laser diode assembly is pushed or screwed into one end of the case. The interior of the case is made of brass or has a brass strip (glued or riveted in place) running down the battery space. The battery space end piece also has an exposed brass area or is made of

1. Cap. 2. Electrically insulative cushion ring. 3. Batteries. 4. Metal barrel. 5. Metal spring. 6. Button. 7. Switch. 8. Spring plate. 9. Laser hole. 10. Metal barrel. 11. Lens barrel. 12. Laser module. 14. Locating frame. 15. Electrically insulative plate.

brass. When this end piece is pushed or screwed into the case, it contacts the other side of the batteries to complete the electrical circuit that allows electricity to flow from the batteries to the laser diode assembly.

6 The case also has a switch button (a piece of plastic sticking through a hole cut in the side of the case) that must be pushed and held for the laser pointer to work. When this button is pushed, the switch on the circuit board closes, electricity flows from the batteries to the laser pointer assembly, and the laser pointer produces a beam of light.

7 After the laser pointer is assembled and tested, a safety label is added. This label describes the rating of the laser in terms of power output, notes which regulations govern its use, and warns the user to avoid direct eye exposure.

Quality Control

A semiconductor manufacturer uses highly controlled processes that have been developed in laboratories and then transferred to the fabrication facility. Laser diodes are tested to make sure that they work after fabrication as well. Each other component is also tested to make sure that it works. Most manufacturing facilities will randomly test their products and use statistical control methods to provide quality products.

When the laser diode assembly or the laser pointer is finally assembled it will be powered and tested with a light detecting device, such as a photodiode, to measure its power output. Laser pointers are Type IIIA laser devices and must produce 5 mW (milliwatt, one thousandth of a watt) of power or less for the United States market. Laser pointers for the European market are typically Class II laser devices and must produce less than 1 mW. These restrictions are for safety purposes.

Byproducts/Waste

Laser pointers contain metals, plastics, and electronic parts. Each of those industries has specific waste byproducts (solvents, halocarbon gases, lead, chemicals), but laser pointer assembly has no specific wastes until the laser pointer is disposed of. A laser pointer contains small amounts of hazardous materials, such as lead and some toxic semiconductors. Like other electronic assemblies, it may be safer for the environment in the long term to recycle the components, though this is expensive and there are few programs in place to recycle or reuse electronics. This may change in the future.

The Future

Red laser pointers are the least expensive and most common today. Green laser pointers have more complicated laser diode assemblies and cost hundreds of dollars. Blue and violet laser pointers will be available soon at a higher price. Newer laser diode types come down in price as production volumes increase in order to keep up with demand, and as production processes improve. Laws that restrict laser pointer use may counteract this trend by causing a drop in demand as laser pointers are banned from public places.

Where to Learn More

Books

Gibilisco, Stan. *Understanding Lasers.* Blue Ridge Summit, PA: Tab Books, Inc., 1989.

Other

CORD Web Page. December 2001. <http://www.cord.org/lev3.cfm/48>.

Laser Focus World Web Page. December 2001. <http://lfw.pennnet.com/home.cfm>.

Andrew Dawson

Lawn Sprinkler

Background

The lawn sprinkler is a mechanism through which water is distributed in a spray so that a residential lawn or garden is irrigated. Sprinklers may take extraordinarily large forms, such as the irrigation systems used by professional farmers to water crops in the field. Many serious gardeners employ landscaping firms to develop and install expensive, permanent sprinkler systems in the ground. However, the average American simply purchases an inexpensive plastic and metal sprinkler and attaches it to a garden hose so that the sprinkler disseminates the water evenly. Residential use lawn sprinklers take a wide variety of forms and are available in a wide variety of materials and associated prices. However, the oscillating sprinkler, with a metal arm that sprays out a fan-shaped curtain of water in an area approximately 600 ft^2 (55.7 m^2), is likely the most popular. Such oscillating sprinklers sell for well under $10, but can be over $40. The permanency of the materials generally determines the price.

The less expensive, plastic and aluminum oscillating sprinkler is very simple in construction and in operation. Most include only a base that enables the piece to seat securely on the lawn, an oscillating arm, a bracket with regulating cam (it controls the width of the fan spray as it moves the spray arm), the mechanical motor, and the connectors to the garden hose. The oscillating sprinkler works on the principle that water provides the power to move the elliptical cam (or heart-shaped cam in some models) which moves the sprinkler arm. Water spins a simple turbine which must be attached to a series of gears stacked up (called a gear train) to slow down the speed of the water.

Without the insertion of the gear train the cam and the oscillating arm would move much too quickly. The gear train reduces the speed of the incoming water so that the cam moves at only about one mile per hour.

History

The advent of the residential lawn sprinkler is inextricably bound with Americans' desire to cultivate the yard for fun and aesthetic reasons and not for farming purposes. In fact, Americans of the seventeenth and eighteenth centuries did not think much of developing bucolic spaces in and around their homes. The town square in New England, now gorgeous green pastures, was a scruffy piece of land in which cattle could graze. In the nineteenth century attitudes toward the cultivation of the lawn change for two reasons. First, more Americans moved out of the city to the evolving suburbs as rail transportation enabled them to live further from places of business. As they moved out where there was space for gardens there was the urge to work the lawn as a hobby. Beautifying the home and garden were thought to strengthen the family. Secondly, Victorian Americans began to long for the rural New England they thought they were losing while others wanted to emulate the pastoral estates developing in England. Notable landscape architects urged Americans to beautify their property and published treatises and designs to help them achieve this goal.

Lawn sprinklers require a city water delivery system because they are generally used with a hose. Such systems were not in place in large measure until about 1870. The very wealthy used fountains to water the lawn and dazzle neighbors. Patents on lawn sprin-

The oscillating sprinkler has a metal arm that sprays out a fan-shaped curtain of water in an area approximately 600 ft^2 (55.7 m^2).

klers follow in short order; the first patent on an American lawn sprinkler was issued in 1871. Hose reels, nozzles, and sprinklers were advertised by 1900. Sprinklers took various forms. Large contraptions that rolled on carts with immense yardage of rubber hose were popular with some in the early twentieth century. By about 1930 the mechanisms took other forms such as cast metal with a three or four-arm rotating head that shot water as it spun around. Some were formed in the shape of animals with a rotating arm on the top. Plastic-bodied sprinklers followed in the later 20th century.

Raw Materials

The oscillating lawn sprinkler may be made from a variety of materials based on price point of the sprinkler. The best selling models are made from very inexpensive materials. These include aluminum tubing, plastic, and rubber. The aluminum forms the oscillating arm. The base of the sprinkler is generally of injection molded plastic. The motor or the mechanical head that forces the sprinkler to oscillate is generally of plastic gears as well. The o-rings that fit onto various joints are of rubber. Most oscillating sprinklers also have steel washers to ensure a tight fit at important junctures.

The Manufacturing Process

The oscillating sprinkler is manufactured in batches or smaller steps that are generally referred to as sub-assemblies. The sub-assemblies put together various larger portions at one time and then manufacture entails the assembling of all these sub-assemblies.

1 First, the mechanical motor that drives the oscillating sprinkler is assembled. The motors are made in huge quantity and set aside for inclusion in any one of the sprinkler models. The parts of the motors are generally injection molded. This entails melting plastic pellets and then forcing the melted, viscous plastic into a hollow mold at high speed and using great force. Once formed, the plastic parts of the motor are carefully assembled. The motor essentially works like a water-powered turbine. It includes gears built into a stack. The small, plastic water-driven motor is assembled by hand. Parts are loaded on an assembly line

and they come to the operator who presses them into place by hand. The motor is set aside until needed.

2 In addition to the small plastic motor parts, other parts of the sprinkler that are injection molded include a regulating cam that is attached to a bracket on one end of the oscillating arm, plastic fittings that attach to the hose, and the plastic base that enables the sprinkler to sit upon a flat platform. The base may be heat foiled stamped, a process by which a brand name or decal is embedded in the plastic base according to the model number and whether the piece is being sold under a retailer's name. These parts are put away until needed for assembly.

3 Next, hollow aluminum tubing that has been shipped to the factory must be configured to form the oscillating arm of the sprinkler. The hollow aluminum tube, already the appropriate length for the sprinkler, is carefully bent into an arch on a hydraulic press that gently exerts pressure on the thin tubing.

4 Immediately following the bending of the tube, a punch comes down and performs two important functions. First, the punch pierces approximately 20 holes into the hollow tube. Then, the punch inserts plastic jets into these holes. These plastic jet heads are simply pressure fit, meaning they are not soldered or glued in place but simply are forced into place. The hollow tube must have these holes and jets so that the water may cleanly spray from the oscillating arm once the hose, with running water, is attached to the sprinkler.

5 One end of the hollow tube—the oscillating arm—is slightly expanded by a machine so that it may receive the bracket, which includes an elliptically-shaped cam, that regulates the width of the water spray. The machine inserts the bracket and cam onto the tube.

6 Then, simple o-rings and washers are manually put on the end of the hollow tube to keep the water from leaking out there before it gets to the spray jets.

7 The tube—with bracket and cam—is now ready to have the small, plastic motor attached to it. A worker attaches the motor to the tube and embeds the entire as-

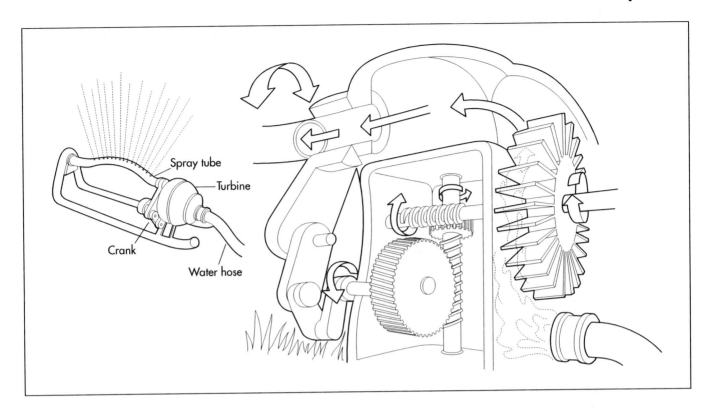

Spray tube

Turbine

Crank

Water hose

sembly into a completed plastic base. The sprinkler is now complete.

Quality Control

All materials that come to the factory are inspected to ensure they are made according to manufacturers' specifications. A few key components of the oscillating sprinkler may come already fabricated at another factory such as the hollow tubing for the oscillating arm or the planetary gears within the mechanical motor. In these cases, incoming inspection of these parts is essential as they are the most important parts of the sprinkler.

During fabrication and assembly the operators are careful to monitor all aspects of production and physical assembly. Visual inspection is expected of all employees and this occurs generally when sub-assemblies are complete. One sprinkler manufacturer uses part-in-place eye machines that have a sensor that will stop the assembly if it notices that a piece is missing. Perhaps most important is that the mechanical motor works as it is advertised. One manufacturer simulates a jet of water coursing through the completed motor mechanism before installation by performing an air test in which air shoots through the mechanism to see if it is working properly.

While it does not exactly duplicate the water stream the manufacturer does discover faulty motors using the air test.

Byproducts/Waste

There is little waste in the fabrication of simple oscillating sprinkler. When there are errors in the injection process, or if a sub-assembly made from plastic is found to be faulty, the plastic parts may be put into a bin and recycled into pellets and reformed. Improperly formed oscillating arms, made from hollow extruded aluminum, may also be recycled if not suitable for use. Some lawn products are sprayed in colors and the powder coating requires some regulation upon its use; however, unless the base is very colorful or unusual the sprinkler is generally not powder coated or layered with additional color.

The Future

The oscillating sprinkler is quite effective and extraordinarily low priced. Like other inexpensive consumer products, many are made out of the country but there are a surprising number still made in the United States. The use of in-ground systems installed by professionals at great cost has become more popular recently but the market for simple sprinklers such as the oscillating sprinkler is still strong. Some manufacturers make them so inexpensively so that each year a consumer who leaves the sprinkler out over the winter does not think twice about buying another the following spring.

Sprinkler use of any kind has caused some concern in the arid Western states as drought in a concern. Some communities restrict the use of sprinklers according to the day of the week or the address of the gardener.

Where to Learn More

Books

Jenkins, Virginia Scott. *The Lawn: A History of An American Obsession.* Washington, D.C.: Smithsonian Institution Press, 1998.

Tice, Patricia M. *Gardening in America, 1830–1910.* Rochester, NY: Margaret Woodbury Strong Museum, 1984.

Other

L. R. Nelson Company Web Page. December 2001. <http//:www.lrnelson.com>.

Oral interview with Mike Simpson, Production Unit Manager at L. R. Nelson Company. Peoria, IL. November 2001.

Nancy EV Bryk

Lighter

History

The discovery of tobacco in the New World in the sixteenth century and the opening of a worldwide market created the need for a portable way to make fire. Pieces of flint and steel struck against each other and modified pistols were early devices. In 1903, Austrian chemist Carl Auer von Welsbach made a hand-held lighter with a striking wheel. During World War I, soldiers made their own using empty cartridges. In New York City in 1886, Louis V. Aronson opened a company for "artistic metal wares" for smokers and patented an automatic lighter after World War I.

In 1931, George G. Blaisdell of Bradford, Pennsylvania, saw a friend trying to light his cigarette with an awkward lighter, but one that worked. Blaisdell acquired the American distribution rights for the Austrian product. He redesigned the case for comfort, improved the chimney (or wind hood) around the wick to make the lighter windproof, and modified the fuel chamber. Blaisdell named his lighter "Zippo" because he liked the sound of the word "zipper," which was another new and publicly acclaimed device. He began manufacturing his lighter in 1933.

Blaisdell's timing was poor and the lingering Great Depression nearly bankrupted the company. His luck changed during World War II, when soldiers found that Zippo lighters worked in all weather conditions. Crews of Navy ships have used these lighters with their own logos since World War II, and custom lighters for soldiers have been distributed during every war through Desert Storm. The lighters had may uses in wartime; GIs heated powdered rations in their helmets with the lighters and were able to start fires in all types of weather.

Collectors seek out Zippo lighters primarily because of the commemoration of large and small events on their cases. Advertisements from the 1940s and 1950s are valuable collectibles, as are lighters marking the 1969 moon landing, sports teams, many corporate clients, and a range of other historic events, personalities, and special interests. In 1997, 9,000 different images were used on the lighters.

Today, Zippo is the only manufacturer of pocket lighters in the United States and produces 50,000 lighters a day. Other production has shifted to Europe and Asia where smoking is more popular. Other large manufacturers of pocket lighters, like Ronson and BIC, have facilities in Austria, France, and Asia. BIC's contributions to the lighter are a childproof metal shield over the spark wheel and disposable, mini-sized lighters. BIC, Scripto, and others also make utility (fireplace) lighters with long tube shapes that are fueled with butane gas. Other firms make cigar lighters.

Raw Materials

The entire bottom case and the parts of a lighter are called the outer case assembly, and the inside case (containing the fuel and sparking action) and its parts is called the inner case assembly. The raw materials used for lighter manufacture are mostly metals. The outer case is made of cartridge brass, a material that was developed for rifle cartridges. A specialized mill makes brass sheets of the proper thickness and cuts them to the width required by the lighter manu-

George G. Blaisdell began manufacturing his Zippo brand lighter in 1933.

215

facturer. The brass is wound on large rolls or spools that are delivered to the factory.

A metal mill processes stainless steel for the inside case in a similar manner. It also arrives on large spools, and each holds enough metal to produce several thousand lighters.

The manufacturer produces the majority of the parts in a lighter. Most of the smaller parts are also made of brass or steel, depending on the purpose and location of the part in the lighter. The flint tube and spring tip (contained in the fuel chamber) are brass, as is the screw that holds these in place and exits the bottoms of both cases.

Other small parts inside and connected to the inner case are stainless steel. The cam, cam rivet, and plate are attached to the back top of the inner case and hold the lighter closed. In the chimney (wind hood) area and on the front, the cam spring and eyelet and the rivet for the flint wheel hold the fire-making parts. Specialized contractors make the eyelets and fasteners.

Three non-metallic components are also parts of the inner case assembly; these are wicking, balls made of a cotton-type substance, and felt. The wick and balls are placed in the fuel chamber. A piece of felt is fixed to the bottom of the inner case, but its front end can be lifted to allow lighter fluid refills and new flints.

Other metals, including nickel, chromium, and gold, are used to plate the brass cases upon customer request. Nonmetallic conducting fluids are used in the electroplating baths for the cases. Many methods are used to add decorations to the right faces of the lighters: three-dimensional (relief) emblems typically representing organizations can be attached to the lighters, designs can be etched or engraved in the metal, and lasers can draw detailed designs that are colored by any of several methods. Epoxies and special inks and powders are needed for these images.

Design

The basic design and operation of the pocket lighter have changed little since the 1930s, but the use of new technology has drastically altered many aspects of production. In manufacturing, "design" includes not only the

product and its parts but also the introduction to and interaction of machines with assembly and other production processes. One new machine or technique may require others.

Consumer interest has also changed. Lighters were once essential pocket tools, but are now often prized as "pocket art." Sophisticated techniques such as laser engraving and technigraphic printing are some of the means of dressing lighter cases. These require skilled artists and engineering expertise in artistic and production considerations, as well as time and cost limitations.

The Manufacturing Process

1 The manufacturing process begins at the factory's receiving dock, where raw materials and parts from suppliers are received. Plant personnel check the quantity and quality of materials and components. They also review blueprints and specifications to confirm that materials and parts meet the design engineer's requirements.

2 The coils of metal used for the outer and inner cases are rolled through presses in a process called "deep drawing." The presses punch in the edges of each lighter (as if it were unfolded and flattened) as well as key details. The holes in the lighter chimney are punched all the way through the steel of the inside case. The manufacturer's name and date codes are pressed into the bottoms of the brass outer cases.

3 The case pieces are trimmed. They are moved to the fabrication area where machines bend and fold them into their box-like shapes, and are then spot-welded together. The welding machines are highly accurate and can spot-weld a number of different positions in a case at the same time.

4 Meanwhile, small metal parts are moved to assembly stations in preparation for receiving complete cases. The lighter manufacturer fabricates most of the specialized parts, including the case hinge and the brass parts that hold and push up the flint. A precision machine fabricates and welds the hinges connecting the case lid and bottom to both parts using a strong welding process called "resistance welding." Other fabrica-

tion machines produce the components of the flint tube.

5 The brass finishes of the outer cases may remain unplated, but often the exteriors are finished in other metals, including nickel, gold, and chromium. This is done in a process called electroplating, in which a small electrical charge is applied to lighters suspended on a moving row of hangers that pass through a liquid bath. This bath contains a conducting solution that is non-metallic as well as a small piece of the plating metal. The opposite charge is applied to the bath, and atoms of the plating metal are drawn from that metal piece to the charged lighters. In this coating process, a thin layer of atoms is electrically bonded to all the surfaces of the cases.

6 Regardless of metal type, all cases receive final finishes. Many are polished to a sparkling luster. Others are given a brushed look or a texture. Assembly line workers apply the final finishes, inspect the lighter cases, and put them in fitted boxes for transfer to the next assembly station.

7 Machined parts are then fixed to the top of the inner case assembly. The cam, a finger-like projection from the inner case, applies enough pressure on the lid of the outer case to keep it closed. When the owner pushes up the front of the lid to operate the lighter, the thumb pressure overcomes the pressure that the cam applies, and the lid pops open easily.

8 The cam plate, which supports the cam, is riveted into place, and the cam is also fastened to the case with a rivet. Holes for these rivets were punched in the case when it was first deep-drawn from the stainless steel strip. A cam spring is added to the base of the chimney, which is also the top of the box-like portion of the case. An eyelet screw through the cam spring fastens it down and also provides the opening for the wick.

9 Elements of the inner case assembly are inserted in the welded shell. Several small balls of cotton-type material are placed inside the fuel chamber that will contain the lighter fluid. A length of wicking is inserted and will be pulled through the eyelet in the chimney later.

10 The sparking or flint wheel, which rubs against the flint to make the

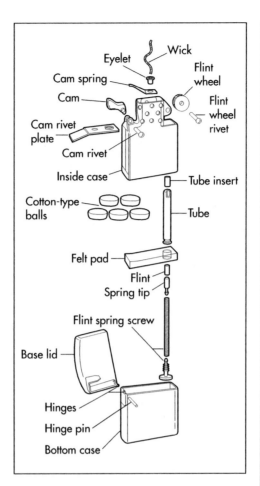

The internal components of a Zippo lighter.

spark and light the fuel, must also be firmly fixed to the top of the inner case assembly. A machine rivets the flint wheel to the case. Connections for the wheel on the lighter were also pre-punched during deep drawing.

11 Flints for producing the spark must be held in position next to the flint wheel and raised as they become worn. A brass tube is inserted in a hole in the bottom of the inner case. A flint, a spring tip that is directly in contact with the base of the flint, and the flint spring are pushed through the brass tube. The spring tip is made of brass, but the flint spring is hard spring steel that withstands wear.

12 A felt pad with a pre-cut hole for the flint and lighter fluid is stamped on the bottom of the inner case. The wicking is pulled through its eyelet, and the inner case assembly is installed in the outer case assembly.

13 Some lighters are sent to separate workstations to be decorated in any number of ways. Some have relief (three-dimensional) emblems attached to the right

A Zippo lighter.

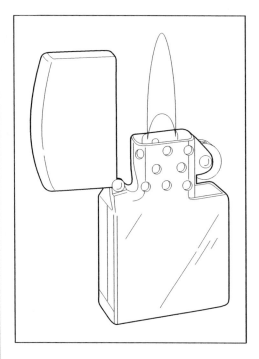

sides (as the lighter openings face forward). Diamond-drag rotary engraving cuts geometric patterns or monograms into some cases. Two types of lasers are used to engrave the outlines of line drawings, photos, company logos, and other designs on the sides of lighters. Still others have designs etched in them with computer-controlled etching machines.

14 Many designs can be filled with color using a painting method. To color more elaborate designs, a sublimation process transfers color by heat and pressure directly into the surface coating. In an example of this technigraphic method, a presentation box displays a design that extends over multiple lighters. A jungle scene that covers four lighters may have designs of animals and plants that stand alone on a single lighter, but may also form a puzzle or mural across the set.

15 The art department also designs packaging for maximum marketing effect. The artwork on the packages changes with the product and with customer orders. The boxes also showcase their contents. Plastic boxes, blister-card plastic containers, deluxe wood boxes, and custom-made collector tins with exterior designs that complement the enclosed lighters are examples. These may also have custom wrappings and may be lined with velour, felt, or other fabrics.

Quality Control

Quality control is subject to the same engineering detail as any of the plant operations. A quality method called Statistical Process Control (SPC) builds controls into all design aspects, from product conception to management. Lighters may not seem like highly sophisticated products, but their manufacture requires advanced technology and equipment to be cost-competitive in today's market.

Programmable logic controls (PLCs) allow machines to operate using information from an extensive database. Data acquisition is readily available plant-wide. Man-machine interfaces (MMIs) correct problems as they happen, and lighters with even the tiniest flaws are pulled from production immediately. These interfaces also keep all machines operating so that the maintenance or repair of one does not shut down others.

Mechanical quality is maintained by seeking the best new technologies, including fabrication processes and robotics. Finally, the personal touch cannot be replaced. Assemblers are responsible for monitoring the product in their area and for alerting supervisors if details are not perfect.

Byproducts/Waste

The processes required for lighter manufacture have been vastly improved to limit waste. Solvents were used in the past to degrease machine parts, but today's operations use only soap and water. Fewer hydrocarbons are used in deep drawing metals, and some plants have a complete water treatment system that returns creek water to the environment in a cleaner-than-natural state.

Lubricants are used as mists to penetrate the fine workings of screw machines, but the mist is fully contained to protect the ozone layer and employees. Fabric dust is vacuumed and contained, and metal and paper wastes are recycled.

The Future

Lighter makers have largely fled the United States for Europe and Asia where smoking is more accepted. However, lighters still have a promising future. They are handy as small light sources for finding lost keys and keyholes in the dark, and their wind resistance helps users determine wind direction because high winds or poor weather will not extinguish the flames.

Metal, reusable lighters are competitive against disposables because of their durability, reliability, quality, and sentimental value. Quality lighters are considered luxury items, however, and the competition for consumer dollars in this area is high. Lighter makers add artwork for uniqueness and adapt the outer cases to other personal accessories with the same convenience and quality.

Where to Learn More

Books

Schneider, Stuart, and David Poore. *Zippo: The Great American Lighter.* Atglen, PA: Schiffer Publishing Ltd., 2000.

Schneider, Stuart, and Ira Pilossof. *The Handbook of Vintage Cigarette Lighters.* Atglen, PA: Schiffer Publishing Ltd., 1999.

Periodicals

Dininny, Paulette. "Keepers of the Flame: After Big Sales in World War II and Parts in Old Movies, Zippos are Still Around, Often as Hot Collector's Items." *Smithsonian Magazine* (December 1998).

Other

BIC Corporation Web Page. December 2001. <http://www.bicworldusa.com>.

International Vintage Lighters Exchange Web Page. December 2001. <http://www.vintagelighters.com>.

"Lighter." *Discoveries and Inventions Web Page.* December 20001. <http://www.quido.cz/objevy/zapalovac.a.htm>.

Zippo Web Page. December 2001. <http://www.zippolighter.com>.

Gillian S. Holmes

Manhole Cover

Construction began on the Chicago sewer system in 1856. New York City had only 200 m (320 km) of sewer line laid by 1870, compared to 6,200 m (10,000 km) today.

Background

The subsurface of a major city teems with subsurface utilities: sewers, storm drains, steam tunnels, and utility corridors. Access ways, called manholes are dug down to these subsurface conduits at regular intervals to allow maintenance workers to reach them. Manholes are required so that people can clean, inspect, or repair the subsurface utilities. Manholes can be quite shallow or as deep as 70-stories in the third New York City water supply tunnel. Manhole covers are the round iron plates sunk into streets and sidewalks that keep passers-by from falling into manholes.

Manhole covers must be a minimum of 22 in (56 cm) in diameter, but can be as much as 60 in (1.5 m) in diameter. The average cover weighs between 250 and 300 lb (113–136 kg). It is important for sewer manhole covers to be heavy as sewers can produce methane gas that could push lightweight covers out of the way, letting noxious gases up into the street.

History

As soon as people began to live in cities the problem of what to do with human waste became an issue. The first cities were built along great rivers that served as open sewers. This was hardly satisfactory due to the periodic plagues that resulted from too much human waste in close contact with people. Roman civil engineers solved this problem with the invention of the underground sewer. The Roman sewers, dug by hand and lined with brick, collected a city's waste and deposited it far downstream. The Romans constructed access ways to these sewers to allow for periodic cleaning. The stone manhole covers that capped these access ways can still be seen in the old Roman City of Jerash in Jordan.

It would be quite a while until modern civilization rose to the level of the Romans. Construction began on the Chicago sewer system in 1856. New York City had only 200 mi (320 km) of sewer line laid by 1870, compared to 6,200 mi (10,000 km) today. The first manholes with covers were probably constructed in the early nineteenth century, not for sewers but for water or town gas pipelines. None of the covers for these manholes are known to survive to the present.

Raw Materials

Manhole covers are made out of cast iron. Cast iron means that the iron is melted and then poured, or cast, into a mold. Typical manhole covers are cast using gray cast iron. Ductile cast iron, because of its greater strength, is used for special manhole covers, like those that would be found near airplane terminals. Gray cast iron consists of the element iron and the alloying elements carbon and silicon. The alloying agents, chiefly carbon, give cast iron its strength and durability. Ductile cast iron is produced by adding manganese to the molten iron. The manganese causes the carbon in the iron to form nodules instead of flakes, giving ductile cast iron its greater strength and malleability.

Besides iron, the other raw material required to make manhole covers is green sand, which is sand bound together with clay. The green sand is used to produce the molds into which the molten iron is poured. The sand mixture consists of about 90% silica sand, 4–10% clay, 2–10% organics (e.g., coal), and 2–5%

water. The sand is not colored green. Green refers to the fact that it is allowed to remain wet during the casting process.

Design

Every manhole cover, from the simplest to the most ornate, is first modeled in wood or aluminum. The model is used to make the mold into which the molten cast iron will be poured. The designs that have been created for the surface of the manholes are as varied as the skilled artisans who created them. All manhole covers are round because a round object cannot be dropped into a round hole of the same diameter. This is vital since the weight of the manhole could easily kill a worker standing underneath it. Round manhole covers are also easier to move around on the surface as they can be rolled. There are rectangular utility box covers, but they are not installed over manholes.

The Manufacturing Process

All castings, including manhole covers, are made in large factories called foundries. Scrap steel comes into the foundry, is melted and alloyed, and leaves as iron casting. Cast iron is everywhere. A typical home in the United States contains around 2,000 lb (900 kg) of iron castings, mostly as pipe and pipe fittings, but also in furnaces and air conditioners. The casting process consists of five steps, pattern making, mold preparation, melting/pouring, and cooling and finishing.

Pattern making

1 Manhole patterns are either carved out of wood or machined out of aluminum. Aluminum models are used for large production runs because of their greater durability. Patterns are designed to be slightly larger than the finished manhole cover to allow for shrinkage as the castings cool. Two patterns, one for the top half of the cover and the other for the lower half, are required for each manhole. The top half of the pattern is usually provided with a decorative design, though the design is usually limited to a basic waffle, basket weave, or concentric circle pattern in modern times. Prior to 1950, the patterns could have been anything from shooting stars to city skylines. The bottom half of the mold may simply be flat,

or may be designed in a three-dimensional spider web pattern to provide much greater strength without increasing the cover's weight to a degree that would make moving the cover impractical.

Mold preparation

2 The sand molds are created by placing the two halves of the manhole model into boxes called flasks so that the models form the base of the box. The upper flask is known as a cope while the bottom flask is known as a drag. Green sand is tightly packed into the flasks to create the two molds. The upper mold contains holes (known as risers or sprues), into which the molten iron will be poured, and vents that allow gases to escape from the mold. For a manhole cover, these risers and vents can be created by simply placing a piece of wood vertically into the flask and removing it once the sand has been packed into the flask. The riser does not usually lead directly into the mold. The riser connects to runners, horizontal channels at the "parting line" (the plane where the two halves of the mold are joined). Using runners allows the molten metal to be fed into the mold at more than one location which helps prevent voids from forming in the final casting.

3 Once the patterns are removed, the bottoms of the flasks are then a hollow image of the upper and lower halves of the manhole cover. The bottom and top halves of the mold are then assembled in a "drag flask," a large metal frame.

4 Some castings are made with sand bound together with a chemical resin that is thermoset, which means it must be heated to become fixed. This process has some advantages in that the molds can be constructed very quickly and require less labor. These types of molds are ideal for automation when large numbers of casting are to be made. However, manhole covers are not usually produced in the quantities that would justify automation.

Melting/pouring

5 Cupola, electric arc, reverberatory, induction, and crucible furnaces are commonly used to melt the scrap steel that most foundries use to produce cast iron. The

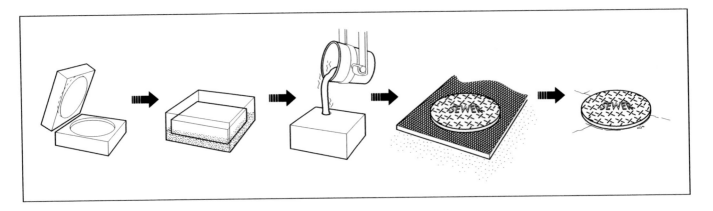

The manufacturing of a manhole cover using a sand mold.

scrap steel is placed into the furnace and melted at about 2,700°F (1,500°C).

6 Any required alloying metals and flux are then added to the molten iron. The purpose of the flux is to bind with any impurities creating a waste product called "slag." Because the slag is lighter than iron, it floats to the top of the molten iron and can be removed.

7 The molten iron is collected into a large metal ladle. Working from a distance to avoid being splashed with molten iron, foundry workers tip the ladle so that the iron pours into the sand mold through the riser (or sprue). The riser is designed to hold extra molten iron. As the casting cools and shrinks, the excess metal fills in the mold. Because the temperature of the molten iron is much higher than the autoignition temperature of the organic materials in the green sand, the organic materials burn and use up all the oxygen present in the mold. This prevents oxidation of the manhole cover. Foundry workers watch for the exhaust products jetting from the mold to make sure the gases are not trapped in the mold where they might cause bubbles in the casting.

Cooling

8 It takes about an hour and a half for the metal to cool sufficiently so that it can be removed from the mold. Complete cooling takes about a day.

9 In large foundries, the cooled casting and mold are placed on a vibratory grate and shaken until all of the sand has been shaken off. In a small foundry, the same process might be accomplished by a worker with a wire brush.

10 Handling the used sand from the molds can be a major headache for foundry personnel as enormous quantities of it can be generated during the casting process. After each use, the sand is sorted in a cyclone to remove any that is too fine to be reused and to sift out all of the metal slag that might be present.

Finishing

11 While finishing can be a large part of the casting process for intricate castings, manhole covers do not require a lot of finishing. For the most part, all that is required is to remove the runners, gates, and risers (the channels into which the molten iron was poured become little stalagmites on the finished manhole covers), shotblast the surface, and then machine the bearing surfaces to assure that the cover will lie flat in its frame.

Quality Control

Cast iron is usually made with scrap steel. As the raw materials are not controlled, casting houses must carefully analyze the molten metal before it is used to assure that it contains the proper percentages of iron, carbon, and alloying metals. After casting, the strength and ductility of the cast irons must be tested to assure that the manhole covers made from the iron will perform as designed. Strength and ductility is assessed by casting bar test specimens from the same metal used to cast the manhole covers. The bars are placed into a tensiometer which pulls on their ends until they either break or elongate past their elastic limit–the elastic limit is the point that the bar can be pulled to and still regain its original shape if the tension is released. Ductile cast iron can usually withstand between 2% and 10% elongation before it will break. Gray cast iron is brittle, and hence will break before it elongates significantly. Engineers who design products made with gray cast iron must always bear in mind that because the product is brittle, it will break without providing any warning if it is overloaded. As this could cause a disaster, cast iron components usually have much higher factors of safety than ductile iron components.

Byproducts/Waste

Gaseous emissions, such as carbon monoxide, hydrogen sulfide, sulfur dioxide, nitrous oxide, and benzene are produced when the molten iron contacts the green sand. In the past, lead was often used as a binder in some types of molds. Landfilling these used molds created heavy metal pollution problems. Resin bound castings produce volatile organic compounds when they are baked to set them.

Most of the sand in green sand can be recycled in new castings. However, a certain percentage of the sand becomes too fine during the casting operations and must be discarded.

The Future

It is unlikely that the production process for manhole covers will change much in the future. Nor is it likely that alternative materials will be used to produce manhole covers as cast iron is extremely economical. The exciting prospect for manhole covers involves computer-aided design and computer-aided manufacturing (CAD-CAM). With CAD-CAM, manhole cover designers can produce intricate patterns that can be cut out of plastic molds by automatic machinery. It will not be necessary for a highly-paid artisan to spend days or weeks creating particularly intricate models for special manhole covers. Once the design is ready, the model can be cut in just a few minutes. A golden age of manhole cover design may be at hand. Rather than dull, utilitarian circles, manhole covers may once again add a touch of artistry to city streets and sidewalks.

Where to Learn More

Books

Baumeister, Theodore, et al. *Marks's Standard Handbook for Mechanical Engineers.* 8th ed. McGraw Hill Book Company, 1979.

Davis, J. R., ed. *ASM Specialty Handbook, Cast Irons.* ASM International, 1996.

Melnick, Mimi. *Manhole Covers.* Cambridge: The MIT Press, 1994.

Samokhin, V. S., ed. *Design Handbook of Wastewater Systems.* New York: Allerton Press, Inc., 1986.

Other

Architectural Iron Company Web Page. 28 September 2001. <http://www.archironco.com/AIC/castingiron.asp>.

Sewers of the World Unite. 28 September 2001. <http://projects.artinfo.ru/sewers>.

Jeff Raines

Maracas

Maracas are used as musical instruments, and they are usually oval or egg-shaped.

Background

One of the most recognizable of the percussion instruments is the maracas, a pair of rattles made from gourds. Maracas are essential to Latin and South American orchestras and bands, and other musical forms that have adopted the rhythm of the maracas.

Maracas are used as musical instruments, and they are usually oval or egg-shaped. The family of musical instruments is divided into groups depending on how sound is produced. Solid or sealed objects that have full, distinctive sounds are classified as "idiophones." Maracas are part of a further subgroup of instruments that are shaken rather than struck. Idiophones that are struck include cymbals, castanets, and the xylophone.

The most universal form of construction of maracas uses dried gourds with beads, beans, or small stones inside. A handle is attached to each gourd, and the handle not only can be used for shaking but also seals in the noisemakers. The manufacturing process has evolved from one using only natural materials including gourds or other plant pods, wood, and leather to using plastic and fiber. It also features more sophisticated machinery to fashion wood handles.

History

Percussion instruments, especially drums, existed as long ago as the Stone Age. Maracas may have originated among several ancient civilizations at almost the same time. African tribes are known to have played drums and a wide variety of rattles and similar instruments from the traditions that have been carried down through the ages. South Pacific Islanders also developed a wide range of rattles by using plants that produced gourd-like seed pods; rattles without handles were even made from coconuts that had been dried out. In South America, maracas linked music and magic because witch doctors used maracas as symbols of supernatural beings; the gourds represented the heads of the spirits, and the witch doctor shook the gourds to summon them.

Just as maracas are essential to today's Latin and South American ensembles, the history of the maracas is best traced through the artwork of pre-Columbian Indians, especially the tribes in Colombia, Venezuela, Brazil, and Paraguay. The word *maraca* is believed to have been given to the instrument by the Araucanian people of central Chile. It is used for all gourd rattles although some also have more specific names. In the region of West Africa along the Atlantic Ocean called Guinea, native people tell the legend of a goddess making a maraca by sealing white pebbles in a *calabash*, a hard gourd that is also shaped into cooking utensils. Natives of the Congo in Africa and the Hopi Indians in America share the tradition of using turtle shells and baskets for rattles; when settlers brought European goods to America, native Americans collected empty shell cartridges, metal spice boxes, and cans to make rattles.

Players of maracas in the countries and regions in South America favor gourds of different varieties as well as unique playing customs. The "typical" maracas are played in Colombia, but musical ensembles in the Andes Mountains play smaller maracas called gapachos because they are filled with seeds from the gapacho plant. In Colombia's Llanos region, instrumentalists play clavellinas, which are similar to gapachos. In Paraguay, the porrongo gourd is used to make maracas, but only the men play them. Venezuelan ensembles use the maracas to set basic rhythms, but only the singers in the groups play them.

Some maracas relatives have beads on the outside. The gourd is larger than those typical of the maracas; the calabash is most common. The end is cut off but farther from the round body of the gourd, so the neck can be used as a handle. Strings of the same length are cut and tied to a center circle of string. Beads are strung along the lengths and tied again to a circle around the neck. Shaking this instrument rattles the loose strings and beads against the outside of the hollow gourd.

In modern times, many rhythm and percussion bands playing all styles of music use maracas. Composers have even written parts for them in classical pieces; for example, Prokofiev's *Romeo and Juliet*, written in 1935, calls for maracas in the fiery portions of this ballet. Maracas are even pounded on the heads of drums for interesting effects in classical music. Leonard Bernstein wrote the *Jeremiah Symphony* in 1942 and scored music for maracas used as drumsticks.

Raw Materials

Materials for the three major parts of the maracas are needed for manufacture. The hollow oval top is called the bell. It can be made of almost any kind of gourd or seedpod that can be dried or hollowed out. Traditional construction also uses leather that is cut into two parts, shaped to make the bell, and stitched along the seams. Plastic or fiber can be molded into round or oval shapes for the bells. Plastic maracas with small, round bells made in bright colors are toys and teaching tools for children and are marketed under catchy names.

The pellets that make the sound when the maracas are shaken are traditionally the dried seeds from inside the gourd. Other seeds, beans, beads, metal pellets, and even shells and buttons can be used inside maracas. Changing the type of material and the number of beans inside will change the sound.

The handle is made of wood or plastic. Wood is the traditional material and was carved or whittled to fit the opening in the gourd and to make an attractive shape and one that was comfortable to hold and shake. Today, wood is still used, but it is shaped with a lathe to make a uniform, attractive handle. Makers of maracas prefer Caribbean wood, mostly for its beauty; but soft to ultra hard wood is chosen depending on the size, shape, and appearance of the maracas.

Lesser materials include heavy thread that is used to stitch the halves of leather maracas together, and thick string that is wound around the top of the handle and the base of the gourd, then glued to hold them together. Cloth bindings (much like hem tape) can also be wrapped around the join of the handle and gourd.

Design

The design of maracas has assumed a traditional shape even though there are many variations within the family of rattles. Maracas have an oval top or bell in a hollow, outer shell and contains bean-sized objects that rattle against the shell when the instrument is shaken. To shake the maracas, a handle is attached.

Within this basic description, the materials used to make the bell, beans, and handle can vary in type of material, shape, and size. Traditional maracas are gourds or stitched leather with wood handles. However, modern technology has produced hard fibers and plastics for the bell as well as plastic noisemakers. Machinery like lathes can be used to shape handles that precisely fit the bell. Machines stitch the parts of the bell when they are made of leather. Modern glues are also a technical improvement that assures the long-lasting fit of bell, handles, and binding. Manufacturers use climate-controlled rooms to dry the gourds carefully. If they are dried too quickly, the outer skins will shrink and shrivel.

The designs on the outsides of the bells are also varied and made of different materials. Most gourds are painted on the outsides with bright colors from their native homes or with colors suited to the instrumental group or musical style. Red, yellow, and green are a vibrant, popular color combination, but images in dark brown on the yellow show instruments, native peoples, or beaches and trees or other scenes. Hawaiian dancers play gourds that have feathers suspended from the binding around the handle, and the feathers sway with the dancers.

The Manufacturing Process

1 Manufacturers of percussion instruments typically make many varieties of instruments ranging from kettledrums (called tympani) to xylophones and maracas. They purchase natural materials from a suppliers but tool their own machines to their standards for cutting and shaping these materials. To produce maracas made of gourds, they buy gourds from a local supplier, often a farmer. The manufacturers cut off the narrow end of each gourd with a thin-bladed band saw. Knives or spoons with long handles and narrow bowls are used to scrap the membranes and seeds out of the gourd. The membranes are disposed, but the seeds are washed and saved.

2 After the insides of the gourds are cleaned, the gourds and their seeds are dried in a climate-controlled room. For some styles and if the necks of the gourds are long enough, the necks are also dried and kept with the gourds they came from. Gourds are usually dried for months (and sometimes as much as a year) in these controlled conditions so that the interior will dry completely and the exterior will not wrinkle.

3 If the necks of the gourds are not used, wooden handles are cut to appropriate lengths and general shapes. By using a lathe, the handle can be shaped with rounded ridges to make it easy to grip. In some cases, more of the neck end of the gourd is cut off to speed removal of membranes and seeds, as well as drying. In this case, the end of the handle that will be attached to the gourd is cut into a funnel-like shape that will fit the gourd. Some handles are much simpler and resemble rod-like pieces of wood called dowels. The various types and shapes of handles are stored in boxes for assembly when the gourds are ready.

4 When the gourds and seeds are dried, the outsides of the gourds may be sanded to smooth any irregularities. Each gourd is then partially filled with seeds. Other noise-makers like beans or small stones may be used for different sound effects, and the quantity of seeds or other materials also influences the sound. Manufacturers usually

have their own special formula for filling the gourds or maracas made of other materials. Percussionists also request certain sounds, so some maracas are custom made.

5 The handles are then attached to the gourds. Those with flared tops are matched to the gourds with larger openings, and the handle tops are shaped to suit the gourds. The fitted handles are then glued to the gourds. After they have dried, the gourd-handle join may be sanded so that the join can barely be seen. Long necks saved with their gourds are also glued in place with high-strength, long-lasting glue.

When handles are not matched to the bells of the maracas, a transition join between the handle and each gourd may be needed. Some styles of maracas use a round piece of wood that is glued to both sections and wrapped with binding for an attractive finish. For other styles, twine soaked in glue is wound around the top of the handle and lower end of the bell. A second layer of twine binding is added to smooth the appearance.

6 When assembly of the maracas is complete and glue has dried, the maracas are painted. In some cases, they are left in their natural finish. Bright enamels are used to paint the maracas by hand; usually, several layers are applied to create an even finish and blend the edges of the colors. When the paint has dried, the instruments are coated with shellac that is also dried.

7 As a final step, the maracas are individually packaged as a pair in a box. Wrapping of the individual rattles prevents them from knocking against each other. The pairs in boxes are then packed in bulk for shipment to distributors or instrument shops.

Quality Control

Although maracas are relatively simple, they are still musical instruments that require care in manufacture. Skilled crafters complete all the steps in making maracas, and handcrafting is essential to many steps. Manufacturers oversee the process, but the workers themselves are the true quality control experts because pride in their work demands skill and attention. Workers also test the sound quality of the instrument. If the filling material is stuck together, the maraca must be discarded.

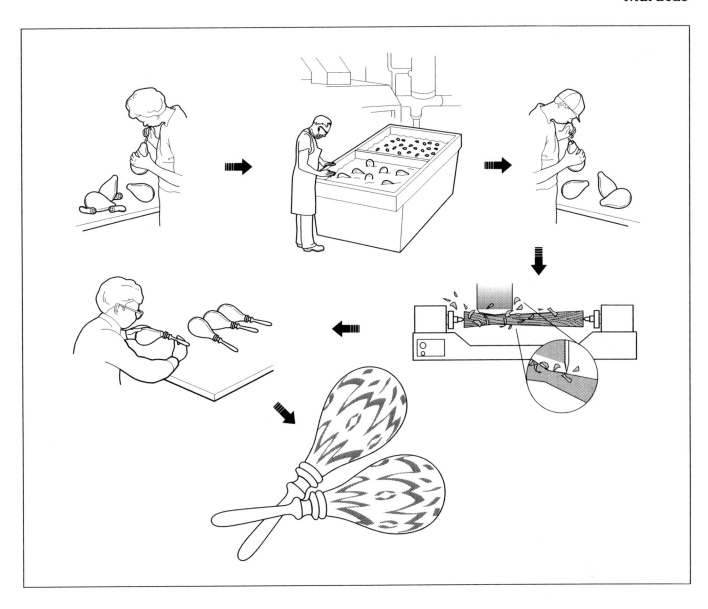

Byproducts/Waste

Manufacture of maracas does not generate any byproducts although many styles may be made in the same facility. Waste is also very limited. Membranes and seeds from the gourds can be disposed as green waste that can be composted. Wood shavings and saw dust and trimmings from other components are minor in volume.

The Future

Maracas have a long past and a promising future because of their rhythmic sound. They are often first instruments for children and so have happy associations. Musicologists, who preserve the history of musical styles that may not have been written down, are recording and documenting ethnic music using maracas in many parts of the world so this musical heritage will not be lost. In modern music, these percussion instruments have found comfortable homes in many musical styles. The recent and increasing popularity of Latin music has brought maracas great attention, and ethnic music demands the essential sound of the maracas. Recordings spread this fascination, building a larger and larger audience for the maracas.

Where to Learn More

Books

Baines, Anthony, ed. *Musical Instruments Through the Ages.* New York: Walker and Company, 1976.

Maracas are typically made out of dried gourds filled with dried seeds.

Buchner, Alexander. *Folk Music Instruments.* New York: Crown Publishers, Inc., 1972.

Hunter, Ilene, and Marilyn Judson. *Simple Folk Instruments to Make and to Play.* New York: Simon and Schuster, 1977.

Sadie, Stanley, ed. *The Grove Dictionary of Musical Instruments.* Vol. 2. London: Macmillan Press, 1984.

Other

Jansky, Charlotte. "Allegro Music." *AllegroMusic-Fremount.com Web Page.* December 2001. <http://www.allegromusic-fremont.com>.

Mambiza Drums & Percussion Web Page. December 2001. <http://www.mambiza.com>.

Gillian S. Holmes

Microphone

Background

A microphone is a device that converts mechanical energy waves or sound into electrical energy waves. Speaking into a microphone excites (moves) a diaphragm that is coupled to a device that creates an electrical current proportional to the sound waves produced.

Microphones are a part of everyday life. They are used in telephones, transmitters for commercial radio and television broadcast, amateur radio, baby monitors, tape recorders, motion pictures, and public address systems. There are many different types of microphones—the design depending upon the application. Sound recording, radio and television, and motion picture studios use ribbon or condenser type microphones because of their high quality reproduction of sound. Public address systems, telephones, and two-way radio communications systems can use carbon, ceramic, or dynamic microphones because of their versatility and low cost.

History

The first microphone was invented as a telephone transmitter by Alexander Graham Bell in 1876. It was a liquid device that was not very practical. In 1886, Thomas Alva Edison invented the first practical carbon microphone. The carbon microphone was used for radio transmissions and extensively in telephone transmitters until the 1970s when they were replaced by piezoelectric ceramic elements.

The carbon microphone had a limited frequency range, and would not reproduce music effectively. In 1916, the condenser microphone was developed by E. C. Wente

of Bell Laboratories. The condensor microphone required an amplifier built within the microphone to pick up the faint signals. Condensor microphones were used for radio broadcasting and the first generation of sound motion pictures.

A major breakthrough in microphone technology would come in 1931 with the invention of the moving-coil or dynamic microphone by Wente and A. C. Thuras of Bell Laboratories. The dynamic microphone has a lower noise or distortion level than that of the carbon microphone and required no power to operate. The dynamic microphone is in extensive use today in all areas of communication and entertainment.

In 1931, the ribbon microphone was introduced by RCA, and became one of the most widely used microphones for the vocal recording and broadcasting industries. It was considered by many as the most natural sounding microphone ever made. The ribbon microphone was very heavy, about 8 lb (3.6 kg), and could easily be damaged by shock or blowing into it. Variations of the ribbon microphone are still use today.

The ceramic or crystal microphone was invented 1933 by the Astatic Corporation when C. M. Chorpening and F. H. Woodworth found that they could make a microphone out of Rochelle salts or piezoelectric crystals. They found that when sound waves struck these crystals, they vibrated and created an electrical current.

Raw Materials

Depending upon the type of microphone, raw materials may vary. Permanent magnets are generally made from a neodymium iron

The first microphone was invented as a telephone transmitter by Alexander Graham Bell in 1876.

Jim Morrison.

James Douglas Morrison was born December 8, 1943 in Melbourne, Florida. After finishing high school in Alexandria, Virginia, Morrison took classes at St. Petersburg Junior College and Florida State University before traveling to California in 1964. By 1966, Morrison was enrolled at UCLA. There, he met organist Ray Manzarek and, soon after, guitarist Robbie Krieger and drummer John Densmore, forming the Doors.

Hard rock, mysticism, lyrical poetry, and theatrics merged in the group's music. Some critics dismissed Morrison as a self-indulgent vocalist who sold out to the demands of the pop music market after becoming popular. Others praised Morrison as a powerful singer and poet and believed the Doors' unique sound represented a brilliant fusion of jazz, rock, blues, and pop.

Late in 1970, Morrison became disillusioned with his celebrity status. He settled in Paris to work on poetry and a screenplay. Morrison died suddenly on July 3, 1971, at the age of 27. Official reports stated he suffered a heart attack while bathing, but his body was only seen by a doctor and Morrison's common-law wife. A legend arose that Morrison was not really dead. His tomb is in the Poets' Corner of the Pere-Lachaise cemetery in Paris, near the graves of Balzac, Moliere, and Oscar Wilde.

Morrison remains a cult figure as a poetic messiah whose uncompromising vision led to an early death. Today fans continue to visit Morrison's grave, buy his records, and read his poetry. Elektra Records, the Doors recording company, still sells over 100,000 Doors records, cassettes, and compact discs every year. July 3, 2001 marked the thirtieth anniversary of Morrison's death and over 20,000 people visited the gravesite.

boron compound. The voice coil and cable are made from copper wire. Plastic is used for cable insulation. The case is usually made from aluminum sheet and sometimes plastic.

Design

The dynamic or moving-coil microphone consists of a thin plastic diaphragm attached to a voice coil. The voice coil consists of many turns of very small diameter insulated copper wire wound on a bobbin. Surrounding the voice coil is a permanent magnet. Sound causes the diaphragm to vibrate, which causes the voice coil to move on its axis. This movement induces a voltage in the coil and creates a varying electrical current proportional to the sound to flow through the coil. This induced current is the audio signal.

The condenser or capacitor microphone consists of two metal plates spaced slightly apart. These two plates act as a capacitor. A capacitor is a device that stores an electrical charge. The front plate acts as a diaphragm. As the diaphragm vibrates, an electrical current is induced to the attached wires creating an electrical signal between the two plates.

A carbon microphone consists of lightly packed carbon granules in an enclosure. Electrical contacts are placed on opposite sides of the enclosure. A thin metal or plastic diaphragm is mounted on one side of the enclosure. As sound waves hit the diaphragm they compress the carbon granules, changing its resistance. By running a current through the carbon, the changing resistance produced by the sound changes the amount of current that flows in proportion to the sound waves.

The diaphragm of a ribbon microphone uses a thin corrugated aluminum ribbon about 2 in (50 mm) in length and 0.5 in (2.5 mm) wide suspended in a strong magnetic field. As sound pressure variations displace the ribbon, it cuts across the magnetic field. This induces a voltage and produces a current that is proportional to the sound striking it.

Ceramic or crystal microphones use a quartz or ceramic crystal. Electrodes are placed on either side of the crystal. When sound pressure variations displace the crystal an electrical current is created that is proportional to the sound striking it.

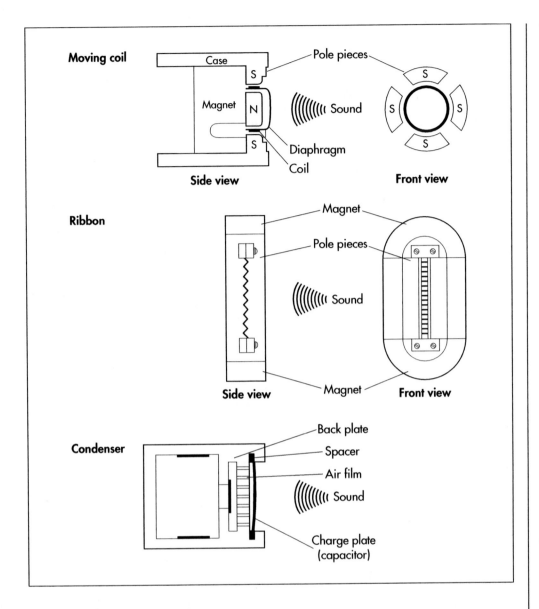

Moving coil

Case — Pole pieces

Magnet

N

Sound

Diaphragm
Coil

Side view

S S S

S

Front view

Ribbon

Magnet

Pole pieces

Sound

Side view

Magnet

Front view

Condenser

Back plate
Spacer
Air film

Sound

Charge plate
(capacitor)

The Manufacturing Process

While the manufacturing process will vary depending upon the type of microphone and how it is used, all microphones had three common parts—a capsule containing the microphone element, internal wiring, and a housing. The following process describe the construction of a moving-coil or dynamic microphone.

1 The case is formed from thin sheet aluminum or mold injected plastic. The aluminum sheet is placed in the die of a punch press. The die is an inverted replica of the desired case shape. The hydraulic punch is release and forces the aluminum into the die. Any excess material is trimmed and dis-

carded. If the case is to be made of plastic, the plastic pellets are fed into a hopper and melted. The liquid is poured into an injection molding machine. The machine feeds the liquid into a closed mold. Once the mold is filled and the plastic has cooled, the mold is opened and the plastic case is taken out. If a switch is required, it is mounted in position in the case and secured with small screws and nuts or rivets.

2 The voice coil is made by winding very fine enameled copper wire onto a plastic bobbin. The wire is secured to the bobbin with glue.

3 The permanent magnet is made from a neodymium iron boron compound. It is formed by sintering the powder (the powder

is placed in a high pressure die and heated, the metals combine and becomes a solid) or by bonding it with plastic binders.

4 The pre-cut plastic diaphragm is placed in a holding fixture. The voice coil bobbin is then glued in the exact center of the bobbin. After the glue has cured (about 24 hours), the assembly is lowered into the permanent magnet assembly and glued together.

5 A coaxial audio signal cable is selected and cut to length. Insulation is stripped from all leads at both ends of the cable. Then, an audio connector is soldered to one end of the cable. The open end to the cable is left free.

6 The open end of the audio cable is inserted through its hole in the bottom of the case. The cable is pulled out through the top of the case a sufficient length to allow the wires to be soldered to the switch and voice coil.

7 A foam rubber spacer is placed around the voice coil assembly and the assembly is lowered into the case. It is secured into proper place with a grille and cap.

8 The microphone is then packaged and shipped to the distributor.

Quality Control

The microphone is tested by placing the voice coil assembly in a test station. The test station emits a white noise signal, which contains all audible frequencies at one time. The frequency response is then measured to ensure that the microphone is within specifications.

Byproducts/Waste

Scrap metal or plastic from the case can be recycled and remolded. Exotic materials such as neodymium iron boron must be disposed of according to government chemical regulations.

The Future

The industry is constantly experimenting with raw materials to improve microphone sound quality, sensitivity, and frequency response. As technology advances, microphones are becoming more and more common. They are now standard with any new computer system, giving the user the opportunity to talk to friends and family over the Internet. Depending on their use, microphones are constantly being redesigned to incorporate the different needs of the customer.

Where to Learn More

Books

"Amplitude Modulation." December 2001. <http://www.tpub.com/neets/book12/48j.htm>.

"History of the Microphone." December 2001. <http://users.belgacom.net/gc391665/microphone_history.htm>.

"Microphones." December 2001. <http://hyperphysics.phy-astr.gsu.edu/hbase/audio/mic.html#c3>.

"Microphone History." December 2001. <http://history.acusd.edu/gen/recording/microphones1.html>.

Ernst S. Sibberson

Mop

Background

Mops are classified in two main divisions as either wet or dry mops. Wet mops are commonly used to clean kitchen and bathroom floors. They usually have sponge or cloth heads that can be put in water with a detergent or other cleaner (under the general term surfactant) and rinsed when cleaning is finished. Wet mop heads can be easily cleaned themselves, and this should be done regularly to make them efficient in cleaning and absorbing dust. Wet mops should be dried thoroughly before they are stored, and those with cotton strings that fray at the ends should be trimmed occasionally. Mop heads are replaceable when they begin to wear.

The dry mop is also called the dust mop and is characterized by a large, flat head that can be pushed easily over a floor surface. The strings making up the head pick up dust, lint, and hair as the mop glides across the floor. A swivel at the point where the mop head joins the handle allows the mop to be pushed under beds and in other places with limited access. A dry mop can be shaken outdoors to remove dust, but, if the dust clogs the mop, it should be soaked in soapy water overnight. A detachable mop head can be machine-washed. Treating the dry mop with dust mop oil after washing also preserves it and helps the dust cling to the mop head.

History

The mop is a patented invention that is part of social history as well as the evolution of house wares. Thomas W. Steward, an African-American inventor, was awarded Patent Number 499,402 on June 13, 1893, for inventing the mop. His creation joined a long list of household equipment invented by African-Americans. The roster includes the eggbeater, yarn holder, ironing board, and bread-kneading machine. Steward's deck mop, made of yarn, quickly became well used for household and industrial cleaning. A wringing mechanism made the process of mopping and cleaning the mop easier and faster.

Another pair of inventors, brothers Peter and Thomas Vosbikian, fled Europe just before World War I and patented over 100 inventions in 30 years. In 1950, Peter Vosbikian developed a sponge mop that used a lever and flat strip of metal to press against the wet mop and squeeze it dry. This automatic mop eliminated the need to bend over and wring the mop repeatedly by hand. Its development was aided by the many technological improvements in the plastics industry that grew out of World War II and made absorbent plastic mop heads possible.

Other modifications have made mops even more adaptable to different cleaning chores. In 1999, Scotch Brite released a new wet mop made of natural cellulose and reinforced with internal polyester net. The cellulose does not leave lint like a cloth mop and absorbs 17 times its dry weight.

Raw Materials

Dust (dry) or wet mops consist of the same three basic parts: the mop head including a frame, a mechanical attachment (linking the head and handle) that may be fixed or may swivel, and the handle. The head of a dust mop is typically made of yarn consisting of natural or synthetic fibers like cotton or nylon. The yarn is attached to a carrier substrate, which is almost rigid and holds the

Thomas W. Steward, an African-American inventor, was awarded Patent Number 499,402 on June 13, 1893, for inventing the mop.

shape of the mopping surface. The carrier substrate is fabric, vinyl, or molded plastic. Heads for wet mops are either made of loosely woven yarn or sponge. Like dry mops, the yarn for wet mops may be made of natural or synthetic materials. Sponge mops usually have rectangular heads made of a natural material like cellulose or a synthetic such as polyurethane foam.

The mechanical attachment fixes the mop head to the handle, but the attachment varies widely depending on the type, shape, and use of the mop. The mechanical attachment for a dust mop is made of steel wire, plated metal, or plastic that supports the shape of the head and carrier substrate. It also usually supports a swivel, also made of metal or plastic, that fastens to the frame and handle. Plastic is the most common material for mechanical attachments and swivels on household dust mops, and the plastic attachments are made of durable resins that are injection-molded.

The frame for the wet mop is also made of stamped metal. Steel is commonly used, but it is plated with zinc to protect it from water damage. The mop head does not swivel, but the mechanical attachment linking it to the handle may be a single plate, a double hinged plate that folds like a butterfly to squeeze the mop dry, or a roller mechanism that squeezes the head between two rollers. The mechanism is integrated into the frame along the major axis (the widest portion) of the sponge and has a lever that parallels the handle so the person operating the mop can activate the hinge to squeeze the mop without bending down. Attachments on wet mops also allow for removing and replacing the mop heads when they get dirty.

Handles for dust and wet mops are similar. Dust mops are made with tubular steel or wood handles. Sometimes fiberglass or aluminum is used, but these are less common and much more expensive. Historically, wood handles have also been used in making wet mops, but tubular steel coated with plastic or chrome-plated is the preferred material today.

Design

New designs for mops are driven by changes in technology, consumer demand for products with specialized functions, or internal resources within the manufacturing company. The basic shapes of mops are well suited to their uses, but they do not have to be unattractive to perform their functions. Some mop makers focus on color schemes and other fashion trends in designing new products. New types of fibers and lightweight components are technical improvements that are built into new mop designs.

Since the mid 1990s, static cleaners with disposable cloth covers have been heavily marketed, and they have had some effect on the mop industry. Mops, however, are much more durable and can be cleaned many times before the heads must be replaced. Nylon mops also hold static electricity charges and are just as effective as the static cleaner cloths in attracting and holding dust and hair.

The Manufacturing Process

1 The manufacturer's receiving department accepts raw materials and components made by subcontractors. All materials are inspected, accepted or returned to the supplier, logged in, and stored until they are needed on the production lines. Most of the components for mops are made by outside suppliers specializing in producing plastics, metal products, wood handles or tubular metal handles, and sponges. Yarn fibers that are used to make dust and wet mops are ordered in bulk quantities and processed by the manufacturer.

2 The manufacturer's production area includes many specialized workstations. Yarn for the mop head is cut to a specific length, sewn together by industrial weight sewing machine, and attached to the carrier substrate also by machine stitching. The substrates of molded plastic, vinyl, or heavy fabric are made to the manufacturer's specifications by outside suppliers. Assembly line workers pull frames from bins next to their workstations and fix them to the carrier substrates. Usually, the substrate is fitted to the frame, and the worker stitches it in place. While the substrate forms the general shape of the mop, the frame holds that shape rigid.

3 At the next work station, the wire metal frames are connected to the mechanical attachments. For dust or dry mops, the attachments are usually swivel devices. The connections between the swivels and the

Sponge mop with
built-in wringer

Wet mop

Dust (dry) mop

frames are designed to fit together securely. Usually, connections that are stamped in the metal pieces guide them in place and clip or snap features fix them snugly together.

4 The other end of the connection is designed to fit the handle. Before the two are connected, the handles are inspected and finished. Wood handles may be sanded or smoothed and painted, and tubular steel is inspected for rough edges or irregularities. Plastic-coated steel handles for wet mops are also checked carefully to make sure that the coating is uniform so water will not damage exposed metal. Mechanical attachments and handles are joined together; again, metal stampings usually lock them in place.

5 Finished mops are carefully bundled together and taken to the shipping department. Preprinted labels and card paper wrappers or preprinted plastic bags are fastened around the mop heads. Specified numbers of units are packed into shipping containers for bulk distribution to retailers.

Quality Control

Mop manufacturers use an inspector-based system of quality control. At various steps in assembling mops, the inspectors look at materials, methods or processes, and the products themselves to see if there are any visual defects. Mops that have scratches, cracks, or loose threads are discarded. Assembly line workers are not responsible for

quality control. The inspectors typically check raw materials and materials from suppliers as they are received. They observe and control processes on the production line and audit quality and quantity of work. Finally, the inspectors check the finished goods including the mops that are packed and ready for shipment.

Byproducts/Waste

Mop manufacturers do not generate byproducts, but they usually make a wide variety of dust and wet mops in different sizes and shapes to suit residential and commercial consumers. Mops are also one category of product that may be made by manufacturers of many other small house wares and cleaning products.

Waste is only produced in tiny quantities. The inspection and quality control process prevents inferior mops from being made on the production line. Most waste consists of yarn trimmings and loss of yarn fiber that generates dust. The waste is controlled and disposed, and workers are protected from the dust in keeping with the requirements of the Occupational Safety and Health Administration (OSHA). Mop production does not result in any hazardous wastes.

The Future

Both dust and wet mops are trusted fixtures in most homes and businesses because their usefulness is proven and they are inexpensive cleaning tools. Manufacturers are always looking for innovative methods of improving their products, especially in the efficiency of performing what most people consider a mundane task. Wet mops and dust or dry mops are surface-specific, so as new types of flooring are developed, manufacturers adapt mops to them. Technical developments are also ongoing in fabrics used as mop fibers and in cleaning media like dust protectants, detergents, and polishes. With such long-standing reliability, mops surely will continue to be needed.

Where to Learn More

Books

Aslett, Don. *Do I Dust or Vacuum First?* Cincinnati: C. J. Krehbiel Co., 1982.

Moore, Alma Chestnut. *How to Clean Everything.* New York: Simon and Schuster, 1977.

Vare, Ethlie Ann, and Greg Ptacek. *Mothers of Invention: From the Bra to the Bomb: Forgotten Women & Their Unforgettable Ideas.* New York: William Morrow and Company, Inc., 1988.

Other

OCedar Brands, Inc. Web Page. December 2001. <http://www.ocedar.com>.

Quickie Manufacturing Corporation Web Page. December 2001. <http://www.quickie.com/aboutbody.html>.

Robinson, Maisah B. "African-American History. 19th Century African American Inventors—Part 1." Feb. 26, 2001. December 2001. <http://www.suite101.com/article.cfm/african_american_history/61415>.

Gillian S. Holmes

Movie Projector

History

The inspiration for the development of motion pictures and projectors can be traced to a variety of sources including theaters, circuses, and magic shows. Another important factor was the understanding of the phenomenon of persistence of vision. While the process was known for hundreds of years, it was only in the early nineteenth century when Roget introduced the underlying theory in an article that it developed popular interest. In short, persistence of vision is the phenomena in which the brain retains an image that is observed by the eyes for slightly longer than it is actually seen. Movies take advantage of persistence of vision to create the illusion of motion. When successive still frames are viewed, the brain "connects" the image and they appear to move.

During the early 1800s, hundreds of novelty devices based on this principle were introduced. Some of the most influential include the Thaumatrope and the Phenakistiscope. Dr. John Ayrton Paris is generally regarded as having invented the Thaumatrope in 1825. This device was a toy with a simple design that took advantage of persistence of vision. It consisted of a small round board with a picture on both sides. The original toy had a bird on one side and a cage on the other. The board was held at the side by two strings and when spun it appeared as though the bird was in the cage.

The Phenakistiscope was introduced in 1832 by Joseph Antoine Ferdinand Plateau. This toy was a disc with a fixed center that allowed it to be spun freely. Various images were drawn on the outer edges of the disk depicting sequential movement. The pictures were spaced evenly and slits were cut in conjunction with each. The toy was held between the user and a mirror and images were viewed reflected by the mirror. The persistence of vision created the illusion of movement. Plateau was the first to realize that there had to be a resting period between images for a perfect illusion and determined that 16 images per second was the optimum number. Other inventors introduced similar devices. In 1853, Baron von Uchatius invented a projecting Phenakistiscope by adding a lantern. This was the earliest known moving picture.

One of the most important early moving picture devices was the Zoetrope that was invented by William George Horner in 1834. This device was a rotating drum that had slits cut in its side. A strip of paper, which contained the images, was affixed to the inside and the top of the drum was open. When the drum was spun, the images appeared to move. This was by far the most popular of all these animation toys. It had the added advantage of being able to change the pictures by putting in a different strip of paper with new images. The next device that advanced the technology of animation toys was the kineograph, invented in 1868. This was essentially a flipbook that had drawings or pictures of sequential movement. When the pages were flipped, the illusion of motion was created.

In 1891, Thomas Edison introduced a mechanized version of the Zoetrope he called the Kinetoscope. While similar in principle, it had significant changes. Instead of being moved by hand, the device had a motor attached for automated movement. Also, instead of simple paper images, it utilized a film with that had pictures on it. The film

Some companies are working on a system by which movies are produced on computer hard drives. Stored in this way, movies promise to be much less expensive to distribute and display.

Louis Jean Lumiére.

Auguste Marie Louis Lumiére was born October 19, 1862, in Besançon, France. His brother, Louis Jean, was born October 5, 1864. In 1894, they began looking for ways to project motion pictures, expanding on Thomas Edison's ideas. In 1889, Edison created the kinetograph, which used strips of photographic paper to take motion pictures. Edison produced the kinetoscope in 1893, allowing a single person to view a moving image. The Lumiéres' goal was to improve Edison's ideas and project motion picture films for a larger audience.

Louis realized the problem of projection was creating continuous movement of the film. He realized the "presser foot" mechanism of a sewing machine could be adapted to move small sections of film quickly across the lens, allowing a short period of time for each frame to be stationary for exposure. This machine, the Cinematograph, could create the negatives of an image on film, print a positive image, and project the results at a speed of 12 frames per second.

The Lumiéres arranged to bring these films to the public. On December 28, 1895, the Grand Café in Paris held the first public show of projected moving pictures. An approaching train shot from a head-on perspective frightened people in the audience who, in a panic, tried to escape; others fainted.

On June 6, 1948, Louis died in Bandol, France at the age of 83. Auguste lived to the age of 91, dying in Lyons, France, April 10, 1954. The Lumiéres are symbols of technological creativity and growth. They are remembered for bringing technology to a wider marketplace, a value seen in their contributions to the motion picture industry, which has become a popular form of entertainment around the world.

was moved past a fixed light source that projected an image on the wall of a closed booth. When it was found that people would gather to watch these moving pictures, a new industry was born. In 1895, the Lumiére brothers, Auguste and Louis, introduced the Cinematograph. This device was a camera that could take pictures, process it into film and project the image. In 1896, they introduced the Vitascope, which was similar to the Kinetoscope. The primary difference was that the image could be projected onto a much larger screen.

During the course of the twentieth century, movie projector design became more complicated and sophisticated. Spools were added to make it easier for film to move past the light source. The length of movies was significantly increased, and by the 1920s sound was available. In the 1930s, color movies were introduced. The industry was revolutionized in the 1960s by the introduction of the platter that made it possible to show a long movie using a single projector. During the 1970s and 1980s, digital sound was developed. Today, movie projectors are much more impressive and functional than the early counterparts, but the basic principal by which they work remains the same.

Design

Movie projectors consist of four primary sections including the spool assembly, lamp assembly, lens assembly, and audio assembly.

Spool assembly

The primary purpose of the spool assembly is to move the film through the projector. While the motion appears continuous, there is actually a slight pause after each frame. This allows light to be passed through the image and projected on the screen. The spool assembly is made up of all the parts related to storing and moving the film. The platter, which is located on the side of the projector, consists of up to four large discs about 5 ft (152 cm) in diameter vertically stacked between 1–2 ft (30–60.1 cm) apart. Each disc is large enough to hold the length of an entire film. Since every second of film requires 24 frames, a two-hour movie can be as long as 2 mi (3.2 km) when stretched out. Therefore, films are provided to movie theaters on numerous reels that must be spliced together before being loaded on the platter.

A payout assembly on the side of the platter moves the film from the feed disc through the lamp and lens assembly and back to the receiving disc. The film has small holes on its edges that allow it to be held by specialized gears called sprockets. An electric motor turns the sprockets that cause the film to be pulled through the device. Spring-loaded rollers, called cambers, provide tension to keep the film from slipping out of the sprockets. Intermittent sprockets have been developed to pull the film one frame at a time and pause before moving again. They are timed to show 24 frames per second. The film is also stretched between two bars as it passes in front of the lens to keep it tight and aligned. Depending on the projector design, the film is passed through a sound decoding system that is located above or below the lens.

Lamp assembly

The lamp assembly includes all of the parts related to illuminating the image on the film. The key element is the light source. Modern movie projectors use a xenon bulb because they burn brightly for thousands of hours. A xenon bulb is constructed with a quartz outer shell, a cathode, and an anode. When current is applied, the bulb burns bright and hot. The bulb is located in the center of a parabolic mirror which is mounted in the lamphouse. The mirror focuses the light and reflects it onto the condenser. The condenser consists of two lenses that focus the light further and direct it to the main lens assembly. The whole setup not only intensifies the light but also the heat which is why film quickly melts if it is suddenly stopped moving through the projector. Most projectors have a cooling system because of the heat generated by the lamp.

Lens assembly

The light is next passed through the picture head and lens assembly. At the start of this section is the shutter which is a small plate that is rotates 24 times a second. Its movement is synchronized with the advancing film so that dark spaces between the frames are not seen. If the shutter was not in place, the film would appear to flicker. To further reduce flickering, some movie projectors are designed with double shutters. The light is then passed through a small metal frame called the aperture. This ensures that light only shines on the part of the film with the image and not on the sprocket holes.

Light passing through the film causes the image to be projected. The main lens first focuses this image. On most movie projectors, lenses can be removed and changed for different movies. There are primarily two types of lenses available: flat and CinemaScope. A flat lens is more suitable for comedies and dramas while a CinemaScope lens is designed for action movies. Flat lenses typically are between 1.5–1.8 in (37–45 mm) long while CinemaScope lenses are 2.8–3.3 in (70–85 mm). Some movie projectors have a turret system which contains multiple lenses that can be automatically moved into place as needed.

Audio assembly

The audio assembly is the part of the projector that gives the film sound. Two types of technologies can be used: optical or magnetic. Optical systems are the most common. They consist of a light source and a photocell. On one side of the film, a transparent line is recorded. The line varies in width depending on the frequency of sound. As it passes by the light source, varying amounts of light are passed though. A photocell located on the side of the film opposite the light source, picks up the transmitted light. This light is then converted to an audio signal that is then amplified before being sent out to the speakers. Magnetic systems have a recorder head that is in direct contact with the film. The differences in the magnetic field on the film are then converted to the audio signal. Magnetic sound systems are not used as much because they have disadvantages such as being easily damaged, more expensive and a shorter life span.

Raw Materials

Numerous raw materials are used in the manufacture of a movie projector. Aluminum alloys and hard plastics are primarily used to make the housings, sprockets, gears and other structural components. Xenon gas is used for the light bulb. Xenon is a so-called inert gas that creates a tremendous amount of light when it is exposed to an electric current. Quartz is also used to make movie projector light bulbs because it can maintain its structure at high heat better than glass. Other materials used in the construction of a movie projector include rubber, stainless steel, and glass.

The Manufacturing Process

The major components of a movie projector, including the spooling system, the projector console, the audio reader, and the lenses, are produced by different manufacturers and typically assembled on-site at the movie theaters.

Making the main body

1 The main body of the movie projector is basically a rectangular box that houses the lamphouse, the lenses, the picture head, and the audio head. It is made from steel that is loaded onto a conveyor belt. The sheets are then placed in a die with the desired shape of the housing. A hydraulic press is then released. The punch forces the steel sheet to assume the shape of the die. The body is then removed and fitted with an adjustable base that can be changed to modify the angle of viewing.

Making the picture head

These parts are all assembled separately and then put together as a whole.

2 The picture head is the area between the lamp and the lens through which the film moves. Its housing is first formed from steel in a punch press process similar to the production of the body.

3 Then a series of sprockets and roller pads are manually screwed above and below the framing aperture. An intermittent sprocket is placed below the aperture. This sprocket is then connected to the motor which causes it to start and stop at a frequency of 24 frames per second.

4 On the other side of the film, across from the aperture, is the film gate that provides pressure to hold the film in place as the image is projected. The film gate is also formed by the punch press process. Behind the film gate is the shutter blade. This is a

small metal device with blades like a fan. It rotates in front of the light condensers at a controlled rate. It is synchronized with the moving film so that dark spaces between frames are not seen.

5 The lens turret is placed in front of the aperture. This is a rotating device into which the lenses are placed. It can be moved when a different lens is desired.

6 One side of the picture head frame is fitted with a door that can be opened so that film can be loaded. The separate pieces are then assembled and the entire picture head is bolted to the main projector housing.

Making the audio head

7 The audio head is constructed in much the same way as the picture head. It is composed of a variety of sprockets and film rollers. On one side of the film path a light source is affixed. This device emits light at a specific wavelength and intensity. On the other side of the film path is a photocell that detects the amount of light that travels through the film. It is connected to a series of amplifiers that are then wired to the theater speakers. The audio head can be located above or below the picture head depending on the design of the movie projector. Like the picture head it is bolted to the main projector body.

Making the lamphouse

8 The lamphouse consists of a frame and xenon bulb. Producing a xenon bulb can be a difficult process. Since these bulbs can produce a tremendous amount of heat, their outer housing is made from quartz instead of glass. First a quartz tube is heated and air blown to create the shape required for the bulb. A metal cathode is attached on one end and an anode on the other. Air is replaced in the quartz envelope by xenon gas and the whole unit is vacuum-sealed. The rarity of xenon and the difficulty in construction makes these bulbs expensive, running anywhere from $700 to $2,000 each.

9 The bulb is then mounted in the center of an aluminum parabolic shaped mirror. This assembly is then manually attached to a metal frame. The frame has an exhaust pipe and several fans to help remove the large amount of heat generated by the bulb. Wires

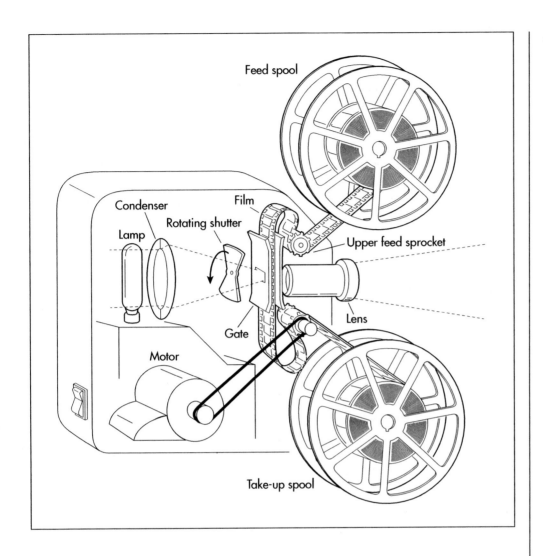

are hand soldered to the anode and cathode which are then connected to the power supply line. The light assembly is then put in the top of the main projection body. Within the body are the condenser lenses that help focus and intensify the light.

Making the lens

10 Lenses are produced from glass. Each movie projection lens is really composed of a number of small glass lenses that have a different magnifying effect. For each component lens, the glass is first cut to the manufacturer's specifications. The glass is then placed on an assembly line and workers polish each piece to the appropriate thickness, then treat it with a special anti-reflection coating. As many as seven component lenses might be used for a single lens system.

11 The component lenses are then fitted with metal and then placed into the lens barrel at specific intervals. This is a highly precise process done by specialized workers because the distances between the glass lenses have a profound effect on the image quality. The inside of the barrel is coated with a dark, non-reflective material. The lenses are then attached to the projector body by being screwed into the lens turret.

Making the spool assembly

12 The spool assembly begins with the construction of the solid metal frame. The typical frame consists of a tall pillar with two foot bars. Each component is placed on a conveyor belt and passed under a hydraulic punch. This punch is equipped with a sharp metal saw that will cut the proper dimensions from solid steel bars. The foot bars are then manually welded perpendicularly to the bottom of the main pillar. They are positioned such that there is about a 45-degree angle between

them. A smaller metal pipe is welded between the foot bars at their midpoints to provide a more stable structure. Finally, metal plates with rubber bottoms are welded to the bottom of the pillar and foot bars to ensure that there is minimal movement during operation.

13 Separately, the support arms and accompanying parts are assembled. To one end of the steel support arm a metal bearing is attached. This bearing can spin freely. To the other end of the support arm a hole is drilled through it and an electric gear motor is fitted through. At the end of the motor is a small rubber wheel that spins. It is the motion of this wheel that creates the spinning of the platter that moves the film.

14 The support arm assemblies are then attached to the main pillar at set intervals. The arms are welded to a metal plate that is then bolted and secured to the pillar. At specific points on the main pillar, the rollers that hold and guide the film to and from the projector are attached. Sensors that monitor the speed of the platters are bolted to the pillar above each support arm to synchronize the motion of the moving film. The electronic wires are fed into a control box located at the junction between the foot and the main pillar.

15 The platters are then placed on the support arms. The platters are made of lightweight aluminum alloy. They can be cut from thick sheets of the metal. A typical dimension is 5 ft (152 cm) in diameter and 0.5 in (1.3 cm) thick. They have a circular cut in the middle that can accommodate the centerpiece. This is a circular device complete with rollers and tension bars that accepts the incoming and outgoing film. A hole is also drilled directly in the center of the platter so it can be held and moved by the bearings on the support arm.

Final assembly

16 The main projector console and the film spooling systems are delivered to the movie theatre. They are connected through an electrical cable so that they move the film in a highly controlled motion. They are then ready to be loaded with film and show a movie.

Quality Control

At each step during the production process quality control testing is done to ensure that a working movie projector is produced. Each manufacturer has their own tests specifically related to the part of the projector that they make. These tests include both visual inspection and physical measurements. For example, the lens manufacturer uses computerized laser calipers to measure the thickness of each lens produced. The lamp producers measure various characteristics of the lamps that are produced such as luminosity, heat and power consumption. The components of the main projector are then assembled, and manufacturers run a sample movie through to adjust and pacing of all the moving components and determine whether the device works properly. Even after the movie projector is assembled in the theater, technicians constantly check and adjust parts as necessary.

The Future

The future of movie projectors looks to change dramatically in the coming years. With significant advances in electronic storage mediums, film may not be used to show movies. In fact, some companies are working on a system by which movies are produced on computer hard drives. Stored in this way, movies promise to be much less expensive to distribute and display. Fewer workers could run movie theaters and the movie images will be much clearer and crisper. Currently, theaters are hesitant to adopt the new technology, but it is just a matter of time before computerized digital projectors replace movie film projectors.

Where to Learn More

Books

Barclay, S. *The Motion Picture Image: From Film to Digital.* Focal Press, 1999.

Case, D. *Film Technology in Post Production.* Focal Press, 1997.

Other

Boegner, Ray F. "Everything You Wanted to Know About Xenon Bulbs." *Xenon Bulb Web Page.* December 2001. <http://www.cinemaequipmentsales.com/xenon1.html>.

Boegner, Ray F. "Film Technology in Post Production." *Scientific American Web Page.* 1998. December 2001. <http://www.sciam.com/1998/1098issue/1098working.html>.

Harrigan. *Movie Projection Lens.* United States Patent 6,317,268. November 13, 2001.

Perry Romanowski

Night Scope

The United States military established a night vision technology development program by the late 1940s, and by the 1950s had come up with viable infrared viewing systems.

Background

Night scopes, or night vision devices, are used to intensify human sight under very low light conditions. There are several types of night vision scopes. Infrared imaging systems, also referred to as "active" night vision devices, focus infrared light on a scene. Infrared is beyond the light spectrum visible to humans, so the beam itself is undetectable. Image-converting technology transforms the scene illuminated by the infrared into a visible image. Thermal imaging systems work in a similar way, converting the pattern of heat emitted by objects, people, or animals into a visual image. The night vision devices perfected for wartime use and also available commercially today are called "passive" night vision systems. These systems amplify images picked up in minimal light, such as starlight, into visible images. The view through a passive night vision device may be from 20,000 to 50,000 times brighter than what the unaided eye could see.

Night vision devices were developed for military use, where seeing in the dark is an obvious tactical advantage. The United States used night vision devices in the Vietnam War and to great effectiveness in the Persian Gulf War. Night vision devices are also used by both urban and rural police forces. In the late 1990s, night vision devices were finding more commercial outlets. They began appearing in some high-end cars and are being marketed directly to consumers for recreational use.

History

Research into night vision devices began in the 1940s. The United States military established a night vision technology develop- ment program by the late 1940s, and by the 1950s had come up with viable infrared viewing systems. This was an active technology, meaning it used a directed beam of infrared light. Though the beam itself was invisible to the unaided eye, opponents armed with equivalent technology could easily pick up the beam. The infrared viewers used in the 1950s and 1960s are referred to as "Generation 0" technology.

ITT Corporation (now ITT Industries, Inc.) in Roanoke, Virginia, began producing night vision devices for the United States military in 1958. The United States Department of Defense founded its own Night Vision Laboratory in 1965, dedicated to improving the existing technology. During the 1960s, scientists developed the first workable passive night vision systems. These devices were called "Starlight" systems because they were able to pick up and amplify images seen only by starlight. They are also known as "Generation 1" devices. They actually worked best in moonlight. Generation 1 night vision devices were used in combat for the first time during the Vietnam War.

Improved technology developed shortly after the war led to smaller, less bulky night vision devices with better resolution. These more reliable instruments were called "Generation 2." The United States military continued to develop and refine night vision technology during the 1970s and 1980s, fitting weapon sights with night vision targeting devices and training pilots in night vision goggles. Passive Generation 2 devices were able to produce a good visible image in very low light situations.

"Generation 3" technology was developed in the late 1980s. These new night vision devices used gallium arsenide for the photo cathode material inside the image intensifier tube. This produced better resolution even in extremely low light situations. United States forces used night vision devices extensively in the Persian Gulf War, where the technology allowed troops to see not only in the dark but through dust and smoke as well. By the late 1990s, the Department of Defense had reduced its funding for night vision development, and some manufacturers began searching for consumer markets for the gear. Individuals may buy night vision devices in the United States, but their export is still restricted.

Raw Materials

The image intensifier tube, which is the main working component of a night vision device, is made up of millions of hair-fine fibers of optic glass. The glass used is a particular formula that preserves its desired characteristics when heated and drawn. Optical quality glass is used for the eyepiece and output window. (The output window is an ocular lens, like the eyepiece of traditional binoculars.) Other materials used in the image intensifier tube are phosphor and gallium arsenide. The tube body is composed of metal and ceramic, and the metals used may be aluminum, chromium, and indium.

Design

Passive night vision devices work by sending light through a lens, an image intensifying tube, and another lens. Light enters through a lens called the objective lens, which is similar to a fine camera lens. The lens focuses the light into the image intensifier tube.

The tube is the most complex piece of the night vision device. It is handmade to exact specifications. The tube is a vacuum tube with a photo cathode, a power source, a microchannel plate, and a phosphor screen (the screen emits light when excited by electrons). The cathode absorbs light (photons) and converts the photons into electrons. The electrons are multiplied thousands of times as they pass down the tube, by a wafer-thin instrument called a microchannel plate.

A standard microchannel plate is 1 in (25 mm) in diameter and about 0.04 in (1 mm) thick—about the size of a quarter. Incorporated into this plate are millions of microscopic glass tubes, or channels. The latest night vision microchannel plates contain over 10 million channels. These channels release more electrons as the electrons bounce through the tubes. The channels must be uniform in diameter and spacing on the plate in order to produce a clear image. The electrons then hit a phosphor screen. The phosphor screen reconverts the electron image into a light image, and focuses it on the output window.

The entire image intensifying tube may vary in size, but the finished tube can be small enough to fit into a gun sight or into a pair of military goggles. For example, a current product available from ITT is a Generation 3 monocular that is 4.5 in (11 cm) long, 2 in (5 cm) wide, and 2.25 in (5.5 cm) high, including both lenses. The entire instrument weighs 13.8 oz (0.4 kg).

The Manufacturing Process

The manufacturing process for night vision devices is complex. Over 400 different steps are needed to make the core component, the image intensifier tube. Manufacturers carry out several major process steps simultaneously in different sections of the plant.

1 The first major step is making the photo-cathode. The manufacturer may buy preformed rounds of glass for the photocathode plate from a subcontractor. Workers drop a wafer of gallium arsenide onto the glass and heat it. This begins to melt the gallium arsenide to the glass.

2 Then the part is put into a press, which firmly binds the gallium arsenide substrate.

3 Workers then grind and polish the part.

4 Meanwhile, the glass microchannel plate is formed using a system known as the two-draw process. This begins with a cast or extruded ingot of special formula glass. The ingot is ground into a rod with a diameter of several centimeters. The rod is fitted into a hollow tube of another type of glass. This is called the cladding. The cladding glass will

later be etched away, but it gives the fibers more uniformity in the drawing process.

5 Now the glass is drawn for the first time. The ingot is hung vertically at the top of a furnace. The furnace may be several feet tall. The furnace has very fine temperature control, so that different points along its length can be held at different temperatures. The ingot is heated at the top of the furnace to about 932°F (500°C). A globule of glass forms at the bottom of the ingot, like a drip coming out of a faucet. As the globule falls, it pulls down a single strand of glass, about 0.04 in (1 mm) in diameter. This strand cools as it stretches. Farther down the furnace, the strand is gripped on either side by a traction machine, which rolls along the fiber, forming it to the precise desired diameter. Cutters clip the fiber to a uniform length (about 6 in [15 cm] long) and pass it down into a bundler. Several thousand fibers are bundled together into a hexagon. This hexagonal bundle is then drawn for a second time, giving the two-draw process its name.

6 The second draw looks much like the first, with the hexagonal bundle suspended at the top of a zone furnace and heated. The fiber is drawn into a hexagonal shape about 0.04 in (1 mm) in diameter. Because the special glass keeps its cross-section properties, the fiber from this second draw is geometrically similar to the larger bundle, with the glass tubes' honeycomb structure still intact, and the whole structure is just reduced in size. (The space between the individual glass tubes has now been reduced to a few hundredths of a millimeter.) The fiber that results from this second draw is also cut and bundled, similar to the first draw.

7 The resulting bundle of fibers is heated and pressed under a vacuum, which fuses the fibers together. At this point, the fiber bundle is known as a boule. To make the microchannel plates, the boule is cut at a slightly oblique angle into wafer-thin slices. The slices are ground and polished. The plates are then finished with an acid etch to remove the softer cladding glass. Removing the cladding glass opens channels throughout the plate. Each plate is then coated with nickel-chrome.

8 Next, a film of aluminum oxide is set onto both surfaces, so that each channel can carry electrical charge. This finished microchannel plate can vary in diameter depending on its designated use, but the thickness remains at about 0.04 in (1 mm). The standard size for finished microchannel plates is 0.9 in (25 mm) in diameter, but they can be as large as 4.9 in (12.5 cm) in diameter.

9 Next, the phosphor screen and tube body are assembled. The screen itself is a small fiber optic disk that may be supplied by a subcontractor. The image intensifier manufacturer must bond the screen to the metal parts that will hold it in the tube, and then apply the phosphor. The screen is dropped into a flange and bonded to it with a ring of a fusable material called frit. Frit is a special glass compound that welds to metal and glass under high heat. Other metal parts are fitted over the screen, making a small, round part. This part is sent on a track through a furnace, which melts the frit, bonding all the components together. After the part is cooled, cleaned, and polished, the phosphor is sprayed or brushed onto the part. A solution of phosphor in water is poured in. The phosphor settles on the screen, and then the water is drained out.

10 Workers assemble the tube body by fitting together a series of small metal and ceramic rings. Each ring has a precise function, supporting the different parts that will be loaded into the tube. Insulators and conductors are also added at this time. Some sections of the tube body are made of a soft metal called indium. The assembled tube is run through a furnace, and the indium parts melt and fuse, holding the tube together.

11 When all the main components are manufactured, they are loaded by hand into the tube body. This is extremely delicate work done in a special clean room environment—in the clean room facilities, the workers wear laboratory suits, gloves, and the work stations are protected by plastic sheeting. The parts mechanically lock into place. First the microchannel plate is locked into the body. Then workers tack-weld electrodes to the parts that will carry voltage.

12 The partially assembled unit is taken next to a piece of equipment called the exhaust station. In the exhaust station, air is removed from the tube, leaving a vacuum. Under the vacuum, the cathode is inserted into place and activated. Once this is

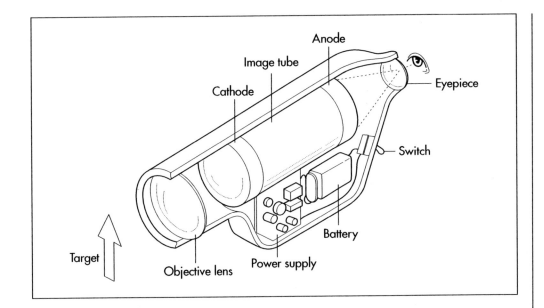

Target

Cathode — Image tube — Anode

Objective lens — Power supply — Battery — Switch — Eyepiece

The internal mechanisms of a simple night vision scope. The anode is fluorescent, and will emit light. (The text refers to the fluorescent anode as a phosphor screen.) This scope does not use a microchannel plate to improve the image quality. In a more complex scope, the microchannel plate would be between the cathode and the anode.

done, the body, cathode, and screen are pressed together. Under high pressure, indium interfaces between the parts fuse, joining all the elements permanently.

13 Next, the image intensifier tube goes through several testing stages to make sure it is activated and working within expected parameters. When the tube is shown to be functioning correctly, workers wire it to its power supply. Then the tube is set into a piece called a "boot," which resembles a simple plastic cup. This boot forms a housing that encapsulates the tube to protect it. The boot is closed and sealed under a vacuum. Now the image intensifier tube is complete. It undergoes several more rounds of testing. The tests may vary depending on the intended use. Thoroughly tested components then move to a final assembly process. Here they are fit into a casing for goggles, gun sights, binoculars, or whatever the end night vision product is.

Quality Control

Quality control at every step of the manufacturing process is essential for the image intensifier tubes to work correctly. Large manufacturers have honed the process so that each step is tested or gauged, and workers are unable to move the part onto the next step if the part has not met the quality control requirements. Manufacturers use sophisticated calibration equipment to measure such things as the diameter of the glass fiber, the thickness of the microchannel plates, and the temperatures in the various

furnaces. Materials supplied by subcontractors are checked as they come into the plant. The calibration equipment used for testing is itself tested frequently for accuracy.

The final product is tested in various ways to ensure that each device works as it should. Each device is checked for its visual action. Other tests may show how tough the device is under adverse conditions. Finished night vision devices may be tested for how they respond to shock and vibration, and there may be a drop test. For some military requirements, the devices may be subjected to days of extreme heat and humidity.

Byproducts/Waste

The manufacture of night vision devices can result in some hazardous waste, as many chemicals are used in cleaning and etching. However, some manufacturers have been able to substitute less toxic or nontoxic chemicals for harmful ones, and in general the manufacturing process now is cleaner than it was when the technology was first developed. Image intensifier tubes are expensive and arduous to produce, so manufacturers try to salvage as much scrap as possible. If a tube is built that does not function, it would be disassembled and the parts reused.

The Future

The night vision industry is making itself available to the non-military consumer market. While prices are still high, as demand increases, the price may decrease until the technology is fairly affordable. The technology is already being used by law enforcement and search-and-rescue teams. As the products become more in the price range of consumers, and because the images viewed can be recorded by video cameras or as photographs, more photographers, wildlife watchers, boaters, campers, and many others may begin to use night vision technology in more innovative ways.

Where to Learn More

Books

Palais, Joseph C. *Fiber Optic Communications.* Upper Saddle River, NJ: Prentice Hall, 1998.

United States Army CECOM. *Night Vision and Electronic Sensors Directorate.* Fort Belvoir, VA: US Army CECOM, 1997.

Periodicals

Justic, Branco, and Peter Phillips. "Night Vision Scopes." *Electronics Now* (October 1994): 57.

Lampton, Michael. "The Microchannel Image Intensifier." *Scientific American* (November 1981): 62–71.

Rhea, John. "The Feedback Loop of Night Vision Devices." *Military & Aerospace Electronics* (February 2000): 8.

Angela Woodward

Oxygen Tank

Background

Oxygen (atomic number, 8; atomic weight, 16) is essential for all living things and has the ability to combine with almost all other elements. When elements fuse with oxygen, they are labeled as being oxidized. Oxygen is the most plentiful element in the world, comprising about 90% of water (hydrogen makes up the other 10%) and 46% of the earth's crust (silicon, 28%; aluminum, 8%; and iron, 5%; among others). Oxygen's melting point is -360°F (-218°C) and its boiling point is -297°F (-183°C). In its free state, oxygen is odorless, colorless, and tasteless. At temperatures below -297°F (183°C) oxygen takes on a pale blue liquid form.

Two-thirds of the human body is composed of oxygen. In humans oxygen is taken in through the lungs and distributed via the blood stream to cells. In the cells, oxygen combines with other chemicals, making them oxidized. The oxidized cells are then distributed where they are needed, providing the body with energy. The waste products of respiration are water and carbon dioxide, which are removed through the lungs.

Pressurized oxygen therapy is used to treat numerous medical aliments such as emphysema, asthma, and pneumonia. This medicinal form of oxygen is typically kept in medium-sized aluminum canisters equipped with pressure regulators and release valves. Large amounts of oxygen are kept in large, insulated steel tanks pressurized at 2,000 lb/in^2 (141 kg/cm^2).

History

The discovery of oxygen has generally been attributed to Joseph Priestley, an English chemist. In 1767, Priestly believed that air mixed with carbon was able to produce electricity. He called this carbonized air, mephitic air. Priestly went on to conduct experiments concerning air, and in 1774 he used a burning glass and solar heat to heat mercuric oxide. While doing this, he noticed that the mercuric oxide broke down under the extreme temperature and formed beads of elemental mercury. The mercuric oxide also emitted a strange gas that facilitated flames and opened the respiratory tract, making it easier to breath when inhaled. This gas was named dephlogisticated air by Priestley, based on the popular thought of the time that phlogiston was needed for material to burn. The phlogiston theory was deemed false by Antoine-Laurent Lavoisier, a French chemist.

Lavoisier had been conducting his own experiments with combustion and air in the mid- to late-eighteenth century. It was in 1774, that he met Priestley who told Lavoisier of the discovery of dephlogisticated air. Lavoisier began to conduct his own experiments on Priestley's pure form of air. He observed that the element was part of several acids and made the assumption that it was needed to form all acids. Based on this incorrect thought, Lavoisier used the Greek words *oxy* (acid) and *gene* (forming) to coin the French word oxygene—translated to oxygen in English—sometime around 1779.

There is yet a third man who is credited for his involvement in the discovery of oxygen in about 1771. Carl Wilhelm Scheele, a Swedish pharmacist and chemist, discovered that a certain element (Scheele also thought it to be phlogiston) was needed in order for substances to burn. Scheele called this element "fire air" due to it being needed for

The standard medical E tank holds 680 l and can provide up to 11.3 hours at 1 liter per minute (lpm). This tank weighs 7.9 lb (3.6 kg) empty.

combustion. During these experiments with fire air, Scheele also discovered "foul air," now known as nitrogen. Despite the fact that Scheele had isolated oxygen before Priestley, Priestley published his findings first.

Raw Materials

The raw materials to produce an oxygen tank are liquid air and aluminum. The aluminum starting stock is cast 6061. The liquid air is condensed and heated until pure oxygen remains then distributed into the aluminum tanks. A compressible Teflon ring is used to form the o-ring, which is placed in the o-gland forming a seal between the valve and the cylinder. The o-ring gland is a precision depression machined in the top of the cylinder. When the valve is screwed in the cylinder and when completely seated, it compresses the o-ring and completes the airtight seal between the valve and the cylinder.

Design

Oxygen tanks vary in size, weight, and function but the manufacturing process is very similar. The typical medicinal oxygen tank contains pure oxygen and has a green top with a brushed steel body.

The Manufacturing Process

Formation of the cylinder

1 Oxygen tanks are manufactured from a single sheet of 6061 aluminum. The starting material is called a cast billet, which is approximately 18 ft (5.5 m) long and shaped like a log.

2 The cast billet is placed on a conveyor belt and cut to the desired size by an automated saw. The sawn piece is called a slug and is almost the same weight and diameter as the finished product.

3 The slug is then placed inside a die in a backward extrusion press. The press forces a punch against the slug. The metal of the slug flows backwards around the punch forming a large, hollow, cup-shaped product called a shell.

4 The shell is then inspected for defects and gauged.

5 Next, the shell is put through a process called swaging. The open end of the shell is heated and forced into a closing die to close the open end of the cup. Now the general shape of a seamless cylinder is finished.

Heat treating the cylinder

6 The cylinder is conveyed through a two step thermal process called solution heat treat and artificial aging.

7 The first thermal process, solution heat treatment, begins when the cylinder is placed in a solution furnace. In this process the alloying elements of the aluminum are put into the solution. The cylinder is heated to about 1,000°F (538°C). A cylinder that has been subjected to this thermal process is labeled as being in the T-4 temper.

8 The second thermal cycle, artificial aging, consists of the cylinder being conveyed through an age oven where it is heated to about 350°F (177°C). This allows the alloying elements to precipitate out of the solution and into the grain boundaries, strengthening the cylinder. A cylinder that has completed both thermal processes is labeled as being in the T-6 temper.

The neck configuration

9 The threads, o-ring gland, and top surface are the sealing surfaces and are machined into the cylinder. The cylinder is placed in a milling machine (a drill press able to move in three directions). Under the direction of Computer-Aided Design (Auto-CAD) software, a hole is milled in the center of the cylinder's neck.

10 The top surface, o-ring gland, and threads (in that order) are machined into the cylinder using a form tool. The form tool has the form of the top of the cylinder, the o-ring gland and the thread relief are under the o-ring. The form tool spins as a drill bit and is lowered into the cylinder machining the form into the cylinder neck.

Finishing

11 The tank is then subjected to hydrostatic testing. During this test the tank is pressurized equal to five-thirds its service pressure. If the tank expands greater than a

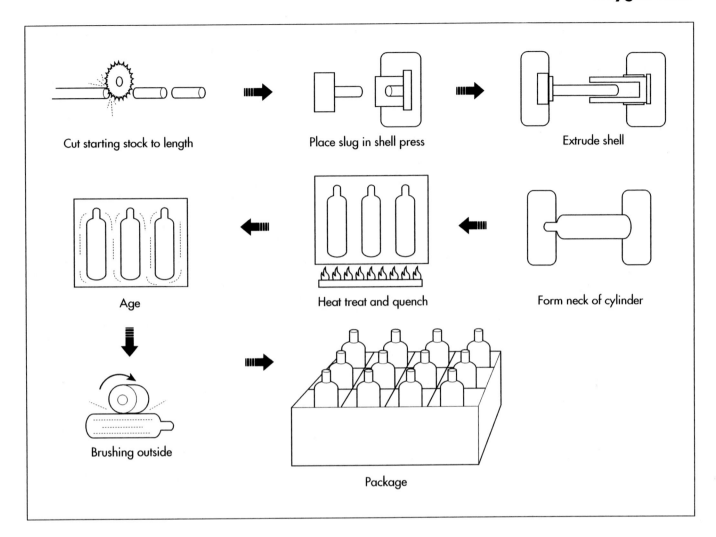

Cut starting stock to length

Place slug in shell press

Extrude shell

Form neck of cylinder

Heat treat and quench

Age

Brushing outside

Package

The manufacturing of oxygen tanks.

specified amount within 30 seconds, it is rejected.

12 Identification marks are stamped onto the tank via a pneumatic stamper. These marks identify the specifications to which the cylinder was manufactured, service pressure, serial number, manufacturer's name or number, and the manufacturing date of the tank.

13 The tanks used for medical purposes generally have a brushed body. The tank is placed horizontally on the conveyor belt and rotated under an automatic sander.

14 The top of the tank is manually painted green then the entire tank is sprayed with a clear powder coat and cured in an oven.

15 The finished tank is then either capped or equipped with a valve depending on the customer requirements.

Filling the tanks

1 Commercial pressurized oxygen is distilled from liquid air in large batches. Air becomes liquid at -297°F (-183°C). The air supply is compressed, then passed through a compartment equipped with a piston (expansion engines).

2 As the air expands, the pistons move, increasing the volume of the compartment and decreasing the pressure and temperature of the air.

3 The air is then rotated through several expansion engines until liquefied. The liquid air is then transported to huge insulated holding tanks.

4 The liquid oxygen is then boiled to get rid of the nitrogen, since nitrogen has a lower boiling point (-320°F; 195°C). The liquid air is then mostly oxygen (97–100%)

and transported to large insulated tanks until dispersed into oxygen cylinders.

Quality Control

During the manufacturing process, the cylinders are inspected and cleaned numerous times. After the tank is sold and put into service, it must be put through hydrostatic and visual retesting every five years. The testing is conducted in accordance to the Compressed Gas Associations requirements. If the tank is not damaged and wear is minimal, there is unlimited service life.

DOT-3AL is the marking identifying the specification in which the cylinder was manufactured in compliance. The Department of Transportation (DOT) regulates the transportation of all goods. The transportation of compressed gases falls into this category.

Byproducts/Waste

In the manufacturing process nearly 93% of the starting material (the cast billet) is used in the final product. There is less than 7% manufacturing scrap of the starting material. After production is complete, any cylinders that are damaged to the point of being condemned are stamped through the "DOT-3AL" marking on the crown. If the tank has been pressurized, it is depressurized, the valve is removed, and the cylinder is sawn in half and recycled. The condemned, sawn cylinders can and should be recycled.

The Future

As the medical use of oxygen tanks increases, the tanks are getting smaller and more maneuverable. The standard medical E tank holds 680 l and can provide up to 11.3 hours at 1 liter per minute (lpm). This tank weighs 7.9 lb (3.6 kg) empty. One of the smaller oxygen tanks is an M9 tank. This tank holds 240 l of oxygen lasting for four hours at 1 lpm or two hours of continuous flow. There are accessories such as carts or bags that allow the user to transport the full tank easily.

Where to Learn More

Other

Catalina Cylinders Web Page. 8 November 2001. <http://www.catalinacylinders.com>.

Tri-Med, Inc. Web Page. 8 November 2001. <http://www.trimed.freeservers.com>.

Deirdre S. Blanchfield

Pacifier

Background

A pacifier is a form of an artificial nipple on which the baby or child sucks. Fluids do no pass through the pacifier, rather, the action of sucking on the nipple is thought to soothe or calm the baby, quieting the baby, and even alleviating the burning and itching of the gums during teething.

Pacifiers generally have three parts: the nipple; the guard, which rests on the baby's lips; and the ring attached at the center of the guard. Artificial nipples are always made of a material that closely simulates a mother's nipple. It is usually of latex or silicone and is occasionally of hard plastic. The guard is firmly attached to the nipple and prevents ingestion of the nipple by the child. Increasingly, pacifier manufacturers believe that the nipple and mouth guard should be of one material and molded together so that the two do not have to be fused during the manufacturing process. There is evidence that when the pacifier is made of two pieces it is at greater risk for failing at that juncture and creating a potential choking hazard. This guard must have holes in it to ensure that in the event of ingesting the guard the holes permit air to pass through to the windpipe. Finally, the ring at the center of the guard must be present in order to pull the pacifier forcibly from the mouth in the event of ingestion.

Manufacturers of pacifiers must comply with extensive government regulations developed to prevent pacifier failures from choking infants and young children. Pacifiers made in the United States must undergo testing to ensure that they comply with these regulations. Even with these regulations pacifiers are occasionally recalled be-cause of product failure. Manufacturers must test each new pacifier design extensively before production, and then keep compliance reports on hand in the event that the information is requested at any time.

History

The early history of pacifiers is inextricably bound to the development and use of the baby rattle. The two were often attached to both amuse and assuage a crying baby. Rattles were used for centuries by primitive adults for ceremonial and musical or dance use and could be of sticks, teeth, shells or pods. In addition, primitive man used rattles with bells to ward off evil spirits and even children wore bells. Romans gave children peony wood bead necklaces that were pulled into the baby's mouth for teething and to ward off illness. Such soothing bead necklaces were used for many centuries. Some rattles were constructed with a handle that had inserted into it a smooth piece of stone or bone that was used for the baby to suck upon and teethe. This end of the rattle was known as the gum stick and could be made of rock crystal, ivory, agate, carnelian, mother-of-pearl, bone, or coral, all of which felt cool upon the gums.

Coral was one of the most effective materials thought to protect children from spirits as well as witchcraft, enchantment, and epilepsy. Coral necklaces were given to babies at birth for this reason. (In some cultures a coral necklace is still an appropriate baby gift.) It did not take long to combine the coral and rattling bells into a single baby toy that was often referred to as coral and bells. These devices had a branch of coral at one end and a metal shank with bells at-

Romans gave children peony wood bead necklaces that were pulled into the baby's mouth for teething and to ward off illness.

253

tached at the other end. The coral was sucked on, and being fairly soft and knobby, could soothe the baby and rub against sore gums as the bells sounded and warded off evil spirits. Fairly expensive coral and bells were generally crafted by silver or goldsmiths and offered for sale in the Colonies by 1700. Sticks of cane sugar or candy were given to babies to suck upon as well. Less expensive rattles used for sucking included gum sticks of wood with small vegetable gourds or wicker or willow rattles that could be put into the mouth. The baby's own fingers or fists have been the pacifier of choice for many babies as well.

When the sap from rubber trees was viably used in manufacture of household products by the middle of the nineteenth century, nipples for nursing bottles and simple soothers were devised. In the twentieth century, with better refinement of rubber and associated materials, many companies entered into the pacifier market. Hard plastics and silicone molders produced the product without much regulation. However, the United States Consumer Product and Safety Commission soon saw catastrophic failure with unregulated two-piece and hard plastic pacifiers and has since extensively regulated the industry.

Raw Materials

Pacifiers are manufactured from a soft, pliable material that closely resembles a mother's nipple. Increasingly, doctors and safety experts are urging manufacturers to produce pacifiers as a single unit, of one material, that cannot separate and pose a choking hazard. Thus, the product is preferably molded from a single material. In 2001, pacifiers are primarily constructed from either latex or silicone, although other soft plastics can be used in their manufacture. Unfortunately, manufacturers are finding that softening these other plastics requires chemical additives that have proven to be harmful to small children, and pacifiers have been recalled by the federal government not for failures of construction but because harmful additives were used in the processing of the raw materials.

Latex is technically referred to as natural rubber latex and is produced by the rubber tree *Hevea brasiliensis*. The milky latex from this tree is harvested by tapping the tree. The sap oozes from the tree when it is scored, then collected and processed for manufacture of thousands of household products. Most of the latex trees tapped for sap are in Southeast Asia and South America. Chemicals are added to raw latex to increase elasticity and strength. Proteins are found in this natural material that have caused severe allergic reactions in some consumers. Most manufacturers are either eliminating latex from their products lines as a result, or are treating the latex to counteract the effect of these proteins on consumers. In addition, latex does not survive repeated "boil and cool tests" in which the product is boiled, cooled, and then assessed for its ability to retain shape and perform successfully. Latex disintegrates more quickly than other substances during these tests. This is of concern as pacifiers cannot be repeatedly boiled or washed for sterilization without risk of disintegration.

The raw material of choice for American pacifiers is silicone, more expensive than latex but of superior performance. Silicones are synthetic polymers that are relatively chemically inert, stable at high temperatures, and resist oxidation. Thus, silicone is able to survive the boil and cool cycle far better than latex and therefore superior for products requiring sterilization. Some pacifier manufacturers buy silicone that is "certified," meaning it is of a superior grade for use.

Design

The design of pacifiers is an extremely important part of its manufacture. The pacifier manufacturer's director of research gathers information that assists with schematic design. He or she looks for the latest information on babies' upper and lower mouth configuration, the role of the tongue in sucking, and how a mother's nipple works and feels during nursing. Some consult directly with medical professionals about what they believe babies need in a pacifier. When the information is gathered, the research department works with the manufacturing department to produce a pacifier that conforms to recommendations made as a result of research.

Of paramount importance is the research department's responsibility to stay on top of the United States Consumer Product Safety Commission's *Requirements for Pacifiers*. These guidelines spell out precisely the

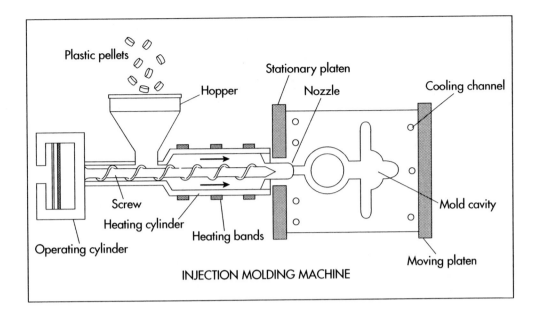

INJECTION MOLDING MACHINE

standards to which the manufacturer must comply. As of 2001, the "pacifier rule," as it is referred to, requires that: the mouth guard not be so small or flexible that it can be sucked into a baby's mouth; that a pacifier have no handles or other protrusions that might force the pacifier into a baby's mouth if he or she should fall on it face first; pacifiers are labeled to warn caregivers not to tie the pacifier around the child's neck; and that the pacifier not come apart into smaller parts when vigorously tested. It is the manufacturer's responsibility to design all new pacifiers with these guidelines in mind. Testing of the product is also specified in these guidelines.

The Manufacturing Process

The manufacturing process for pacifiers is very simple. Since most are made of one-piece construction, the preferred method for safety's sake, the production occurs in one stage using liquid injection molding. Liquid injection molding was derived from metal die casting, but unlike molten metals, polymers have a high viscosity and cannot be poured into a mold. Instead, a large force must be used to inject the polymer into a hollow mold. A great deal of melt, the name given to molten polymers, must be forced into the cavity as there is some shrinkage upon cooling.

1 First, the pellets of polymer must be melted at very high temperatures, usually 360–420°F (182–216°C). Pellets are fed from a hopper into a machine that liquefies the pellets. The polymer is now molten and ready for injection.

2 Next, the melt is rammed into the mold at very high speed and under intense pressure, approximately 300–700 psi. The melt fills the mold, then a bit more melt is added in order to compensate for the contraction due to cooling and solidification of the polymer.

3 Once the polymer has cooled, it is separated from the mold. This solidified part is simply ejected from the mold automatically, and the melt and injection cycle is ready to begin again. Each cycle takes between 10 and 100 seconds depending on the time it takes the polymer to cool and set.

4 The one-piece pacifier is then inspected and packaged for shipment. Pacifiers are not considered sterile when packaged but are considered clean.

Quality Control

Perhaps the most important part of quality control happens at the very beginning and very end of the manufacturing process. First, the materials acquired for use in the product are often certified for cleanliness and to ensure that no unsafe chemicals are present. Then, prototypes are extensively tested to ensure the product complies to specifications set forth by the United States

A plastic pacifier.

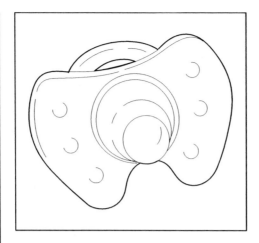

Consumer Product and Safety Commission. This federal agency specifies tests and how they are to be conducted. Testing that must be performed includes the "pull test" in which intense pressure is used to pull on the nipple to see how hard it is for the baby to ingest the nipple. Ten pounds (4.5 kg) of pressure are exerted for a period of time. Also, measurements for the size of the product are meticulously checked and re-checked. In addition, choke holes in the mouth guard must be a certain diameter according to the commission; the factory must check the size of the holes after production to ensure the holes have not changed in size or shape after manufacturing or with extensive mold use. Boiling tests are conducted to determine whether the material can withstand repeated use and whether the configuration changes after boiling and cooling. Some companies contract with an independent laboratory to conduct these tests. Results of these tests are generally held by the company in the event that anyone challenges the safety of the product.

Byproducts/Waste

When making silicone pacifiers there is little waste as the raw material is so expensive that the company seeks to re-use all flash or waste. In a mold with multiple cavities, the melt flows to each cavity using runners, which are long channels. The runners sometimes contain bits of polymer referred to as flash. If the runners are allowed to cool and solidify, the runners are separated from the part and must then be reground and transformed into pellets for reheating and melting. In hot runner molds, the runners are reheated and flow back into the system automatically. While the hot runner system virtually eliminates all waste, the mold system is very expensive.

The Future

Manufacturing pacifiers as a single unit device is the primary trend for the future. This design is preferred because it is less likely to pose a choking hazard for the child. Improvements in the types of raw materials used are also being researched. It is important that the rubber or plastic materials that become the pacifier do not contain harmful chemicals that could be transferred to the infant as it sucks on it.

Orthodontic problems often occur in children who use pacifiers for long periods of time and after they have developed their primary teeth. Research and development workers analyze the structure of children's mouths and their sucking patterns in order to develop pacifiers that will minimize the long-term effects of pacifier use.

Where to Learn More

Books

Henry Francis DuPont Winterthur Museum. *Kids! 200 Years of Childhood.* Hanover, New Hampshire: University Press of New England, 1999.

Weiss, Harry B. *American Baby Rattles.* Trenton, NJ: 1941.

Other

"Requirements for Pacifiers, 16 C.F.R. Part 1511." *U.S. Product and Safety Commission Web Page.* December 2001. <http://www.cpsc.gov>.

Oral interview with Paul Dailey, Director of Research and Development of Children's Medical Ventures, Inc. Norwell, MA. July 2001.

Nancy E.V. Bryk

Paper Clip

Background

The paper clip is a nearly ubiquitous device, used worldwide to temporally hold papers together. The technology for manufacturing paper clips evolved in the early years of the twentieth century, and has remained virtually unchanged since the 1930s. Paper clips come in several forms, but the one most often seen in common use is called the Gem clip. The origin of the term "Gem" is supposed to have originated from a British firm that began exporting them at least as early as 1907. The term has come to stand for the iconic shape of the oval-within-oval design. Any clip of this shape is called a Gem clip, regardless of the manufacturer. Another type of paper clip sometimes used by archivists and librarians is called the Gothic clip. It has a rectangular shape, with a triangular inner loop. Other distinguishing marks of paper clips are the overall size, the thickness, and quality of the wire, and whether the clip is corrugated or smooth. Most paper clips in the United States are made domestically by a few firms that specialize in their manufacture. These manufacturers put out roughly 20 million lb (9 kg) a year of paper clips.

History

The paper clip evolved to fill a specific need. A large amount of paper could be bound into a book in order to hold the leaves together. Binding was not a viable solution to keep together a few sheets, such as a short set of records or receipts. Though paper was invented in China sometime in the first century A.D., and was widespread in Europe by the thirteenth century, people made do without anything like the modern paper clip until the end of the nineteenth century. People used two general methods to fasten together a few sheets of paper. They could slit the pages in the corner, making two short parallel cuts. Then ribbon or string could be threaded through the slits, the ends tied and often sealed with wax. The second method was to take a common straight pin and pin the sheets together. Machinery to make cheap and uniform quality pins developed early in the nineteenth century. Business people bought boxes of loose pins, sold by the pound and called "bank pins," to use in offices. Both the slit method and the pin method had the same drawback: the paper had to be pierced. Pin holes caused less wear on the paper than slits, yet if pages needed to be unpinned and repinned many times, the pinned corner was subject to a lot of wear, leading to the drooping "dog ear." Pins also had the disadvantage of being sharp enough to prick fingers and tended to rust if left in place for any length of time.

Designs or patents for early paper clips date to the middle of the nineteenth century, but none of these early devices seem to have worked well enough to have made a lasting impression. Some of these were closer in form to what is known today as a binder clip or bulldog clip, and others enfolded the entire corner of the paper within teethed overlays of thin metal. This type of paper fastener was made from stamped sheet metal. Wire forming technology advanced in the mid-nineteenth century, and from about 1870 on, paper clips of various designs competed in Europe and the United States.

The earliest recorded patent for a paper clip was granted by the United States Patent Office to a Pennsylvanian, Matthew Schooley, in 1898. Schooley's patent application mentions other devices already on the market of

In the 1990s a Pennsylvania company began marketing what looks like essentially a giant Gem clip, which can hold more than one hundred sheets of paper at one time.

a similar design, so it would not be accurate to name him or any other individual as the father of the paper clip. A Norwegian, John Vaaler, is often credited with inventing the paper clip in 1899. His patent application included several possible paper clip shapes, including one that is similar to the modern Gem clip. Because of Vaaler, the paper clip became a symbol of Norwegian nationalism during World War II. There is even a 23 ft (7 m) high statue of a paper clip in Norway. It was set up to commemorate the solidarity the Norwegians represented against the Nazis by pinning paper clips on their lapels.

A Massachusetts inventor, Cornelius Brosnan, received a patent for a paper clip design in 1900. Again, his application spoke of the product as an improvement over other paper clips already in existence. His clip was marketed as the Konaclip. The Konaclip was an oval loop of wire with an inner arm terminating in a rounded eye. At least these three clips, the Schooley, the Vaaler, and the Brosnan designs, existed by the turn of the century. In 1899 a Connecticut inventor, William Middlebrook, applied for a patent for a machine to make paper clips. Middlebrook's patent application was not for making any particular type of clip, but the one pictured on the application's illustration looked like the archetypal Gem. Gem clips were imported to the United States from England by at least 1907. The Gem was advertised as a fine English product, superior to all others. Though paper clips of differing designs continued to be made for several years, by about the mid-1930s, the Gem had become the most commonly used.

Raw Materials

Paper clips are generally made from galvanized steel wire. The wire diameter depends on what size and quality clips are being made from it. Paper clips can be made from light, cheap steel, or from better quality steel, depending on the manufacturer. The material used, however, has to fall within certain physical parameters to make satisfactory paper clips.

Design

The Gem clip is often held up as a paragon of modern design. It is simple, elegant, and surpassingly functional. Yet leaving the iconic shape aside, a paper clip designer must consider a host of mechanical and engineering questions. The material used to make a paper clip must possess certain properties. The wire needs to be stiff enough to hold its shape in use, but not so stiff that it is difficult to open. Engineers also consider a quality called yield stress when designing a paper clip. Yield stress is the amount of stress needed to permanently reshape the wire. If the wire has too low yield stress, it will stay bent open and not hold the papers tightly. Engineers also must consider the cost effectiveness of the material used. Using a cheaper, thinner wire may save the manufacturer money. Yet the material must also perform well in the manufacturing process, not leaving sharp burrs at the cut ends and resisting cracking or breaking. The material used also should be non-corrosive. The finished appearance of the clip is also a design consideration. The clip can have various finishes, smooth or slightly serrated, shiny or dull, and it can be made in many different sizes. So even though the basic Gem design has survived primarily unchanged for about a hundred years, manufacturers still confront design and materials options when making new paper clips.

The Manufacturing Process

The manufacturing process for paper clips is fairly simple, using a specialized wire forming machine. Moreover, the process has not changed much since the 1930s.

1 The process begins with a huge spool of galvanized steel wire. A worker feeds the end of the wire into the paper clip machine. A finished paper clip has three bends. The machine forms the wire into these three bends by cutting it and passing it by three small wheels. The wheels are slightly roughened, and catch the length of wire as it passes.

2 The first wheel turns the wire 180 degrees, making the first bend, the second makes the next bend, and the third wheel makes the final turn. The entire process is so quick, the machine can churn out hundreds of clips a minute.

3 The finished paper clips fall into open boxes. The boxes are shut and sealed. Depending on the size of the factory, many

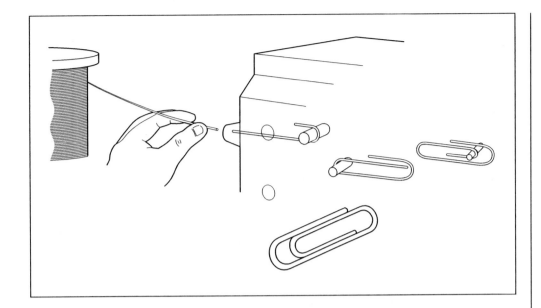

paper clip machines may be operating at once. Automated controls allow one worker to monitor dozens of machines.

Quality Control

Quality control is not a particularly important issue in paper clip manufacturing. Visual inspection of the product is enough to identify a problem with the process. No special tests are needed. The manufacturing equipment must be maintained in order to work properly. Some machines still in use today in the United States were built in the 1930s or even earlier. Trained workers check the equipment for wear and defects that might affect the quality of the finished clips.

Byproducts/Waste

Though paper clips are re-usable, many are thrown away. Some office paper recyclers ask that paper clips be removed before paper is put in recycling bins. Some recyclers use metal detecting equipment that can separate out staples and paper clips, so this material can be recycled separately. One paper clip industry study estimated that the vast percentage of paper clips were never used as intended—to hold paper—but were bent and destroyed by people, used as cleaning or prying instruments, etc. Since paper clips are inexpensive, both to manufacture and to buy as a retail item, most are not re-used or recycled but simply thrown away.

The Future

The Gem clip has held sway against other contenders in paper clip design for a very long time. All-plastic paper clips came on the market in the 1950s, to some success, followed by plastic-coated clips. In the 1990s a Pennsylvania company began marketing what looks like essentially a giant Gem clip, which can hold more than one hundred sheets of paper at one time. None of these developments differs markedly from the turn-of-the-century design consumers are so familiar with. This leads to the question of whether the Gem clip is already a perfect design, thus leaving no room for improvement.

Where to Learn More

Books

Kalpakjian, Serope. *Manufacturing Engineering and Technology.* Reading, MA: Addison-Wesley, 1992.

Petrovsky, Henry. *The Evolution of Useful Things.* New York: Alfred A. Knopf, 1992.

Periodicals

Allen, Frederick. "How Do You Make Paper Clips?" *American Heritage Invention & Technology* (Summer 98).

"Now This Is a Paper Clip!" *Managing Office Technology* (April 1997): 16.

Angela Woodward

Pizza

Background

A pizza is a round, open pie made with yeast dough and topped with tomato sauce, cheese, and a variety of other ingredients.

History

Flatbreads or rounds of dough with various toppings can be found throughout the history of civilization. What is known as pizza today can be traced to Naples, Italy in the Middle Ages. The Italians are also credited with coining the term pizza, although its origin is not clear. It could have derived from the Italian word for point, *pizziare*, meaning to pinch or pluck, or a verb meaning to sting or to season.

Early toppings may have included cheeses, dates, herbs, olive oil, and honey. Tomatoes or tomato sauce were not introduced until the sixteenth century when New World explorers brought the red fruit back from South America. The wealthy classes regarded the tomato as a fruit to be avoided; indeed many thought it to be poisonous. But in the peasant neighborhoods of Naples, residents were enjoying it with the rounds of dough that constituted their primary staple. Somehow the news of this tomato pie spread, and open-air pizza parlors began to do a brisk business. It was also not unusual to see the pizza marker, or *pizzaioli* plying his wares through the streets.

Just as the tomato made its way to Europe, the pizza traveled to the United States with the large influx of Italian immigrants in the latter part of the nineteenth century. One of the earliest known pizzerias was opened by Gennaro Lombardi in New York City in 1905. The thin-crust pie served featured a layer of tomato puree, mozzarella cheese, and various toppings such as sausage and pepperoni. In 1943, Ike Sewell created a deep-dish version at his Chicago restaurant, Pizzeria Uno. The deep-dish pizza combines the sausage, pepperoni, mushrooms, and such with the cheese, which is then poured into a high-sided crust. A layer of tomato sauce is then ladled over the top.

By the end of the 1940s, Frank A. Fiorello was packaging and marketing the first commercial pizza mix. Frozen pizzas were introduced in 1957. By the 1990s, one out of every 20 meals eaten American homes each week was pizza. From its humble beginnings as a staple of the peasant diet, pizzas now sport everything from shrimp to pineapples to barbecued chicken. The manufacturing process, however, remains virtually the same.

Raw Materials

Flour is ground from grain. All grains are composed of three parts: bran (the hard outer layer), germ (the reproductive component), and endosperm (the soft inner core). All three parts are ground together to make whole wheat flour. To make white flour, the bran and the germ must be removed. Since bran and germ contain much of the nutrients in grain, the white flour is often "enriched" with vitamins and minerals. Some white flour has also been fortified with fiber and calcium.

Yeast is a single-celled fungus. The variety *Sacchromycetais cerevisae* is cultivated for use in fermentation to produce alcoholic beverages and bread. Yeast enzymes allow its cells to extract oxygen from the starch in flour and produce carbon dioxide. The carbon dioxide then causes the flour to rise.

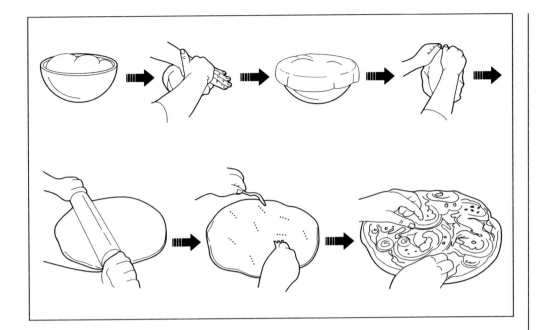

The dough must be kneaded for about 10 minutes until it is allowed to rise.

Baker's yeast is sold fresh in cakes or dried in powder form.

Mozzarella cheese was originally made from buffaloes' milk. In the Italian regions of Latium and Campania, it is still made this way, but the vast majority of mozzarella cheese is now made from cows' milk. It is stored in salted water or whey to keep it moist. For use as a pizza topping, the mozzarella is shredded.

Pizza sauce is made from pureed tomatoes seasoned with a variety of spices including garlic, oregano, marjoram, and basil. Both fresh and dried spices can be used.

The preparation of olive oil is as old as, if not older, than the that of pizza. Olives are gathered from orchards of olive trees and pressed to release their oil.

The list of pizza toppings is exhaustive. Meats include sausage, pepperoni, bacon, chicken, and pork. Vegetables include mushrooms, spinach, olives, broccoli, onions, green peppers, and artichokes. Some of the toppings may be partially cooked before being added to the pizza.

The Manufacturing Process

Making the pizza crust

1 A small amount of baker's yeast, about 1 tbsp, is mixed with a cup or so of warm water. It is left in a warm place until the mixture becomes foamy.

2 Several cups of sifted flour are poured into a bowl. The yeast and water mixture along with 1 tbsp of olive oil is poured into a well made in the center of the flour. The liquids are mixed into the flour with the hands and then kneaded on a floured surface until smooth and elastic. The kneading time is approximately 10 minutes.

3 The kneaded dough is formed into a ball, dusted with flour and then placed in a bowl and covered with a damp kitchen towel. The bowl is placed in a warm place until the dough has doubled in size. This occurs in approximately one to two hours.

4 The dough is kneaded again for about one minute and then rolled out onto a floured surface into a circle. The standard pizza is approximately 10 in (25 cm) in diameter. The edges of the circle are raised by pushing up on the dough with the thumbs.

Filling the pizza

5 A half cup or so of tomato sauce is spooned over the pizza dough. The sauce is spread over the surface of the pie to within 0.5 in (1.3 cm) of the rim. The shredded cheese may be added before the toppings or on top of them.

Baking the pizza

6 Using a wide metal pizza peel, a long-handled flat shovel, the pizza is eased onto a metal pan or clay stone. Pizza pans feature a flat, circular bottom set into a round metal frame. After the pizza is baked, the outer frame is removed. Pizza stones are made of a clay similar to that of old-fashioned brick ovens. Because the clay is porous, it absorbs moisture. The thickness of the stone, usually about 0.75 in (2 cm), radiates heat evenly.

7 The pizza is baked at 450°F (230°C) for about 15 minutes or until the cheese is bubbling. The pan or stone is removed from the oven with the peel. The pizza is allowed to sit for approximately five minutes before cutting it into slices with a pizza wheel. Slice shapes, like the placement of the mozzarella cheese, differs from region to region. In some cities the pizza is sliced into pie-shaped pieces. In other cities, the pie is cut into squares.

The Future

A staple since the beginning of human civilization, the pizza shows no sign of diminishing in popularity. So-called gourmet pizzas, made with pastry dough, goat cheese, and escargot, can be found on the menus of upscale restaurants. And in spite of increased awareness about cholesterol levels and fat content, a slice of pizza oozing with cheese and pepperoni is a favorite item in storefront pizzerias and shopping mall food courts.

Where to Learn More

Books

Anderson, Kenneth N., and Lois E. Anderson. *The International Dictionary of Food & Nutrition.* New York: John Wiley & Sons, 1993.

Lang, Jenifer Harvey, ed. *Larousse Gastronomique.* New York: Crown Publishers, 1998.

Scicolone, Charles, and Michela Sciolone. *Pizza: Any Way You Slice It.* Broadway Books.

Other

Stradley, Linda. "History and Legends of Pizza." *What's Cooking America Web Page.* 2000. December 2001. <http://www.geocities.com/familysecrets/History/Pizza/PizzaHistory.htm>.

Mary McNulty

Pop Up Book

Background

The pop up book is a book with paper elements within the pages that may be manipulated by the reader. Many refer to such a book as a moveable book. Pop up books include text, illustrations, and folded, glued, or pull-tab elements that move within the pages of the story. The pop up book is primarily marketed to children.

The moveable paper elements within the pop up book require the expertise of a paper engineer to effectively design these elements. This paper engineer is part engineer and part creative designer, constantly seeking new, fun elements to design into pop up books while ensuring they will inexpensive to produce and successfully manipulated. The paper engineer communicates to the printer/publishers how the moveable elements are die cut and then assembled. The die cutting of these elements is expensive and complex. Even more expensive is the extraordinary amount of hand work the moveable elements require as many must be cut, folded and pasted by hand (some books include 100 elements that require hand manipulation). In fact, pop up books are becoming somewhat collectible because of the extraordinary amount of hand assembly that goes into the construction of each book. The pop up book has been the subject of at least two art museum exhibits in which the art of the illustrator and the design of the paper engineer have been highlighted.

History

Moveable books are hardly new. About 700 years ago people used simple books with moving parts to teach about anatomy or make astronomical predictions. Even for-tune-telling used moveable books. The pop up book was the domain of adults until the late 1880s when *metamorphoses* books, also called turn-up books, included fold-out illustrations within the pages of children's books. By the nineteenth century such moveable books were published in some quantity in England. By mid-century a British firm was happily producing such books for children, and by the twentieth century they had published over 50 titles.

An American firm named McLoughlin Brothers of New York city produced the first moveable books in the United States about 1880. They were large plates that unfolded into multi-layered displays. As Europeans found cheaper papers and booksellers sought to enlarge their markets cheaper and more inventive pop up books were developed in the early twentieth century. By the 1960s American Waldo Hunt created advertising inserts and premiums inspired by Czechoslovakian works. Hunt began to produce his own moveable books for popular consumption and is believed to have popularized the moveable book in the mid-twentieth century. Today, pop up books are enormously popular with children because of the novelty of the moveable elements greeting cards and advertisements include pop up elements. Recent pop up books for children are written on topics as diverse as astronomy, geology, meteorology, children's classics, and dinosaurs.

Raw Materials

The typical pop up book uses heavy gauge paper for the pages and the moveable elements of the book, heavy board cover in front and back, glue for securing the cover, and glue for the attachment of the pop up elements. Inks of a wide variety may be used in the printing, from soy-based inks to more traditional oil-based inks. Many pop up books are coated with a coating on the page to make them sturdier and dirt-resistant. These coatings include oil-based varnishes that render a shiny surface on the page. Some companies use aqueous or water-borne varnishes. Other pop up books use a plastic film that is put over the pages as a laminate.

The Manufacturing Process

1 The author and the editor of the pop up book take ideas about the movement of elements within the book to a paper engineer. A paper engineer is trained to understand how paper may be folded in order to render certain effects, and he or she designs how the paper may be cut and folded in order to create pertinent pop up book elements. A paper engineer's primary job is to design elements that are to move within the book, including cut parts, parts with tabs, and so forth. The paper engineer must fold, cut, and paste elements and create by hand in rough draft form for examination and approval. This page-by-page understanding of the moveable parts of the pop up book is put together in an all-white paper model referred to as a white dummy. This white dummy is meant to highlight how the elements will work together through the manipulation of paper elements.

2 The white dummy is an important element for the author, editor and paper engineer to examine and assess. Any changes in gluing, paper movement, size of pull tabs, or folded elements are assessed and finalized at this point. The paper engineer must do two things to ensure the book can be produced. First he or she must produce a digital file or some other template that will allow the printer to create dies in order to produce the moving paper elements. Secondly, the paper engineer also must lay out or nest all the pages so that the moveable pieces will all fit onto the size sheet that will be run through the press.

3 Next, the editor works with a graphic designer who will work with the paper engineer, author, and illustrator to lay out each page, element by element. The graphic designer will produce a flat lay out. This flat lay-out determines the relative positions of the text, the illustrations and the pop up elements. This process is highly collaborative. In some cases, the art work is place first and the text accommodates the art; in other cases the art is placed according to moving paper elements and text.

4 Once the flat lay out is agreed upon by all, the illustrator must create the flat art in full color to fill in the areas that he or she is given to work in. When the art work is complete, the illustrations are sent back to the graphic designer.

5 The designer takes the art work and creates a mechanical, which is an electronic file that shows the printer where the art work, is inserted into the page. Then, the text is placed on the mechanical. The mechanical and the art work is sent to printer.

6 Now, the pop up book is ready for printing. The digital file of the art work and text for each page is output to a film. The film is then used to make a plate for printing. Four different films are generally used to print children's pop up books including blue, yellow, red, and black. Each plate carries a different color. The individual pages are off set printed, meaning an inked blanket deposits color onto the page.

7 The printed pages are proofed, which means they have an initial printing that must be assessed and approved by the editor before production can continue. Once they are approved, final printing can occur. The pages are printed and await the attachment of the moveable paper elements. Some pages receive a coating that is applied to the surface of the paper. Shiny pages receive a varnish coating as opposed to a matte finish.

8 Meanwhile, the moveable elements must be created. The digital files help create the dies in some publishing companies; thus the computer generates these dies automatically. However, in some publishing companies, especially those overseas, the dies are often made by hand using sophisticated machinery by tool and die makers.

9 The dies are used to cut out the moveable paper parts. About 10 sheets of paper are placed on a die. A hydraulic press forces the

cut die through the paper to produce the desired shape.

10 The moveable parts of the pop up book are of often cut out by hand and are folded and glued by hand upon the printed pages.

11 The cover is glued or sewn to the lining. Front and backs are often made up from board, which is just a heavier gauge paper than is used for the pages. However, some pop up books do not have covers that are thicker or heavier than the inner pages but are of precisely the same material as the pages within.

12 The book is then packaged and sent to the distributor.

Quality Control

There are many checkpoints in the production of a pop up book as there is much at stake if small mistakes are discovered after the book goes to publication. Publishing is very expensive and few publishers are able to make a large amount of money on these books. The eyes of the designer, illustrators, paper engineer, and author must look for mistakes prior to production.

Byproducts/Waste

There is some waste in the production of pop up books. Paper waste may be discarded by the manufacturer; however, it is more likely the paper is recycled or re-used by another company. Solvent-based coatings on the pages are fairly common but are being phased out because of problems with disposal. Companies increasingly prefer aqueous varnishes, which are water-based and

A pop-up book.

not oil-based. Dies are rather expensive to make and bulky to keep so many dies are melted down and re-used quickly.

Where to Learn More

Other

Carvajal, Doreen. "Boing! Pop-Up Books are Growing Up." *New York Times Web Page*. 27 November 2000. December 2001. <http://www.nytimes.com/learning/students/quiz/articles/27POPU.html>.

Designmation, Inc. Web Page. December 2001. <http://www.pop-ups.com/new/process.html>.

Oral interview with Greg Witt, Customer Service Representative at Leo Paper Company. Seattle, WA. November 2001.

Oral interview with Sarah Ketcherside, editor at Candlewick Press. New York, NY. November 2001.

Nancy EV Bryk

Punching Bag

By the beginning of the first century A.D., boxing had been forbidden and would not be seen again in the sports world until the eighteenth century in Great Britain.

Background

A punching bag is a round or cylindrical piece of athletic equipment used by professional boxers for training and by amateurs for exercise. The bags come in a variety of sizes for a variety of uses. The largest, known as the heavy bag, is used to develop footwork and power. The timing bag, usually suspended from the ceiling and floor by bungee cords, develops timing and hand-eye coordination. The small speed bag develops hand speed, coordination, and rhythm.

History

The sport of boxing dates to the ancient Olympic Games. The first gold medal in boxing was awarded to a fighter named Onomastos in the twenty-third Olympiad. The earliest boxers were trained as if soldiers for war. There was little sophistication in the sport. Participants boxed bare-handed. One of the earliest boxers, Eurydamus, was known for his fierceness; he was reported to have swallowed his own teeth during a match rather than admit that he had been severely injured.

By the beginning of the first century A.D., boxing had been forbidden and would not be seen again in the sports world until the eighteenth century in Great Britain. The first recorded English boxing champion was James Figg, who fought in the early part of the 1700s. Figg was followed by Jack Broughton in the mid century. Broughton was considered a master of blocking, parrying, and hitting on the retreat. In 1743, Broughton created a boxing code of conduct called the London Prize ring rules. (The rules were later revised and became known as the Revised London Prize Ring rules.) When Broughton's patron, the Duke of Cumberland, asked him to teach some of his well-to-do friends to box, Broughton devised special gloves, or mufflers, so that these "gentlemen" would not injure their hands.

The sport was further refined by the establishment of the Queensberry Rules, which were created under the sponsorship of another patron of the sport, John Sholto Douglas, the eighth marquess of Queensberry. These rules, still in effect today, set round limits, established glove weights, and created fighter classes by weight. They also forbade hits below the belt, to the back of the head, to the neck and to the kidneys.

It is not certain when the punching bag became part of the boxer's training regiment. The United States Office of Patents and Trademarks awarded a patent for the punching bag to Simon D. Kehoe in 1872. Since that time, others have made improvements to better simulate the human body.

With the increased interest in physical fitness and the advent of health clubs in the 1970s, amateurs became more interested in boxing as a form of fitness. By the beginning of the twentieth century, large numbers of people in the United States were enrolled in some type of boxing class.

Design

Paper patterns are created for the various panel sizes. The patterns are placed on sections of leather. Using chalk or grease pencils, a worker traces the shape of the pattern onto the leather.

A striking bag is usually constructed of two leather balloons, inserting one inside the other, then inflating the inner balloon with air to create a resilient ball.

Raw Materials

The earliest striking bags were made from kangaroo skin. However, goatskin is now more commonly used for small punching bags. The animal skin is dipped in strong chemicals to remove the hair prior to being cured in salt water. After curing, the skin is stretched and dried and ready for use in manufacturing.

The larger, heavy bags are constructed of polyvinyl or canvas. Canvas is a heavy cotton material. Polyvinyl is a plastic material developed during the Second World War. It is made from the byproducts of petroleum and coal.

A heavy, coated synthetic thread, typically nylon and polyester, is used to stitch the pieces of leather together.

While the striking bag is inflated with air, the heavy bags are filled with sand or finely shredded wood clippings.

Snaps, hooks, zippers, chainlink, and cord lacings are used to close the bags, attach them to the rebound board, and/or attach them to other bags.

The Manufacturing Process

The manufacture of striking and training bags are accomplished through a combination of manual and mechanical steps.

Constructing the striking bag

1 The leather for the bag is usually derived from goatskin. The skin must be tanned by placing it in rotating drums filled with a salt and water solution. The salt ingredient is typically chromium. In approximately eight hours, the chromium soaks through the skin. The chromium is then "fixed" to the skin by the addition of an alkaline chemical such as sodium carbonate or bicarbonate.

2 The tanned skin is run through a machine that shaves it to the desired thickness. After that, it is passed through a wringer to remove excess moisture.

3 Once the leather has been dried and prepared, it is ready to be cut. First, patterns that are the exact size and shape of each panel are traced onto the leather using chalk

Joe Frazier throwing a punch at Muhammad Ali.

Joe Frazier was born on January 17, 1944. Growing up in the rural South with his 12 brothers and sisters, Frazier rigged a punching bag from a burlap sack, rags, corncobs, brick, and Spanish moss. Leaving school at 14, he worked as a delivery man and then as a construction worker in South Carolina. Arriving in Philadelphia, he got a job at a slaughterhouse and developed a habit that would be immortalized in the movie *Rocky*: Frazier practiced his punches on hanging sides of beef.

Frazier's boxing career began in 1964 after wining the Olympic gold medal in Japan, and peaked when he became the first American Olympic heavyweight champion to also win the heavyweight title of the world. When he was champion, Frazier held the highest knockout percentage in history, and while he had been knocked down a few times, he had never been knocked out. Frazier was involved in "The Fight of the Century" with Muhammad Ali in 1971 for the world heavyweight title, which Frazier held. The Frazier-Ali fight was the first of three and set indoor boxing records for attendance and revenue.

Frazier underwent cataract surgery on his left eye in 1975. While the cataract was removed, it was too late—he was legally blind in his left eye and now wore contacts to fight. A rematch with George Foreman in June of 1976 was stopped in the fifth round, and Frazier knew his career was over. He came out of retirement in December of 1981 to fight Floyd Cummings. Even though the bout was a draw Frazier had to admit it was time to hang up the gloves for good.

or a grease pen. Then, these sections are manually cut by workers using leather-cutting shears or knives.

Some manufacturers employ automated cutting machines in order to create the panels. In these cases, a very sharp metal component is constructed in the exact size and shape of the panel. As the leather passes

A boxer practicing with a speed bag.

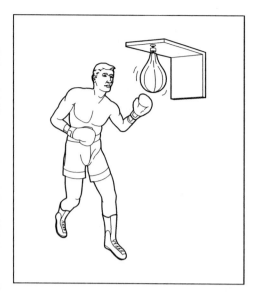

under this metal die, the die cuts down into the fabric similar to a cookie cutter.

4 The next step is to sew the leather panels together. The bags are constructed with either four or six panels, depending on the desired size. When sewn together, the panels create a pear-shaped balloon. Heavy duty nylon or polyester thread is used to stitch the pieces to one another. The stitching is performed by a worker operating a sewing machine.

5 Once the two bags have been sewn together, the inner bag needs to be inflated with air. An inflation device called a collar is inserted into the inner bag just before the two bags are completely stitched. A valve on the inflation collar regulates the amount of air that is blown into the outer bag. Inflated air is measured in terms of pounds per square inch, or psi. The striking bag is inflated to a measure of 4–4.5 psi. A cap on the collar then clamps the bag wall to the collar skirt.

6 The final step involves manually connecting the striking bag to an overhead rebound board by a piece of flexible metal combined with a ball-and-socket joint. The rebound board may be constructed of wood or a thick, durable plastic.

Constructing the training bag

1 Training bags are usually made of vinyl or canvas. The vinyl or canvas is cut from patterns using a garment knife or die cutting

machine, just as the leather was cut for a striking bag. The round top and bottom pieces are cut on a punch press.

2 Once all of the material has been cut to the specified size, the pieces are sewn together. Again, strong nylon or polyester thread is used. The pieces are stitched together by a worker operating a sewing machine. The top is left open.

3 In order to fill the bag, it is placed over a sleeve attached to a hopper. The stuffing materials, such as shredded wood clippings or sand, are loaded into the hopper. Forced air fills the bag with the stuffing materials.

4 The top of the bag is sealed by sewing tabs of vinyl or canvas to the top edges of the bag. A round length of tubing, called a torus, is threaded through the tabs. Four metal rings, attached to four chains, are fitted around the torus. The top is then closed with laces or with zippers.

Quality Control

In order to ensure that bags have been properly constructed, manufacturers will periodically test the final product. Either manually or using automatic robotic devices, the bags undergo rigorous tests that determine the durability of the seams, the outer material, and the hanging components. To test the seams, the fabric may be pulled in opposing directions at measured forces or repeatedly tugged and timed. Punching the bag in a manner similar to how it will be used is also another quality control measure. It is important to ensure that the final product can withstand excessive wear and tear and that the primary components will remain in tact after use, especially considering the purpose of this product.

Byproducts/Waste

Scraps of leather, vinyl, canvas, thread, and fill material are the excess waste produced through the manufacture of punching bags. Depending on the size of the scrap fabrics, they will either be reused for other goods produced by the company or, more likely, disposed of in the trash. Any pieces of thread will also be thrown away. Fill material may be recollected for the same use if it is still in good condition. Excess wood clippings could be safely incinerated and sand can be disposed of at an appropriate dump site if not adequate for reuse.

The Future

Manufacturers continue to make improvements to punching bags in an effort to better simulate the human body and its reactions to strikes. A patent was issued in 1998 by the United States Trademark and Patent Office for a training apparatus that combined a head-sized striking bag to a hanging training bag.

Other innovations are designed to attract the amateur boxing enthusiast. One such version is the Soc-o-Mac, developed by Howard "Mack" McConnell in 1976. The Soc-o-Mac weighs about 450 lb (204 kg) and sits in a weighted steel pan on the floor. The steel pan's curved bottom keeps the bag upright and allows it to roll over and back up when hit.

In the early 1980s, Tom Critelli, a former deputy sheriff invented a water-filled bag that can be drained and refilled. Another invention in the last decade of the twentieth century, was the SoloSpar, an automated heavy bag that moves when punched and talks back when struck in certain spots.

Where to Learn More

Books

Carpenter, Harry. *Boxing: A Pictorial History*. Chicago: Henry Regnery, 1975.

Periodicals

Lidz, Franz. "A Fighter Would Have to be All Wet." *Sports Illustrated* (20 June 1983): 12.

A boxer practicing footwork with a heavy bag.

McDonnell, Terry. "Punchline." *The Business Journal (serving Phoenix and Valley of the Sun)* (19 June 1989): 1.

Millman, Chad. "Automated Attitude." *Sports Illustrated* (21 February 1994): 89.

Other

"U.S. Patent 5142758: Punching Bag Construction and Suspension." *Delphion Web Page*. December 2001. <http://www.delphion.com>

"U.S. Patent 5769761: Striking Bag Training Apparatus." *Delphion Web Page*. December 2001. <http://www.delphion.com>.

Mary McNulty

Pyrex

In Mesopotamia, archeologists have uncovered clay tablets that contain ancient "instructions" for making glass in furnaces.

Background

Pyrex glass is a borosilicate glass first produced by The Corning Glass Works company. It is made by heating raw materials like silica sand and boric oxide to extremely high temperatures for extended periods of time. The molten material is then processed into different types of glassware. First formulated during the early twentieth century, Pyrex has become an important material for a variety of applications that require heat and chemical resistance.

To understand how Pyrex is unique it is important to understand the nature of glass itself. Glass is a state of matter that has characteristics similar to both crystalline solids and liquids. On a macroscopic level glass appears to be like solids. It is rigid and remains in one piece when removed from a container. However, on a molecular level, glasses are more like liquids. In crystalline solids, molecules are arranged in an orderly fashion. In liquids they are randomly arranged. This random arrangement is also a characteristic of glass.

Glass is typically made by heating crystalline compounds to temperatures high enough to melt them. Melting breaks the ordered molecular structure, leaving them in a disordered state. When the melted material is cooled, the molecules become locked in place before they can reform in the ordered crystalline structure. The properties of a specific glass such as hardness, brittleness, clarity, and chemical and thermal resistance are dependent on its chemical composition.

When Pyrex was being developed, scientists were trying to create a glass composition that had a high thermal resistance. At some point it was discovered that glass compositions with boron could be heated to high temperatures without breaking. Boron, which is the fifth element on the periodic chart, has the unique ability to create a variety of chemical bonds. When bonded with oxygen it can create a three dimensional structure that is strong. In a glass composition, this extra strength gives it thermal and chemical resistance that makes it useful for cooking applications, thermometers, and laboratory equipment. Pyrex also has a low alkali content that gives it high corrosion resistance.

History

While the exact date people found that sand could be combined and melted with other materials to produce glass is not known, its discovery was likely accidental. Formal processes for glassmaking have been known for over 3,000 years. In Mesopotamia, archeologists have uncovered clay tablets that contain ancient "instructions" for making glass in furnaces. Throughout history, glass production technology became more sophisticated. People steadily discovered the best proportions to combine the raw materials and also learned manufacturing practices like glass blowing.

During the early twentieth century, kerosene lanterns were widely used for streetlights and railroad signaling devices. Unfortunately, the glass used for making these lanterns was sensitive to the heat of the flame and would often break. Scientists began searching for glass formulas that could withstand heat.

The first experiments led to the discovery that when boric acid was present in the raw materials, the glass was more heat resistant.

These early formulas were chemically weak however, often breaking down in water. Work proceeded to find the right proportions of silica sand and boric oxide that would continue to be heat resistant and chemically stable. In 1912, an adequate formula was found. These glasses, called borosilicates, were then introduced into lantern production. One of the original types of borosilicate glass introduced by the Corning Glass Works Company was brand-named Nonex.

The potential for this product in the area of cooking was discovered in 1913 by Dr. Jesse T. Littleton who worked at Corning. He gave his wife a casserole dish made out of Nonex, the precursor to Pyrex. It worked as well as a ceramic cooking dish and a new era in cooking ware had begun. The Nonex glass formula was revised to remove lead, and the ovenware was given to the Philadelphia Cooking School for more testing. A series of successful tests there led to the introduction of Pyrex ovenware in 1915. This same year the Corning Glass Works Company patented the formula and gave it the trademarked name Pyrex. It has been suggested that the term Pyrex was either a derivative of the word "pie" (referring to its original use) or the Greek "pyra," which means hearth. In both cases, the "ex" suffix was used to give it brand-name similarity to Nonex.

When World War I broke out, scientists who relied on German glass products found that the new Pyrex material met their needs for beakers, test tubes, and other laboratory glassware. Borosilicate glass has steadily been made more chemical, heat, and shock resistant. It has also been applied to numerous products such as eyeglasses, telescopes, and electronic components.

Raw Materials

Three classes of materials are used in making Pyrex including formers, fluxes, and stabilizers. Formers are the main ingredients in all glassmaking. These are crystalline materials that, when heated high enough, can be melted and cooled to create glass. Fluxes are compounds that help lower the temperature required to get the formers to melt. Stabilizers are materials that help keep glass from crumbling, breaking, or falling apart. They are needed because fluxes typically destabilize glass compositions.

New tests prove that FOODS ACTUALLY BAKE BETTER THIS WAY

An ad for Corning Pyrex.

Eugene G. Sullivan established Corning Glass Works' research laboratory in 1908 and set out with William C. Taylor to make a heat-resistant glass for railroad lantern lenses. The problem was that flint glass (the kind in bottles and windows, made by melting silica sand, soda, and lime) has a fairly high thermal expansion but poor heat conductivity. Both causing the glass to break. Two solutions were possible: improve thermal conductivity or reduce thermal expansion. The formulation that Sullivan and Taylor devised was a borosilicate glass—a soda-lime glass with borax replacing the lime—with a small amount of alumina added. This gave the low thermal expansion needed and also had good acid-resistance, leading to use for the battery jars required for railway telegraph systems and other applications. The glass was marketed as "Nonex" (for nonexpansion glass).

Jesse T. Littleton joined Corning in 1913. A physicist, Littleton knew glass absorbs radiant energy well, while metal mostly reflects it. Littleton took a cut-off battery jar home and asked his wife to bake a cake in it. He took it to the laboratory the next day. Littleton developed variations on Nonex and the result was Pyrex, patented and trademarked in May of 1915.

Initial sale of Pyrex took place at the Jordan Marsh department store in Boston in 1915. By 1919 more than 4.5 million pieces were sold. In 1915, Pyrex was introduced into the laboratory. Laboratory glassware came from Germany but World War I cut off the supply. Corning filled the gap with Pyrex glassware, which worked so well that Pyrex replaced most other items. Today, Corning-style glassware is found in laboratories all over the world.

The primary formers used for making Pyrex include silica sand and boric acid. Silica sand is also known as silicon dioxide. It is a crystalline material and was probably the major component of the first glass used by humans. In a typical Pyrex glass composition, silicon dioxide makes up about 60–80% by weight.

Pyrex has a droplet in matrix phase structure. The silicon dioxide creates the basic matrix. The borate material creates the droplets within that structure. The borate former can come from a material like sodium tetraborate. Prior to manufacture, this compound is chemically reduced with sulfuric acid to create boric acid. When boric acid is mixed with silicon dioxide and heated, it oxidizes into boric oxide. Boric oxide is responsible for the unique Pyrex molecular structure. Boric oxide makes up anywhere from 5% to 20% of Pyrex glass.

Secondary ingredients used in glass production include fluxes, stabilizers, and colorants. Fluxes are included in glass mixtures because they reduce the melting temperature of the borosilicate glass. Fluxes that can be used in manufacture include soda ash, potash, and lithium carbonate. They make up about 5% of a Pyrex glass composition.

Unfortunately, fluxes also cause the glass to be more chemically unstable. For this reason stabilizers such as barium carbonate and zinc oxide are included. In Pyrex manufacture, about 2% aluminum oxide is added to make the glass stiffer when it is molten. Finally, to produce glass with different colors, silver compounds can be added.

The Manufacturing Process

The manufacturing process can be broken down into two phases. First, a large batch of molten glass composition is made. Next, the glass is fed into shaping machines to create different types of glassware. The process moves at tremendous speeds and is quite efficient.

Batching

1 Large batches of Pyrex glass are produced in a specified compounding area of the production plant. Here, glassmakers follow formula instructions and add the required raw materials in the correct proportions into large tanks. Prior to use, the raw materials are pulverized and granulated to a uniform particle size. They are stored in batch towers. The materials are mixed together and heated to temperatures over 2,912°F (1,600°C). This high temperature melts the ingredients and allows them to thoroughly mix to create

molten glass. However, the mixture typically requires longer heating—up to 24 hours—to remove excess bubbles that can lead to a weaker structure.

Forming

2 The batch tanks are designed so that the molten glass will flow slowly toward the working end of the tank. This end of the tank is connected to continuous feed forming machines. As the glass moves from the tank it looks like a thick, red-orange syrup. The forming machines work the material quickly because as it cools it becomes rigid and unworkable. Typical glass processing machines blow, press, draw, and roll it into various shapes.

3 The forming process used depends on the final product. Glass blowing is used to create thin-walled products like bottles. A bubble of the molten glass is put inside a two-piece mold. Air is forced into the mold, which presses the glass against its sides. The glass cools inside the mold and conforms to the shape. Glass pressing is used to create thicker pieces of glass. The molten glass is put into a mold and a plunger is lowered which forces the glass to spread and fill the mold. Drawing is used to create tubing or rods. In this process molten glass is drawn down over a hollow cone called a mandrel. Air is blown through it to keep the tube from collapsing until the glass becomes rigid. For glass sheets, like windows, a rolling process is used.

4 After the product is formed, it is cooled and polished. It may then be decorated with various printing or markings and fitted with plastic pieces if necessary. The glass product is then checked for imperfections, put in protective boxes and shipped out to customers. Depending on the size of the batching tank, as much as 700,000 lb (317,520 kg) of glass product can be produced in one year.

Quality Control

Since the quality of the glass depends on the purity of the raw materials, manufacturers employ quality control chemists to test them. Physical characteristics are checked to make sure they adhere to previously determined specifications. For example, particle

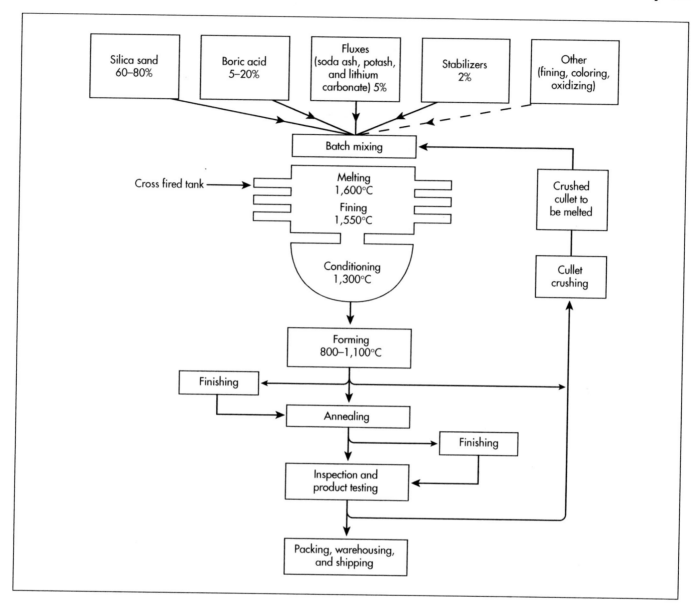

| | | | | | |
| Silica sand 60–80% | Boric acid 5–20% | Fluxes (soda ash, potash, and lithium carbonate) 5% | Stabilizers 2% | Other (fining, coloring, oxidizing) | |

Batch mixing

Cross fired tank →

Melting 1,600°C

Fining 1,550°C

Conditioning 1,300°C

Crushed cullet to be melted

Cullet crushing

Forming 800–1,100°C

Finishing

Annealing

Finishing

Inspection and product testing

Packing, warehousing, and shipping

A diagram of the production of Pyrex.

size is measured using appropriately meshed screens. Chemical composition is also determined with an IR or GC. Other simple checks that are done on the raw materials include color checks and odor evaluations. During production of a glass product, inspectors watch the glass products at specific points on the manufacturing line to ensure that each product looks correct. They notice things such as cracks, flaws or other imperfections. For certain products, the thickness of the glass is measured.

Byproducts/Waste

Since Pyrex is made from compounds that become oxides when heated, air pollution is a potential problem. A variety of byproducts may be released during manufacture including nitrates, sulfates, and chlorine. These chemicals can react with water to form acids. Acid rain has been shown to cause significant damage to manmade structures as well as natural ecosystems. One method glassmakers use to reduce pollution is by making glass compositions that have lower melting temperatures. Lower temperatures reduce the amount of volatilization thereby reducing the amount of gaseous pollutants. Another pollution control is the use of precipitators that are installed in chimneys. These devices help reduce air pollution by filtering out solids that persist in smoke and vapor created by the melting process. Waste-disposal drains are monitored to ensure that only allowable amounts of factory waste are released into the environment. This helps prevent water pollution.

An additional method of pollution control is the use of ventilators. These devices are also called regenerators because they help recover and recycle heat energy consumed during manufacture. This has the double effect of reducing air pollution and lowering production costs. Other cost reducing and environmentally sound techniques employed include the use of electric heat instead of gas heat, and the incorporation of broken recycled glass during the production of new glass.

The Future

In the future, borosilicate glass manufacturers will concentrate on increasing sales and improving the production process. To increase sales, glass manufacturers will be involved in finding and promoting new applications for their products. This could require new glass formulations that have a range of characteristics from clarity, melt point, and shatter resistance. From a production standpoint, future improvements will focus on increasing manufacturing speeds, minimizing chemical waste, and reducing overall costs.

Where to Learn More

Books

Bansal, N. P., and R. H. Doremus. *Handbook of Glass Properties.* New York: Academic Press, Inc., 1986.

Kirk-Othmer Encyclopedia of Chemical Technology. Vol. 12. New York: John Wiley & Sons, 1994.

Mazurin, O. V. *Handbook of Glass Data.* New York: Elsevier Science Publishing Co., 1991.

Rogove, S. T., and M. B. Steinhauer. *Pyrex by Corning: A Collector's Guide.* New York: Antique Publications, 1993.

Other

Corning Museum of Glass Web Page. 1 October 2001. <http://www.cmog.org>.

United States Patent 4,075,024. Colored Glasses and Method. 1976.

Perry Romanowski

Radio

Background

The radio receives electromagnetic waves from the air that are sent by a radio transmitter. Electromagnetic waves are a combination of electrical and magnetic fields that overlap. The radio converts these electromagnetic waves, called a signal, into sounds that humans can hear.

Radios are a part of everyday life. Not only are they used to play music or as alarms in the morning, they are also used in cordless phones, cell phones, baby monitors, garage door openers, toys, satellites, and radar. Radios also play an important role in communications for police, fire, industry, and the military. Although there are many types of radios—clock, car, amateur (ham), stereo—all contain the same basic components.

Radios come in all shapes and sizes, from a little AM/FM "Walkman" to a highly sophisticated, multi-mode transceiver where both the transmitter and receiver are combined in one unit. The most common modes for a broadcast radio are AM (amplitude modulation) and FM (frequency modulation). Other modes used by ham radio operators, industry, and the military are CW (continuous wave using Morse code), SSB (single sideband), digital modes such as telemetry, radio teletype, and PSK (phase shift keying).

History

Guglielmo Marconi successfully sent the first radio message across the Atlantic Ocean in December 1901 from England to Newfoundland. Marconi's radio did not receive voice or music. Rather, it received buzzing sounds created by a spark gap transmitter sending a signal using Morse code.

The radio got its voice on Christmas Eve 1906. As dozens of ship and amateur radio operators listened for the evening's traffic messages, they were amazed to hear a man's voice calling "CQ, CQ" (which means calling all stations, I have messages) instead of the customary dits and dahs of Morse code. The message was transmitted by Professor Reginald Aubrey Fessenden from a small radio station in Brant Rock, Massachusetts.

In the years from 1904 to 1914, the radio went through many refinements with the invention of the diode and triode vacuum tubes. These devices enabled better transmission and reception of voice and music. Also during this time period, the radio became standard equipment on ships crossing the oceans.

The radio came of age during World War I. Military leaders recognized its value for communicating with the infantry and ships at sea. During the WWI, many advancements were made to the radio making it more powerful and compact. In 1923, Edwin Armstrong invented the superhetrodyne radio. It was a major advancement in how a radio worked. The basic principles used in the superhetrodyne radio are still in use today.

On November 2, 1920 the first commercial radio station went on the air in Pittsburgh, Pennsylvania. It was an instant success, and began the radio revolution called the "Golden Age of Radio." The Golden Age of Radio lasted from the early 1920s through the late 1940s when television brought in a whole new era. During this Golden Age, the radio evolved from a simple device in a bulky box to a complex piece of equipment housed in beautiful wooden cabinets. People would gather around the radio and listen to the lat-

Guglielmo Marconi successfully sent the first radio message across the Atlantic Ocean in December 1901 from England to Newfoundland.

275

est news and radio plays. The radio occupied a similar position as today's television set.

On June 30, 1948 the transistor was successfully demonstrated at Bell Laboratories. The transistor allowed radios to become compact, with the smallest ones able to fit in a shirt pocket. In 1959, Jack Kilby and Robert Noyce received the first patent for the integrated circuit. The space program of the 1960s would bring more advances to the integrated circuit. Now, a radio could fit in the frame of eyeglasses or inside a pair of small stereo earphones. Today, the frequency dial printed on the cabinet has been replaced with light emitting diodes or liquid crystal displays.

Raw Materials

Today's radio consists of an antenna, printed circuit board, resistors, capacitors, coils and transformers, transistors, integrated circuits, and a speaker. All of these parts are housed in a plastic case.

An internal antenna consists of small-diameter insulated copper wire wound around a ferrite core. An external antenna consists of several aluminum tubes that slide within one another.

The printed circuit board consists of a copper-clad pattern cemented to a phenolic board. The copper pattern is the wiring from component to component. It replaces most of the wiring used in earlier radios.

Resistors limit the flow of electricity. They consist of a carbon film deposited on a cylindrical substrate, encased in a plastic (alkyd polyester) housing, with wire leads made of copper.

Capacitors store an electrical charge and allow alternating current to flow through an electrical circuit but prevent direct current from flowing in the same circuit. Fixed capacitors consist of two extended aluminum foil electrodes insulated by polypropylene film, housed in a plastic or ceramic housing with copper wire leads. Variable capacitors have a set of fixed aluminum plates and a set of rotating aluminum plates with an air insulator.

Coils and transformers perform similar functions. Their purpose is to insulate a circuit while transferring energy from one circuit to another. They consist of two or more sets of copper wire coils either wound on an insulator or mounted side-by-side with air as the insulator.

Transistors consist of germanium or silicon encased in a metal housing with copper wire leads. The transistor controls the flow of electricity in a circuit. Transistors replaced vacuum tubes used in earlier radios.

The integrated circuit houses thousands of resistors, capacitors, and transistors into a small and compact package called a chip. This chip is about the size of the nail on the little finger. The chip is mounted in a plastic case with aluminum tabs that allow it to be mounted to a printed circuit board.

Design

Radios consist of many specialized electronic circuits designed to perform specific tasks—radio frequency amplifier, mixer, variable frequency oscillator, intermediate frequency amplifier, detector, and audio amplifier.

The radio frequency amplifier is designed to amplify the signal from a radio broadcast transmitter. The mixer takes the radio signal and combines it with another signal produced by the radio's variable frequency oscillator to produce an intermediate frequency. The variable frequency oscillator is the tuning knob on the radio. The produced intermediate frequency is amplified by the intermediate frequency amplifier. This intermediate signal is sent to the detector which converts the radio signal to an audio signal. The audio amplifier amplifies the audio signal and sends it to the speaker or earphones.

The simplest AM/FM radio will have all of these circuits mounted on a single circuit board. Most of these circuits can be contained in a single integrated circuit. The volume control (a variable resistor), tuning knob (a variable capacitor), speaker, antenna, and batteries can be mounted either on the printed circuit board or in the radio's case.

The Manufacturing Process

There is no single process for manufacturing a radio. The manufacturing process depends upon the design and complexity of the radio.

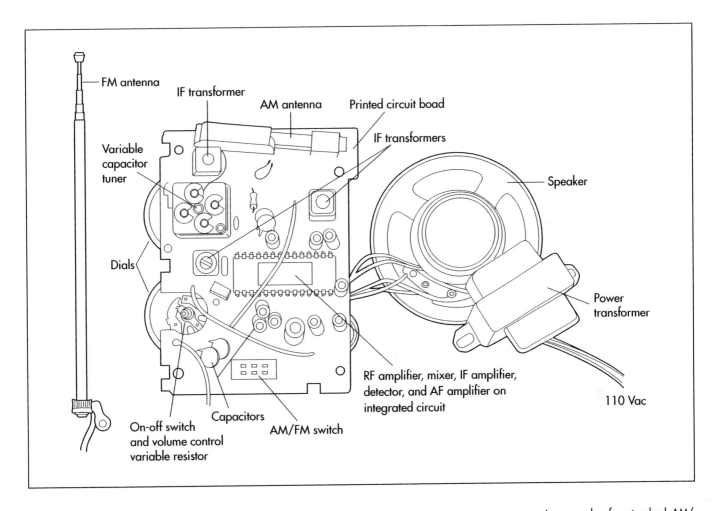

The simplest radio has a single circuit board housed in a plastic case. The most complex radio has many circuit boards or modules housed in aluminum case.

Manufacturers purchase the basic components such as resistors, capacitors, transistors, integrated circuits, etc., from vendors and suppliers. The printed circuit boards, usually proprietary, may be manufactured in house. Many times, manufacturers will purchase complete radio modules from an vendor. Most of the manufacturing operations are performed by robots. These include the printed circuit boards and mounting of the components on the printed circuit board. Mounting of the printed circuit board and controls into the case and some soldering operations are usually done by hand.

1 The blank printed circuit board consists of a glass epoxy resin with a thin copper film cemented to one or both sides. A light sensitive photoresist film is placed over the copper film. A mask containing the electrical circuitry is placed over the photoresist film. The photoresist film is exposed to ultraviolet light. The photoresist image is developed, transferring the image to the copper film. The unexposed areas dissolve during etching and produce a printed circuit on the board.

2 Holes are drilled in designated locations on the printed circuit board to accept the components. Then, the board is pre-soldered by dipping it in a bath of hot solder.

3 Smaller electronic components such as resistors, capacitors, transistors, integrated circuits, and coils are installed in their designated holes on the printed circuit board and soldered to the board. These operations can be performed by hand or by robots.

4 Larger components such as power transformer, speaker, and antenna are mounted either on the PCB or cabinet with screws or metal spring tabs.

5 The case that houses the radio can be made either of plastic or aluminum. Plas-

An example of a standard AM/ FM radio.

tic cases are made from pellets that are melted and injected into a mold. Aluminum cases are stamped into shape from sheet aluminum by a metal press.

6 External components not mounted on the printed circuit board can be the antenna, speaker, power transformer, volume, and frequency controls are mounted in the case with either screws, rivets, or plastic snaps. The printed circuit board is then mounted in the case with screws or snaps. The external components are connected and soldered to the printed circuit board with insulated wires made of copper and plastic insulation.

Quality Control

Since most of the components or a radio are manufactured by specialized vendors, the radio manufacturer must rely on those venders to produce quality parts. However, the radio manufacturer will take random samples of each component received and inspect/test them to ensure they meet the required specifications.

Random samples of the final radio assembly are also inspected to ensure quality. The overall unit is inspected for flaws—both physical and electrical. The radio is played to ensure it can select radio frequencies it's design to receive, and that the audio output is within specifications.

Byproducts/Waste

Today's environmental awareness dictates that all waste be disposed of properly. Most byproducts from the construction of a radio can be reclaimed. The etching solutions used in the printed circuit board manufacture are sent to chemical reclamation centers. Scraps from the leads of electronic components are sent to metal waste recovery centers where they are melted to create new products.

The Future

Radios are being combined with computers to connect the computer to the Internet via satellites. Eventually radios will convert from analog to digital broadcasting. Analog signals are subject to fade and interference, digital signals are not. They can produce high quality sound like that found on a CD.

Digital radios can be programmed for specific stations, types of music, news, etc. Eventually, radios will have mini-computers built in to process sounds in numerical patterns "digits" rather than an analog waveform. This will allow listeners to program their radios for favorite radio stations, music type, stock quotes, traffic information, and much more.

Where to Learn More

Books

Carter, Alden R. *Radio From Marconi To The Space Age.* New York: Franklin Watts, 1987.

Floyd, Thomas L. *Electric Circuit Fundamentals.* Columbus: Merrill Publishing Company, 1987.

The American Radio Relay League. *The ARRL Handbook for Radio Amateurs.* Newington, CT: ARRL, 1996.

Other

Canadian Broadcasting Company Web Page. "The Future of Digital Radio.: December 2001. <http://radioworks.cbc.ca/radio/digital-radio/drri.html>.

UC Berkley Web Page. December 2001. <http://www.cs.berkeley.edu/~gribble/cs39c/Comm/radio/radio.html>.

Ernst S. Sibberson

Radio Collar

Background

Pet owners have long struggled with adequate measures of pet containment. Inadequately confined pets run the risk of damaging property and endangering the animal. Physical barriers such as fences can be quite expensive and often not aesthetically pleasing. For over 20 years a number of pet containment systems have been developed to keep animals to predefined areas. Imbedding a wire perimeter and connecting to a transmitter system has been a popular method. The pet wears a collar that has a receiver and a motivation system for providing a stimulus to the pet as it approaches a boundary. Limitations have been the cost associated with the wiring and the fact that the pets are unable to reenter the boundary area if they escape. There is also the problem that a lawnmower can damage wires not buried properly. To overcome this limitation, wireless systems have also been developed. The receiver only produces a stimulus when there is a loss of signal from the pet.

In radio collars there is a safe area in which the pet can roam without receiving a warning stimulus. When the pet moves away from this area the collar/receiver gives a warning signal (a shock, loud sound, or other stimulation) to the pet. Therefore the pet learns to stay within the safe area. The collar unit comprises a multiplicity of radio signal receivers each having a receiving antenna. The system transmitter continuously transmits a radio frequency (RF) signal and a mobile receiver assembly mounted in the collar unit on the dog. The receiver assembly receives the RF signal and measures the intensity of the received signal.

History

Early patents for underground pet containment systems were granted in the early 1970s. Documentation as to who the original inventor was is not available but technology from electronic transmission had been in use for Global Positioning System (GPS) since the early 1960s to track animals in the wild. The desire to keep pets safe and away from restricted areas in private homes (such as gardens and children's play areas) moved the technology into the home market.

Raw Materials

The materials for the collar range from nylon to leather with a metal buckle and D-ring to hold the collar in place. Suitable wiring and electronic components consist of oscillators, transistors, capacitors, and resistors that are incorporated in the receiver and transmitter devices. Systems using buried wire containment systems use heavy duty 18-gauge wire.

Design

Some radio collars use a wire imbedded in the ground to transmit signals to the receiver. Other forms use only a transmitter box and the collar, avoiding the wire. Depending on the device's strength, it can cover anywhere from 50 to 500 yd (46–457 m). The level of intensity will also vary depending on the size and temperament of the dog. There are separate devices for smaller, larger, or more stubborn dogs.

The Manufacturing Process

The collar

1 A dog collar can be made out of many materials, though nylon is the most pre-

The first patents for underground pet containment systems were granted in the early 1970s.

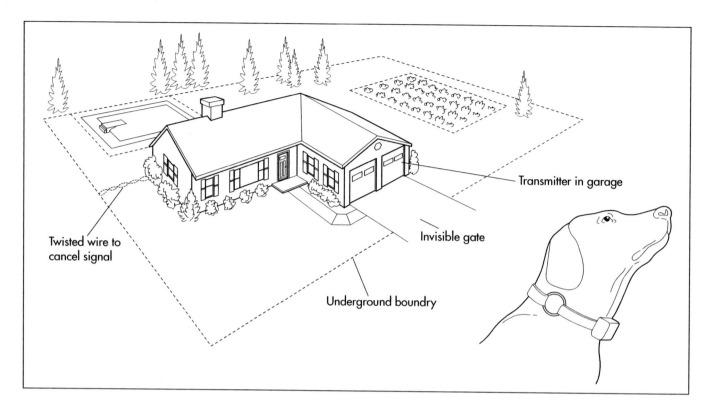

A layout of a yard using an underground pet fence system and an example of a collar on a dog.

ferred due to its strength and durability. Sheets of woven nylon are shipped to the manufacturing plant and loaded onto a conveyor belt. They are then conveyed to an automatic cutting press, which cuts the material to size (usually about 3 ft [0.9 m] long and 1–1.5 [2.5–3.8 cm] wide).

2 The edges of the collar are heat sealed to prevent loose threads.

3 The material is then taken to a press that punches round metal rivets through the collar, simultaneously removing the material in-between the rivets. The rivets enable the buckle to hold the collar together.

4 Next, the collar is attached to a preformed metal buckle. The buckle generally arrives at the plant already formed. The fabric of the collar is fed through the bottom of the buckle and a D-ring. It is then folded back onto itself and manually stitched using an industrial sewing machine, securing the buckle in place. The collar is now ready for the attachment of the RF receiver.

5 The RF receiver arrives at the plant preformed. It is put on an assembly line and then manually placed in plastic cases. The cases have been made by heating plastic pellets and injecting them into a liquid mold.

6 A 6-volt battery is attached to the radio receiver. The RF receiver is mounted on the lower portion of the collar along the strap by mechanical fasteners (screws, bolts, rivets, etc.) that pass through the strap and clamp the plate from the housing of the transmitter.

The transmitter

7 The parts of the transmitter also arrive at the plant preformed. The transmitter consists of a printed circuit board that is connected to a correlation point of the capacitors and to the ground through a resistor. The resonator circuit may be a Liquid Crystal resonator circuit or a circuit including an electromechanical converter element (an element that converts electrical energy to technical energy), such as a Surface Acoustic Wave Resonator (SAW), a crystal resonator, a lithium-tantalite resonator, or a tantalite-niobate resonator.

8 In this transmitter, an oscillation loop is formed by a closed circuit including the resonator circuit, the antenna, and the base and emitter of the transistor, which constitutes the oscillator.

9 The collector of a transistor is connected to the ground through a series of circuits

including bias resistors as well as to the power voltage.

10 The base of the transistor is connected to the ground through a series circuit including capacitors. The base is also connected to the ground through the connecting point where the resistors are connected, and where a series circuit of an antenna and a resonator circuit is also connected.

11 A flexible antenna consists of two copper wires wrapped in a form of plastic installation. It typically extends out of the transmitter and is directed upward by a guide attached to the side of the device.

12 The bottom of the antenna includes a through-hole through which a threaded stud is attached to the transmitter, used to couple the antenna to the transmitter and broadcast signals to the receiver.

13 The transmitter is then placed in an injection molded plastic case, manufactured in the same manner as the receiver case on the collar.

14 The wire used to mark the borders of the electric fence is a roll of 18 gauge multi-stranded wire. The wire is insulated within a plastic coating and arrives at the plant in large rolls. The wire is then automatically cut into 500 ft (152.4 m) rolls.

15 Once the transmitter is finished, it is packaged with the wire rolls and collar. The packages are then shipped to either the distributor or directly to the customer.

Quality Control

Skilled technicians at work stations perform calibration tests such as soldering inspection, high and low voltage vibration, and humidity thermal cycling. The receivers are tested to ensure that the correct amount of voltage or level of sound is distributed when activated. The collar is also subjected to tests such as strength and durability.

Byproducts/Waste

Waste materials (discarded wire and electrical components) are disposed through industry waste removal. Any plastic that can be reused is melted down and remolded. When the collar material is defective, it is discarded.

The Future

Future systems are being designed that use orthogonally positioned antennas to ensure boundary signals are detected. The risk of inadvertent shocks from nonboundary RF signals has been greatly reduced by encoding the transmitter with a preselected signal eliminating errant shocks from abhorrent RF signals. Additionally trapped animals will no longer continue to receive corrections until the battery is drained of power. The shocks are suspended but the circuitry continues to monitor for the boundary signal. Microprocessor circuitry is used to analyze the received signals and to control the annoyance signals. The microprocessor circuitry is also be powered down when not in use and then turned back on when a signal is received to be analyzed to further conserve power. Additionally, the circuitry may include one or more motion sensors that allow the power-draining circuitry to be energized in response to movement of the animal such that when the animal is at rest and not trying to cross the boundary, the battery is not wasted trying to detect a boundary signal that should not be present.

Where to Learn More

Books

Radio Fence Distributors, Inc. Web Page. December 2001. <http://radiofence.com>.

United States Patent and Trademark Office Web Page. December 2001. <http://www.uspto.gov/patft/index.html>.

Periodicals

Woolf, Norma Bennett. "Hidden Fences, Out of Sight May not Mean Peace of Mind." *Dog Owner's Guide* December 2001. <http://www.canismajor.com/dog/fences1.html>.

Bonny P. McClain

Revolving Door

The exact first use of revolving doors is unknown. However, it is known that they have been in use since about 1790 in Chicago.

Background

A revolving door is used to control traffic or heating and air conditioning in a building. The revolving door structure consists of individual door panels (or wings), a center shaft with the hardware needed to support the door wings, a circular structure called a "rotunda" or "drum" that is usually fitted with glass, and the ceiling (supported by the rotunda) that contains either a mechanical braking device (used to control the speed of the doors) or an electronic device that uses a motor to drive the doors automatically.

The main benefit of a revolving door is that it is always closed and always open. This means that the design of the system is such that there is at least one door wing sealing the opening at all times reducing the amount of heating volume and air conditioning (HVAC) that escapes from a building and these savings in energy costs can be considerable. The revolving door achieves these savings because the curved walls of the rotunda allow the seals to fit tightly as the door wings rotate.

Each wing is fitted with a rubber and felt weather seal. With time and use, these weather seals on the door wings must be replaced because their effectiveness diminishes with age. This process is also a part of the design of the revolving door system.

History

The exact first use of revolving doors is unknown. However, it is known that they have been in use since about 1790 in Chicago where they are still widely used today. Revolving doors solved the problem of how to automatically close the door opening in order to keep from losing heating or cool-ing. Early revolving doors were manufactured using wood because the technology and materials needed to economically manufacture them from metal was either not yet available or prohibitively expensive.

Raw Materials

The materials used in the construction of a revolving door consist of aluminum extrusion, steel tubes, machined steel hardware pieces designed to attach the doors to the structure, marine grade plywood, glass (curved and flat), felt and rubber, and the mechanical devices used to control speed or a motor driven operator device to move the doors in response to a signal from a sensor.

Design

Each basic revolving door design is modified (with respect to dimensions, number of door wings, whether the function is manual, automatic or security, and finish or color) to suit the users requirements. The basic design of the revolving door system is not changed, but it is available in several configurations. When an architect specifies a revolving door, he or she is primarily interested in the best design that will conserve energy but also one that will bet suit the traffic requirements of the finished building. From this basic idea, the architect specifies the revolving door system and configuration, and manufacturers can adapt their designs to comply with the specifications.

The Manufacturing Process

1 All aluminum members are cut to length depending on the overall final dimen-

sions of the revolving door. Aluminum extrusions are manufactured utilizing a very large press device called an extrusion press. The press forces pre-heated aluminum cast billets through a steel die that has a shape cut to the requirements of the user. Much the same as a cake decorator who uses a sack of icing forced through a small nozzle to create designs and shapes, an extrusion takes the shape of the cut out in the die and these shapes can be made to very exact tolerances. These extrusions are the main components used in revolving door manufacturing—they are designed to lock together and provide a very solid framework to support the rest of the system.

Revolving door systems can also be manufactured using exotic metals like brass or stainless steel to enhance their appearance. These systems do not use aluminum extrusions. Instead, sheets of metal that are cut and formed to wrap around a steel skeleton (sub-frame) are welded in place. They can easily cost as much as three times that of a standard aluminum revolving door.

2 The circular rotunda walls are formed from aluminum extrusions that are repeatedly fed through a bending machine until the correct radius is achieved. The radius depends on the required size of the revolving door. The most common size (width) is 8 ft (2.4 m), but these systems can be manufactured to much larger widths. Larger systems are usually motorized because the structure would be very difficult to push manually.

The bending machine consists of three rollers through which the straight piece of aluminum is fed. After each bend, the rollers are moved closer together and the material is fed through again until the correct radius is achieved. This process is critical because the door panels must not only fit inside the rotunda, but they must also evenly rotate 360 degrees without hitting the rotunda walls. The seal between the door panels and the rotunda must also be maintained to reduce the amount of hot and cold air entering and leaving through the door.

3 All aluminum parts (doors, frames, etc.) are welded together and prepared to receive the hardware components that make up the emergency breakout system.

4 The circular ceiling is cut from 1 in (2.54 cm) thick marine grade plywood and then laminated with 0.125 in (0.318 cm) thick aluminum sheet to form the interior ceiling.

5 The ceiling is prepared for the light fixtures (cut-outs) and the center shaft hole is cut at the very center.

6 Steel angle is welded together in a grid pattern to provide the support for the ceiling and to secure the mechanical speed control device. The grid is installed on top of the ceiling.

The speed control device is a system of spring loaded brake shoe assembly that rotates inside a drum. As the rotational speed of the revolving door increases, so does the brake shoe assembly and the pressure of the brake shoes against the drum slows the revolving door down preventing the "freewheeling" effect. The speed control device can be mounted in the ceiling or in the floor. Floor mounted speed controls are used on systems that have glass ceilings or have insufficient clearance at the top of the door system.

7 The center shaft is fabricated and fitted with the balance of the breakout hardware. Revolving door systems (in the United States) must be fitted with a "panic collapsing mechanism" or "breakout system" that permits the door frames to fold against one another. This bookfold position permits an unobstructed exit from the interior to the exterior of the building in the event of a fire or other emergency. The Uniform Building Code (adopted by most municipalities in the United States) requires the breakout system together with another manual or automatic swinging or sliding door next to the revolving door for emergency exit and handicap access.

The breakout mechanism consists of two breakout plates that are attached to each end of the center shaft. They are slotted because the door frames are fitted with pivot bars that allow the door frame to fold out of position as described above. The breakout plates have removable access gates that allow the door frame to be hung to the assembly using a pivot bar assembly.

The revolving door panels rotate with the center shaft assembly and the center shaft

An aerial view and a cut-away of a revolving door.

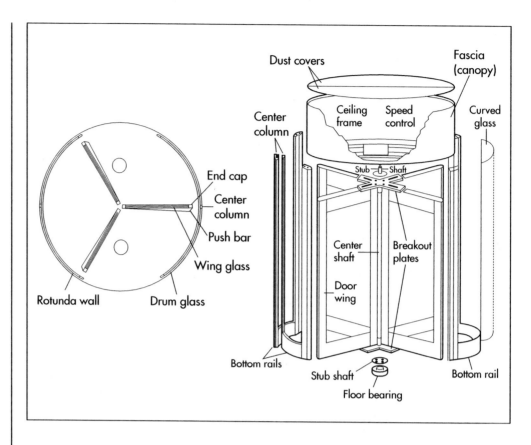

assembly is, as previously discussed, fitted with breakout plates at the top and bottom. There are additional hardware devices at each end of the center shaft and these devices are the actual center shaft pivot points. They are referred to as stub shafts.

The bottom of the center shaft assembly sits inside a bearing assembly that is secured to the floor of the building. The bearing permits the revolving door to rotate smoothly and is necessary to prevent mechanical breakdown due to wear.

8 The entire revolving door is assembled at the factory for testing. Once cleared for shipment, the revolving door is disassembled and wrapped in protective plastic. A wood crate is built and the revolving door components are packed and secured for their journey to the job site.

Quality Control

When the revolving door assembly is erected, a series of tests and measurements are made in order to verify that the product conforms to the requirements of the customer. The system is checked for correct height and width, all functions of the revolving door are tested (speed control, breakout system, weather stripping, etc.) and adjustments are made where needed. In the case of security systems or automatic systems, each computer function or mode of operation is checked and re-checked for proper operation. The quality control inspector must sign off on an inspection form before the door system can be disassembled.

Byproducts/Waste

All waste materials are recyclable and consist of wood, aluminum and steel. All waste material is stored and sent to a recycling center. There are no harmful chemicals used or wastes in the manufacturing process.

The Future

Designs are being examined that would incorporate enhanced security features for specialty applications such as airports and prisons to include metal and chemical detection systems as well as new video recognition systems. Although the revolving door basic design remains the same after over 200 years of use, some new door systems are very large (20 ft [6.1 m] in diameter) and their design incorporates elaborate and exotic electromechanical devices designed to allow several people to transverse the opening at once, where most revolving door designs accommodate one or two people at a time. The mechanical limitations of the equipment make the large door systems expensive to own and troublesome to maintain and repair. Revolving doors will always be the best way to control HVAC loss in areas that experience extremes of climate but they will also remain effective for access control and traffic control well into the future.

Where to Learn More

Other

Manfredi, Bob. *Revolving Doors: An Open And Shut Case.* December 2001. <http://www.greendesign.net/bcnews/SEPOCT96/rvlvgdrs.htm>

Sierra Automatic Doors, Inc. Web Page. December 2001. <http//:www.autodoors.com>.

O. Harold Boutilier

Rolling Pin

Over the centuries, rolling pins have been made of many different materials, including long cylinders of baked clay, smooth branches with the bark removed, and glass bottles.

Background

A rolling pin is a simple tool used to flatten dough.

History

The first civilization known to have used the rolling pin was the Etruscans. These people may have migrated from Asia Minor to Northern Italy or may have originated in Italy. They established a group of city states (called Etruria) and were a dominant society by about the ninth century B.C., but their civilization was cut short after attacks from the Greeks, the growing Roman Empire, and the Gauls (tribes that lived in modern day France). The Etruscans' advanced farming ability, along with a tendency to cultivate many plants and animals never before used as food and turn them into sophisticated recipes, were passed to invading Greeks, Romans, and Western Europeans. Thanks to the Etruscans, these cultures are associated with gourmet cooking.

To prepare their inventive foods, the Etruscans also developed a wide range of cooking tools, including the rolling pin. Although written recipes did not exist until the fourth century B.C., the Etruscans documented their love of food and its preparation in murals, on vases, and on the walls of their tombs. Cooking wares are displayed with pride; rolling pins appear to have been used first to thin-roll pasta that was shaped with cutting wheels. They also used rolling pins to make bread (which they called *puls*) from the large number of grains they grew.

Natives of the Americas used more primitive bread-making tools that are favored and unchanged in many villages. Chefs who try to use genuine methods to preserve recipes are also interested in both materials and tools. Hands are used as "rolling pins" for flattening dough against a surface, but also for tossing soft dough between the cook's two hands until it enlarges and thins by handling and gravity. Tortillas are probably the most familiar bread made this way.

Over the centuries, rolling pins have been made of many different materials, including long cylinders of baked clay, smooth branches with the bark removed, and glass bottles. As the development of breads and pastries spread from Southern to Western and Northern Europe, wood from local forests was cut and finished for use as rolling pins. The French perfected the solid hardwood pin with tapered ends to roll pastry that is thick in the middle; its weight makes rolling easier. The French also use marble rolling pins for buttery dough worked on a marble slab.

Glass is still popular; in Italy, full wine bottles that have been chilled make ideal rolling pins because they are heavy and cool the dough. Countries known for their ceramics make porcelain rolling pins with beautiful decorations painted on the rolling surface; their hollow centers can be filled with cold water (the same principle as the wine bottle), and cork or plastic stoppers cap the ends.

Wood has always been the material preferred by cooks and craftsmen in the United States. Pine was probably the wood of choice from colonization to the mid-1800s, but the pine forests in the northern states were already being depleted by this time. Rolling pin manufacturers started using other hardwoods like cherry and maple for their wooden kitchenware, which also included ladles and butter molds. Late in the nineteenth century, J. W. Reed invented the rolling pin with handles connected to a center rod; this is similar to the tool we know today, and it prevents cooks from putting their hands on the rolling surface while shaping pastry. Reed invented new versions of the dough kneader and dough roller; his contributions are notable, not only because he eased the cook's tasks, but also because Reed was one of many African-Americans who developed and patented improvements to household items.

Raw Materials

Some 600,000–750,000 rolling pins are manufactured and sold in the United States every year. By far, the majority of these are made of wood with handles to rotate them around central spools. Wood from maple or ash trees is the most common raw material, depending on availability and customer preference. Hard woods like rock maple are the high-end materials found in bakeries, cooking schools, and retail stores selling fine cookware. Less desirable and softer woods are ash or soft maple. Soft maple and birch form the rolling pins for sale in discount and other mass-marketing stores. Matching woods are used to produce handles.

Rolling pins turn on stainless steel center rods and ball bearings; these are held in place with nylon bushings. Specialty suppliers provide these parts to the rolling pin manufacturers based on their requirements. Handles used to be painted or lacquered, but this practice is out of fashion. Manufacturers no longer use paints or other applied finishes.

Design

Designs for most rolling pins follow long-established practices, although some unusual styles and materials are made and used. Within the family of wooden rolling pins, long and short versions are made as well as those that are solid cylinders (one-piece rolling pins) instead of the familiar style with handles. Very short pins called mini rolling pins make use of short lengths of wood and are useful for one-handed rolling and popular with children and collectors.

Mini pins ranging from 5 to 7 in (12.7–17.8 cm) in length are called texturing tools and are produced to create steam holes and dec-

orations in pastry and pie crusts; crafters also use them to imprint clay for art projects. These mini pins are made of hardwoods (usually maple) or plastic. Wood handles are supplied for both wood and plastic tools, however.

Blown glass rolling pins are made with straight walls and are solid or hollow. Ceramic rolling pins are also produced in hollow form, and glass and ceramic models can be filled with water and plugged with stoppers. Tapered glass rolling pins with stoppers were made for many centuries when salt import and export were prohibited or heavily taxed. The rolling pin containers disguised the true contents. The straight-sided cylinder is a more recent development, although tapered glass pins are still common craft projects made by cutting two wine bottles in half and sealing the two ends together so that the necks serve as handles at each end.

Tiny rolling pins are also twisted into shape using formed wire. The pins will not flatten and smooth pastry, and the handles do not turn. The metal pins are popular as kitchen decorations and also to hang pots, pans, and potholders.

The Manufacturing Process

1 Production of wooden rolling pins starts with the selection of the wood. Trees are selected by log buyers in approved forests, and are then cut and hauled to sawmills. There, they are sawn into squares of either 1.5 in (3.8 cm) or 2 in (5.1 cm); both sizes of squared wood are cut into lengths of 48 in (1.2 m). The square pieces are then kiln-dried.

2 The prepared wood lengths are brought into the rolling pin plant and fed through a specialized machine called a hawker. The hawker produces a large, rounded dowel by taking off the corners of the squares and about 0.25 in (0.6 cm) of wood all around the length. The trimmed lengths are inspected. The 4 ft (1.2 m) length may be free of defects for its full length. If 3 ft (2.7 m) of the length are acceptable, the imperfect portion is trimmed off. These long, perfect lengths are called clear dowels and are sold to the dowel market primarily for use in furniture manufacture.

Rolling pins are typically used to smooth out dough.

3 The long dowels containing defects like knots, mineral deposits, or major color changes are clipped to appropriate lengths for rolling pins. The standard lengths are 12, 15, and 18 in (30.5, 38.1, and 45.7 cm, respectively); a high-quality large rolling pin can weight 12 lb (4.48 kg). Typically, one or two rolling pins of different sizes can be clipped from acceptable parts of the dowels. Shorter pieces than those suitable for standard rolling pins can be further trimmed in diameter and length and made into handles or mini rolling pins.

4 From the clipping station, the rolling pin lengths are transferred to the next workstation, where they are deep bored with the holes that will hold the rods. The lengths are chamfered (machined with beveled or gently angled edges) and counter bored. The wood length is then considered to be completely machined and is termed a pin blank. Handles are shaped with a different woodworking machine. They are made from short lengths of wood and are turned on spool lathes to produce rounded, uniform handles.

5 At the next station, the pin blanks and handles are given their fine outer finishes on a machine with motors and belts that is something like a sander capable of a series of tasks. The outer surface of the rolling pin (or handle) is sanded with two or more types of sandpaper, from a coarse 80-grit paper to a very fine 150-grit paper. The machine then waxes the surface and buffs it to an attractive polish.

6 When the machining and treating of the wood is completed, the rolling pins and their handles are put on carts and taken to the assembly area. Until recently, the assembly work was done by hand, but it is now fully automated. The assembler inserts rods and ball bearings in the bores through the rolling pins and adds nylon bushings that will keep the rods centered in the pins. Wood handles are fitted on the rod ends, and the assembly machine taps the handles firmly in place.

7 Labels are applied to complete rolling pins. Each is boxed in a pre-labeled box, and the boxes are packed into bulk cartons for storage or shipping to retailers.

Quality Control

Product quality begins with wood of outstanding quality. The log buyers are highly skilled in choosing fine-quality timber. When the dried, trimmed, and squared lengths are ready to be clipped into standard rolling pin lengths, they are carefully inspected for any defects. About 95–97% of the wood is perfect.

During finish machining and assembly, the chamfering, deep boring, and sanding machines and the lathes may occasionally introduce a small flaw. The operators recognize these conditions and pull the affected pin or handle from production and toss it into a recycling bin. During assembly, the workers also monitor the quality of the wood and their own processes. They have the authority to reject substandard work.

Byproducts/Waste

Rolling pin manufacture generates byproducts when a length of dowel is not suitable for a rolling pin. For example, if a 9.5 in (22.9 cm) piece is too short for a 12 in (30.5 cm) rolling pin, it can be clipped to make an 8 in (20.3 cm) pin. These useable lengths of quality wood can also be sold to other manufacturers to make dowel-based products such as legs for stools.

The machinery in a rolling pin plant generates a lot of heat that manufacturers use to heat other operations and save other resources. All wood trimmings and waste are clipped up into sawdust that is used to make paper or is sold to farmers as barn flooring.

The Future

Like the cliché of "building a better mousetrap," it would seem difficult to improve a device as elegantly simple and durable as the rolling pin. However, in 2000 at an inventors' show in Geneva, Switzerland, South African native Yvonne Bekker introduced a newly patented rolling pin that is perforated to release a steady sprinkling of flour. Bekker had grown increasingly frustrated with pastry sticking to the rolling pin, and this prompted her bright idea. Chrome rolling pins are also experiencing a revival, except that the new versions have coatings of Teflon to limit sticking.

Yet the familiar, reliable rolling pin appears ready to take on all newcomers. One possible threat to its future exists in ready-made food that eliminates the need for rolling pins; pre-rolled piecrusts are already on the market. The quality of the pins themselves seems to discourage new production. At least 10 major and 20 significant manufacturers of rolling pins in the United States produce 600,000–750,000 a year, and these sales figures are steady but also unexplainable, given the pins' longevity.

Not only are rolling pins kept in families, but they also are gaining popularity as kitchen collectibles. The aluminum and chrome pins that were once produced are now sought after. Wooden pins can be dated by checking the connection of the rod and pin; plastic bushings are characteristic of modern pins. Wooden rods through the handles and pins, metal bushings, or no bushings at all are indicators of collectible rolling pins. Lacquer and different colors of paint on the handles also help date rolling pins.

Where to Learn More

Books

Editors of Consumer Guide. *The Cook's Store: How to Buy and Use Gourmet Gadgets.* New York: Simon and Schuster, 1978.

Field, Carol. *The Italian Baker.* New York: Harper & Row Publishers, Inc., 1985.

Mauzy, Barbara E. *The Complete Book of Kitchen Collecting.* Atglen, PA: Schiffer Publishing, Ltd., 1997.

Schat, Zachary Y. *The Baker's Trade: A Recipe for Creating the Successful Small Bakery.* Ukiah, CA: Acton Circle Publishing, 1998.

Other

Bethany Housewares Web Page. December 2001. <http://www.bethanyhousewares.com/products10.htm>.

Somé, Lucio. "A Salute to the Etruscan Origins of Tuscan Cuisine." *Castello Banfi Web Page.* December 2001. <http://www.castellobanfi.com/features/story_salute.html>.

Koppel, Naomi. "Bright Ideas on Display: Everyday Problems Solved at Inventors' Fair." *Cnews Web Page.* 13 April 2000. December 2001. <http://www.canoe.ca/CNEWS Features0004/13_inventors.html>.

Gillian S. Holmes

Rubik's Cube

At its peak between 1980 and 1983, 200 million cubes were sold world wide. Today sales continue to be over 500,000 cubes sold world wide each year.

Background

Rubik's cube is a toy puzzle designed by Erno Rubik during the mid-1970s. It is a cube-shaped device made up of smaller cube pieces with six faces having differing colors. The primary method of manufacture involves injection molding of the various component pieces, then subsequent assembly, labeling, and packaging. The cube was extremely popular during the 1980s, and at its peak between 1980 and 1983, 200 million cubes were sold world wide. Today sales continue to be over 500,000 cubes sold world wide each year.

The Rubik's cube appears to be made up of 26 smaller cubes. In its solved state, it has six faces, each made up of nine small square faces of the same color. While it appears that all of the small faces can be moved, only the corners and edges can actually move. The center cubes are each fixed and only rotate in place. When the cube is taken apart it can be seen that the center cubes are each connected by axles to an inner core. The corners and edges are not fixed to anything. This allows them to move around the center cubes. The cube maintains its shape because the corners and edges hold each other in place and are retained by the center cubes. Each piece has an internal tab that is retained by the center cubes and trapped by the surrounding pieces. These tabs are shaped to fit along a curved track that is created by the backs of the other pieces. The central cubes are fixed with a spring and rivet and retain all the surrounding pieces. The spring exerts just the right pressure to hold all the pieces in place while giving enough flexibility for a smooth and forgiving function.

History

Puzzle makers have been creating problems for people to solve for centuries. Some of the earliest puzzles date back to the time of the ancient Greeks and Romans. The Chinese have a ring puzzle that is thought to have been developed during the second century A.D. This was first described by Italian mathematician Girolamo Carolano (Cardan) in 1550. When the printing press was invented, complete books of mathematical and mechanical problems designed specifically for recreation were circulated.

From these early riddles and word problems, toy puzzles were naturally developed. In 1857, the Irish mathematician Sir William Hamilton invented the Icosian puzzle. Sometime around 1870, the famous 15 Puzzle was introduced, reportedly by Sam Lloyd. This puzzle involved numerical tiles that had to be placed in order and became extremely popular in the early twentieth century. In 1883, French mathematician Edouard Lucas created the Tower of Hanoi puzzle. This puzzle was made up of three pegs and a number of discs with different sizes. The goal was to place the discs on the pegs in the correct order.

There are various puzzles that involve colored square tiles and colored cubes. Some early precursors to the Rubik's cube include devices such as the Katzenjammer and the Mayblox puzzle. The Mayblox puzzle was created by British mathematician Percy MacMahon in the early 1920s. In the 1960s, Parker Bothers introduced another cube puzzle type toy called Instant Insanity. This toy achieved a moderate level of popularity in the United States. The early 1970s brought with it a device called the Pyraminx, which was invented by Uwe Meffert. This toy was a pyramid that had movable pieces that were to be lined up according to color.

Erno Rubik, an architect and professor at the University of Budapest developed the first working prototype of the Rubik's cube in 1974. He received a Hungarian patent in 1975. Apparently, it was also independently designed by Terutoshi Ishige, an engineer from Japan, who received a Japanese patent in 1976. Professor Rubik created the cube as a teaching aid for his students to help them recognize three-dimensional spatial relationships. When he showed the working prototype to his students, it was an immediate hit.

Over the next few years, Rubik worked with a manufacturer to allow production of the cube on a mass scale. After three years of development, the first cubes were available on toy store shelves in Budapest. While the cube remained popular in Hungry, the political atmosphere of the time made it difficult for it to be introduced in the United States. The two men who were most responsible for making the cube an international success were Dr. Laczi Tibor and Tom Kremer of Seven Towns Ltd., London. Seven Towns licensed the Rubik Cube invention from Professor Rubik for worldwide distribution. Dr. Tibor worked within Hungry to convince bureaucrats to allow the technology out of the country. Kremer found a United States toy maker, the Ideal Toy company, who was willing to help market the product. The product was an immediate hit, and during the 1980s, over 200 million cubes were sold. Around 1983, the frenzied popularity of the cube began to wane and sales slowed drastically. It remained in small scale production until Seven Towns took over the marketing, and licensed the Rubik Cube to the Oddzon Company for the United States market in 1995. Since that time sales have steadily increased to over 500,000 units a year.

Design

The most important part in the manufacture of a Rubik's cube is designing the mold for the various pieces. A mold is a cavity carved into steel that has the inverse shape of the part that it will produce. When liquid plastic is put into the mold, it takes on the mold's shape when it cools. The creation of the mold is extremely precise. The cavity is highly polished to remove any flaws on the surface. Any flaw would be reproduced on each of the millions of pieces that the mold will produce. In the manufacture of the cube parts, a two piece mold is typically employed. During production, the two mold pieces are brought together to form the plastic part and then opened to release it. The tool includes ejector pins that release the molded parts from the tools as it opens. All the parts are molded with auto gating tools that automatically remove the parts from the sprue as it is ejected. The molds are also produced with a slight taper, called release angle, which aids in removal. Finally, when molds are designed, they are slightly bigger than the pieces that they ultimately will produce. This is because as the plastics cool, they shrink. Different plastics will have a different shrink rate, and each tool must be specifically designed for the material that will be used.

The commercial cube is composed of six fixed cubes, eight movable cubes on the corners and 12 movable cubes on the edges. Each cube is one of six colors. The Rubik's cube has red, yellow, blue, green, white, and orange colors. In its solved state, each color is on only one face. When the cube is rotated, the edges and corners move and the cube becomes scrambled. The challenge of the puzzle is to restore each cube to its original position. The cube is extremely challenging because there are slightly more than 43 quintillion (4.3×10^{19}) possible arrangements, and only one solution.

The standard Rubik's cube has sides of about 2.2 in (5.7 cm) per square. Various other sizes have also been produced such as a 1.5 in (3.8 cm) mini cube, a 0.8 in (2 cm) key chain micro cube, and a 3.5 in (9 cm) giant cube. While the standard cube is a $3 \times 3 \times 3$ segmentation other types have also been introduced. Some of the more interesting ones include the $2 \times 2 \times 2$ cube, the $4 \times 4 \times 4$ cube (called Rubik's Revenge) and the $5 \times 5 \times 5$ cube. The shape has also been varied and puzzles in the form of a tetrahedral, a pyramid, and an octahedral are among types that were produced. The Rubik's cube also led to the development of game derivatives like the Rubik's cube puzzle and the Rub it cube eraser.

Raw Materials

The individual pieces that make up the Rubik's cube are typically produced from plastic. Plastics are high molecular weight materials that can be produced through various chemical reactions called polymerization. Most of the plastics used in a Rubik's cube are thermoplastics. These compounds are rigid, durable, and can be permanently molded into various shapes. The plastics used in the Rubik's cube are acrylonitrile butadiene styrene (ABS) and nylon. Other plastics that might be used include polypropylene (PP), high impact polystyrene (HIPS), and high density polyethylene (HDPE).

For decorative purposes, a colorant is typically added to the plastic. The pieces of a Rubik's cube are typically black. During production, colored stickers are put on the outside of the cube to denote the color of a side. The plastics that are used during production are supplied to the manufacturer in a pellet form complete with the filler and colorants. These pellets can then be loaded into the molding machines directly.

The Manufacturing Process

The manufacture of the first Rubik's cube prototypes was by hand. During the late 1970s, methods for mass production were developed and continue to be used today. Typically, production is a step by step process that involves injection molding of the pieces, fitting the pieces together, decorating the Rubik's cube, and putting the finished product in packaging.

Molding

1 When production is initiated, the plastic pellets are transformed into Rubik's cube parts through injection molding. In this process, the pellets are put into the hopper of an injection molding machine. They are melted when they are passed through a hydraulically controlled screw. As the screw turns, the melted plastic is shuttled through a nozzle and physically forced, or injected, into the mold. Just prior to the arrival of the molten plastic, the two halves of the mold are brought together to create a cavity that has the identical shape of the Rubik's cube

part. This could be an edge, a corner, or the center piece. Inside the mold, the plastic is held under pressure for a specific amount of time and then allowed to cool. While cooling, the plastic hardens inside the mold. After enough time passes, the mold halves are opened and the cube pieces are ejected. The mold then closes again and the process begins again. Each time the machine moulds a set of parts is one cycle of the machine. The Rubik's cube cycle time is around 20 seconds.

2 After the cube parts are ejected from the mold, they are dropped into container bins and hand inspected to ensure that no significantly damaged parts are used. The waste sprue material is set aside to be reused or scrapped. Waste material can be ground up and melted again to make new parts, however reground material can degrade and cause poor quality parts. Rubik's cubes are always made from virgin material and never use reground waste plastic.

Parts assembly

3 The Rubik's cube parts are taken to an assembly line. In this phase of production, the individual cube pieces are put together. Starting with the nylon core, each ABS center cube is riveted to the core with a spring spacer. The rivet is carefully controlled with a depth stop to ensure the spring is compressed just the right amount. Each center cube has a plastic cover that is glued on to hide the rivet. One of the six center cubes is left until the last part of the assembly. The ABS edges and corner pieces are individually stacked around the core. The cube is built from the bottom up and the last piece to be assembled is the final center cube which is again riveted into the core with a spring spacer and the final cap is glued on.

Labeling

4 Next, the Rubik's cube faces need to be labeled. The labels are made from sheet polypropylene material that is printed with the colors. The printed sheet PP is then laminated with a clear PP protective covering. The material is then die cut with the labels wound onto rolls. The labels are made with all nine squares of each face exactly aligned. This way the labels can be perfectly aligned when they are applied to the cube.

Packaging

5 After all the labeling is completed, the cubes are put in their final packaging. This can be a small box that has an instruction booklet included or a plastic blister pack with a cardboard backing. The package serves the dual purpose of protecting the Rubik's cube from damage caused by shipping and advertising the product. The Rubik's cube packages are put into cases and moved to a pallet. The pallets are then loaded on trucks and the products are shipped all over the world.

Quality Control

To ensure that each toy will be a high quality product, quality control inspectors check the product at each phase of production. The incoming plastic pellets are chemically tested to determine whether they meet certain chemical specifications. These include checks on appearance, color, melting point, toxicity, and molecular weight.

The quality of the individual parts are also inspected just after exiting the mold. Since thousands of parts are made daily, a complete inspection would be difficult. Consequently, line inspectors may randomly check the plastic parts at fixed time intervals and check to ensure they meet size, shape, and consistency specifications. This sampling method provides a good indication of the quality of the overall Rubik's cube production run. Things that are looked for include deformed parts, improperly fitted parts and inappropriate labeling. While visual inspection is the primary test method employed, more rigorous measurements may also be performed. Measuring equipment is used to check the length, width, and thickness of each part. Typically, devices such as a vernier caliper, a micrometer, or a microscope are used. Just prior to putting a cube in the packaging it may be twisted to ensure that it holds together and is in proper working order. This can be done by hand or by a turning machine. If a toy is found to be defective it is placed aside to be reworked later.

The Future

While the extreme popularity of the Rubik's cube subsided around 1984, it has recently made a significant come back. This has been a result of impressive marketing efforts by Seven Towns. In the future, this marketing effort should continue to increase sales of the Rubik's cube. In addition to the cube, other derivative puzzles have been introduced including the Rubik's snake, Rubik's triamid, and the Rubik's magic folding puzzle. It is expected that new variants will also be introduced in the near future.

Where to Learn More

Books

Chabot, J. F. *The Development of Plastics Processing Machinery and Methods.* Brookfield: Society of Plastics Engineers, 1992.

Othmer, Kirk. *Encyclopedia of Chemical Technology.* Vol. 22, 1992.

Rubik, E. *Rubik's Cubic Compendium.* Oxford University Press, 1987.

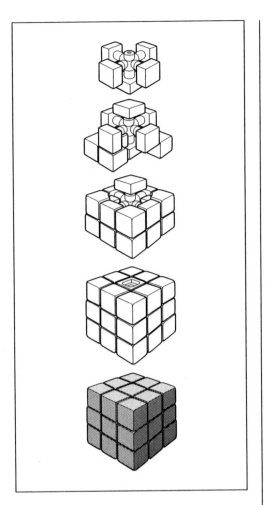

The Rubik's cube starts with a nylon core. The plastic squares are then attached from the bottom up with the labels being attached last.

Seymour, R., and C. Carraher. *Polymer Chemistry.* New York: Marcel Dekker, Inc., 1992.

Other

Seven Towns Ltd. *Rubik's Online Web Page.* 27 September 2001. <http://www.rubiks.com>.

United States Patent no. 4378116.

United States Patent no. 4471959.

Virtual Puzzle Museum Web Page. 27 September 2001. <http://www.virtualpuzzle museum.com>.

Perry Romanowski
Steven Perrin

Scale

Background

The traditional bathroom scale is used to measure a person's body weight. It is based on a spring system that uses the weight of the person to depress a lever, which in turn rotates a sprocket attached to the dial. The dial rotates until it stops, and a plastic marker marks the person's weight. A home-use bathroom scale has a margin of error of ±0.25 lb (±0.12 kg). Scales for bathrooms or kitchens are generally designed as spring balances.

History

Units of measurements have been used for all of antiquity. People have always used some type of set standard for trade. The first known measurement device was used by the Romans 2,000 years ago. They devised an equal beam scale which was shaped like the letter T with both arms measuring 7.4 in (18.8 cm) wide. Attached to each arm were metal pans that were typically 1.5 in (4 cm) in diameter.

The first known unit of weight was the wheat seed. The ancient Romans and Greeks used this standard to measure any other object against, generally for barter or trade. For instance, farmers would bring their crops to sell and they would be weighed against the known standard of wheat grain. X-amount of produce was equal to X-amount of grain needed to maintain the equilibrium of the balance. The Arabs improved on these techniques and established weight standards for gold, silver, and gems.

By the thirteenth century trade had become much more widespread, but people in different parts of the world (or even within the same country) used different standards of measurement. King Edward I of England established a base standard of measurement to which objects or materials could be compared to. This standard soon traveled through trade and became somewhat acceptable in other parts of the world.

In 1793, the French government devised a system based on a line running along the ground through Paris that measured the distance from the North Pole to the Equator. The French called this the metric system. People were unfamiliar with this system and it was not fully enforced until 1837 when it became the standard in European countries.

Scales themselves continued to evolve to meet both the distributor and customer's needs. Customer's wanted to be able to count on the accuracy of the distributor's scales to make sure that they were not being cheated. The first scales used a simple balance beam to weigh an object against a known standard.

The first spring balance was brought into widespread use in the eighteenth century. In Bilston, England Richard Salter began making what is today known as a fisherman's scale which used a spring balance to measure weight. The Salter brand was also the first company in England to marked bathroom scales. Modern home scales have evolved from these early industrial prototypes. Today, the scale is based on the same spring balance idea.

Raw Materials

The case of the spring scale is manufactured from stainless steel or aluminum. The interior is composed of metal springs, pins, gears,

The first known unit of weight was the wheat seed. The ancient Romans and Greeks used this standard to measure any other object against, generally for barter or trade.

and plastic. The gears can be made from aluminum, copper, brass, bronze, stainless, steel, nickel silver, monel, zinc, iron, or plastic. The non-slip mat is formed from a mix of polyvinyl chloride and rubber.

Design

There are many different types of scales; solar, electronic, digital, and spring to name a few. The scales may also differ on what they measure. Some scales are able to measure a person's body fat ratio. The color and size of scales vary greatly to meet all customer needs.

A typical spring scale is comprised of weight transmitting levers, a weight sensing mechanism, and dial enclosed in a metal casing. Generally, the scale is equipped with a non-slip pad on the platform so that the person does not slip and fall off the scale.

The Manufacturing Process

1 Aluminum is melted until molten and then fed into a die that has the desired shape of the scale casing. The aluminum is cast in a cold chamber process at temperature of 1,202°F (650°C) so it will not bond with the steel die.

2 The aluminum is then cooled and ejected from the mold. Both the top and bottom of the scale body are manufactured using this process.

3 The top of the case is manufactured with a slot missing that will serve as the window through which to view the correct weight. This plastic covering is made from molted plastic fed into an injection molding machine. The plastic is then injected into a mold of the cover and left to cool. After cooling, the cover is removed and manually inserted into the top casing.

4 There are four levers used to distribute a person's weight through the scale. The levers are manufactured from thin sheets of aluminum or steel that is delivered to the plant. The sheets are then placed on a conveyor belt to be laser cut. A laser beam that is 0.008 in (0.2 mm) in diameter focuses 1,000–2,000 watts on the aluminum sheet. The laser gets the outline of the lever and in-structions from a Computer Aided Drafting and Design (CAD) drawing.

5 The springs, brackets, and gears arrive preformed at the plant. They are inspected for quality and then distributed to work stations along an assembly line.

6 The dial is formed from a coining method. In this process the aluminum is placed in a set of dies that close, exerting up to 200,00 psi (1,375 mpa) depending on the level of detail on the dial.

7 The dial is then extracted and automatically painted, typically white with black numbers.

8 The non-slip pad is made from a mixture of polyvinyl chloride (PVC) and rubber. The resin is measured and mixed, then poured into the mold of the non-slip mat. The mold is then cured and the finished slip mat is removed.

Putting it all together

9 The case is then placed on a conveyor belt and the metal plate that holds the main lever and the spring is fitted through a slot that was molded into the side of the case.

10 The main lever that runs vertically through the middle of the scale is then rested upon the plate and hooked over the base of the casing.

11 The other levers are then hooked over the corners of the case bottom and hooked to the main lever.

12 The tooth rack and pinion are then manually connected to two compression springs.

13 Next, the metal dial is fixed to the main vertical lever. It rests on the rack and pinion system that will turn the dial based on the person's weight.

14 Four brackets that will secure the levers in place are connected to aluminum latches that were molded into the top cover of the scale.

15 The two halves of the case are then manually fitted together. The top cas-

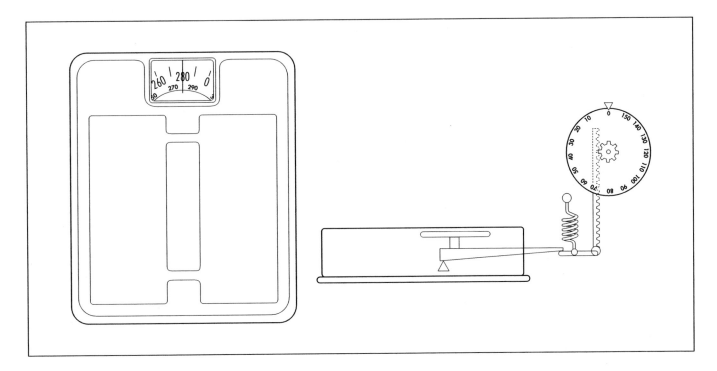

The inner mechanisms of a bathroom scale.

ing is a little larger than the bottom, but will ensure a snug fit.

16 The scale is tested for correct calibration. Then it is packaged and shipped to the distributor.

Quality Control

The parts used to manufacture the scale are checked for defects. Any defective parts that can be salvaged are removed and reused. Parts that are extremely damaged are discarded or recycled. The workers check the calibration of the scale against a known weight before it is packaged. Typically the scale should be able to detect weight within 0.25 lb (0.12 kg).

Byproducts/Waste

Any excess or defective parts are assessed for quality and then either reused or discarded.

The Future

As technology advances, so does the accuracy and application of scales. Today scales can measure not just weight but also body fat. These scales send a mild electrical current through the person's feet and up the rest of the body. The more quickly the signal travels through the body, the less fat. Software is also being developed that allows the scale to keep track of a person's weight loss or gain. Some are even able to track the weights of more than one person. These systems will be able to hook up to software on the home computer to better track weight loss or gain.

Where to Learn More

Other

Bodytrends.Com Web Page. December 2001. <http://www.bodytrends.com>.

ReallyFit.Com Web Page. December 2001. <http://www.reallyfit.com>.

Salter Scales Online. December 2001. <http://www.salterhousewares.com>.

Deirdre S. Blanchfield

Ship In A Bottle

Models of human and heavenly figures were put in bottles as early as about 1750 and may have originated in monasteries, when many quiet hours were available for crafts.

Background

The ship is obviously much larger than the opening in the bottle. Many people think the underside of the bottle is cut away; the ship, however, is made of wood and its sails and rigging are paper and thread. The secret is that the ship's hull is small enough to fit through the bottle's neck, but the sails and spars (the masts and sail supports) are collapsible and can be pulled into position using controlling threads.

History

The history of ships in bottles is the history of the two major components. Sailors on ships of all sizes and types have used scrap wood, cloth, and rope to make model or toy boats to pass long hours at sea. This model-making dates back perhaps 4,000 years. The Egyptians buried miniature ships with their mummified masters, and the Phoenicians, Etruscans, and Greeks produced models that are shown in wall murals.

The merging of model ships with bottles is a much more recent development, due largely to the poor quality of early bottles. Models of human and heavenly figures were put in bottles as early as about 1750 and may have originated in monasteries, when, again, many quiet hours were available for crafts. Character and puzzle models were put in bottles of flawed glass and of shapes that help date them. When techniques of manufacturing glass improved, glass bottles were clearer, less distorted, and free of bubbles and heavy seams. Today, minor distortions, soft tints, and the antique appearance of hand-blown bottles are seen as advantages.

Model ships were not bottled until about 1850 when the great clipper ships plied the seas from port cities in England and America. These ships had as many as seven masts and many sails for the speeds needed to cross oceans and deliver products and profits. They were also equipped with guns and the large crews of sailors for manning the rigging and weapons. The date of the first construction of a ship in a bottle is unknown; but the patience needed to fold the masts in the bottle was a challenge, and the bottle protected the model. Most of the classic sailing ships have been preserved in bottles and in maritime museums.

Raw Materials

The wood for the hull and the glass bottle should be chosen after the model is selected; the proportions of the ship are better suited to some bottles, and measurements of the ship parts are controlled by the inner diameter of the neck of the bottle. Usually, the wood is a hardwood such as spruce or fir, and it should be close-grained with no flaws. Bottles with flat sides rest on shelves or tables easily. Three-sided bottles with "dimples" also display attractively. Round bottles require stands or supports for stability. Ships with more than three masts look sleek in slender, elongated bottles. Sloops, schooners, and other ships with one or two masts fit shorter bottles well.

Other wood supplies include bamboo cocktail skewers or small-diameter dowels for spars, popsicle sticks that can be carved into deckhouses and lifeboats, wood matching the hull for a display stand, and larger blocks of scrap wood for a raised work stand. Sandpaper in grades from about

120–200 for smoothing the hull and other wood is also essential.

Thread, wire, glue, and clear nail polish are used for the rigging, metal trim like rails, and gluing pieces together (the nail polish is also used as glue). Beeswax helps seal the fine wisps of thread together. Paint thinner, model enamels in a variety of colors, and fine paintbrushes are the materials and tools for painting the ship parts. Medium-weight white bond paper is cut into the shapes of the sails, and seams are drawn on with a pencil that is also used to curve the sails.

The "sea" beneath the ship can be made with one of two materials. Linseed oil putty and artists' oil colors, especially white, shades of blue, and some green, are the materials for one method. In the second technique, Plasticene (artists') clay is used to shape the sea. The clay is manufactured in a wide range of colors, so the model builder can choose the best sea color or combination and use white clay slivers for whitecaps.

The method of sealing the bottle should also match the style of the ship and bottle, and the method dictates the materials. Corks, red sealing wax, and cotton fishing line tied around the neck of the bottle in Turk's Head or other sailors' knots are a common combination. Finally, the underside of the bottle (or back of a display board) should be inked with white enamel describing the model.

The hobbyist also needs a selection of simple tools like Exacto knives, a hobby drill with fine bits, and miniature screwdrivers, saws, and a vise. Some tools have to be made for the specific bottle and model size. These include wire tweezers, scoops, and tampers for reaching the back of the bottle and for scooping and tamping putty or clay into place. Clothes hangers can be cut and shaped into long handles for these tools, and pieces cut from a tin can should be soldered to the wire to finish these.

Design

The ships that are featured in bottles are historical subjects, and part of the modeler's skill is recreating a miniature version of the original including the colors it was painted, the carving of its figurehead, and the national flag at the time the ship sailed. Design as-

pects of the ship are the modeler's choice of which ship to build and his or her depth of research. Crafters should begin with a simple model and learn some of the basic nautical terms for sails, rigging, and parts of a ship.

Apart from the ship, other aspects of the display are the crafter's choice. These include the type of bottle and display stand or wall mount, decorations like rope edging and sailors' knots, and other touches inside the bottle.

Bottles may be chosen for size, shape, color, character, or eccentricity. Sizes can range from 3 in^3 (50 cm^3) to 2.7 qt (3 l). A ship can be finished in a large bottle with a companion version in a tiny bottle. Pairs of identical bottles and ships have been sealed together mouth-to-mouth and mounted on an elevated display stand to emphasize the unusual construction. Ships have also been sealed in light bulbs from large, clear globes to Christmas tree bulbs.

Ultimately, the most successful designs balance creativity and faithfulness to historical accuracy and realism. Small vessels should sail on green or greenish blue near coastal water, rather than the deep blue of the open ocean. Similarly, ships do not confront violent seas in full sail, so the modeler needs to show restraint in painting whitecaps. Proportions of masts and rigging to hulls, deck houses, lifeboats, and flags should be as true as possible because some errors will be obvious even to someone who has never seen a ship in a bottle before.

The Manufacturing Process

1 When the ship and its bottle have been selected, all measurements of both should be checked and double-checked. The ship and its collapsed parts must fit through the neck of the bottle and must not hit the top or sides of the bottle when the masts are erected. The modeler cannot forget to add in the thickness of the planned sea under the ship. The bottle should be cleaned and dried.

2 If the sea is to be made of putty, this is the next step. The putty is mixed with oil paint in a shade suiting the sea; partial blending of several colors will create the right tint and add variety to the ocean's col-

ors. The sea should also be dark so the ship and waves are visible. The crafter uses a custom-made scoop to spoon the putty in the bottle. A wire tamper is then used to spread the putty and shape some waves and a flat area for the base of the ship's hull. Smudges should be cleaned from the sides of the bottle, and it must be left open until the putty is dry. Whitecaps, wakes, and waves should be touched with white paint when the putty is dry. If putty is used, the sea should be made before the model is carved and finished; if Plasticene clay is the ocean water, the same process should be followed after the ship is constructed and when the crafter is ready to tamp the ship into the clay.

3 Construction of the ship begins with carving the hull. The block of wood should be gripped in a vise until the basic shape, curved sides, and deck are cut out. Chiseling out the extra wood makes raised parts of the deck and the bulwarks around the edge of the deck. The bow and stern (front and rear ends) of the ship are shaped next, and the hull is cut away from the host block of wood. The hull should be sanded with increasingly fine sandpaper and coated with clear nail polish that will seal the wood and "varnish" the deck. The outer hull is then painted with two coats of enamel of the correct colors. Thread is used to mark straight lines showing gun ports and other lines.

4 The deck is finished, but details can be added by cutting lifeboats, hatches, and deckhouses from wood skewers or popsicle sticks. Other trim like metal rails, stanchions (posts), and davits supporting the lifeboats can be made from wire and inserted in holes drilled with a fine bit. Clear nail polish again glues these details in place and coats the wire. Tiny hitches can be tied on the stanchions.

5 The wooden supports for the sails are collectively called spars. The spars include masts, the bowsprit (a single spar projecting from the bow or front of the ship), yards (spars that hold square sails and cross a mast), booms (spars along the bottom edges of fore-and-aft sails), and gaffs (spars along the top edges of fore-and-aft sails). The spars and rigging must be true to the ship being modeled or the model will not appear authentic. Spars usually have to be on the order of 0.06 in (0.16 cm) in diame-

ter. Birch doweling or bamboo skewers are used to make the spars but are larger in diameter and should be sanded to be more slender and round. Masts are larger in diameter than other spars and, ideally, should taper from bottom to top. Completed spars are coated with clear nail polish to prevent the wood or bamboo from splitting and to add a glossy finish. Holes for rigging lines and holes at the bases of the masts for wire pivots are drilled next.

6 No. 30 gauge wire is fed through the base of each mast and bent in a U-shape with the two arms projecting down to construct a pivot. After the pivot ends are fastened into matching holes on the deck, the pivots will act as hinges to lower the masts and raise them again inside the bottle.

7 The bowsprit is the first spar that is glued or drilled into the foredeck. The threads that will be used to raise the masts will surround the bowsprit, so it has to be fixed securely to the hull. A fine drill point is used to drill holes through the spars for the rigging. Sewing thread tipped with nail polish to stiffen the ends is appropriate for all rigging.

8 Each mast with its set of spars including yards, gaffs, and booms is assembled as a unit. Types of knots have to be chosen carefully because some lines of rigging run fore and aft and others side to side. Rigging that is tied in the wrong direction, with the wrong knots, or too tightly will prevent small and fragile pieces from folding to fit into the bottle and from being erected inside the bottle.

9 After each mast and its spars are complete, its position on the deck should be marked and holes should be drilled for the pivot wire or hinge. The pivot wires for all masts should be inserted in the holes and checked to confirm that the masts will lie almost parallel to the deck and that the spars will also turn to parallel to the long axis of the ship. When the spars are proven to move freely, the masts can be glued in place. Later, after the sails are fixed in place, the tips of the spars will be painted white for visibility.

10 The stays or controlling lines that will be used to raise and lower the spars, rigging, and sails are tied to the masts using clove hitch knots or running them through

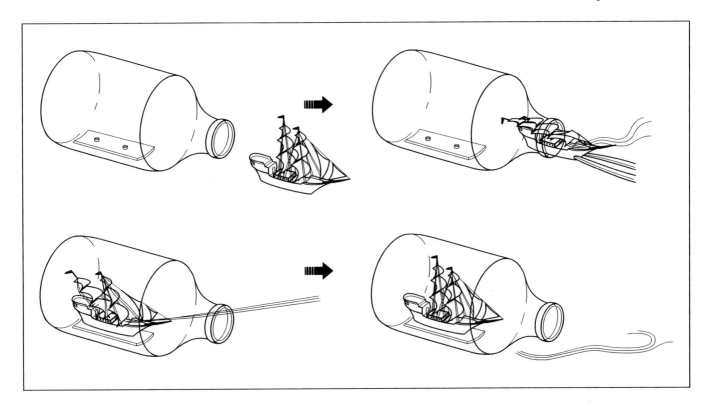

holes and the masts, hull, and bowsprit. All of the knots and holes must be positioned above other spars on the masts so the lines will not hang up on the sails. The lines have to extend at least 18 in (46 cm) below the hull and bowsprit for enough working length to erect the masts later. This excess should be tied to hooks or tacks in the work stand supporting the model.

11 The masts are stabilized on the sides with shroud lines that are attached to the bulwark, which is the rail-like edge of the hull extending above the deck. The shroud lines have to extend from immediately to the sides of each mast or to the bulwarks aft of the masts so they will not prevent the masts from being folded aft to pass through the mouth of the bottle. An 18-in-long (46-cm-long) piece of thread is knotted at one end, pulled through the inside of the forward hole on the starboard (left) side of the bulwark until the knot stops at the hole and further threaded through the hole in the mast. The thread is then run outside the port side of the bulwark, through the outside of the forward hole to the inside then wound on the inside of the bulwark through to the next hole on the port side and back through the mast. This process continues until the shroud lines are complete, the mast stands at

the correct position, and the lines are tight. Nail polish is then painted on the thread on the inside and outside of the bulwarks; when the polish dries, the thread should be cut. Shroud lines are attached to each mast by the same method. Enamel should be painted over the shroud line holes on the hull to blend with the existing paint.

12 A square-rigged vessel requires further rigging called lifts and braces. They are tied to the ends of the yards and passed through holes in the masts. Each lift rises from one end of a yard through a hole in the mast above the yard and back to the other end of the yard. Each brace attaches to one end of the yard, passes through a hole in the mast behind the mast supporting the subject yard, and is tied to the opposite end of the yard. In other words, the lift rises perpendicular to the deck, and the brace parallels it. This rigging allows the yards to be raised and lowered and moved fore and aft, like those on a true square-rigger. All rigging knots should be touched with clear nail polish to seal them.

13 Medium-weight bond paper is excellent for sails because it can be easily marked and curled. Cloth can also have weaves and thicknesses that are too large for

The ship is constructed outside of the bottle, then gently placed inside and raised.

the scale of the model. Soaking it in tea or coffee, drying it, and ironing it can "age" the paper. The sails should be drawn on the paper to match the dimensions on the plan. After cutting out the sails, each one should be held in position against its spar to confirm the fit. The seams and reef points (short lengths of rigging that pull up the bases of the sails) are drawn on the sails with a sharp-pointed pencil. The pencil is also used to curve each sail by wrapping the sail around it. The sails are glued in place with clear nail polish, but some are glued along one side only so the mast and other spars will fold back. The edges that should be glued must be carefully checked.

14 As noted in Step 2 above, the "sea" in the bottle can either be made of putty before the ship model is constructed or from Plasticene when the ship is finished and ready to be pressed into in the bottle. Plasticene has the advantage of providing its own adhesive effect. If Plasticene is used as the sea, it should be added to the bottle at this point in the construction process. Otherwise, the bottle with the putty sea should have glue placed on the flat pad prepared earlier to hold the ship. It will stay wet in the confines of the bottle until the ship is collapsed to fit in the bottle.

15 To collapse the ship, the controlling lines attached to the work stand should be untied. Beginning with the aft (rear) mast, each mast should be lowered, and the spars should be turned to parallel the masts. The sails will extend over the bulwarks and should be wrapped around the hull. The stern of the ship should be inserted in the bottle first. When most of the ship is in the mouth of the bottle, long tweezers should be used to support the rest of the model to guide it into the bottle. With the model gripped with the tweezers and well inside the bottle, the lines (extending outside the bottle) should be pulled gently and in the correct order to raise the masts from fore to aft and to align the spars. The model can then be put on the pad of glue or Plasticene and pushed down with the tamper. The rigging may have tangled during the lowering and raising of the masts and can be untangled when the hull is stuck in place. Similarly, the sail alignments can be corrected. The whole process of inserting the ship in the bottle and unfurling and correcting

its parts must be done carefully so the sails are not torn or other damage is not done.

16 After the sea dries, the masts should be fastened in their permanent positions. The stay or controlling lines should be pulled and taped to the outside of the bottle. They should be secured to the bowsprit with drops of nail polish, then the tape holding the lines should be removed to test the security of the masts. If they keep their positions, the lines can be cut where they pass through to the underside of the bowsprit. Final corrections can be made to the rigging and sails.

17 The bottle is sealed with a cork, but this can be cut off flush with the mouth of the bottle or left partially extended. The bottle can be resealed with its metal screw-on cap, if appropriate. All seals can be anchored with sealing wax. Cotton fishing can be tied into a Turk's Head knot commonly seen on ships and nautical items. A sequence of knots forms a line that can be wrapped around the bottle's neck. The modeler's name, date of construction, and the type and name of the ship can be written on the underside of the bottle or engraved on a metal tag.

18 To provide the finishing touch, a wooden display stand or wall mount can be constructed to complement the model. The ship in the bottle must remain the focal point, and, ideally, the stand will consist of rope, wood, or "period" materials rather than modern choices. The range of possibilities is large, but some research and woodworking techniques are useful in selecting a stand and finishing it elegantly. Displaying the model at eye level and lighting it attractively should also be considered.

Quality Control

The final impact of a ship in a bottle depends on the crafter's skill in every step of research, planning, selecting the bottle, modeling the ship, finishing all details including rigging and sails, erecting the model inside the bottle, and displaying the finished work of art. A ship in a bottle is a work of art and should be treated like the revered craft it is. If a model builder has any interest in learning this craft, he or she must emphasize quality throughout, including the process of checking and double-checking plans and measurements.

Byproducts/Waste

Building ships in bottles produces almost no waste because of the small features of the models and the limited amounts of materials required. Some wood trimmings may result and are easily disposed.

The artist's safety is also reasonably secure. Chisels, hobby knives, pins, wires, and other sharp tools may cause the occasional cut. However, the tools are small and are generally familiar to those who become fascinated with this hobby. Other materials like nail polish and modeling enamels produce fumes, but these are also minor. Adequate ventilation and lighting are best for the hobbyist's safety.

The Future

The hobby of building ships in bottles is not for everyone. Love of research, ships and sea lore, history, woodworking and other skills, and minute details, as well as considerable patience, are required. The finished models are surprisingly durable and are treasured possessions to leave to children and grandchildren. Competitions are held around the world and opportunities to display models, including museum exhibits, are plentiful, so many people can enjoy these creations and purchase and collect them.

This nautical craft thrives because parts can be produced in numbers and sold in kits for hobbyists of all skill levels. The temporary nature of so many modern collectibles and mass production on a far greater scale than the ship-in-a-bottle kits have also encouraged crafters to pursue this relatively unusual interest. Those who appreciate ships in bottles are not likely to grow to huge numbers, but they are intensely loyal to the blend of skill and mystery in these models, insuring a small but stable future.

Where to Learn More

Books

Hubbard, Donald. *Ships-in Bottles: A Step-by-Step Guide to a Venerable Nautical Craft.* New York: McGraw-Hill Book Company, 1971.

Needham, Jack. *Modelling Ships in Bottles.* Wellingborough, England: Patrick Stephens Limited, 1985.

Smeed, Vic. *The World of Model Ships.* Secaucus, New Jersey: Chartwell Books, Inc., 1979.

Other

Nautical Gift Shop Web Page. December 2001. <http://www.nautical-gift.com/enindex.htm>.

Langfords Marine Antiques Web Page. December 2001. <http://www.langfords.co.uk/gallerysb.htm>.

Uptown Sales Web Page. "Authentic Ship-In-A-Bottle Kits." December 2001. <http://www.hobbyplace.com/woodmodels/shipinbotl.html>.

Gillian S. Holmes

Shrapnel Shell

The Shrapnel shell was first used in combat in 1804 in Surinam on the north coast of South America against Dutch settlers.

Background

There has always been a high demand among military strategists for economical means of killing enemy soldiers. Economy is required not so much to save money, but to allow outnumbered soldiers the opportunity to win battles. Prior to the advent of high-powered rifles, soldiers in opposite armies would form ranks preparing for battle in clear sight of each other. However, artillery was generally ineffective against troop formations at long range until late in the eighteenth century.

History

Lieutenant Henry Shrapnel of the British Royal Artillery solved the distance problem in 1784. Shrapnel's contribution was to pack musket balls into a container that could survive being fired out of a cannon. The round case shot was simply a hollow cannon ball that contained musket balls in a gunpowder matrix. A time fuse made out of paper wrapped around more gunpowder, rather like a firecracker fuse, was inserted into the cannon ball and lit. The cannon ball was then fired at the enemy troops. If the cannoneer timed the flight of the ball properly, the ball would explode just as it arrived above the enemy troops, releasing the musket balls.

Shrapnel was largely ignored. However, by 1803 he was a captain and was allowed to demonstrate his invention for the British Army. Shrapnel's invention was instantly recognized to be one of the super weapons of the day, evidenced by the speed with which the British Army put it into production—only two months after Shrapnel first demonstrated it.

The Shrapnel shell was first used in combat in 1804 in Surinam on the north coast of South America against Dutch settlers. The Dutch surrendered after receiving their second round of Shrapnel shells. Shrapnel was promoted to lieutenant colonel in 1804, less than a year after making major.

There were numerous improvements made in the Shrapnel shell between the final defeat of Napoleon and the phasing out of Shrapnel shells during World War I. Shrapnel's round ball evolved into an artillery shell that looked very much like a modern shell and was manufactured in much the same way. It also performed the same function: the delivery of lead balls over long distances in large quantities at high velocities.

Raw Materials

The shell was made out of forged carbon steel. The purpose of the shell was simply to contain the lead balls and funnel them downward toward the target. The shell was not intended to explode into fragments. Cartridge cases were almost always made out of brass. Brass was used because it expands during firing. As the cartridge case expanded, it sealed the gun barrel in a process called obturation. Obturation provides greater thrust to the projectile and also protects the artillerymen against backfire. The Shrapnel balls were made out of lead. Lead was also used in bullets, as it is both heavy and soft. Because lead is soft, it gives up more of its energy to the target (flesh) rather than passing through the target and expending its energy against the landscape. The rotating band was made out of an alloy known as gilding metal, which consists of 90% copper and 10% zinc. The rotating band provided forward obturation (so that none of the propelling charge would blast by the shell in the gun barrel and be wasted) and also imparted a spin on the shell as it moved up the barrel. Spin was induced in the shell by the barrel's rifling—the spiral ridges cut in the barrels of many types of guns. Just as a football that does not spiral will rotate end over end and not go where it was intended, an artillery shell that is not spin-stabilized might end up anywhere.

The base charge for most artillery shells was usually a combination of nitrocellulose and nitroglycerine. Common primer materials used to ignite the base charge include mercury fulminate, lead azide, lead styphnate, and nitromannite. These chemicals are extremely shock-sensitive and will explode when hit sharply. The artillery primer would ignite a booster charge of gunpowder that was inserted into a perforated hollow spike that penetrated most of the length of the base charge. The purpose of the booster charge was to ignite as much of the base charge as possible at the same time. The fuse in Shrapnel shells consisted of a brass plug that screwed into the top of the shell. The brass plug contained hollow channels that contained gunpowder, and the fuse could be adjusted to provide a given delay in firing. The fuses were initiated by the force of the initial acceleration of the shell as it left the cannon barrel. Modern artillery fuses are almost always solid state electronic timers or proximity fuses.

Design

The design of an artillery shell involved determining the purpose of the shell and then matching the purpose to the cannon (modern artillery pieces are mainly howitzers, the distinction being that howitzers fire along parabolic arcs over the horizon whereas cannons fire along a line of sight) from which the shell will be fired. The designer had the specifications for the cannon, and thus knew that the shell must have a certain diameter and could only generate a certain amount of thrust without damaging the cannon. The shell had to be simple enough to allow for rapid firing but intrinsically safe so that a shell dropped in the heat of battle would not explode and kill the wrong people. The fuses of Shrapnel shells were precisely designed so that the shell would explode at exactly the right moment. A Shrapnel shell that went off

too far from the target would do little damage, while a shell that went off after hitting the ground would do no damage.

The main components of a Shrapnel shell were the shell itself, the cartridge case, the lead balls, a base charge to propel the shell to its target, a charge to expel the lead balls from the shell, a primer charge to set off the base charge, and a fuse to set off the expelling charge. Other miscellaneous components included a rosin mixture to hold the lead balls in place and which produced smoke to assist the artillery spotters, a steel push plate between the lead balls and the expelling charge, a rotating band on the base of the shell to spin the shell as it moved up the gun barrel, and a nose cone to reduce aerodynamic resistance of the shell.

The Manufacturing Process

The shell

1 First, the Shrapnel shell was forged. In forging, a cylinder of carbon steel is heated almost to the melting point and then manually beaten into the rough shape of the final product. The rough forging was then machined to the final shape.

2 The cross-section of an artillery shell is slightly smaller than the inside diameter of the cannon barrel except for two locations: the top of the cylindrical portion of the shell and the rotating band. The top band is known as a bourrelet. The bourrelet and rotating band provide a very close tolerance (only a few thousandths of an inch) between the shell and the cannon barrel. A groove is milled into the base of the shell into which the rotating band is pressed. For a Shrapnel shell, the center of the shell is then drilled out to hold the lead balls.

3 Gunpowder was used to expel the lead balls. One to two ounces (28–56 g) of gunpowder was inserted into the shell under carefully controlled conditions to prevent accidental detonation. A cloth disk was inserted into the shell to separate the base charge from the lead balls. A metal diaphragm (push plate) was then placed on top of the cloth separator. The push plate and cloth separator contained a hole into which a steel flash tube was press-fitted prior to in-

A cut-away of a shrapnel shell.

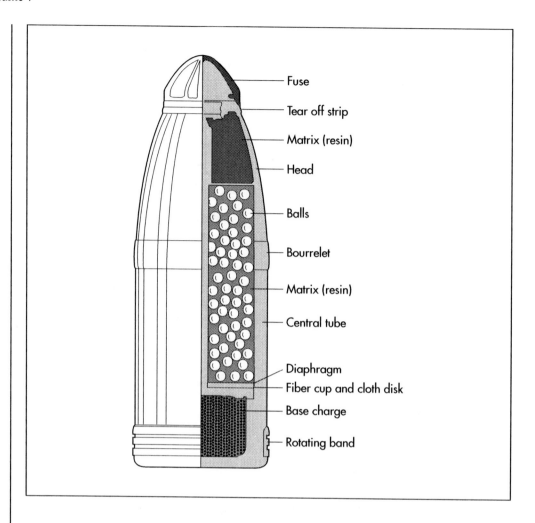

Fuse

Tear off strip

Matrix (resin)

Head

Balls

Bourrelet

Matrix (resin)

Central tube

Diaphragm

Fiber cup and cloth disk

Base charge

Rotating band

sertion. In press fits, the part to be inserted has a slightly smaller diameter than the hole into which it is forced. Large forces are necessary to press the part into its hole, which provides a tight fit that will not come loose. The purpose of the flash tube was to transmit the flame from the primer charge in the fuse at the nose of the shell down to the gunpowder in the base of the shell.

Lead balls

4 The lead balls were manufactured by pouring molten lead through a steel screen. As the molten lead flowed through the screen, it formed spherical droplets, the diameter of which were controlled by the size of the openings in the screen. The molten lead droplets fell against a countercurrent of forced air that solidified the molten lead, and then into running water, which hardened them further. Typical Shrapnel balls were about 0.51 in (13 mm) in diameter, though larger balls were sometimes included to kill horses.

5 The lead balls were mixed with pine rosin (the substance left over after turpentine is distilled out of pine sap) and poured into the shell. The rosin was allowed to harden in the shell. The purpose of the rosin was to prevent the lead balls from rattling around in the shell during flight, which might have caused premature ignition of the gunpowder. The rosin also provided smoke so that artillery spotters could determine if the shells were being timed correctly to explode above their targets.

The fuse

6 The Shrapnel shell fuse was a complex mechanical/chemical device machined out of brass and threaded to fit the shell. Its complicated design and tiny components made for difficult assembly, and a large portion of the munitions industry was devoted to turning out these parts. Brass was chosen as it is non-sparking (if struck, it does not produce sparks that may ignite the powder train and cause premature explosions). The

fuse consisted of two different primer charges separated by two channels containing gunpowder. The connection between the two channels (and thus the speed of ignition) could be adjusted by rotating the bottom portion of the fuse. The first primer was activated by the acceleration of the shell as it was fired. The acceleration drove a plunger against a stiff spring into a gliding metal or aluminum cup containing the primer, which exploded on contact.

The cartridge case

7 The cartridge case was stamped out of brass. Stamping involves placing a flat piece of metal between dies and progressively beating it into the desired form. In the case of cartridge cases, multiple anvils and hammers were used to obtain the final shape.

8 A separate primer was required to set off the base charge in the cartridge case. The primer was contained in a gilding metal or aluminum cup that would be contacted by a steel anvil. The anvil would be pushed into the primer by the howitzer's firing pin. The primer ignited a booster charge of gunpowder that then ignited the base charge. The primer and booster charge were contained in a hollow brass spike that was pressed into a hole in the base of the cartridge case.

9 Final assembly of the shell was accomplished by crimping. A groove semi-circular in cross section is cut into the shell. The cartridge case is fitted over the groove and the entire diameter around the groove is compressed until the brass cartridge case actually flows into the groove and forms a tight bond.

Quality Control

Quality control is extremely important in ammunition manufacturing as faulty ammunition can kill valuable soldiers. All artillery shells were manufactured in lots of specified sizes, usually 2,000–5,000 pieces per lot. The lot number was painted onto artillery shells so that the shells could be tracked down if problems with the lot cropped up later. A certain percentage of shell components were measured to verify that the parts were the correct size. Destructive testing was performed on representative samples to assure that metal components had the proper

strength and that chemical components burned at the proper rate. Fuses were tested for waterproofing. Rotating bands were torn off shells to ensure that they had sufficient strength to withstand firing.

Once it was determined that the shells had been manufactured according to design, the shells were then field tested to determine whether the design had produced a shell that behaved in a predictable fashion. Some shells were deliberately overloaded with base charge and fired to ensure that they would not destroy the gun. Shells with inert fuses were fired, and then recovered, to assess whether the force of the firing would have set off the fuse prematurely. Shells were filled with sand and fired to assess how well the shell held together during flight. And a certain number of shells were fired to assure that the base charges would send the shells where the artillery operators intended them to go.

Byproducts/Waste

The major wastes generated by the production of artillery shells were produced during testing of the shells and the training of artillery operators. There are currently large sections of the United States that will never be able to be used because of the presence of artillery shells that were fired but did not go off. During actual production, the largest waste stream consists of cutting fluids and metal chips produced during machining.

The Future

Shrapnel shells were rendered obsolete during the first world war. They proved to be ineffective against troops protected by trenches, could not clear barbed wire entanglements, and were proven difficult to set so that the shells exploded at the proper height above the enemy troops. The Shrapnel shell was superseded by the high-explosive fragmentation shell, in which the shell casing was filled with an explosive that fragmented into hundreds of deadly pieces upon detonation. The latest technology for killing enemy troops at a distance is the Improved Conventional Munition, or ICM. The ICM is more similar to a Shrapnel shell than a fragmentation shell. The difference is that rather than spilling out simple metal balls, it spits out hand grenades, land mines, or anti-tank bombs. It is inevitable that the ICM will someday be superannuated by something even more efficient and tailored to overcome new defense strategies.

Where to Learn More

Books

Hogg, Ian. *Allied Artillery of World War One.* Great Britain: Crowood Press, 1998.

Other

New Zealand Permanent Force Old Comrades' Association Web Page. December 2001. <http://riv.co.nz/rnza/hist/shrap/index.htm>.

United States Army. *TR 1355-75A Mobile Artillery Ammunition. Ammunition for 75-mm Field Guns, M1897 (French); M1916 (American); and M1917 (British).* 21 November 1927.

United States Army. *TR 1355-155A Mobile Artillery Ammunition. Ammunition for 155-mm Howitzers, M1917 (French) and M1918 (American).* 23 November 1927.

Jeff Raines

Sleeping Pill

Background

A sleeping pill, also commonly called a sleep aid, is a drug that helps a person fall asleep or remain sleeping. Disorders such as insomnia (inability to sleep) are widespread, and drugs to induce sleep have been used since ancient times. Two distinct categories of sleeping pills are sold in the United States: prescription and over-the-counter drugs. Most prescription sleeping pills have a type of drug known as a benzodiazepine (a central nervous system depressant) as the active ingredient. Benzodiazepines include chlordiazepoxide (Librium) and diazepam (Valium). Pharmacists developed non-benzodiazepine hypnotics in the 1990s, such as zopiclone and zaleplon (Sonata). Over-the-counter sleep aids, which can be bought without a prescription, contain antihistamines. Both prescription and over-the-counter sleep aids can cause side effects, such as next-day drowsiness, and an overdose can be hazardous. The manufacturing of sleeping pills is highly regulated and overseen by the Food and Drug Administration (FDA).

History

Sleeping potions were some of the earliest drugs discovered, and sleep aids are still among the most widely used drugs today. The ancient Greeks and Egyptians used the extract of the opium poppy to induce sleep. The Greek god of sleep, Hypnos, was usually depicted holding a poppy flower. The juice of the poppy contains chemicals known as opiates, from which morphine and heroin are distilled. Ancient Greeks and Romans knew several other herbal sleep-inducers. The bark of mandrake, or mandragora, was used as a sleep aid, as were the seeds of an herb called henbane. The juice of lettuce was also used to induce sleep. As early as 300 B.C., Greek doctors were known to prescribe concoctions of these different plant derivatives. Similar prescriptions were also apparently known throughout the Arab world. Apothecaries of the Middle Ages in Europe stocked "spongia somnifera," a sponge soaked in wine and various herbs. Other mixtures were known in England in the Middle Ages and the Renaissance as "drowsy syrups." Plant-based sleep aids were all that were available up until the nineteenth century. The chemist Frederick Seturner synthesized opium in 1805, and other advances in sleep drugs followed by the middle of the century.

Two drugs used in the nineteenth century to induce sleep were bromides and chloral hydrate. Chloral hydrate was synthesized in 1832 by a German chemist, Justus von Liebig. Chloral hydrate is a central nervous system depressant that acts very rapidly. Chloral hydrate alone could send a person into deep sleep in about half an hour. Chloral hydrate works much more quickly in combination with alcohol. Slipped into whiskey, it was the "knockout drops" of the underworld, also called a "Mickey Finn." The class of sleeping drugs called bromides were invented in 1857 by an English chemist, Sir Charles Locock. There are several bromide salts, including sodium bromide, potassium bromide, and ammonium bromide, which all act as central nervous system depressants. Locock first used bromides as an anticonvulsant to treat epileptics. A German doctor, Otto Behrend, discovered in 1864 that potassium bromide was a useful sedative. The various bromides be-

The ancient Greeks and Egyptians used the extract of the opium poppy to induce sleep.

came popular as sleep aids in the late nineteenth and early twentieth century.

The most popular sleeping pills of the early twentieth century were the barbiturates. The barbiturates comprise a huge class of drugs with at least 25,000 known compounds. Of these compounds, about 50 were or are marketed as prescription drugs. The forebear of the barbiturates was actually discovered in the mid-nineteenth century. A Prussian chemist, Adolf von Baeyer, is credited with inventing and naming barbituric acid in 1863 or 1864. He created the acid out of a compound of malonic acid and urea. On the day of his discovery, Baeyer is said to have gone to a nearby tavern to celebrate. Some sources say it happened to be the feast of St. Barbara that day; others say the barmaid was named Barbara. In any case, he named the compound barbituric acid. In itself, barbituric acid was useless. In 1903, a student of Baeyer's, along with another German chemist, produced a new compound out of barbituric acid and a diethyl derivative. The new chemical, given the trade name Veronal, was an excellent sedative and sleep aid. Other researchers came up with more barbituric acid derivatives. The most widely used was phenobarbital. Many European and American pharmaceutical companies came up with new barbiturates in the 1920s and 1930s. The Eli Lilly Company produced the widely used Amytal and Seconal, and Abbott Laboratories invented Pentothal.

Though the barbiturates were effective sleep aids, they also proved dangerous. Barbiturates are addictive, can have a variety of unpleasant side-effects, and their effectiveness is greatly increased when taken with alcohol. Barbiturate sleeping pills taken with alcohol can quickly bring on death, whether through accidental overdose or planned suicide. Scientists developed safer sleeping pills in the 1970s, the benzodiazepines. Early benzodiazepine drugs shared problems with barbiturates. They were addictive and had side effects such as memory impairment. Improved benzodiazepines were developed in the late 1970s and early 1980s. In the 1990s, non-benzodiazepine drugs were developed, which leave the body much more quickly than the older drugs.

The barbiturates and benzodiazepines are ingredients in prescription drugs, given out by a doctor. Until the 1970s, ingredients in over-the-counter sleep medications were not closely regulated in the United States. The FDA began reviewing over-the-counter drugs in the early 1970s, and by 1978 had approved one active ingredient for an over-the-counter sleep aid. This was the antihistamine doxylamine succinate. In 1982, the FDA approved two more antihistamines for non-prescription hypnotics. These are diphenhydramine HICl and diphenhydramine citrate. These three drugs are the only active ingredients approved for non-prescription sleep aids in the United States. Though it is possible to overdose on these drugs, they are not nearly as strong as prescription sleep aids, and in most cases can be used safely when the directions are followed.

Raw Materials

A non-prescription sleeping pill sold in the United States may contain only one of three approved active hypnotic ingredients. As stated above, these are the antihistamines diphenhydramine HICl, diphenhydramine citrate, or docylamine succinate. Some over-the-counter sleep aids also contain other active ingredients for other conditions, such as an analgesic for pain relief. In addition, sleep aids contain inactive ingredients that are used to bind the tablet, coat it, flavor it, color it, and give it the proper consistency. Some common ingredients are sugars, starches, magnesium stearate, various artificial colors, microcrystalline cellulose, and wax. Though sleeping pills are sold under a variety of brand names, many brands are virtually identical. The number of formulas used to manufacture sleeping pills in the United States is actually quite small.

Design

For over-the-counter sleep aids, design is not a terribly important element in the manufacturing process. Sleeping pills are made in a controlled and regulated environment with a limited number of approved ingredients. Chemical engineers design a formula that meets the drug maker's needs, producing a tablet of a particular color and shape, for example. The tablet's designers also need to consider how stable the drug will be over its stated shelf life and how quickly it dissolves or breaks down in the body. Manufacturers would test a new formula as well

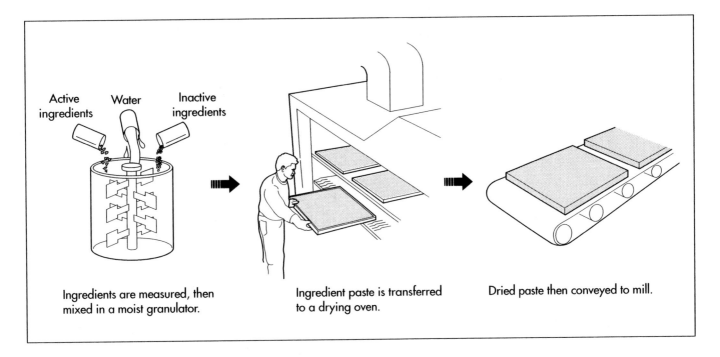

Active ingredients   Water   Inactive ingredients

Ingredients are measured, then mixed in a moist granulator.

Ingredient paste is transferred to a drying oven.

Dried paste then conveyed to mill.

to see how it works in the tabletting machine. Some redesign may be necessary in order to come up with a tablet that can be produced easily.

The Manufacturing Process

1 First, workers perform what is called the "weigh up," measuring out the active and inactive ingredients for the sleep aid formula. A large manufacturer may make a batch of tons of tablets, while a smaller plant may make only several pounds at a time. In either case, workers follow the Standard Operating Procedure in measuring out the specified amounts of each ingredient. Workers are required to work in pairs, so one worker may measure and the second will verify that his partner has measured out the correct amount.

2 After all the ingredients are correctly measured, they are blended. A variety of mechanical devices may be used to blend the ingredients, depending on the manufacturer and the size of the batch. Also, the blending may take part in stages. Workers may place all the ingredients in the blender at once, or groups of ingredients may be done separately. Workers then add a measured amount of water to the blender in order to wet the ingredients, again following exactly the written procedure and checking each other. The wetted powder becomes a thick paste, with the

consistency of damp sand. Workers feed the paste into a hopper that has rotating hammers which crush the paste. This process is called moist granulation.

3 The paste exits the moist granulator machine onto trays, forming a layer about one inch thick. The trays are next conveyed to a drying oven. The size of the oven depends on the manufacturer. It may be a small cupboard-like apparatus, or something much larger. The paste dries in the oven for several minutes to half an hour, depending on the particular formula, type of heat used, and manufacturing protocol. When dry, the material is in chunks about the size of a dime, of a brick-like hardness. The granules making up the material are coarse and uneven. In the next step, more granulation breaks the material into uniform granules.

4 Next the hardened paste is conveyed to another granulator or mill, which breaks the paste into small particles. Various designs of machine can perform this function. A common design uses a screen or sieve with holes the desired size of the finished granules. A mechanical arm moving over a series of screens forces the hardened paste to form smaller and smaller particles. After passing through the final screen, the particles are the correct size for compression in the tabletting machine.

The ingredients are mixed and dried.

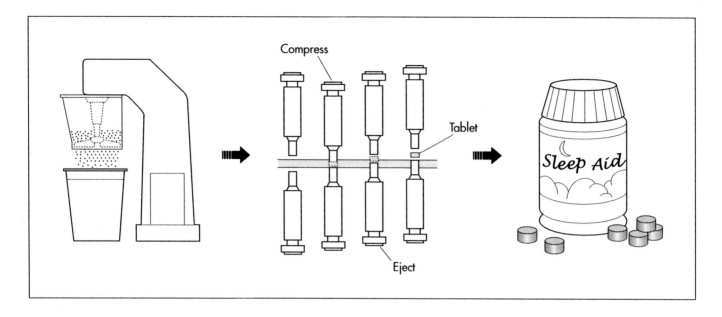

The paste is ground and formed into tablets.

5 At this stage, the material for the sleep aid has been converted to fine granules of uniform size. Workers feed the powder into hoppers above a tabletting machine. This machine has a rotating tray of hollow dies in the shape of half the finished tablet. The powder falls from the hopper into the hollow dies. Then an arm holding an inverted die of the same shape descends and exerts several tons of pressure on the powder. This compresses the powder into tablet form. Depending on the size of the tabletting machine, it may be able to produce thousands of tablets a minute. The tablets fall onto a belt that counts them into sterile packaging. The packages are sealed, and readied for shipping.

Quality Control

Quality control occurs at every step of the sleep aid manufacturing process. All the raw ingredients, the inactive as well as the active, must be tested when they arrive at the factory, in order to make sure each is exactly what it is supposed to be. Workers follow explicit directions in carrying out each step of measuring, blending, granulating, etc., always working in pairs and checking to verify that the procedure is being carried out in the right way. A quality control inspector also follows the manufacturing process, formally "releasing" the tested ingredients so they can be mixed, testing and releasing the mixture, and releasing the final product too, if it is exactly as it is supposed to be. A por-

tion of each batch must be taken aside and subjected to testing before the batch is released. Chemical tests might include ascertaining the correct medicinal strength of the tablet, as well as physical tests for proper texture, correct speed of disintegration, and stability of the product over its shelf life. Quality control inspectors keep complete written records. Each batch is assigned a batch number, which is imprinted on the packaging. If there is any problem later with the tablets, the manufacturer can trace when and where the batch was made.

The manufacture of sleeping pills is also regulated by the FDA. Drug manufacturers need to follow guidelines laid down in an FDA monograph on night-time sleep aids. The monograph details what ingredients are acceptable in the drug formula and how the package can be labeled. Sleep aid packages explain the indications for use and must stick to FDA-approved wording. The FDA also lays out the warnings the package must carry and the dosage. The manufacturer of an over-the-counter sleep aid must follow precise manufacturing steps laid out in federal Good Manufacturing Practice regulations. These regulations spell out what kind of facility is considered adequate for producing drugs, what kind of machinery can be used, how the machinery is to be installed and maintained, how measuring instruments are to be calibrated, and many other rules regarding the physical aspect of the plant and equipment. Good Manufacturing Practice also requires certain quality control steps be taken. Inspectors from the FDA visit the plant at least once every two years in order to ascertain that the Good Manufacturing Practice rules are being followed. A plant making sleep aids must have its Standard Operating Procedures written out and approved. The Standard Operating Procedures explain exactly how each step in the manufacturing process will be carried out. Workers must be trained to follow the Standard Operating Procedures to the letter. This prevents any variation in how the product is made. The FDA inspector will also check the actual operation of the plant against the written Standard Operating Procedures and note any discrepancies. Because of this tight control of every step of the manufacturing process, sleep aids manufactured in the United States should be of highly uniform quality.

The Future

Advances in sleep aids are likely to come from prescription drug makers. Because of the side-effects and danger of overdose of powerful sleeping pills, researchers are investigating new drugs that leave the body quicker. One of the new non-benzodiazepine sleep aids discovered in the late 1990s, zaleplon, has the advantage of leaving the body in as little as five hours. This makes it possible to treat a form insomnia where a person falls asleep, but then wakes up after a few hours, unable to fall back asleep. Drugs like this have what is called a short elimination half-life. If a new drug like this proved effective as a prescription drug, it is possible that the manufacturer would move to petition the FDA to change its status to an over-the-counter drug. Yet since the FDA issued its first monograph on over-the-counter sleep aids in 1978, it has not added any new classes of drugs to this category.

Where to Learn More

Books

Alderborn, Göoran, and Christer Nyströom, eds. *Pharmaceutical Powder Compaction Technology.* New York: Marcel Dekker, Inc. 1996.

Kales, Anthony, ed. *The Pharmacology of Sleep.* New York: Springer-Verlag, 1995.

Periodicals

"New Hypnotic, Zaleplon, Shows Advantages for Treating Insomnia." *Psychopharmacology Update* (August 1998): 1.

"Over-the-Counter Sleep Aids." *Consumers' Research Magazine* (April 1995): 34.

Angela Woodward

Smoked Ham

Background

Smoked ham is a popular serving of meat, cut from the pork leg. It is cured with salt and spices, then subjected to slow and steady heat for varying periods. The smoking is carried out in a special chamber called a kiln.

History

Foodstuffs were originally smoked as a means to preserve them. The practice may have started as early as the Stone Age and was probably discovered by accident when food was left out in the sun. The discovery of fire would have made the smoking of foods more prevalent. Throughout the centuries, until the development of refrigeration, smoking and salting meat for future use was a regular practice.

Chemicals released from the wood during the smoking process slows the growth of microorganisms. Likewise, in curing, salt reduces the amount of available water for bacteria to grow.

Pork has always been a popular meat for many civilizations due to the ease of raising pigs and preserving the meat. People began raising pigs about the same time that they established group settlements. By 600 B.C., pig breeding was a thriving industry. Pigs were brought to the New World by Spanish explorer Hernando de Soto in the sixteenth century and soon became a major commodity here as well.

A number of cultures, such as Orthodox Jews and Muslims, forbid the eating of pork. This food prohibition dates to ancient times when Egyptians only ate pork during the feats of the god Osiris.

Raw Materials

Today, pigs are raised around the world, primarily in areas of temperate climates and dense human populations. China and the United States are the largest producers of pigs. Pig breeding incorporates a combination of pen-rearing and pasture-feeding. Domesticated pigs are fed a diet consisting of corn, grain, roots, and fruits.

Domestic pigs generally reach their market weight of 175–240 lb (79.4–108.9 kg) between the ages of five and 11 months. At that time they are taken to the slaughterhouse. The specific cuts are then created from the carcasses. The ham portion, cut from the leg, is then cured and smoked.

Before smoking, the pork is submerged in a brine solution containing water, salt, and sugar. Pickling spices (mace, allspice, cloves, cinnamon, peppercorns, and bay leaves) and garlic may also be added.

Using the correct type of wood is essential to successful smoking. The wood must be one that burns slowly and steadily. Non-resinous woods, such as beech, oak, chestnut, and hickory are the most common types used for smoking. Aromatic herbs such as juniper, laurel, sage, and rosemary may also be added. Conversely, woods containing resin, such as pine, will impart a bitter taste to the meat.

Smoking kilns are built in a variety of fashions. They can be brick chimney-like structures or stainless-steel drums. The inside can be fixed with racks or hooks. The fuel is loaded into the bottom and covered with a perforated plate so that the smoke can filter through to the ham.

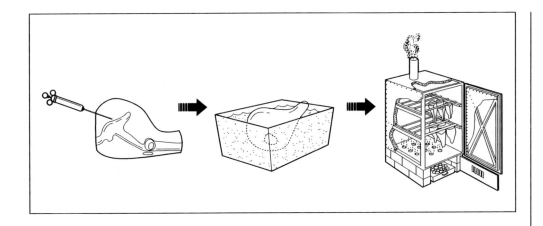

The Manufacturing Process

Brining the ham

1 Before the ham can be smoked, it must be soaked with a salt solution. First the solution is injected from a brine pump into the thickest part of the meat, preferably near the bone. Similar to a hypodermic needle, the brine pump has a long hollow tube with rows of holes down the sides. As the plunger is depressed, the brine is forced through the holes into the meat tissues.

A typical brine solution might be composed of 5 gal (19 l) of water, 5 lb (2.3 kg) of salt, 1 lb (0.5 kg) of white sugar, and 1 oz (28 g) of saltpeter (sodium nitrate). If a mild cure is desired, the ratio is 1 fl oz/lb (28 g/0.5 kg) of meat. For a stronger cure, the ratio is increased to 1.5 fl oz/lb (42.5 g/0.5 kg).

Soaking the ham

2 The ham is then immersed in the remaining brine for a period of 10–16 days depending on weight. The ham and the brine are inspected daily. The brine is overhauled, or turned over, every three days. Overhauling prevents the brine from becoming weak on top and heavy on the bottom. A large wooden paddle is used to remix the brine.

Rinsing and drying the ham

3 After the brining is complete, the ham is rinsed thoroughly with fresh water.

4 It then is hung or loaded onto racks to dry for a period of a few days to several weeks. During the drying period, a crust may form on the ham. This crust is shaved off and discarded.

Smoking the ham

5 Several hams are smoked together. They can be hung from meat hooks in a large kiln or placed on racks in smaller smokers. The smoker is kept at a temperature of 70–80°F (21–26°C). Wood chips are used rather than logs to create a slow burning fire. The chips are stirred, or stoked, regularly. The amount of time that the ham remains in the smoker is determined by the desired intensity of flavor. That period can be as little as 48 hours or as much as six weeks.

6 When smoking is completed, the hams are packaged in shrink wrap and shipped in refrigerated trucks to retail outlets and restaurants.

Quality Control

Pig farming and pork production is closely regulated by government agencies such as the United States Department of Agriculture (USDA). For example, according to USDA standards, a cut of meat marketed as ham must come from the hind leg of a hog. Meat cut from the front leg cannot be called a ham, but rather a pork shoulder. Additionally, the use of hormones in pig farming is illegal.

Pigs are extremely susceptible to several diseases, including foot-and-mouth disease, anthrax, and hog cholera. To prevent the occurrence of these diseases, pigs are often treated with antibiotics. In those cases, a withdrawal period is required between the time the drugs are administered and the time that the animals are slaughtered. The USDA's Office of Food Safety and Technical Service conducts random tests at slaughterhouses to test the pork for antibiotic residues.

Government agencies also set rules for the operation of slaughterhouses. In the United States, the Federal Meat Inspection Act of 1967 includes guidelines set out in the Humane Slaughter Act of 1958. Under these regulations, animals must be rendered unconscious before they are killed. This is accomplished by stunning or gassing the animals.

Byproducts/Waste

Waste material from hog farms are often piped into open-air pits or waste lagoons. This has caused concern among environmental agencies that toxic gases emanating from the bacteria that feeds on the decomposing waste material will make its way into the groundwater. Contaminating this environmental hazard will result in increased safeguards imposed on the farmers, which could impact those farmers economically.

The Future

In the early months of 2001, a severe outbreak of foot-and-mouth disease occurred in the United Kingdom. By mid-year, 2.8 million animals had been affected. The disease, which is characterized by sores and blisters on the hooves and mouths of livestock, is extremely contagious. Although humans cannot contact foot-and-mouth disease, infected animals must be destroyed

The outbreak caused world-wide concern. Imports of meats from the United Kingdom were restricted. A number of tourists sites, such as Stonehenge, were closed to visitors. Travelers returning from the United Kingdom were required to clean their shoes in disinfectant at airport customs points. Whether or not the epidemic can be contained and its long-term effect on pork production is yet to be seen.

At the end of the twentieth century, pork farmers were facing a severe economic downturn. An increase in hog production was offset by the closing of a significant number of slaughterhouses. This caused hog prices to fall from 55 cents a lb/kg to eight cents a lb/kg.

Where to Learn More

Books

Erlandson, Keith. *Home Smoking and Curing.* London: Vermilion, 1977.

Lang, Jenifer Harvey, ed. *Larousse Gastronomique.* New York: Crown Publishers.

Nissenson, Marilyn, and Susan Jonas. *The Ubiquitous Pig.* New York: Harry N. Abrams, Inc., 1992.

Sleight, Jack, and Raymond Hull. *Home Book of Smoke-Cooking Meat, Fish & Game.* Harrisburg, PA: Stackpole Books, 1971.

Other

United States Department of Agriculture Web Page. December 2001. <http://www.fsis.usda.gov>.

Mary McNulty

Speedometer

Background

A speedometer is a device used to measure the traveling speed of a vehicle, usually for the purpose of maintaining a sensible pace. Its development and eventual status as a standard feature in automobiles led to the enforcement of legal speed limits, a notion that had been in practice since the inception of horseless carriages but had gone largely ignored by the general public. Today, no automobile is equipped without a speedometer intact; it is fixed to a vehicle's cockpit and usually shares a housing with an odometer, which is a mechanism used to record total distance traveled. Two basic types of automobile speedometer, mechanical and electronic, are currently produced.

History

The concept of recording travel data is almost as old as the concept of vehicles. Early Romans marked the wheels of their chariots and counted the revolutions, estimating distance traveled and average daily speed. In the eleventh century, Chinese inventors came up with a mechanism involving a gear train and a moving arm that would strike a drum after a certain distance. Nautical speed data was recorded in the 1500s by an invention called the chip log, a line knotted at regular intervals and weighted to drag in the water. The number of knots let out in a set amount of time would determine the speed of the craft, hence the nautical term "knots" still applied today.

The first patent for a rotating-shaft speed indicator was issued in 1916 to inventor Nikola Tesla. At that time, however, speedometers had already been in production for several years. The development of the first speedometer for cars is often credited to A. P. Warner, founder of the Warner Electric Company. At the turn of the century, he invented a mechanism called a cut-meter, used to measure the speed of industrial cutting tools. Realizing that the cut-meter could be adapted to the automobile, he modified the device and set about on a large promotional campaign to bring his speedometer to the general public. Several speed indicator concepts were introduced by competing sources at the time, but Warner's design enjoyed considerable success. By the end of World War I, the Warner Instrument Company manufactured nine out of every 10 speedometers used in automobiles.

The Oldsmobile Curved Dash Runabout, released in 1901, was the first automobile line equipped with a mechanical speedometer. Cadillac and Overland soon followed, and speedometers began to regularly appear as a factory-installed option in new automobiles. Speedometers in this era were difficult to read in daylight and, with no lamp in the housing, virtually illegible at night. The drive cable in early models was attached to either the front wheels or the back of the transmission, but the integration of the drive cable into the transmission housing wouldn't happen for another 20 years. After that improvement was made, the basic technical design of a speedometer would remain untouched until the advent of the electronic speedometer in the early 1980s.

Raw Materials

Materials used in the production of speedometers vary with the type of gauge and intended application. Older mechanical models were entirely comprised of steel and

Early Romans marked the wheels of their chariots and counted the revolutions, estimating distance traveled and average daily speed.

other metal alloys, but in later years about 40% of the parts for a mechanical speedometer were molded from various plastic polymers. Newer electronic models are almost entirely made of plastics, and design engineers continually upgrade the polymers used. For example, the case of a speedometer's main assembly is usually made of nylon, but some manufacturers now employ the more water-resistant polybutylene terephthalate (PBT) polyester. The worm drive and magnet shaft are also nylon, as is the speedometer's gear train and spindles. The glass display lens of the recent past is now made of transparent polycarbonate, a strong, flexible plastic that is resistant to heat, moisture, and impact.

Design

In a mechanical speedometer, a rotating cable is attached to a set of gears in the automobile's transmission. This cable is directly attached to a permanent magnet in the speedometer assembly, which spins at a rate proportional to the speed of the vehicle. As the magnet rotates, it manipulates an aluminum ring, pulling it in the same direction as the revolving magnetic field; the ring's movement, however, is counteracted by a spiral spring. Attached to the aluminum ring is the pointer, which indicates the speed of the vehicle by marking the balance between these two forces. As the vehicle slows, the magnetic force on the aluminum ring lessens, and the spring pulls the speedometer's pointer back to zero.

Electronic speedometers are almost universally present in late-model cars. In this type of gauge, a pulse generator (or tach generator) installed in the transmission measures the vehicle's speed. It communicates this via electric or magnetic pulse signals, which are either translated into an electronic readout or used to manipulate a traditional magnetic gauge assembly.

The Manufacturing Process

Steel components

1 To form molten steel, iron ore is melted with coke, a carbon-rich substance that results when coal is heated in a vacuum. Depending on the alloy, other metals such as

aluminum, manganese, titanium, and zirconium may also be introduced. After the steel cools, it is formed into sheets between high-pressure rollers and distributed to the manufacturing plant. There, the individual parts may be cast into molds or pressed and shaped from bar stock by large rolling machines.

Plastic components

2 The various plastics that arrive in an instrument manufacturing station were first created from organic chemical compounds derived from petroleum. These polymers are distributed in pellet form for use in the injection-molding process. To make the small parts for a speedometer assembly, these pellets are loaded into the hopper of a molding machine and melted. A hydraulic screw forces the plastic through a nozzle and into a pre-cast mold, where the plastic is allowed to cool and solidify. The parts are then gathered and transported to assembly stations.

Assembly

3 The manner of assembly and degree of human interaction depends on the quality of speedometer. Some inexpensive speedometer systems are made to be "disposable," meaning that the instruments are not built for easy disassembly or repair. In this case, the hardware is fastened using a process called riveting, in which a headed pin is inserted and blunted on the other end, forming a permanent attachment. Higher-end speedometer systems consist of two major assemblies attached by screws; the advantage is that the inner hardware of the gauge is accessible for repair and recalibration.

4 The inner shaft and speedometer assembly are then fused into place with rivets or screws. The permanent magnets used in mechanical speedometers are compressed and molded before arrival at the plant, and therefore only require mounting onto the worm drive. In the case of electronic speedometers, fiberglass-and-copper circuitry is also manufactured by vendors, and does require programming before it is screwed into the larger system. These larger components are transported to a separate assembly station, where they are mounted into the housing with stud-terminal or blade-terminal plastic connectors. Beyond its prima-

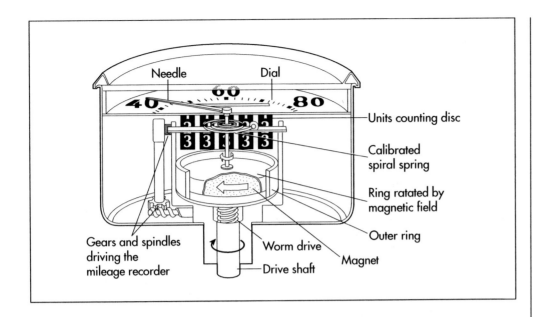

Needle   Dial

Units counting disc

Calibrated spiral spring

Ring ratated by magnetic field

Outer ring

Gears and spindles driving the mileage recorder

Worm drive

Magnet

Drive shaft

ry duty as a protective case, the housing also serves as a platform for attaching exterior features such as the dial face, needle, and display window. Again, these processes require automation due to large output, but human effort is needed at every step to inspect and ensure product consistency.

Calibration

5 Calibration is the process of determining the true value of spaces in any graduated instrument. It is an especially vital process in the manufacture of speedometers because driver safety is reliant on an accurate readout. In a mechanical gauge, magnetic forces produce the torque that deflects the indicator needle. When calibrating this type of gauge, an electromagnet is used to adjust the strength of the permanent magnet mounted in the speedometer until the needle matches the input from the rotating cable. When calibrating an electronic gauge, adjustments are made when calibration factors are written into the memory of the meter. The system can then refigure the balance between input from the transmission and output of the needle. New automated systems for calibrating both mechanical and electronic speedometers are now available, saving an immense number of the man-hours usually required for this process.

Quality Control

Probably the most direct method of quality control is the calibration process. Auto parts manufacturers work under the measurement standards developed by International Organization for Standardization (ISO), which ensures that universal guidelines between gauge manufacturers are used. In-house quality assurance teams develop specifications for each new product before it moves to the assembly line, and the same teams later report whether those guidelines are adhered to on the factory floor. Gradual levels of assembly also involve inspection by factory personnel to make sure that the automation is working smoothly.

Byproducts/Waste

No byproducts result from the manufacture of gauges. Waste materials include scrap metals and plastics, some of which can be reused in later production runs. Because the raw materials involved are prepared outside of the factory, no significant amount of hazardous industrial waste results from manufacture. Emissions from factory automation are government-regulated and surveyed by environmental protection groups.

The Future

Design firms are currently experimenting with improvements in speedometer readout, an effort to eliminate the moment of distraction needed for a driver to look down and gauge his or her speed. Digital readouts projected onto the windshield appear to be the next developmental step. Some prototypes for these speedometers actually make the readout appear as though it is floating over the engine hood. Because this type of display looks as though it is several feet beyond the steering wheel, drivers will be able to continually monitor speed without having to take their eyes off the road. The mirrors and projection devices used in this system could also be adjusted to suit the driver's position, much in the same way that a rear-view mirror does. In addition, speedometer projection systems will eventually be integrated with navigation tools, allowing directional information to appear with gauge readouts.

Where to Learn More

Other

Devaraj, Ganesh, et al. "Automating Speedometer Calibration." *Evaluation Engineering Web Page.* December 2001. <http://www.evaluationengineering.com/archive/articles/1100auto.htm>.

"How a Tachometer/Speedometer Works Using a Magnetic Sensor." Manual. Stewart-Warner Co., April 2001.

"How an Electrical Gauge is Put Together." Manual. Stewart-Warner Co., April 2001.

"How Odometers Work." *Marshall Brain's How Stuff Works.* December 2001. <http://www.howstuffworks.com>.

"Speeding Through Time." *Transport Topics Electronic Newspaper.* November 1998. December 2001. <http://www.ttnews.com/members/printEdition/0000395.html>.

"Speedometer." *Complete Computer Software Web Page.* December 2001. <http://www.iao.com/howthing/Default.htm>.

"The Floating Speedometer." *Siemens.com Web Page.* December 2001. <http://www.siemens.com/page/1,3771,257095-1-999_5_4-0,00.html>.

Kate Kretschmann

Spinning Wheel

Background

A spinning wheel is a machine used to turn fiber into thread or yarn. This thread or yarn is then woven as cloth on a loom. The spinning wheel's essential function is to combine and twist fibers together to form thread or yarn and then gather the twisted thread on a bobbin or stick so it may be used as yarn for the loom. The action is based on the principle that if a bunch of textile fibers is held in one hand and a few fibers are pulled out from the bunch, the few will break from the rest. However, if the few fibers are pulled from the bunch and at the same time are twisted the few pulled out will begin to form a thread. If the thread is let go it will immediately untwist, but if wound on a stick or bobbin it will remain a thread that can be used for sewing or weaving.

Many different kinds of fibers can be spun on a simple spinning wheel, including wool and hairs; bast fibers which come from below the surface of a plant stem including flax (linen), hemp, jute, ramie, and nettle; and seed fibers, particularly cotton. Each of these fibers vary tremendously in length of staple, quality and strength. Different fibers require different kinds of pieces or bobbins placed on the spinning wheel and even call for spinning wheels of different size or configuration in order to spin the specific fiber more efficiently.

History

Man has been spinning fibers for centuries as woven cloth cannot be made without producing yarn or thread. Ancient Egyptians processed flax into linen and surely used the earliest form of spinning apparatus, known as the drop spindle. It was simply a weighted stick onto which yarn was wound with a twist as the spindle was dropped downward, pulling thread from the pack of unspun fibers. There is uncertainty about the development of the spinning wheel as some argue it was developed in China as early as the sixth century for silk and ramie spinning, while others believe it may have developed later in India for cotton. Early Eastern spinning wheels are similar in that the base sat on the ground and the wheel was powered by hand or hand crank. These wheels were rimless and soon spread to the West.

Western examples throughout the Middle Ages were both rimless and had a hoop rim. By the fourteenth century the hoop rim spinning wheels appear to overtake others in popularity. The Flemish who settled in the British Isles brought with them strong textile traditions and with that brought improvements on the traditional spinning wheel. Great wheels with very large driving wheels were known by the sixteenth century in the British Isles for the spinning of wool solely. Endless small variations were made in the wheel to ensure efficient spinning until the present, as some modern manufacturers do not reproduce old examples precisely but make their own wheels that are attractive and effective.

Spinning wheels were one of the first craft tools to be supplanted by modern machinery. Richard Arkwright, an English industrialist, developed a method for machine spinning cotton by the mid-eighteenth century and American Samuel Slater stole the system and brought it to Rhode Island. He began Slater Mill, which began producing the first machine-spun thread in the New World with machinery driven by water power. As machine-spun yarn was commercially available

Perhaps fewer than 1,000 spinning wheels are made yearly for hobbyists within Canada and the Untied States today.

from that point on, fewer used spinning wheels unless it was for small, domestic needs such as the production of wool yarn for knitting from a farmer's sheep fleece. Today, spinning wheels are carved and turned of hardwood and used only by craftspeople for handspun yarns. Spinning wheels are entirely obsolete as large manufacturers use industrial spinners to produce millions of yards of thread or yarn each day. Perhaps fewer than 1,000 spinning wheels are made yearly for hobbyists within Canada and the Untied States today.

Raw Materials

The raw materials for most modern spinning wheels are wood, wood glue, clear lacquer or urethane, and some bits and pieces of metal, primarily used as wire on the wheel. Some spinning wheels use a bit of brass as well. The wheels made on the North American continent are made of native hardwoods. Most consumers are looking for spinning wheels that not only work well but are keepsakes. For this reason, spinning wheels are made from at least three different woods depending on the aesthetic preferences of the purchaser. Maple is easy to acquire and a fine wood to turn and shape but it is not a beautiful wood nor does it take a stain well. For that reason, spinning wheel manufacturers also offer more expensive wheels in woods considered "prize" such as cherry and walnut. Cherry, from the eastern United States, fine-textured and straight-grained. It is favored by many because it may be light pink when first cut but turns a mahogany red as it is exposed to air. It is easily worked and makes a fine spinning wheel; however, it is more difficult to get and more expensive than maple. Walnut, generally American black walnut, is a dark purple-brown in color when applied with a clear lacquer. It works easily and produces a spinning wheel much prized for is beautiful wood.

Design

While there is no single spinning wheel design, the iconic spinning wheel has three splayed legs, a foot treadle connected to a footman that attaches to the driving wheel thus making the wheel go around. A horizontal stock is the wooden plank or bed upon which most of the apparatus rests upon. The driving wheel is perhaps the most prominent feature of the spinning wheel and resembles a wheel with turned spindles in the center (thread is pulled along the outside of that wheel). The bobbin is a grooved wooden spool that fits on the flyer, which gathers the yarn after it is spun. The bobbin fits into a U-shaped flyer (a bracket with hooks on it that guide the yarn onto the bobbin and keep the yarn evenly distributed on it). The distaff and distaff arm hold a batch of unspun fiber. These are only the primary parts of the spinning wheel as the typical wheel has well over 100 small parts that fit together to ensure quality spinning.

The Manufacturing Process

There are different kinds of spinning wheels available for purchase, from small portable spinning wheels, to those that are exact reproductions of early American pieces, to variations on the tradition spinning wheel sometimes referred to as the Saxony Wheel. This essay will concentrate on the manufacture of a modern variation of the traditional wheel with a medium-size driving wheel. It is important to note that this type of wheel is constructed of at least 150 parts; the basic parts' manufacture will only be described below.

1 First, planks of wood enter the factory. These are pieces bundled into planks that 6 in (15.2 cm) wide and about 10 ft (3 m) long. These planks are then rough cut into smaller shapes that can be accommodated on the machinery that will be used to further shape the pieces into more finished parts. The rough cuts are generally performed by hand. Larger pieces are cut into shapes that will become major parts such as the wheel rim, the turnings within the wheel, or the treadle.

2 These rough cut pieces are then clamped into a mechanized cutter and shaper that is computer controlled. This machine is referred to as a computer numerically controlled (CNC) machine, and it may route, shape, or turn these rough cuts into appropriate shapes for the spinning wheel. This machine enables the manufacture to create high quality routed, shaped, or turned pieces for the spinning wheel with minimal hand work. A computer-based program, often programmed by the manufacturer, is fed into the machine and the program moves the machine in order to produce the pieces. A

single machine may be reprogrammed numerous times in order to produce many different parts for the same product. The hub of the wheel or other parts may all be made on the CNC router, shaper, or lathe.

3 The wood pieces are removed from the CNC machine and grouped. These parts are rough to the touch and must be sanded smooth. The turnings (resembling the spokes on a wheel) are put into a lathe and sanded. The flat pieces are then run through either four or five different sanders. These sanders are both mechanical belt and drum sanders, each rendering the piece smoother and readying it for a finish.

4 Some spinning wheels are sold unfinished and are now ready for assembly. Others have a finish applied to the surface. Those that will be finished may be stained, but most consumers prefer the beauty of the natural wood. Thus, a clear lacquer finish may be sprayed upon the surface. A water-based lacquer is applied to these pieces using a spray.

5 All parts are gathered together by hand and readied for assembly. The operator uses furniture-grade wood glue on parts that peg or mortise and tenon through one another. Many of the largest joints are bolted together as well as glued. Sometimes the manufacturer only performs partial assembly, meaning that large parts are assembled together such as the wheel-including the rim, the hub, and spokes or turnings. Thus, the majority of the assembly may be performed by the consumer or the retailer, as shipping an assembled spinning wheel is quite impractical. It is spindly and unstable and could be smashed in shipping. Furthermore, this shipping could be extremely costly as a product of significant volume can be expensive to ship. Thus, spinning wheel manufacturers may well leave the assembly to others at point-of-sale. The company provides simple instructions for the store or consumer.

Quality Control

The quality control issues primarily revolve around the grade of the wood used in the manufacture of the product. North American manufacturers generally negotiate for the delivery of wood from reliable lumber suppliers who can provide goods free of knots, bug damage, and are of the minimum lengths de-

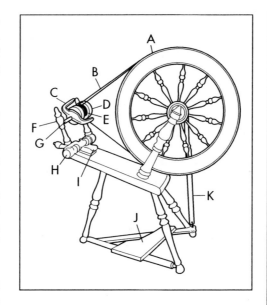

A. Fly wheel. B. Drive band. C. Flyer. D. Flyer whorl. E. Bobbin. F. Maidens. G. Orifice. H. Mother-of-all. I. Tension knob. J. Treadle. K. Footman.

sired. A Canadian manufacturer receives wood from the most economical and reliable suppliers and thus gets different types of woods from all over the continent, from Pennsylvania to distant parts of Canada.

The CNC machine renders parts that are only as good as the programming fed into it. Thus, the manufacturer ensures that the programmer produces a program that is fully compatible with economical manufacture and easy assembly. However, when the program is successfully designed and implemented, the machine is able to make the requisite pieces almost without end. The machines are extraordinarily reliable. Human operator error (problems with clamping or securing the pieces in the machine) or poor-quality pieces of wood (knots or other imperfections) may pose problems but are generally very minor.

Byproducts/Waste

There is a fair amount of wood waste after pieces are routed, shaped, and turned on the CNC machine. A manufacturer may sell the wood chips to a chipboard manufacturer for composite wood for engineered wood furniture. The wood waste may also go for sawdust bedding for animals.

Furniture and other wood products manufacturers are quite concerned about harmful vapors or effluvia that result from wood finishing of their products. Thus, spinning wheel manufacturers may prefer to use the water-based finishes as they do not leave harmful volatile organic compounds, otherwise known as VOCs, the use of which is monitored by the federal government.

The Future

The manufacture of spinning wheels is currently an interesting combination of traditional designs and streamlined manufacturing. The North American manufacturers do not make more than a few thousand pieces a year, and share the market primarily with New Zealanders who have a long history of wool processing and spinning wheel craftsmanship. Thus far, these North American manufacturers do not feel threatened by foreign competitors. However, the viability of the manufacture of wheels rests solely on the vitality of the craft of spinning by those whose hobbies include textile production. Spun yarn is easily available to all cheaply and easily and does not require the use of what is essentially an out-moded spinning wheel in order to acquire spun yarn.

Where to Learn More

Books

Baines, Patricia. *Spinning Wheels: Spinners and Spinning.* New York: Charles Scribners Sons, 1977.

Nylander, Jane. *Our Own Snug Fireside.* New Haven: Yale University Press, 1994.

Other

Lendrum Web Page. December 2001. <http://fox.nstn.ca/~lendrum/instructions.htm>.

Oral interview with Gord Lendrum, owner of Lendrum Spinning Wheels. Odessa, Ontario. October 2001.

"Spinning Wheel." *Encylopedia Britannica CD Edition.* Encyclopedia Britannica, Inc., 1994–1998.

Nancy EV Bryk

Spork

Background

A spork is an eating utensil designed with features of both a spoon and a fork. The overall shape is similar to a spoon complete with a handle and a small bowl-like structure at the end. At the very end of the spork are short tines that are useful for picking up solid food. Also called a runcible spoon, sporks are often provided by take-out restaurants as disposable utensils. They are generally composed of plastic and made through a thermoforming process.

History

Eating utensils have slowly developed over centuries. Early humans used naturally occurring sharp stones to scrape and cut foods. When enough of these were not available, they learned to sharpen dull stones into suitable shapes. These stones represent the earliest known knives. Ancient coastal populations would use shells attached to sticks for eating hot liquids. Other populations used hollowed horns from sheep for similar purposes. These applications of natural products demonstrate the early development of the spoon.

Over the years, the design and production of eating utensils became more efficient. Different materials such as wood, ivory and metal were often used. Knives and spoons were the most common types of eating utensils throughout Europe for years. In fact, the term spoon is thought to come from the ancient Anglo-Saxon word "spon," which means splinter or chip of wood. This reflects the fact that by the fifth century spoons were fashioned from wood.

One of the earliest known forks dates back to the time of the ancient Greeks. These utensils were larger than the modern fork and had only two tines. By the seventh century, Middle Eastern royalty used smaller forks regularly at the dinner table. In Europe, the regular use of forks was slow to be adopted. The attitude of many people was that food should be eaten with the hands and that forks were unnecessary. It was not until the sixteenth century that forks were widely available and used in Italy. In subsequent years, forks were introduced in France and England. By the mid-eighteenth century, forks had achieved the form most commonly used today.

The idea of combining spoons with forks is not new. One of the first patents issued in the United States for such a product was done in 1874. In this patent, a device is described that has a handle, spoon bowl, knife-edge, and fork tines. This is said to be the basic design for all future combination eating utensils. Through the next decades, improved products and materials were patented. One patented design had a deeper bowl and shorter tines that made it useful for eating liquids. Plastics were adopted as the construction material of choice during the 1940s and 1950s. The term Spork was introduced in a patent issued in 1970 to the Van Brode Milling Company.

Raw Materials

Sporks can be made from all types of materials including steel, wood, glass, and plastic. By far the most often used material is plastic, specifically polypropylene and polystyrene. These materials are combined with other additives to create the finished utensil. It is important to note that all the materials used in spork manufacture are regulated by

One of the earliest known forks dates back to the time of the ancient Greeks. These utensils were larger than the modern fork and had only two tines.

the United States Food and Drug Administration (FDA) to ensure that they are safe for contact with food.

Plastics are high molecular weight materials that are produced from monomers through a process called polymerization. These monomers like ethylene and propylene are ultimately derived from oil and natural gas. In a process called the "cracking process" crude oil or natural gas is heated to convert the constituent hydrocarbons into reactive monomers.

For spork manufacture, polypropylene (PP) and polystyrene are often used. PP is produced from a polymerization reaction of propylene monomers. It is said to have excellent chemical resistance and is used for many types of packaging. It is ideal for spork manufacture because it is resistant to degradation by water, salt, and acids, all of which are destructive to metals. Solid polystyrene also shows good chemical and temperature stability. It is made through the polymerization reaction of styrene monomers. Styrene was first produced commercially during the 1930s and was important during World War II as a constituent in synthetic rubber production.

In addition to the base polymeric material, other modifiers are added to change the characteristics of the material, improve the stability, and make manufacturing easier. Since the bulk polymer is typically colorless, colorants are added to make sporks more appealing. These may be soluble dyes or comminuted pigments. To produce a white color, an inorganic material such as titanium dioxide may be used.

A host of other filler materials are added to produce high quality sporks. For example, plasticizers are added to increase the workability and flexibility of the polymer. Plasticizers are nonvolatile solvents and include things such as paraffinic oils or glycerol. Since the plastic is typically heated during production, stabilizers are also added to protect the plastic from breaking down. An unsaturated oil such as soy bean oil may be used as a heat stabilizer. Other protective materials that are added include ultra violet protectors and antioxidants. These materials help prevent degradation of the plastic due to environmental effects. Finally, compounds such as ethoxylated fatty acids or

silicones are used to aid processing during manufacture. These materials make it easier to remove the plastic from the mold.

Design

A spork is an eating utensil with combined elements of both a spoon and a fork. It has a handle portion that is gripped by the user. At the end of the handle a small, curved, bowl-like structure is attached an at the very end of this curved structure are small pointed tines. In some designs, inventors also incorporate a sharp edge in the spork so it can also be used as a knife.

The Manufacturing Process

Creating a plastic sheet

1 At the start of spork manufacturing, a plastic sheet is formed through an extrusion process. In this phase of production, pellets of polymer are fed into a large bin attached to a thin, flattened opening. As the pellets are moved through the extruder, a hydraulic screw crushes and melts them at temperatures between 400–550°F (204–288°C) forming a thick, semi-solid liquid. The polymer is pushed through an opening and extruded as a thinned mass. It is then run under a series of water-chilled rollers to thin it further and increase its width and length. At the end of this process, it can be rolled onto large spools or cut into sheets.

Thermoforming

2 As with most plastic products, a mold is created for production of a spork. The mold is a steel block with a carved cavity that has the inverse shape of the spork. When softened, warmed plastic is pulled into the mold, it acquires the mold's shape upon cooling. To make removing the spork from the mold easier, special silicon release agents may be used to coat the cavity.

When a mold is designed, the cavity is highly polished to remove any flaws on the surface. A single flaw on the surface could be reproduced in the final product thereby ruining an entire production run.

3 The plastic sheets are next loaded onto a thermoforming machine. The sheet of plastic is moved over the spork mold and

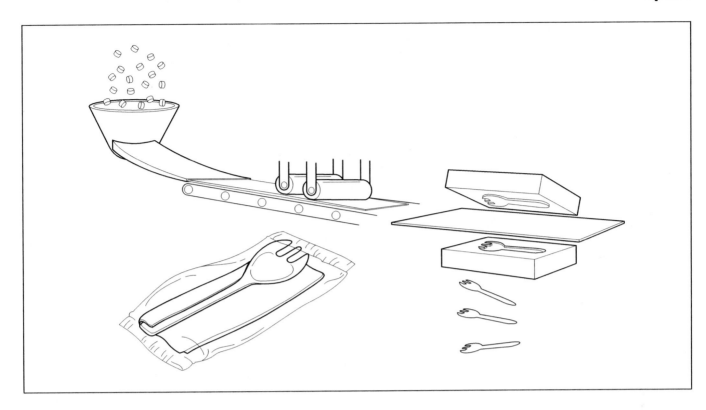

clamped in place. Situated above the mold is a heater that softens the plastic.

4 Air is drawn through a few holes at the bottom of the mold, creating a vacuum. This drop in pressure causes part of the softened plastic sheet to be pulled down into the mold.

5 When the plastic contacts the mold, it cools and hardens. This causes it to maintain the shape of the mold, which in this case is that of a spork. A large metal stamping plate is moved on top of the mold. This cuts the individual spork from the rest of the plastic sheet and also provides an opportunity to emboss a logo or design on the spork handle. The sporks are then ejected from the molds and moved to the next phase of manufacture. The unused plastic is moved to a separate area to be reground and reused.

Assembly and packaging

6 Sporks are sold in a variety of packages. A common method of packaging is to have a spork wrapped in a plastic film with a napkin. In this case, the sporks are moved to a large holding bin that is attached to a placing machine. A continuous, thin plastic film moves under a machine where a napkin is placed on it. After the napkin is set, a spork is placed on it. If salt or pepper packets are to be included in the package, those are also laid down by another device.

7 Next, the filled film is passed under a folding and stamping machine. This device is designed such that the film is folded over the spork and napkin as it enters. It is then stamped on both ends creating a cut, sealed package. The individual package is then transferred to a storage case. When the case is adequately filled, it is sealed with tape and stacked on a pallet for shipping.

Quality Control

Various quality control measures are taken to ensure that each spork produced meets specified standards. These include both laboratory testing and line inspections. Prior to manufacturing, the physical and chemical properties of the starting materials are determined. For example, molecular weight and chemical composition determinations are done on the plastics. Also, visual inspections of the color, texture and appearance are performed. The physical performance properties may be evaluated as well. Stress-strain testing may be done to ensure the plastic is durable. This is particularly important for spork manufacture because the products must be strong enough to pick up solid food. Quality control technicians in a laboratory generally do this testing.

On the manufacturing floor, line inspectors are placed at various points along the production line. They visually inspect the plastic utensils making sure the size, shape and colors are correct. They also check the products in the final packaging ensuring that each spork shipped is of an acceptable quality. When defective sporks are found they are taken out of the production line a set aside for reforming.

Byproducts/Waste

In the manufacture of sporks, the primary waste product is unused plastic. The disposal of this material is managed through a system that includes source reduction, recycling, waste-to-energy conversion, and landfilling. During the manufacture of sporks, source reduction is the primary method for reducing plastic waste. This is achieved by reusing plastic from misshapen products.

The other phases of waste disposal address the finished products themselves. Since sporks are designed to be disposable they are destined to become waste. Some of this waste ends up in the recycling system and gets used for different recycled product applications. Other parts of this waste end up being used in incinerators to convert it from waste to energy. During this energy conversion, a polymer like polystyrene produces carbon dioxide, water vapor, and trace amounts of non-toxic ash. The final resting spot for many disposed sporks is in landfills.

The Future

In the future, spork manufacturers will likely concentrate on improving production efficiencies and increasing sales. From a production standpoint, research is focusing on increasing manufacturing speeds, reducing raw material costs, and minimizing chemical waste. For example, one raw material supplier has introduced a plastic substitute made from wheat gluten resin that is durable enough for spork manufacture but is also biodegradable. Another supplier has introduced a soybean product that has similar characteristics. To increase sales, spork manufacturers will focus on getting more fast food restaurants to use their product. Sporks present an opportunity for these restaurants to cut costs by eliminating the need to carry both spoons and forks.

Where to Learn More

Books

Chabot, J. *The Development of Plastics Processing Machinery and Methods.* Society of Plastics Engineers, 1992.

Giblin, James Cross. *From Hand to Mouth, Or, How We Invented Knives, Forks, Spoons and Chopsticks, and the Manners to Go with Them.* New York: Crowell, 1987.

Petroski, Henry. *The Evolution of Useful Things.* New York: Vintage Books, 1994.

Seymour, R., and C. Carraher. *Polymer Chemistry.* New York: Marcel Dekker, Inc., 1992.

Other

Albanese, Joseph. *U.S. Patent 4,984,367 Combination Utensil.* 1991.

Perry Romanowski

Spray Paint

Background

Spray paint is an aerosol product designed to be dispensed as a fine mist. Compared to conventional brush methods of painting, spray painting is faster and provides a more uniform application. While industrial spray painting relies on special air compressors that break the paint particles into a fine mist, commercial spray paints are self-contained aerosol cans that use liquefied gasses to atomize the paint.

History

The art and science of painting date back more than 30,000 years. Primitive humans painted crude depictions of their lives on cave walls that are still visible today. Over the centuries, as improved methods and materials were developed, painting evolved as both a way of expressing art and as a functional tool. In 1700, the first recorded paint mill in America was built in Boston by Thomas Child. The first ready to use paints for the consumer were developed more than 150 years later by D. R. Averill in Ohio.

While these paints were commercially desired by consumers, they were also very expensive to ship around the country because of their weight. As methods of mass production became increasingly available, manufacturers learned how to make paint more efficiently. Small factories began springing up all across the country. This system of small, decentralized manufacturing plants allowed manufacturers to sell paint across the nation. This system persisted in the industry until the mid twentieth century.

In the 1940s, the paint industry took another step forward with the invention of the aerosol can. Originally developed by the military as a tool to dispense insecticide, aerosol systems were quickly adapted to other product categories including spray paint. In 1948, the Chase Company in Chicago became one of three businesses licensed by the United States Department of Agriculture to make aerosol mosquito repellents. Using similar technology and equipment, a few years later they became the first commercial producers of spray paint.

Since its birth in the 1950s, the spray paint industry has enjoyed considerable success but has also met with many challenges. In the late 1970s, legislators banned paints from using chlorofluorocarbon propellants (CFCs) due to the role these solvents are though to play in atmospheric ozone depletion. In the late 1990s, the California Air Resource Board (CARB) began imposing limits on the amount of Volatile Organic Compounds (VOCs) that can be used in spray paint. VOCs have been shown to contribute to air pollution. These regulatory mandates have dramatically affected the quality of spray paint formulations. In spite of these challenges, spray paints continue to be popular consumer commodities. In 1997 nearly 25 million gal (94 million l) of spray paint where produced in the United States alone.

Raw Materials

Pigments

Pigments are used in spray paint to provide color and opacity. There are four basic types of pigments used in spray paint. White pigments such as titanium dioxide are used to scatter light and make the painted surface more opaque. Color pigments, as the name implies, provide color to the paint mixture.

In 1700, the first recorded paint mill in America was built in Boston by Thomas Child. The first ready to use paints for the consumer were developed more than 150 years later by D. R. Averill.

These include a variety of synthetic chemicals. Inert pigments are used as fillers that alter the film characteristics of the paint. Finally, functional pigments provide extra performance characteristics such as imparting protection from ultraviolet rays.

Pigments must be chosen carefully because they can also affect certain formulation characteristics such as viscosity. If the pigments are not properly dispersed they may agglomerate, that is they may come together to form larger clumps that will settle to the bottom of the container. When this occurs the pigments are not able to be separated into small enough particles to spray through the valve.

Solvents

Solvents are the liquids that carry the rest of paint ingredients. While water is a good solvent for many materials, it is slow in drying and tends to cause corrosion in metal cans. Therefore, non-aqueous, quick drying solvents are used. Solvent selection can also affect the stability of the pigment dispersion. Some solvents may absorb on to the outer layer of the particle and cause it to swell— this interaction helps stabilize the dispersion. Other types of solvents, on the other hand, can have a negative effect upon the pigment dispersion. If the solvents completely cover the surface of the particle they can prevent the interaction of other ingredients and may actually destabilize the formula.

Propellants

Propellants are gasses that force the paint out of the can by expanding rapidly when the valve is opened. Chlorofluorocarbon gasses (CFCs) were originally used as propellants but these were banned from use in 1978 because it was discovered that they deplete the ozone layer. Other gasses like butane and propane were used as replacements for CFCs. These hydrocarbons are classified depending on the amount of pressure they create in the can. Butane 40, for example, is a mixture of butane and propane and has a vapor pressure of 40 psi (2.8 kgf/cm^2) per square inch. Hydrocarbon propellants were used as primary propellants until the 1980's when the California Air Resource Board determined that these chemicals contribute to smog. They passed regulations that limited

the amount that could be used in spray paint. To solve these problems, a new class of propellants known as Hydrofluorocarbons (HFCs) where developed for use in aerosols. These include and 1,1,-difluoroethane (Propellant 152A) and 1, 1, 1, 2,-tetrafluoromethane (Propellant 134A).

Other Ingredients

Other ingredients are included in the formula to stabilize the pigment dispersion, to control pH and viscosity and to prevent corrosion in the can.

Packaging

Spray paints are packaged in tin plated steel or aluminum cans. The can is sealed with a valve that controls how the paint is dispensed. The top of the valve is a button that controls the shape of the spray; it is attached to the valve body that acts as a mixing chamber for the liquid paint concentrate and the propellant. At the bottom of the valve a plastic tube is attached that carries the paint upward from the bottom of the can.

The Manufacturing Process

Batching the concentrate

1 The first step in manufacturing aerosol spray paint is to prepare the liquid concentrate in large metal or glass tanks. This process involves mixing the liquid ingredients such as solvents, corrosion inhibitors, and pH and viscosity control agents with large impeller type mixers driven by electric motors.

Dispersing the pigments

2 The critical step in the manufacturing process is ensuring that solid pigment particles are properly dispersed. Care must be taken to ensure that the liquid displaces all the air surrounding the particles. Simple mixing with a propeller blade is not enough to disperse the pigments, so special mixing equipment such as a ball mill is used. A ball mill is a circular container, like a drum, that is filled with ceramic or stainless steel balls. The dry pigments are mixed with some of the paint concentrate to form a slurry that is poured into this drum. The drum is then placed on a pair of rotating metal rollers; as

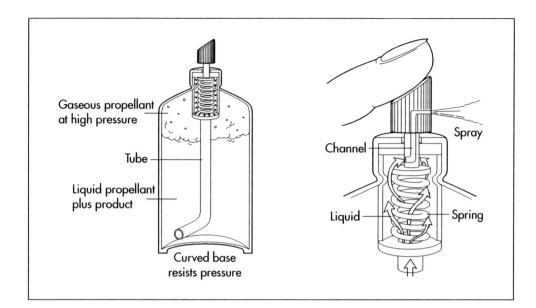

Gaseous propellant at high pressure

Tube

Liquid propellant plus product

Curved base resists pressure

Channel

Spray

Liquid

Spring

it spins around, the balls tumble in the drum and break apart the pigment particles.

Another type of mixer that can be used in this process is a roller mill that consists of two closely spaced rotating metal cylinders. The pigment slurry is passed through the rollers as they rotate against each other. The pigment particles are broken apart by the action of the rollers until only the smallest particles passed through the spacing—the larger aggregated particles are broken apart.

3 Once the pigments have been properly wetted, the slurry can be added to the remainder of the liquid concentrate in the batching tank. This mixture is then stirred until it is homogenous. At this point in the process a sample of the paint concentrate may be taken to check for consistency and color. If the color does not match appropriately additions can be made to the batch to adjust the color. Adjustments can be made to increase the pigment load to boost the color or to add more solvent to dilute it. Once it is known that the batch meets the appropriate specifications it can be transferred to a filling tank.

Filling process

4 The filling processes used for aerosols is highly automated. The empty cans travel down an conveyor belt to reach the filling equipment. Jets of compressed air blast away any dust or dirt that may be in the cans before they are filled with the concentrate by the fill-

ing heads. These heads are a series of nozzles that are connected to tubes that transfer the paint from the filling tank. A piston mechanism controls how much liquid is injected into the can. After filling the cans proceed down the assembly line to a gassing device that injects liquefied propellant into the can and then immediately crimps the valve against the rim of the can to seal it shut.

5 After gassing, the cans travel through a trough of hot water so they can be observed to check for leaks. If the can has a hole in it or if the valve is not sealed properly a small stream bubbles will be visible in the water bath. Faulty cans are removed and discarded. After passing through the water trough, the cans are dried with more compressed air. At the end of the assembly line an overcap is fitted over the valve to protect the aerosol from accidental activation. Finally the cans are packed into cartons and placed on pallets for shipping.

Quality Control

The quality of the spray paint product is evaluated at several stages. During batching, the concentrate is checked to ensure it is the proper shade. This may be done simply by visually comparing a sample of the fresh batch to an approved standard. A small amount of the paint can be spread on a white background to aid in this comparison. In addition, more sophisticated colormetric or photometric instrumental methods of analysis may be used. Analytical test methods, such as the Daniels Flow Point Test, are used to ensure that the paint dispersion will be stable. During the aerosol filling process random samples are pulled from the assembly line to be checked. Critical evaluations include fill weight, the solids concentration, and the pressure of the can. Spray rate (the amount of paint delivered per unit time) and spray pattern (the size and shape of the spray) are carefully evaluated as well. After manufacture is complete, accelerated aging studies may be done to ensure that the cans will spray without clogging and that the inside of the cans remain free from rust.

The Future

The aerosol spray paint industry faces a variety of future challenges involving both marketing and technical issues. As the market has matured, manufacturers struggle to find new ways to market their products. The Krylon company (a division of Sherwin-Williams) is gearing future marketing efforts in two new areas. One new product line is aimed toward woman and children with paints that offer bright new colors, enhanced washability, and a new fresh fragrance. The other line is targeted toward specific home contractor applications such door/shutter paint, vent paint, and tread and grip paint.

In addition to marketing challenges, future paint formulators will have to continue to search for ways to reduce cost or improve performance. Examples of future technology can be found in two new formulation approaches. One deals with two new solvents that improve the appearance of the paint film after drying and the other involves a reduced pigment-fill-to-binder ratio that improves surface coverage. Finally, other future challenges that aerosol spray paints face include environmental regulations meant to control VOC emissions and global warming, legal issues regarding safety labeling, and continued product abuse by graffiti artists.

Where to Learn More

Books

Johnsen, Montfort A. *The Aerosol Handbook.* Wayne Dorland Company, 1982.

Periodicals

Johnsen, Montfort A. "Aerosols—The VOC Challenge Moves into the 21st Century." *Spray Technology* 11 (1999): 21.

Other

"Economic Value of Paints and Coatings." *National Paint & Coatings Association Web Page.* December 2001. <http://www.paint.org/ind_info/value.htm>.

Randy Schueller

Stereo Speaker

Background

A loudspeaker or speaker is a device that converts electrical energy waves into mechanical energy waves or audible sounds. Sound is produced by the vibration of an object. This vibration sets up a series of ripples or waves much like what is seen when a stone is thrown into a pond. Speakers reproduce sound waves (or audio) at various frequencies. The frequency is the rate at which the particles in the air vibrate. Sound that the human ear can hear is from about 20 hertz (Hz) to 20,000 Hz or 20 kilohertz (kHz). Speakers are used in all types of communications and entertainment equipment such as radio and television receivers, tape recorders, telephone answering machines, baby monitors, and stereo home entertainment systems

History

The basic principle of a dynamic speaker has changed little since it was patented by Ernst Siemens in 1874. Siemens described his invention as a means for obtaining mechanical movement of an electrical coil from electric currents that flowed through it. The original intent of his invention was to move a telegraph arm. Alexander Graham Bell applied the principles of Siemens device to the telephone two years later. Thomas Edison is credited with inventing the loudspeaker as it is known today. It consisted of a flexible diaphragm (cone) attached to the throat of an acoustical horn.

The cones of early loudspeakers used various materials such as thin metal sheets, leather, and paper. Paper was (and still is) used in the construction of speaker cones because it is cheap and readily available.

Raw Materials

The dynamic speaker has not changed in decades. The frame is made from stamped iron or aluminum. The permanent magnet is a ceramic ferrite material consisting of iron oxide, strontium, and a ceramic binder. The cone, surround, and spider are made of treated paper coated with an adhesive glue. The voice coil consists of a plastic bobbin with fine gauge insulated copper wire wound around it.

Design

The most common speaker is the dynamic speaker. It consists of a frame, permanent magnet, soft iron core, voice coil, and cone. The frame supports the cone and permanent magnet assembly. The voice coil consists of an insulated wire wound around a plastic bobbin. One end of the bobbin is attached to the cone and the body of the bobbin slides over the soft iron core.

The wires from the voice coil are connected to an audio amplifier. When electrical audio signals from the amplifier are applied to the voice coil, an electromagnetic field is produced around the voice coil. This causes the voice coil to move back and forth along the soft iron core which aids or opposes the magnetic field produced by the permanent magnet. The movement of the voice coil causes the attached cone to vibrate and produce sound.

There are four major types of speakers: full-range, tweeter, midrange, and woofer. The full-range speaker can reproduce most of the audio sound spectrum. However, a single speaker cannot accurately reproduce the entire audio frequency range of the human ear.

Sound that the human ear can hear is from about 20 hertz (Hz) to 20,000 Hz or 20 kilohertz (kHz).

Akio Morita.

Akio Morita was born January 26, 1921 in the Japanese village of Kosugaya. Morita would have been the fifteenth generation heir to his family's 300 year-old *sake* business, but instead entered the prestigious Eighth Higher School as a physics major.

Rather than be drafted into World War II, Morita entered Osaka Imperial University, agreeing to serve in the Japanese Imperial Navy after graduation. There Morita met Masura Ibuka, an electronics engineer. After the war, Morita and Ibuka created the Tokyo Telecommunications Engineering Corporation with only $500 and 20 employees. In 1953, Morita bought the rights to the transistor, a miniature electronics circuit developed by the American company Bell Laboratories and considered to be impractical. Within two years, Morita and Ibuka created the AM transistor radio. In another two years, they began to produce the pocket-sized transistor radio, AM-FM transistor radio, and first all-transistor television set.

In 1958, Morita changed the company's name to Sony Corporation and moved to New York to set up United States operations. Sony became the first foreign-owned business to offer stock for sale in the United States, and in 1970 became the first Japanese company listed on the New York Stock Exchange.

During the 1980s and early 1990s, Morita wrote two books dealing with careers in business and international business trade: *Made in Japan* (1986) and *The Japan That Can Say No* (1991). In 1994, he retired at age 73, after suffering a debilitating stroke. Morita died of pneumonia in Tokyo on October 3, 1999.

Therefore, other speakers were designed to overcome the limitations of the full-range speaker.

The tweeter is designed for the higher audio frequencies or treble sounds in the 4–20 kHz range. They are very small, usually around 2 in (5.1 cm) in diameter or under. The midrange speaker reproduces sounds in the 1,000 Hz to 10 kHz frequency range. Their size ranges from 2 in to 8 in (5.1–20.3 cm) in diameter. The woofer reproduces the bass or very low sounds in the 20–1,000 Hz frequency range. A subwoofer can extend this to as low as 3 Hz. Woofers range from 4 in to 15 in (10.2–38.1 cm) in diameter with 10–12 in (25.4–30.2 cm) in diameter being the most common.

The Manufacturing Process

1 The permanent magnet is constructed by mixing iron oxide with strontium and then milling the compound into a very fine power. The power is mixed with a ceramic binder and then closed in a metal die. The die is then placed in furnace and sintered to bond the mixture together.

2 The frame is constructed from an aluminum or steel sheet. The sheet arrives at the plant preformed. It is then placed on a conveyor belt and transported to a cutting machine that used a hydraulic press to cut holes in the sheet to allow free air movement from the cone. The sheet is then formed by a hydraulic press that forces the sheet into a die of the desired shape. Mounting holes are then drilled at their proper locations.

3 The cone, surround, and spider are individually formed out of composite paper and then glued together as an assembly.

4 The voice coil is built by winding many turns of very fine insulated copper wire on a plastic bobbin. The bobbin and voice coil assembly is glued to the dust cap of the cone assembly.

5 The frame, soft iron core, and permanent magnet are bolted together as an assembly.

6 The cone assembly is then attached to the frame assembly by first manually gluing the spider to the base of the frame and then gluing the surround to the top of the frame.

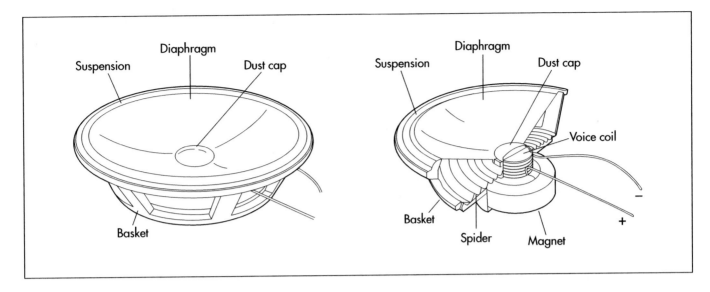

Suspension — Diaphragm — Dust cap

Basket

Suspension — Diaphragm — Dust cap

Voice coil

Basket — Spider — Magnet

−

+

A stereo speaker.

Quality Control

Inspectors monitor all steps of the manufacturing process. The permanent magnet is checked for chips or cracks. The paper cones are inspected for flaws or holes in the material, and proper gluing of the cone assembly. The entire assembly is inspected for overall quality and adherence to specifications.

The final speaker assembly is connected to an audio generator that tests the frequency response and power capabilities of the speaker to insure that in produces sound within the required specifications.

Byproducts/Waste

Scrap metal from the frame and scrap paper from the cone are sent to recycling plants for reclamation. Exotic materials such as strontium must be disposed of according to government regulations.

The Future

The industry is constantly experimenting with raw materials to improve speaker sound quality, frequency response, and power output. New technologies are being developed in the form of a direct-drive speaker that uses no cone.

Where to Learn More

Books

Prompt Publications. *Speakers for Your Home and Automobile*. Indianapolis: 1992.

Other

McIntosh Laboratory Web Page. December 2001. <http://www.mcintoshlabs.com>.

Ernst S. Sibberson

Stereoptic Viewer

In 1999, View-Master celebrated its birthday, having produced more than 1.5 billion reels in 60 years.

Background

The stereoptic viewer is a toy with a relatively simple plastic body, but also a sophisticated lenses for looking at a pair of photographic transparencies mounted, along with six other pairs, in a flat paper reel. Each so-called stereo pair has a photo viewed through the left eyepiece and another viewed through the right. The photos are slightly different. The brain merges the images seen by the eyes to give them depth (also called a three-dimensional or stereo effect).

History

The human urge to see three dimensional (3-D) pictures of the world began with the ancient Greeks. Euclid, the mathematician who established the principles of geometry, proved that the right and left eyes see slightly different views. In the sixteenth century, Jacopo Chimenti, a painter from Florence, Italy, made pairs of drawings—called stereo pairs—that, when viewed together, produced 3-D images. In 1838, Sir Charles Wheatstone patented a stereo viewer that used a complex series of mirrors to look at pairs of drawings. The invention, improvement, and popularity of photography during the period from 1790 to 1840 revived interest in 3-D views because photos can be more easily reproduced than drawings. In 1844, a camera for taking pairs of stereo photographs was created in Germany. Sir David Brewster, the Scottish physicist who also invented the kaleidoscope, used prismatic (mirror-like) lenses to make a compact stereo viewer that became known as the stereoscope.

Sets of stereoscopic slides of the area that was to become Yellowstone National Park were given to members of Congress in 1871, convincing them to approve the first national park. News events were featured on the slide sets, so scenes of the building of the Panama Canal, the World's Fairs in Chicago and St. Louis (1892 and 1904, respectively), and the Great San Francisco Earthquake (1906) could be seen. From 1870 forward, local commercial photographers made slides of stores, farms, and even family gatherings.

The immediate predecessor to the 3-D reel viewer was the filmstrip viewer, developed in the 1920s. The Tru-Vue Company began manufacturing these viewers in 1931 using filmstrips with 14 stereo frames each. Meanwhile, in 1939, William Gruber and Harold Graves invented the View-Master viewer and a system that used reels to hold the stereo photos. Sawyer's, a photofinisher and card manufacturer in Oregon, financed the Gruber-Graves viewer that was introduced in 1940. During World War II, department stores sold the increasingly popular products, and Sawyer's began packaging the reels in three-packs.

Tru-Vue began producing "stereochrome" filmstrips in color in 1951 and acquired the exclusive license to use 3-D images of Walt Disney cartoon characters. Sawyer's bought out Tru-Vue and expanded the reels to include Tru-Vue's Disney characters. In 1966, Sawyer's was purchased by General Aniline & Film Corporation (GAF). Called the View-Master International Group by 1981, the firm bought the Ideal Toy Company and became View-Master Ideal, Inc. (V-M Ideal). In 1989, Tyco Toys bought V-M Ideal. The next merger did not occur until 1997 when Tyco joined Mattel, Inc.; View-Master became a part of Fisher-Price, a Mattel subsidiary.

Raw Materials

The viewer has two basic parts, the viewer itself and the reel with the photographs. The reel also has two primary components, the outside supporting structure and the photos. The outside is paper laminated (layered) with polyethylene film; this patented product is called Lamilux. The paper is delivered to the factory in huge rolls; thousands of reels are stamped from a single roll. Four-color, printed paper labels are also made outside. The labels are backed with adhesive and mounted on rolls; these "crack-and-peel" labels are like self-adhesive postage stamps, and the adhesive remains moveable temporarily and bonds later.

The pictures mounted in the reels are transparencies. A film-processing house mass-produces the transparencies on 16-mm (0.63-in) film.

The viewer is made of three different kinds of plastic. The body is polystyrene, a high-quality plastic that withstands impact, shattering, and other stresses. The advance lever is acetal plastic that is also strong with good dimensional stability and stiffness. The viewer holds four lenses of optical-grade, clear acrylic plastic. Acrylics are also strong and resist change so the lenses remain clear and focused. The three types of plastics are received at the factory in small pellets and are pre-colored.

The viewer contains a metal extension spring that returns the advance lever after each advance of the reel. The extension spring is made of music wire and is a finished part delivered to the factory.

Packaging materials are furnished by outside suppliers and include card and cardboard sheet stock and thin sheets of polyvinyl chloride (PVC) plastic that will be vacuum-formed into "blisters" in the shapes of the products to make display packages. The paper supplier applies heat-sensitive adhesives to the card stock, but printing for packages containing the reel sets is done in the factory.

Design

A representative, basic viewer resembles a small pair of binoculars enclosed in a colorful plastic housing. A slot at the top of the viewer where the focus adjustment for binoculars would be is the opening for the photo reel. A lever extends from the right or the top; it slides down a narrow channel to advance the photo reel and pops back up when the lever is released. The outsides of the lenses on the front of the viewer have looked like recessed binocular lenses. The lens eye openings at the back of the viewer are approximately 0.5 in (1.3 cm) in diameter and are set into eyepieces. The eyepieces are about 1.5–2 in (3–5 cm) in width.

The models of "standard" viewers are typically about 3.5–4 in (9–10 cm) high, 5 in (13 cm) wide including the advance lever, and 3–3.5 in (8–9 cm) deep from the front of the viewer to the user's eyes. The viewers have been made in a variety of colors over the years. Blue and red are the most popular with consumers and have been used the most frequently.

Each reel looks is circular with a ring of photos that are open so they can be seen from both sides. The reels are about 3.5 in (9 cm) in diameter. The coating on the reel is the Lamilux(r) film.

The Manufacturing Process

Reel assembly

The viewer reel complete with photos is called the reel assembly. Production of the photos and the laminated paper portions of the reel begin separately but meet later in the process. The photos are reproduced in mass quantities from originals. The original is a negative and the reproduction, also on film rather than paper, is a positive transparency.

1 At the film processing house, the rolls of 16-mm film are fed through a processing machine. They emerge uncut from the rolls as visible, positive images. Each roll has only one image, but that image is reproduced thousands of times on the roll. It is also either a right or left version of the image, with the versions slightly offset to produce the stereo effect. The large roll of right versions and the companion roll of the left versions are processed at the same time using the same chemicals so the colors will match. The large rolls of identical images are delivered to the viewer maker in cans

just like those used to ship movies and are stored until they are needed.

2 Production of the laminate reels begins with huge rolls of processed paper. A punch press with dies that the tool design engineers have produced stamps reel shapes out of the paper rolls as they are fed through the press.

3 The assembled reels are then printed with descriptions of the photo pairs. Because the assembly machine keeps the reels oriented properly, they are in the correct order for printing captions. After they are printed, they are transferred to the labeling machine, where a pressure-sensitive label is applied to each reel.

4 Assembled and labeled reels are packed in cardboard boxes and dispatched to either of two locations in the plant. Single demonstration reels are included with each viewer that is packaged for sale.

5 Reels that are part of sets are sent to the reel packaging line. Elsewhere in the plant, cards are printed for the packages containing reels. The supplied card may be die cut with openings to allow plastic blisters to be inserted.

6 Package assembly uses a combination of an assembly machine, a conveyor system, and bins supplying packing materials including the preprinted cards with heat-sealing adhesive and clear plastic blister pack. For reel packages, a vacuum-formed blister of clear polyvinyl chloride (PVC) plastic is used to protect the reels and allow them to be easily seen. A blister is loaded through a pre-punched hole in the card that will hold all three reels. The reels rest on their edges and are machine-fed into the blister on the card, the card is folded, and the adhesive is heat-sealed. The packages are boxed for display trays or for display on wire racks in stores and then packed again in "master shippers" for distribution.

Viewer assembly

1 In advance of production of any of the plastic parts, design engineers make highly detailed molds that are contained in a steel box called a tool. The tool is a large box that will be lifted into an injection-molding machine to shape hot liquid plastic into forms that, when cooled, will be the plastic components of the viewer. The tool has two halves that can be locked together to mold a part and then opened to release it.

For the viewer, the tool contains four cavities that look exactly like the front and rear halves of the viewer housing. Two surfaces shape the inside and outside of the rear housing, and the other two are exact images of the inside and outside of the front housing. The outside halves of both front and rear housings are called cavity relief molds, and the inside surfaces are core relief molds. Similar tools for the lenses, reel retainer, and advance lever are designed for manufacture of the viewer.

2 To form the housing, pellets of polystyrene plastic are released from their storage hopper into the injection-molding machine that holds the tool for the housing. The machine melts the pellets until they are liquid plastic, then forces the liquid into the cavities in the tool. Both the front and rear halves of the viewer housing are produced from the same tool so the halves will be the same plastic.

3 The advance lever and the reel retainer (the internal guide for holding the reels) are also produced by injection molding. Molding of the advance lever and the reel retainer produce thin lines of plastic waste called runners. To finish these parts, the runners are hand-trimmed and recycled. The housings and other parts are moved to the assembly stations where they will be transformed into complete viewers.

4 The lenses are injection molded, but the process is different because these critical parts must be of high optical quality. Pellets of clear acrylic plastic are melted and injected into molding tools for the four lenses in the viewer. The time for the mold cycle is two or three times longer than the process for the viewer halves.

5 At the next station, the front housings are positioned exterior sides down in an assembly machine with fittings to hold them securely in the machine. A separate plastic part called the diffuser is mounted in the front housing of the viewer. The diffuser prevents a light source like the image of a light bulb from showing through the picture.

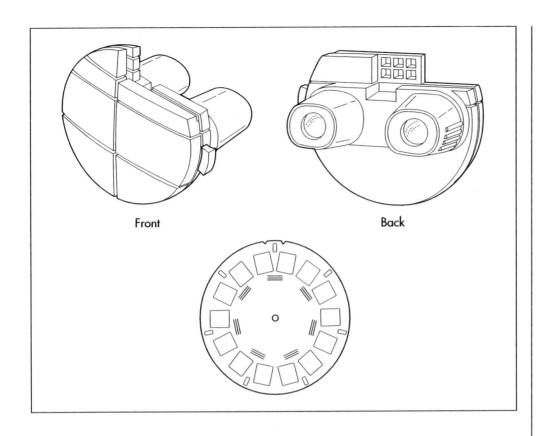

Front   Back

The machine lays the diffuser flat on the inside face of the front housing in an area sized to hold it by surrounding posts that will be used to lock the two housings together. The positions of the posts keep the diffuser from moving from side to side, and, when the viewer is snapped together, the tight fit of the front and rear housings will prevent the diffuser from moving from front to back.

6 Molded features that were designed into the housings help to hold the lenses and other internal parts securely. Later, these features will also be used to seal the housing halves together. Assembly of the rear housing illustrates this better than the front housing because four parts (or a pair in the case of the lenses) are attached to the rear housing. Each part is positioned on posts that have been molded into the rear housing. The parts themselves have holes that pair up with the posts, like those in the lenses that must be precisely positioned in the viewer. Holes in the lenses are placed on posts in the rear housing and are "heat-staked" (heated but not to the point of melting or misshaping) in position. The reel retainer (or internal guide) is heat-staked to posts on the rear housing. A hole in the advance lever (a moving part) is only slid onto its housing

post. Loop-like ends of the spring are pulled over a hook-shaped post on the housing and wired through a hole in lever. As an added means of keeping these internal parts in place, the matching hole and post moldings (called "bosses") in the front and back housings will prevent the parts from moving off of their posts when the housing is closed.

7 The front housing, with the diffuser in place, is loaded on the posts of the rear housing and pressed together. These posts and holes are called "bosses" to differentiate them from those supporting the lenses and internal parts of the viewer. The bosses fit together tightly, and, when the halves are pressed together under high pressure, the grip of the holes on the posts is as strong as any glue. Engineers describe this process as "interference fitting." It capitalizes on physical properties of the plastic (such as friction value), dimensions of the objects (the precise sizes and shapes of the bosses), and applied properties (mechanical pressure) to make the housing halves snap together but not loosen or pull apart.

8 The completed viewers are ready to be packaged and are moved to the packing assembly line. This assembly line is not ma-

chine-fed or autofed like the packaging for the reels but is staffed with employees called operators. The operator inserts a demo reel into the viewer, advances the reel, and cycles through all the views as a final check of the operation of the viewer and the quality of the lenses.

9 The operator fills a tray with four viewers. Blisters are dropped in the tray, and the worker positions each viewer, with the advance lever exposed, face down in a blister in the face of one of the four packages. Posts on the tray help position the cards that form the backs of the packages square to the blisters. When a tray is full, the operator shuttles the tray into the sealing machine. A plate inside the machine drops down and seals the four packages using heat and pressure. When the plate lifts from this tray, a worker on the other side of the machine inserts another full tray into the machine, and the first of the two is shuttled back out. The process is continuous.

10 Packages are boxed in different containers that give stores options for displaying the packages on racks, in cardboard trays, or in larger boxes. A number of smaller containers are packed in master shippers, and each large box also serves as a master shipper.

Quality Control

Quality control steps begin during conceptualization and design of a new product or part, redesign, and trials of new materials. During the first run of a new product such as a viewer, tests are done in the manufacturer's laboratory and include operation of the viewer and drop tests. The viewer must work 10,000 times for the product to be accepted. Each drop test includes 14 different drops, with one drop on each side and each corner of the viewer. If the lever breaks off, for example, the design and materials are modified to correct the faulty part.

Quality control throughout manufacturing is part of a product integrity process that is mandated the manufacturer. During assembly of the reels, the positions of the film chips in the reels are critical to producing the 3-D effect. A machine checks the images, and, if the alignment is incorrect, the reel is rejected. The machine operators are responsible for confirming

quality and rejecting products throughout the reel assembly process. During production of the viewer parts, some machines are instrumented to provide continuous feedback on operating temperatures, pressures, and other parameters. During viewer assembly, quality checks range from simply looking through lenses to confirm that they are clear to measuring dimensions with precision instruments and comparing the measurements with those in design drawings and specifications.

Byproducts/Waste

Viewer manufacture is largely free of waste. Plastic parts like the mold runners are recycled back into the injection-molding machine, reground, and used to form other parts. Plastic of different colors can be blended; the red and blue wastes from the viewers are mixed with other colors to make black plastic for other products. Acrylic for the lenses is an exception. It cannot be reground for use in future lenses, but it can be recycled for other acrylic parts. Other wastes are minor considerations. Dust, for example, is routinely vacuumed or sucked away from specific operations by exhaust systems.

The Future

The future of the stereoptic viewer is secure despite apparent competition from computers and other high-tech, rapid operation toys. Public interest, as well as company commitment, is a strong motivator for improving products and developing dynamic new product lines. View-Master's sales have tripled since the last change in ownership in 1997. Because designs of viewers and reels are well established, the major channels of change will be new processes and materials and availability of film, cartoon, and other entertainment properties that can be licensed.

Appeal to collectors is also a key to a stable future. Stereoptican viewers sold for about $2,500 in the late 1980s. Viewers and reel sets are highly collectible, and early viewers sold for $100 with sets of reels priced from $5 to $100, also in the late 1980s.

Where to Learn More

Books

Sell, Mary Ann, and Wolfgang Sell. *View-Master Viewers—An Illustrated History*

1939–1994. Mission Viejo, CA: Berezin Stereo Photography Products, 1995.

Sommer, Robin Langley. *I Had One of Those: Toys of Our Generation.* New York: Crescent Books, 1992.

Other

Baird, Keith. *A Look at View-Master History.* December 2001. <http://www.3dstereo.com/vmhist.html>.

History of View-Master(r). Press packet, Fisher-Price, Inc., 1999.

International Stereoscopic Union Web Page. December 2001. <http://www.stereoscopy.com/isu>.

Gillian S. Holmes

Stirling Cycle Engine

The first practical engine was the steam engine patented by James Watt in 1769. Watt's engine converted energy into work using steam from coal-fired boilers.

Background

An engine is a machine that converts energy into useful work: burning coal to turn the drive shaft of a power plant generator, for example. The most common engine in production today is the gasoline-powered automobile engine. Other common engines are the diesel engine used in heavy trucks and some passenger cars, the steam turbine that generates electricity in power plants, the jet engine used to propel aircraft, and the two-stroke gasoline engine used to power smaller appliances like lawnmowers. Each of these engines converts heat generated by burning a fossil fuel into useful work.

Energy is the capacity to do work. The two quantities are related and have the same units, but energy cannot be completely converted into work. If used to fuel a stove, for example, 1 gal (3.8 l) of gasoline contains enough chemical energy to boil approximately 14 gal (53 l) of water under standard conditions. However, if that same gallon of gasoline were to be put into a portable generator (which would convert the gasoline into work and then the work into electricity) and if the electricity were then used to boil water on an electric stove, it is unlikely that more than 3 gal (11.4 l) of water could be boiled before the generator ran out of fuel.

The reason the electric stove cannot boil as much water as a gasoline-fueled stove is that engines are not 100% thermally-efficient in converting heat into work—thermal efficiency meaning the amount of useful work produced divided by the energy provided to the engine. That's why a gas range or clothes dryer is cheaper to operate than the equivalent electric appliance. In the portable generator's case, some of the gasoline's en-ergy would end up in the engine's exhaust gases, some would be wasted heating the generator, and some would be wasted internally as the moving parts inside the generator rubbed together converting mechanical energy into frictional heat.

The science that studies how heat is cycled in an engine to create work is called thermodynamics, from the Greek *therme* (heat) and *dynamis* (power). A cycle that converts heat into work is known as a thermodynamic cycle. A gasoline-fueled automobile engine uses the Otto Cycle. A diesel-fueled engine uses the Diesel Cycle. A steam engine, or steam power plant, uses the Rankine Cycle. None of these cycles can be used to completely convert energy into work. This is because all of them have to reject heat into the environment. A power plant or steam engine has to condense steam in order to send the water back to the boiler (losing energy). An automobile engine must reject the hot exhaust gases, containing a considerable amount of energy, out the tailpipe. The most thermally-efficient practical cycle for converting heat into work is the Stirling Cycle. The Stirling Cycle is the most thermally-efficient engine because it wastes (or rejects) the least amount of heat to the environment for the amount of work it produces of any engine. An engine that uses the Stirling Cycle is known as a Stirling Cycle engine. A Stirling Cycle engine can be used to power a car, truck, or airplane, or to generate electricity. It will do this work for less energy input than a comparable Otto, Diesel, or Rankine Cycle engine could.

History

The first practical engine was the steam engine patented by James Watt in 1769.

Watt's engine converted energy into work using steam from coal-fired boilers. The Watt engine consisted of a boiler, a piston contained in a cylinder, a water-cooled condenser, a water pump, piping and conduits to move the water and steam around the engine, and linkages which converted the up and down motion of the piston into circular motion on a drive shaft. The drive shaft could be put to any number of uses, such as powering a mill or pumping water out of a coal mine.

Watt's engine used a four step thermodynamic cycle to create work. The cycle began with a valve opening to allow steam under pressure to flow into the cylinder. As the steam expanded in the cylinder, it depressed the piston, producing useful work. When the piston reached the bottom of the cylinder, the valve allowing steam to enter the cylinder was closed and a valve between the cylinder and the condenser opened. Because the condenser was at a much lower pressure than the cylinder it literally sucked the steam upward into the condenser. As the steam was pulled out of the cylinder, the piston was drawn up along with the steam, returning the piston to its starting location where it was ready to create more work. Once the steam in the condenser had been completely turned back into water, the water was pumped back to the boiler where it was converted back into steam, completing the cycle.

The thermal-inefficiency in this cycle is that there is still a great deal of energy left in the steam when it is sent to the condenser. However, hardly any of this energy can be reclaimed because steam cannot not be pumped back into the boiler without performing a large amount of work on it; often more work than the heat that is lost in the condenser. The steam must be converted to water before it can be pumped to the boiler. Thus, a great deal of the heat supplied by the burning coal is lost.

The steam engine made the modern industrial world possible, but it was not without drawbacks. The mixing of cold water and steam in conjunction with primitive metallurgy led to frequent boiler explosions. The resulting loss-of-life was the motivating factor that led the Reverend Robert Stirling (in addition to being one of the foremost engineers of his day, he was also an ordained minister of the Church of Scotland) to develop an engine that used air instead of steam to drive its piston. As a by-product, the Stirling's engine was much more thermally-efficient than Watt's engine, principally because it did not require that steam be condensed during the cycle. Although Stirling's engine was much safer, the technology of the time did not allow for the manufacturing of Stirling engines of more than a few horsepower (kilowatts).

Stirling's engine never caught on in the nineteenth century. Fossil fuels were plentiful and metallurgy improved to the point where steam engines were no longer quite as hazardous. Thus, the inherent thermal-efficiency advantage of the Stirling Cycle was not enough of a motivator to overcome the significant design challenges that faced engineers wishing to build more powerful Stirling Cycle engines. In the twentieth century, the internal combustion engine—running on the Otto Cycle—dominated the industrial world because it was less expensive to construct than a Stirling Cycle engine and because fossil fuels were still reasonably priced and plentiful. However, engine designers have never forgotten that the Stirling Cycle is the most thermally-efficient possible thermodynamic cycle and have continued to design engines that utilize it. Today, Stirling Cycle engines are used to produce most of the liquefied air made in research laboratories. They are also used in weather and spy satellites and by the Swedish Navy to power some of its submarines.

Raw Materials

The Stirling Cycle engine can be made from a variety of metals. The engine block is usually made from cast ductile iron or a cast aluminum alloy (aluminum and silicon, typically). Many of the internal parts (cranks and pistons) are also made from cast ductile iron or aluminum, but some of the components that require higher strength can be fabricated out of high strength S-7 tool steel. Gaskets and seals are made out of Lexan, Neoprene, or natural rubber. The engine is filled with pressurized helium or air, which is referred to as the working fluid. The component that transfers heat from the heat source to the working fluid is required to withstand very high and constant temperatures. It can be made out of high-strength steel or a ceramic composite material such as silicon carbide (SiC).

Design

Stirling Cycle engine design is a complex fusion of thermodynamics, heat transfer analysis, vibratory analysis, mechanical dynamics, strength of materials, and machine design. Thermodynamics is used to size the engine and select the temperature at which it will operate. Heat transfer analysis is required to determine how heat will be transferred from the heat source to the working fluid and how the engine components will be designed to withstand this heat flow. Vibratory analysis is used to balance the engine for smooth operation. Mechanical dynamics is required to calculate the induced stresses in the individual engine components. Strength of materials analysis is required to determine the size of the individual components in the engine so that they can withstand the induced stress. Machine design is required to translate the thermodynamic cycle into a working engine. Each of these design requirements involves tremendous amounts of analysis.

The Stirling Cycle engine is similar to a steam engine. Both have pistons and cylinders, and both are external combustion engines as the fuel burning takes place outside the engine. The first major difference between the two engines is that the Stirling Cycle engine uses a gas (air, hydrogen, or helium, usually) instead of water and steam as the working fluid, the fluid that moves the piston and creates work. Another important difference is that the Stirling Cycle engine has two cylinders, or spaces, one for working fluid expansion and one for working fluid compression while a steam engine has only one cylinder. However, the most important difference between the two engines is that, instead of wasting its excess heat in a condenser, the Stirling Cycle engine completes its thermodynamic cycle by storing its excess heat for use in the next cycle. Because of this, the Stirling Cycle engine is not just the most thermally-efficient engine there is, it is the most thermally-efficient engine there can be. A typical automobile has a thermal efficiency of about 30%. A coal-fired power plant might be 45% effi-

cient. A very large diesel engine might have a thermal efficiency of 50%. The theoretical maximum thermal efficiency of a Stirling Cycle engine operating at a combustion temperature of 2,500°F (1,370°C) would be about 78%. Of course, no one has been able to build a Stirling Cycle engine with anything near that thermal-efficiency. To date, engineers have not been able to overcome the significant design problems posed by the realization of Stirling's cycle.

In a steam engine, heat is applied to a boiler to create steam, which is then used to drive pistons. In a Stirling Cycle engine, heat is applied to the outside of the engine's main cylinder, which heats the air within the cylinder. This hot air expands, driving the engine's power piston. One of the major advantages of an external combustion engine over an internal combustion engine is that the working fluid in an external combustion engine is never exposed to combustion products, and thus stays much cleaner. Also, because the heat can be created in a controlled manner outside of the rapidly cycling engine, the Stirling Cycle engine produces less than 5% of the smog-creating nitrous oxides produced by an internal combustion engine for the same output of work.

The Stirling Cycle consists of four steps, just like the steam engine's Rankine cycle. However, instead of moving the working fluid from boiler to cylinder to condenser to boiler, the Stirling Cycle engine moves the working fluid from an expansion space at a high temperature to a regenerative heat exchanger to compression space at low temperature and back. The working fluid is moved because of the temperature differences between the hot and cold sides of the engine. The hot side is heated, by burning waste for example. The cold side is simply the side that is not heated, it is only cold relative to the hot side. The key to the process is the regenerative heat exchanger. It is called regenerative because it stores heat in one part of the cycle and then gives it back in the next.

Starting at the beginning of the power stroke, the four steps of the Stirling Cycle are: The working fluid is all contained within the expansion space, it adsorbs heat from the external heat source, which causes it to expand, depressing the power piston and the displacer, producing work; the power piston is stationary while the displacer, a piston that shuttles the working fluid between spaces in the engine, but does no work, moves up, pushing the working fluid from the expansion space into the compression space. On the way, most of the heat remaining in the working fluid that was not converted to work, is transferred to the regenerative heat exchanger; with the working piston fixed at the top of the main cylinder, the working fluid is compressed in the compression space back to the original volume, which requires rejecting some heat to the cold side of the engine, a source of lost heat, and thus lost thermal efficiency; the working fluid is passed back through the regenerative heat exchanger, where it reclaims a large portion of the stored heat, and into the expansion space where it is ready to be expanded again by the external heat source to perform work.

The various movements of the power piston and the displacer (at times, they move together for constant volume processes, while at other times one is stationary while the other moves for compressions and expansions) are controlled by a rhombic drive.

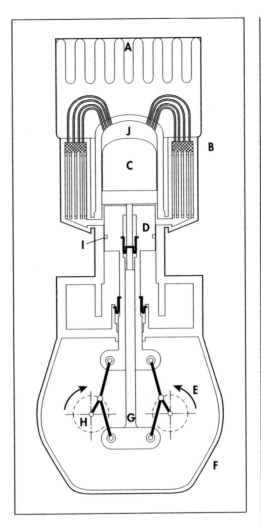

A. Heat source. B. Regenerative heat exchanger. C. Displacer. D. Working piston. E. Rhombic drive. F. Crank case. G. Displacer connecting rod. H. Crankshaft. I. Gas seals. J. Expansion space where working fluid is heated.

The Manufacturing Process

Component manufacturing

1 Engine blocks and pistons are manufactured as casting. Molten steel or aluminum is poured into a hollow mold that is shaped like the final desired product and allowed to cool. A Stirling Cycle engine block requires room for two piston cylinders, one for the power piston and one for the displacer piston (which moves the working fluid back into the power cylinder), a regenerative heat exchanger, a crankshaft, a combustion chamber, and various passageways for the working fluid to move back and forth between the two cylinders.

2 Once the casting has cooled, any extraneous material is ground off of it. It is usually necessary to finish engine blocks by drilling out holes that could not be economically cast (because of their size or complex geometry) and reaming out the cylinders to the final desired diameter. Reaming is nec-

essary because of the fine tolerance required between the piston and cylinder.

3 The crankshaft and connecting rods are manufactured by forging or casting. In forging, a piece of metal stock—called a billet—is placed between two dies (dies are molds made out of very high strength tool steel). A hammer, weighing several tons, is then dropped onto the dies. The billet is usually heated almost to the melting point prior to forging. The forging process may be performed in several steps using different dies to obtain the final shape. A casting is made by filling a hollow mold with molten metal. The metal can be ductile iron, steel, aluminum, or an alloy.

4 The engine components are machined to achieve the final tolerances. The final shaping is usually done on a lathe. The crankshaft/connecting rod/piston is spun while cutters (made of a harder material

than the part being machined) are advanced onto the spinning part to remove excess metal. Modern, computer-controlled lathes can easily attain tolerances of 0.0001 in (0.0025 mm). A milling machine, in which the part is stationary and the cutting tool rotates, is used to cut any required holes, slots, or channels in the final part.

5 The regenerative heat exchanger is manufactured by inserting thousands of fine steel wires through a steel plate. As the working fluid moves from the main cylinder to the auxiliary cylinder, it gives up its heat to these pins. The pins closest to the main cylinder will be the hottest. The pins closest to the auxiliary cylinder will be the coolest.

6 Any heat source, solar or combustion, can be used to power a Stirling Cycle engine. Regardless of source, the heat is concentrated in a chamber directly adjacent to the working piston. The cycling of the engine removes some of this heat and coverts it to work. Because the Stirling Cycle engine is an external combustion engine, the input heat can be constant (unlike the heat generated in an internal combustion engine that is produced in a series of explosions). However, because the heat is constant, the engine components in contact with the heat source must be designed to accommodate high temperatures for a long period of time.

Assembly

7 The crankshaft is inserted into the engine block and held there with bearings. Bearings allow the crankshaft to turn within the engine block without generating excessive frictional heat. The bearings are fixed to the engine block by pressing (the outside diameter of the bearing is slightly larger than the inside diameter of the hole in the engine block). By forcing the bearing into the engine block, the bearing is firmly fixed to the block.

8 The pistons and connecting rods are dropped into the cylinders and attached to the crankshaft from below using high strength bolts and lockwashers. The bolts are tightened using a predetermined torque.

9 The regenerative heat exchanger is inserted into the conduit that flows between the main cylinder and the auxiliary cylinder and bolted in place.

10 The cylinder heads are bolted to the top of the engine, an access cover is bolted to the bottom of the engine. Gaskets are used between the engine block and the covers to provide an effective seal. The heat source chamber is built into the main cylinder cover.

11 The working fluid is pumped into the engine. The working fluid is usually pressurized helium.

Byproducts/Waste

The Stirling Cycle engine produces much more useful work than does an internal combustion engine for the amount of greenhouse gases and smog-producing chemicals it emits. The engine can also be used to reclaim heat that would otherwise be wasted, such as landfill gas that it simply burned to get rid of it. Thus, on the whole, the engine is environmentally friendly. By harnessing solar heat in Stirling Cycle engines, electricity can be produced in areas without access to the electrical grid without the need for photovoltaic cells.

The Future

The future of the Stirling Cycle engine is very bright. If engineers can design and mass-produce a small, reliable Stirling Cycle engine, there would be no need for nuclear power or fossil-fuel burning power plants. Most of the electrical power used in homes could be generated on the premises. The engine could cool the house in the summer without use of ozone-depleting refrigerants and heat it in the winter. Unfortunately, there are serious practical design difficulties that must be overcome before the Stirling Cycle engine can come into wide use. The most significant engineering obstacle is the design of the engine combustion chamber. Because the Stirling Cycle engine operates at very high temperatures, the combustion chamber cannot be built out of the same inexpensive materials used to produce automobile engines. The use of high strength stainless steel or ceramic composites, in addition to being expensive, make manufacturing of the engine extremely difficult. Other non-trivial design obstacles include designing a reliable gearing mechanism to translate the Stirling Cycle piston motions (which are very complex compared to a standard Otto Cycle automobile engine) into crankshaft motion and designing seals capable of keeping the working fluid contained within the engine.

Where to Learn More

Books

Moran, Michael J., and Howard N. Shapiro. *Fundamentals of Engineering Thermodynamics.* 4th ed. John Wiley and Sons, 2000.

Organ, A. J. *Thermodynamics and Gas Dynamics of the Stirling Machine.* Cambridge University Press, 1992.

Walker, Graham. *Stirling Engines.* Oxford University Press, 1980.

Walker, Graham, Graham Reader, Owen R. Faubel, and Edward Bingham. *The Stirling Alternative, Power Systems, Refrigerants and Heat Pumps.* Gordon and Breach Science Publishers, 1996.

Other

Griessel, Eugene. Home Page. "Animation of a Stirling Cycle." 27 September 2001. <http://www.dynagen.co.za/eugene/stirling htm>.

"Stirling Cycle Frequently Asked Questions." *American Stirling Company Web Page.* 27 September 2001. <http://www. stirlingengine.com/FAQ.asp>.

Jeff Raines

Straight Pin

Prehistoric people used thorns as pins. In ancient Egypt, pins were crafted of bronze with decorative heads. The clothes of medieval Europeans were adorned with pins of many materials including bone, ivory, silver, gold, and brass.

Background

A straight pin is a small length of stiff wire with a head at one end and a point at the other end. It is used to fasten pieces of cloth or paper together.

History

Since their ancient beginnings, human beings have devised methods for securing cloth together. Prehistoric people used thorns as pins. In ancient Egypt, pins were crafted of bronze with decorative heads. The clothes of medieval Europeans were adorned with pins of many materials including bone, ivory, silver, gold, and brass.

The use of iron wire, still applied during modern times, began as early as the fifteenth century in France. The craft of tailoring was also well-established by this time. Descriptions of a tailor's equipment from Spanish books dating back to this period included the mention of pins. A "paper of pins" became a familiar cultural phrase, signifying the possessions of the simplest nature.

At the dawn of the Industrial Revolution in the eighteenth century, noted economist Adam Smith employed the imagery of a pin factory as the perfect example of the intricate division of labor. In his book, *Wealth of Nations*, published in 1776, Smith described how one worker drew out the wire, another straightened it, a third cut the wire, the fourth sharpened one end, and another worker ground the opposite end for the attachment of the head. At the end of the process, the pins were polished and inserted into paper packets. These early pin factories produced just under 5,000 pins per day.

Attaching the heads presented a particular challenge. In the early to mid-1800s, American inventors Seth Hunt and John Ireland Howe and British inventors Lemuel Wright and Daniel Foote-Taylor patented machines that produced pins with a solid head from a single piece of wire. American Samuel Slocum also invented a similar machine but did not patent it. In spite of not having an official claim to this invention, the pins manufactured in Slocum's Poughkeepsie, New York factory became known as Poughkeepsie pins.

A physician by profession, Howe also liked to tinker with machinery. After watching the inmate/patients at the New York Alms House laboriously make pins by hand, he began to explore ideas for a pin-making machine. Howe enlisted the help of a printer press designer named Robert Hoe. Howe obtained a patent for his machine in June of 1832. After the machine was exhibited at the American Institute Fair in New York City, Howe was awarded a silver medal for his contribution to manufacturing.

In December of 1835, Howe formed the Howe Manufacturing Company, which was soon turning out about 70,000 pins daily. However, the packaging step slowed down the process. Workers had to manually insert the pins into paper or cards. In 1843, with the help of his employees, Howe developed a machine that crimped the paper and then inserted the pins.

Although electroplating was invented in the mid-1800s, the process was not perfect. The nickel coating would flake off and the pins would rust. To combat this flaw, tailors and seamstresses cleaned the rusted pins by pushing them back and forth into a bag of

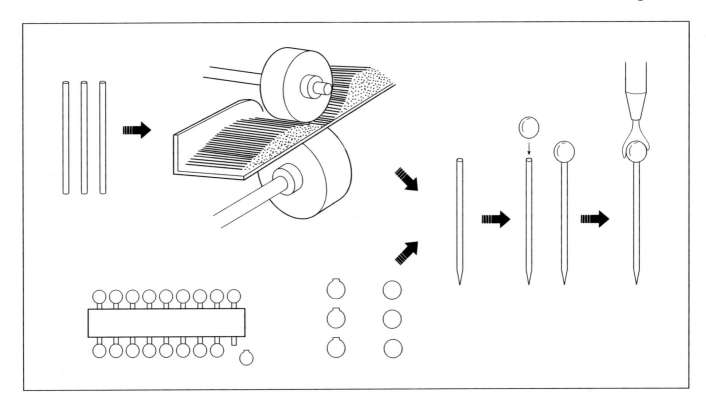

emery grit. Today this bag is known as the pin cushion.

Raw Materials

Blunt wire with an international steel regulation of ISR 9002 is generally used to make straight pins. To create the wire, a bar of steel is heated to a temperature of 2,200°F (1,200°C), rolled into a long thin rod, coiled, and then allowed to cool. The heating causes an oxide coating to form on the wire. To remove this coating, the wire is immersed in an acid bath, then rinsed in water.

The cleaned wire is inserted into a drawing block that pulls the wire through a die whose opening is smaller in diameter than that of the wire. Thus the wire is reduced in diameter and increased in length. The drawing is undertaken several times until the desired diameter is obtained. As it comes off of the last stage of drawing, the steel is coiled. The dies are coated with grease or soap to protect the wire as it passes through. This lubrication process also removes defects and gives the wire a smooth finish.

Nickel is a silvery chemical element extracted from the earth's crust. It is combined with sulfate to create a solution for coating the pins to keep them from rusting.

Design

The straight pin was designed to provide a simple function, secure two or more objects together. The design is relatively simple and unchanged. The sharp tip allows penetration through materials such as cloth and paper. The head of the pin stops the entire body from slipping through the hole created by the tip, thus creating a temporary bond.

The Manufacturing Process

In the modern pin manufacturing plant, hundreds of thousands of pins are produced daily. Although several United States companies produce and sell straight pins, virtually all of the manufacturing plants are in Asia.

1 One-hundred-foot rolls of steel wire are unwound by means of a roll straightener. The end of each roll is threaded into the straightener, which pulls the wire flat. Rotating blades cut the wire into pre-set lengths, usually between 1–1.25 in (2.5–3.2 cm) long.

2 The cut wire travels via conveyer belt to the next station where the heads are "stamped" on. One end of the wire is slammed against a block. This sharp blow

Steel wire is cut and sharpened at one end. Raw plastic peal heads are then stamped onto the blunt end.

causes the end to mushroom out from the shank of the wire and create a flattened head.

3 The pins are loaded into a circular cavity where they are hung by their heads over large grinding wheels. The grinding wheels spin the pins around to sharpen them.

4 To ensure a strong bond between the pins and the plating solution, the pins are cleaned by dipping them in an acid solution. They are then are placed in a rack and lowered into large electroplating tanks filled with a plating solution such as nickel-sulfate. The pin rack is connected to the negative terminal of an external source of electricity. A second conductor is connected to the plating solution. A steady, direct electrical current of a low voltage, between one to six volts, is passed through the tanks. This causes the plating solution to coat the pins and give them a shiny finish.

5 The shined pins are mechanically packed in pre-ordered amounts into plastic clamshell boxes or blister packs. Bar codes are mechanically affixed to each box or pack. The individual containers are then hand-packed into cartons for shipping.

Quality Control

When the wire arrives at the straight pin factory, it is inspected for physical properties such as tensile strength and brightness. Machinery is regularly checked while they are running. Samples are drained off from each electroplating bath and sent to on-site labs. A sample of pins is pulled from each batch of 20,000 and checked again for tensile strength and brightness. The count and weight of each batch is checked before and after processing.

Byproducts/Waste

The electroplating bath creates a toxic waste product that has serious implications for the environment. Companies that employ electroplating techniques are strictly regulated by the United States Environmental Protection Agency (EPA). The concentrations of metal in the electroplating bath must be removed or disposed of in a prescribed manner. The used solution or waste water cannot be emptied into a septic system or storm water sewer. Companies must use an autho-

rized waste transporter and the containers must meet federal United States Department of Transportation packaging standards

Chromium, one of nearly 200 toxic chemicals regulated by the federal Clean Air Act, is released into the air during the electroplating process. Plants must apply for and be granted an air pollution control permit. To qualify for the permit, plants must meet standards that regulate the amount of emissions that are allowed per day, work practices, performance testing, monitoring, record keeping, and reporting.

One way in which manufacturers reduce the incidence of these pollutants is to limit the dragout and the use of water. Dragout is any solution that escapes from the electroplating solution. By allowing the pins and pin rack to drain completely over the bath, the dragout can be substantially cut. Drain boards between the process tanks and the rinse tanks catches any solution still remaining on the parts and the product.

Spray rinses installed over the baths wash the dragout directly back into the bath. Employing spray rinses rather than continuously flowing water also reduces the incidence of pollutants.

The Future

The straight pin has gone relatively unchanged for years and few improvements have been made. The head of the pin can now be either solid metal or plastic. Its use to bind paper together has been replaced by the stapler, but despite its simplicity, the straight pin is the first choice for a temporary way to bind cloth.

Where to Learn More

Books

Kane, Joseph Nathan, Steven Anzovin, and Janet Podell. *Famous First Facts.* New York: H. W. Wilson, 1997.

Petroski, Henry. *The Evolution of Useful Things.* New York: Alfred A. Knopf, 1992.

Travers, Bridget, ed. *World of Invention: History's Most Significant Inventions and the People Behind Them.* Detroit: Gale Research Inc., 1998.

Other

"Watch Your Waste." *Illinois Environmental Protection Agency.* 1 October 2001. <http://www.epa.state.il.us/small-business/ electroplating-shops>.

Mary McNulty

Sushi Roll

As early as 500 B.C., people living in the mountains of southeast Asia wrapped fish in rice as a means of pickling and fermenting.

Background

A sushi roll is a food of Asian origin that features rice and seafood wrapped in seaweed (nori). Until the end of the twentieth century, sushi rolls were only available in restaurants. Today, a number of companies prepare them for retail sale in grocery stores. Although a few of these companies use mechanical sushi makers—called robots—to shape the rice and add condiments, the finest quality sushi rolls are still handmade. An expert sushi chef, a shokunin, can roll and cut six to eight sushi rolls in a matter of moments. It is not the desire for faster production that has led some companies to use the sushi robots; rather it is a shortage of accomplished chefs.

Sushi-making is a time-honored tradition in Japan. This craft is a matter of intense national pride, and it is often noted that the red of the fish and the white of the rice symbolize the red and white of the Japanese flag. This tradition also extends to the types of utensils used. Although some substitutions are acceptable, in the traditional Japanese kitchen the following utensils would be found: hangiri, a small tub made of cypress bound with copper hoops and used for cooling vinegared rice; shamoji, a flat, wooden, rounded spatula used to turn and spread sushi rice; uchiwa, a hand-held fan made of bamboo and covered with paper or silk used to remove moisture from sushi rice; and makisu, mat constructed of thin bamboo strips woven together with string used for rolling sushi.

History

As early as 500 B.C., people living in the mountains of southeast Asia wrapped fish in rice as a means of pickling and fermenting.

In Japan, alternating layers of carp and rice were placed in a covered jar and left for up to a year. During this time, the fermenting rice produced lactic acid, thus pickling the fish. When the jar was opened, the carp was eaten, but the rice was discarded.

A Japanese legend holds that a kindly husband and wife placed rice in an osprey nest. When they later checked on the bird, they found a fish nestled in the rice, which they took as a token of the bird's appreciation. As they ate the thank-you gift, they noted that the fermented rice had imparted a distinctive taste to the fish.

In the seventeenth century, the people of culinary-rich Edo (now Tokyo) began the practice of adding vinegar to the rice so that it would ferment in just a few days. Before long, sushi shops were popular sites on the streets of Tokyo. One of the earliest, Sas Maki Kenukesushi, opened in 1702 and was still in business at the turn of the twentieth century.

Although the Japanese have eaten seaweed, or nori, since the eighth century, it was not until the late seventeenth century that it was regularly cultivated in inlets and estuaries up and down that nation's coasts. The nori was harvested in December and January when it had reached its maturity. It was not an easy task because the nori disappeared during the summer months.

In the 1940s, a British scientist named Kathleen Drew-Baker began to investigate what happened to nori spores in the summer. Drew's studies were published in a paper in 1949, which concluded that nori spores burrow into the pores and crevices of seashells, where they grow into pink thread-like organisms. When the weather turns cold, the organisms detach themselves and then adhere to other surfaces where they grow to maturity.

After Drew's conclusions were published, the Japanese quickly developed a cultivating system and nori production increased tenfold from 1950 to 1980. In 1963, nori farmers erected a bronze statue in Drew's honor overlooking the Bay of Shimbara. On April 14 of every year, a ceremony is enacted in which Drew's cap and gown are placed on the statue, a Union Jack is raised, and farmers placed a tribute of nori from the current crop at the statue's feet.

Raw Materials

Today the cultivation of nori is a very prosperous industry in Japan. Miles of bamboo nets are submerged in inlets on the Japanese coast to provide a growth field for the nori spores. At the end of the growing season in early April, the healthiest spores are selected from the nets and transported to Prefectorial Seeding Centers. There, they are mixed with a liquid suspension and sprayed onto clean oyster shells. It takes 1.5 tons of porphyra seeds to fill 20,000 shells.

The shells are suspended from ropes draped over bamboo sticks over large tanks of water that are held at 50–60°F (10–15°C). The walls and roof of the seeding centers are lined with curtained windows so that the heat intensity can be monitored. The seeds are left to germinate throughout the summer and early fall. The plants are harvested, washed with sea water, then with fresh water. They are then dried into sheets.

Although variety of seafood is used in sushi rolls, including shrimp (ebi), crab (kani), and salmon (sake), tuna (maguro) is by far the most popular. The bluefin tuna market is very competitive. Tokyo's Tsukiji market sets the market price and the day's catch is auctioned to the highest bidder. Prospective buyers extract small samples from the flesh of the fish to test for color and fat content. In order to be considered for sushi, the tuna must meet "kata" or ideal form requirements pertaining to color, texture, fat content, and body shape.

While Japan remains the center for tuna fishing, it is a also major industry in the North Atlantic and in the Mediterranean Sea. However, Japanese techniques are so highly revered that experts from that country are often recruited to advise on matters of catching, handling, and packing. Special Japanese paper is used for wrapping the fish before it is placed on ice. The fish is shipped whole to Japan to be sliced and trimmed. It is not unusual for a tuna to be caught in New England, shipped to Japan for processing, and then shipped back to a restaurant in Boston.

The vinegar used in sushi and sushi rolls is made from fermented rice. It is then poured sparingly into the rice to be used in the sushi roll.

Wasabi, also known as Chinese horseradish, is a common ingredient in sushi rolls. Difficult to cultivate, it grows best on the northern sides of shaded mountain valleys near cold running streams. Wasabi can take two to three years for the edible roots to mature. It is prepared fresh, powdered, and/or as a paste.

Soy sauce is made from fermented soybeans, toasted wheat, barley, salt, and water. It may be purchased from an outside supplier, or processed at the same plant that produces the sushi rolls.

Fresh ginger root is one of the most common spices used with the preparation of sushi rolls. It can also be purchased from an outside source or cultivated in-house.

Vegetable ingredients are as varied as the seafood, but can include cucumber, avocado, and spinach. The vegetables can be purchased from outside vendors.

The Manufacturing Process

1 A half sheet of nori is spread onto the makisu. About 0.25 in (6 mm) of vinegared rice is spread onto the nori. A groove is made down the center of the rice with the shamoji.

2 Strips of seafood and/or vegetables are laid into the groove. Wasabi is distributed evenly on top the seafood and/or vegetables.

3 The makisu is used to roll the nori around the rice and other ingredients. After rolling, it is pressed manually into a square shape.

Sushi filing and preparation techniques vary depending on the shokunin.

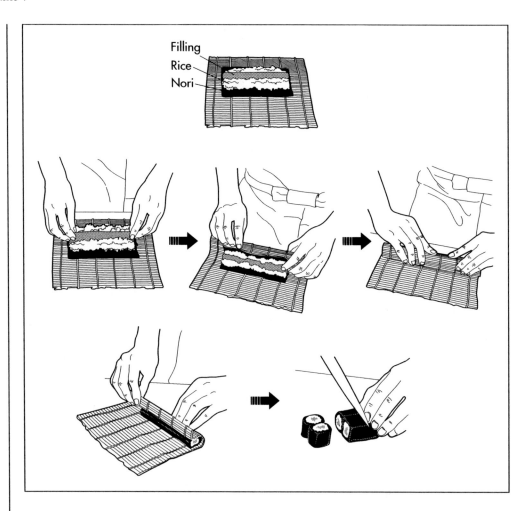

4 The sushi roll is removed from the makisu and sliced into 1.5-in (3.8-cm) pieces.

5 The shokunin places the finished sushi rolls on small, wooden tables. Fresh slices of ginger are usually also placed on the table along with a side of wasabi. If the sushi is to be shipped to grocery stores, factory workers manually place the sushi rolls in plastic cartons, usually in groups of six or eight. Packets of soy sauce are added. Plastic covers are attached to the cartons and labels are affixed. The cartons are loaded onto refrigerated trucks and shipped immediately to grocery stores.

Quality Control

As with any foodstuff, ensuring the health of the consumer is of utmost concern. Sushi raises particular concerns because of the existence of parasites in raw fish. The Centers for Disease Control in Atlanta, Georgia, and the United States Food and Drug Administration (FDA) both recommend that raw fish be flash frozen at -4°F (-20°C) for three to five days to kill parasitic worms. This process of flash freezing the fish usually takes place at sea. At the wholesalers, the frozen fish is sliced into small, rectangular sections and wrapped in plastic. Still rock hard, the sliced fish is then shipped to the factory.

In Japanese sushi bars (both in Japan and in the United States) raw fish is the norm. These restaurants contract to have fresh fish flown in from around the world. In Japan, the Ministry of Agriculture, Forestry & Fishers (MAFF) administers a voluntary product quality and labeling mark called the Japanese Agriculture Standard (JAS). Over 300 JAS standards exist for agri-food imports. The Japanese government's regulations are particularly restrictive on imports of sushi and rice. The use of American rice in Japanese cuisine is strictly forbidden.

Nori used by the food processing industry is often roasted rather than dried. At the processing plant, the nori is tested for the presence of heavy metals, herbicides, pesticides, *Escherichia coli* (E. coli), yeast, and molds.

Byproducts/Waste

Although much of the waste material from fish processing is returned to the sea, a lucrative fish byproduct industry exists. In the United States alone, this amounts to 2 million pounds annually. Organs, bones, and scales are used to make fish meal and bait. Bones are also used to produce fish stock and soups. Fish skins are used in the production of some leather products. The medical community uses fish oil to produce food supplements.

The Future

The popularity of sushi is expected to continue to grow in the twenty-first century. The challenge will be keeping up with the demand. Tuna is especially in danger of being over-fished. At the end of the twentieth century, some American restaurants were refusing to serve tuna in an effort to end the depletion of the species.

The International Commission for the Conservation of Atlantic Tuna (ICCAT), based in Madrid, is responsible for assigning quotas for bluefin tuna in the North Atlantic and in the Mediterranean Sea. However it is difficult to regulate a fish that can swim up to 50 miles per hour (80 km/hr) across multiple jurisdictions. The result is that ICCAT and various wildlife agencies, including the United States National Academy of Sciences and the National Audubon Society, disagree on exactly how many bluefin tuna migrate over the North Atlantic.

The proliferation of salmon farming is also causing controversy. Farmed salmon are fed a diet of dead fish and are often treated with anti-bacterial chemicals. In addition, coloring agents are added to food pellets to give the salmon the pink tint that would normally

be gotten from eating krill and shrimp in the wild. Opponents argue that the waste material that is discharged as a result constitutes a serious environmental hazard. An outbreak of Infectious Salmon Anemia in 1998 in Scotland has also been blamed on poor conditions in the fish farming industry.

Where to Learn More

Books

Davidson, Alan. "Sushi Roll." In *The Oxford Companion to Food*. Oxford University Press, 1999.

Omae, Kinjiro, and Yuzui Tachibana. *The Book of Sushi*. Tokyo: Kodansha International, Ltd., 1981.

Trager, James. *Sushi in The Food Chronology*. New York: Henry Holt, 1995.

Yoshii, Ryuichi. *Sushi*. Boston: Periplus Editions, 1998.

Periodicals

Bestor, Theodore C. "How Sushi Went Global." *Foreign Policy* (November 2000).

"Bowing Out." *Progressive Grocer* (January 1998).

"Entrepreneurs Try Fast-Food Techniques to Feed Growing U.S. Appetite for Sushi. *Wall Street Journal* (23 August 2000).

"Sewage with Your Salmon, Sir? Salmon Farming: More Trouble over Salmon Farms." *The Economist* (23 June 2001).

"Smells Fishy." *The Economist* (9 December 2000).

Thornton, Emily. "The Sushi-Matic." *Fortune* (9 September 1991).

Mary McNulty

Suture

Background

A surgical suture is used to close the edges of a wound or incision and to repair damaged tissue. There are many kinds of sutures, with different properties suitable for various uses. Sutures can be divided into two main groups: absorbable and non-absorbable. An absorbable suture decomposes in the body. It degrades as a wound or incision heals. A non-absorbable suture resists the body's attempt to dissolve it. Non-absorbable sutures may be removed by a surgeon after a surface incision has healed.

Sutures are made from both man-made and natural materials. Natural suture materials include silk, linen, and catgut, which is actually the dried and treated intestine of a cow or sheep. Synthetic sutures are made from a variety of textiles such as nylon or polyester, formulated specifically for surgical use. Absorbable synthetic sutures are made from polyglycolic acid or other glycolide polymers. Most of the synthetic suture materials have proprietary names, such as Dexon and Vicryl. The water-resistant material Goretex has been used for surgical sutures, and other sutures are made from thin metal wire.

Sutures are also classified according to their form. Some are monofilaments, that is, consisting of only one thread-like structure. Others consist of several filaments braided or twisted together. Surgeons choose which type of suture to use depending on the operation. A monofilament has what is called low tissue drag, meaning it passes smoothly through tissue. Braided or twisted sutures may have higher tissue drag, but are easier to knot and have greater knot strength. Braided sutures are usually coated to improve tissue drag. Other sutures may have a braided or twisted core within a smooth sleeve of extruded material. These are known as pseudo-monofilaments. A suture can also be classified according to its diameter. In the United States, suture diameter is represented on a scale descending from 10 to 1, and then descending again from 1-0 to 12-0. A number 9 suture is 0.0012 in (0.03 mm) in diameter, while the smallest, number 12-0, is smaller in diameter than a human hair.

Suture manufacturing comes under the regulatory control of the Food and Drug Administration (FDA) because sutures are classified as medical devices. Manufacturing guidelines and testing for the industry is provided by a non-profit, non-governmental agency called United States Pharmacopeia, located in Rockville, Maryland.

History

Physicians have used sutures for at least 4,000 years. Archaeological records from ancient Egypt show that Egyptians used linen and animal sinew to close wounds. In ancient India, physicians used the heads of beetles or ants to effectively staple wounds shut. The live creatures were affixed to the edges of the wound, which they clamped shut with their pincers. Then the physician cut the insects' bodies off, leaving the jaws in place. Other natural materials doctors used in ancient times were flax, hair, grass, cotton, silk, pig bristles, and animal gut.

Though the use of sutures was widespread, sutured wounds or incisions often became infected. Nineteenth century surgeons preferred to cauterize wounds, an often ghastly process, rather than risk the patient's death from infected sutures. The great English physician Joseph Lister discovered disinfecting techniques in the 1860s, making surgery much safer. Lister soaked catgut suture material in phenol making it sterile, at least on the outside. Lister spent over 10 years experimenting with catgut, to find a material that was supple, strong, sterilizable, and absorbable in the body at an adequate rate. A German surgeon made advances in the processing of catgut early in the twentieth century, leading to a truly sterile material.

Catgut was the staple absorbable suture material through the 1930s, while physicians used silk and cotton where a non-absorbable material was needed. Suture technology advanced with the creation of nylon in 1938 and of polyester around the same time. As more man-made textiles were developed and patented for suture use, needle technology also advanced. Surgeons began using an atraumatic needle, which was pressed or crimped onto the suture. This saved the trouble of threading the needle in the operating room, and allowed the entire needle diameter to remain roughly the same size as the suture itself. In the 1960s, chemists developed new synthetic materials that could be absorbed by the body. These were polyglycolic acid and polylactic acid. Previously, absorbable sutures had to be made from the natural material catgut. Synthetic absorbable suture material is now far more prevalent than catgut in United States hospitals.

The FDA began requiring approval of new suture material in the 1970s. A Medical Device Amendment was added to the FDA in 1976, and suture manufacturers have been required to seek pre-market approval for new sutures since that time. Manufacturers must comply with specific Good Manufacturing Practices, and guarantee that their products are safe and effective. Patents for new suture materials are granted for 14 years.

Raw Materials

Natural sutures are made of catgut or reconstituted collagen, or from cotton, silk, or linen. Synthetic absorbable sutures may be made of polyglycolic acid, a glycolide-lactide copolymer; or polydioxanone, a copolymer of glycolide and trimethylene carbonate. These different polymers are marketed under specific trade names. Synthetic nonabsorbable sutures may be made of polypropylene, polyester, polyethylene terephthalate,

polybutylene terephthalate, polyamide, different proprietary nylons, or Goretex. Some sutures are also made of stainless steel.

Sutures are often coated, especially braided or twisted sutures. They may also be dyed to make them easy to see during surgery. Only FDA approved dyes and coatings may be used. Some allowable dyes are: logwood extract, chromium-cobalt-aluminum oxide, ferric ammonium citrate pyrogallol, D&C Blue No. 9, D&C Blue No. 6, D&C Green No. 5, and D&C Green No. 6. The coatings used depend on whether the suture is absorbable or nonabsorbable. Absorbable coatings include Poloxamer 188 and calcium stearate with a glycolide-lactide copolymer. Nonabsorbable sutures may be coated with wax, silicone, fluorocarbon, or polytetramethylene adipate.

Suture needles are made of stainless or carbon steel. The needles may be nickel-plated or electroplated. Packaging material includes water-resistant foil, such as aluminum foil, as well as cardboard and plastic.

Design

Sutures are designed to meet many different needs. Sutures for abdominal surgery, for example, are different from sutures used in cataract surgery. Since no one type of suture is ideal for every operation, surgeons and medical designers have come up with sutures with varying qualities. One may be more absorbable but less flexible, while another is exceedingly strong but perhaps somewhat difficult to knot. This gives surgeons many options. Designers of a new suture have to take into account many factors. The rate the suture degrades is important, not only along the length of the suture but at the knot. Some sutures need to be elastic, so that they will stretch and not break. Others need to hold tight. Suture manufacturers use specially designed machines to test and study sutures. New suture designs are also tested by subjecting them to chemical tests, such as soaking them in various solutions, and testing on animals.

The Manufacturing Process

The manufacturing of sutures for surgical use is not very different from the production

An example of a person being sutured.

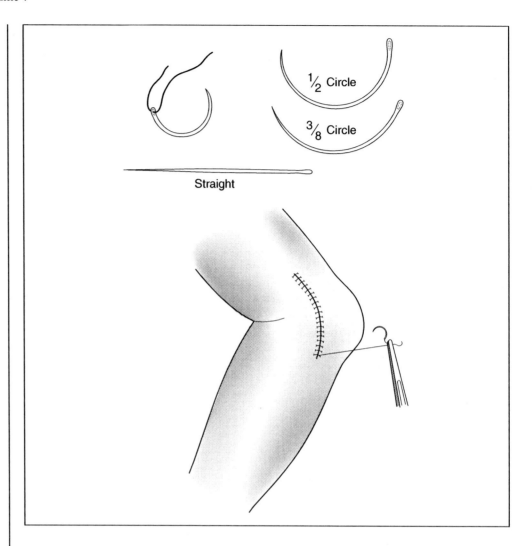

of other synthetic textiles. The raw material is polymerized, and the polymer extruded into fiber. The fiber is stretched and braided on machines similar to ones that might be found in a factory producing polyester thread for the garment industry. The manufacturing process typically occurs at three sites: one plant produces the suture textile, another produces the needles, and a third plant called the finishing plant attaches needles to the sutures, packages, and sterilizes.

1 The first step in suture manufacturing is to produce the raw polymer. Workers measure the chemicals making up the polymer into a chemical reactor. In the reactor, the chemicals are combined (polymerized), forced through a die, and discharged as tiny pellets.

2 Next workers empty the pellets into an extrusion machine. The extruder has a nozzle, looking something like a shower head, pierced with many tiny holes. The machine melts the polymer, and the liquid flows through the tiny holes, forming many individual filaments.

3 After extrusion, the filaments are stretched between two rollers. The filaments stretch to as much as five times their original length.

4 Some sutures are produced as monofilaments. Others are braided or twisted. To braid the suture, the extruded monofilament is wound onto bobbins, and the bobbins are loaded onto an automatic braiding machine. Such a machine is typically of an old design that might also be used in the manufacture of textiles for fabric. The number of filaments braided together depends on the width of the suture made for the particular batch. A very fine suture might braid 20 filaments, a medium width hundreds, and a very thick suture might braid thousands of

filaments. The braiding machine produces one continuous strand of braided material. It works very slowly, and typically the machine is set to run for as long as four weeks at a time. The process is almost entirely automatic. Workers in the plant inspect the equipment for break-downs and reload empty bobbins, but generally the process requires little man-power.

5 After braiding, the suture undergoes several stages of secondary processing. Non-braided sutures will also go through these steps after extrusion and initial stretching. Workers load the material onto another machine that performs another stretching and pressing operation. Unlike the first stretching, this step might take only a few minutes, and adds to the length of the material by only about 20%. The suture passes over a hot plate, and any lumps, snags, or imperfections are ironed out.

6 Next, workers pass the suture through an annealing oven. The annealing oven subjects the suture to high heat and tension, which actually orders the crystalline structure of the polymer fiber into a long chain. This step may take several minutes or several hours, depending on the type of suture being made.

7 After annealing, the suture may be coated. The coating material varies depending on what the suture is made of. The suture passes through a bath of coating material, which may be in solution or may be in a thick, paste-like state called a slurry.

8 All the major manufacturing steps at the processing plant are complete at this point. Now the quality assurance workers test the batch of suture for various qualities. These workers make sure the suture conforms to the proper diameter, length, and strength, look for physical defects, and check the dissolvability of an absorbable suture in animal and test-tube tests. If the batch passes all the tests, it is shipped to a finishing plant.

9 The surgical needles are made at another plant, and also shipped to the finishing plant. The needles are made of fine steel wire, and drilled lengthwise. Workers at the finishing plant cut the suture into standard lengths. The length of suture is mechanically inserted into the hollow in the needle, and the needle is crimped onto the fiber. This process is called swaging.

10 Next, the suture and attached needle are inserted into a foil packet and sterilized. Sterilization differs according to the suture material. Some sutures are sterilized with gamma radiation. In this case, the sutures are packaged completely. The whole package, typically a sealed foil pack inside a cardboard box, is set on a conveyor belt. The sealed package passes under pencil-shaped lenses emitting gamma radiation. This kills all microbes. The suture is now ready for shipment. Some suture material cannot withstand gamma radiation, and it is sterilized in a different process. The suture and needles are packaged in a foil pack, but the pack is left open. The packages move into a gas chamber, which is then filled with ethylene oxide gas. Then the foil packs are sealed, inserted into boxes or other packaging, and readied for shipment.

Quality Control

Sutures, as medical devices, are subject to strict quality control. All the raw materials that arrive at the manufacturing plant are tested to make sure they are what they are supposed to be. Each batch of sutures is tested after the main manufacturing steps for a variety of physical characteristics such as diameter and strength. The suture industry has developed an array of sophisticated instruments for testing special suture characteristics such as knot security and tissue drag. Tests for diameter, length, and strength of the suture are also performed at the finishing plant. The finishing plant must also test how well the needle is attached to the suture. Guidelines for suture quality control are laid down by the independent organization United States Pharmacopeia.

The Future

New sutures are being developed all the time, to better respond to particular surgical needs. While not replacing sutures, scientists have also devised alternative methods of wound closure. The first surgical stapler was invented in 1908, but stapler technology developed considerably in the 1990s. Precise machines are able to place absorbable staples, as thin as four human hairs, beneath the top layer of skin to secure an incision with minimal scarring. A related device, first tested on patients in the United States in 2000, is a surgical zipper. A surgeon can place the zipper over a straight incision and zip the wound closed, eliminating the need for suturing. After the wound heals, the patient can wash the zipper off in the shower. Another surgical closure method that is still evolving is surgical glue. Surgical glue is less painful than sutures if a wound must be closed without anaesthetic. The glue may leave less scarring in some cases, and be easier to care for post-operatively.

Where to Learn More

Books

Mukherjee, D. P. "Sutures." In *Polymers: Biomaterials and Medical Applications.* New York: John Wiley & Sons, 1989.

Planck, H., M. Dauner, and M. Renardy, eds. *Medical Textiles for Implantation.* Berlin: Springer-Verlag, 1990.

Periodicals

"Dermabond 'Super Glue' Receives Mixed Reviews." *Dermatology Times* (October 1999): 1.

Mraz, Stephen J. "From the Jaws of Ants to Absorbable Staples." *Machine Design* (12 January 1995): 70 ff.

"Zip-it-y Doo Dah." *Nursing* (May 2000): 62.

Angela Woodward

Swimming Pool

Background

The most common type of in-ground manufactured swimming pool on the market today is the concrete pool. Although there are a wide variety of manufactured pools on the market (concrete, fiberglass, and vinyl), concrete pools represent 60% of pools being built today. Concrete pools offer limitless options for shape, configuration, and spa features. The excavation site is reinforced with steel and provides a sturdy support for pools of any shape or size. Fiberglass pools are manufactured in a factory and the prefabricated pool arrives at its destination and is set in a previously excavated site. Fiberglass pools constitute 7% of the market and are limited in the variety of shapes available. Vinyl-lined pools arrive in a kit with construction completing at the site and represent 33% of the pool market. A custom fit vinyl liner is installed after the decking is completed and the structure is then filled with water. Previously limited to geometric designs only, free form options are now available.

In the design and manufacturing of pre-fabricated aboveground pools, the cost of fabrication continues to be a major concern. Pools have to be manufactured in a variety of styles and sizes to accommodate the available space requirements of the customer and the variety of components necessary for manufacturing is quite large. Designs that can incorporate this flexibility and reduce the necessary inventory required to accommodate a variety of sizes and structures have begun to arrive in the above ground swimming pool manufacturing marketplace creating easier systems to erect without sacrificing stability.

Above ground swimming pools have an exterior wall of machined sheet metal construction formed into a continuous circular or oval shape. Support or reinforcing posts are located around the perimeter of the pool and serve to strengthen and maintain the pool wall in the desired position. Conventionally, various components make up these post assemblies and may include upper and lower rails for engaging and covering the upper and lower edges of the metal pool wall, some form of plate connectors which serve to mount the vertical posts in position and which may also be used to secure sections of ledges around the top of the pool wall. Some form of cap is usually fastened to finish the top of the post assembly.

Above ground pools consist of a sidewall, a water-impermeable liner, and a frame or superstructure for supporting and reinforcing the sidewall. The frame assembly generally includes a lower rail and a top rail to which the pool sidewall is attached. The liner is attached to an inner surface of the pool sidewall along an upper edge.

History

Swimming as an organized activity goes back as far as 2500 B.C. in ancient Egypt. Actual swimming pools were invented in the first century B.C. by the Romans. In Rome and Greece, swimming was part of the education of young boys. Many of the Roman pools were also heated using water diverted through piping from natural springs. The elaborate bathhouses built with marble and expensive gilding were very popular with the elite society but the majority of people continued to swim in lakes and rivers. These bathing pools were the precursors to modern day swimming pools.

Although there are a wide variety of manufactured pools on the market (concrete, fiberglass, and vinyl), concrete pools represent 60% of pools being built today.

In Europe, many people refrained from building swimming pools due to the fear that infections could be caught from infected swimmers. Europeans formed their first swimming organization in 1837 in London, which by then had six indoor pools. The popularity of swimming pools did not begin to increase until after the first modern Olympic games in 1896.

Raw Materials

Raw materials for manufacturing swimming pools consist of polyvinyl chloride (PVC) plastic, galvanized steel or metal, fiberglass, concrete, and polyurethane foam. Adequate steel bars for reinforcing a concrete pool range in size from 0.38 in (0.97 cm) to 0.75 in (1.9 cm) diameter (these values vary depending on the structural requirements of the design). All piping used for pool plumbing is a minimum of schedule 40 PVC or the equivalent and must be stamped with American Society for Testing and Materials (ASTM) approval. The liners are typically plastics or similar materials.

Design

The variety of designs for swimming pools are a reflection of the unique spatial and economic concerns of the consumer. Concrete pools are the most flexible with the ability to create any shape or configuration. Vinyl-lined pools are evolving into free form designs that can incorporate a larger variety of design options with fiberglass pools having the most restrictive design elements (one-piece construction designed at the factory). Design drawings are completed showing the actual layout of the pool, including shape, elevation, and size. An engineer provides soil and structural analysis assisting in the location and optimal design for a swimming pool installation.

Soil analysis is performed on all surfaces prior to installation to ensure a structurally sound pool. A pool also needs to have a system that circulates the water from the bottom of the pool to the surface to help in the quality of the water. All pools must use grounded wiring and have breakers that protect from shorts in under water lighting systems.

The Manufacturing Process

The following is the manufacturing process of a vinyl-lined concrete pool.

1 The drawings and designs of the pool are completed as the first step in the process of pool installation. After determining the grade level of the pool the site is excavated.

2 The excavation is 3 ft (91.4 cm) outside of the actual pool dimension to allow a working area around the pool. First, stakes are driven into the four corners of the pool area. The stakes are 2 × 3 in (5.1 × 7.6 cm) and sharpened at one end. Next, a second set of stakes is driven, outlining an area 6 ft (1.8 m) larger than the pool.

3 An experienced worker uses a bulldozer to excavate the earth in the site of the pool. The excavation needs to be level so a transit level is used. During excavation 2 in (5.1 cm) is allowed for bottom material. The bottom is smoothed using sand, vermiculite, cement, grout mix (sand and cement), or stone dust. The bottom must be a smooth surface as this is where the liner rests. The size of the pipe is determined by calculating the volume of water in the pool and varies depending on the holding capacity of the swimming pool.

4 The wall of the in-ground swimming pool is made from a continuous length of machined fiberglass sheeting. The walls are lowered into the excavation manually beginning with the corner sections. The subsequent wall panels will be self-supporting as each additional panel is added. They are installed working in two directions placing panels with machined holes for light and skimmer in their appropriate places.

5 A steel rod driven through the hole in the bottom of the panel fixes the positions.

6 The bottom circulation or main drain is connected to the suction side of the pump and motor. Two 1.5 in (3.8 cm) plugs are installed into the bottom threads of the main body of the drain. Duct tape is used to cover the drain to keep dirt and/or bottom fill from getting into the gaskets or faceplate.

7 Next, the pipe is dug below the existing grade and under the panel wall. A small

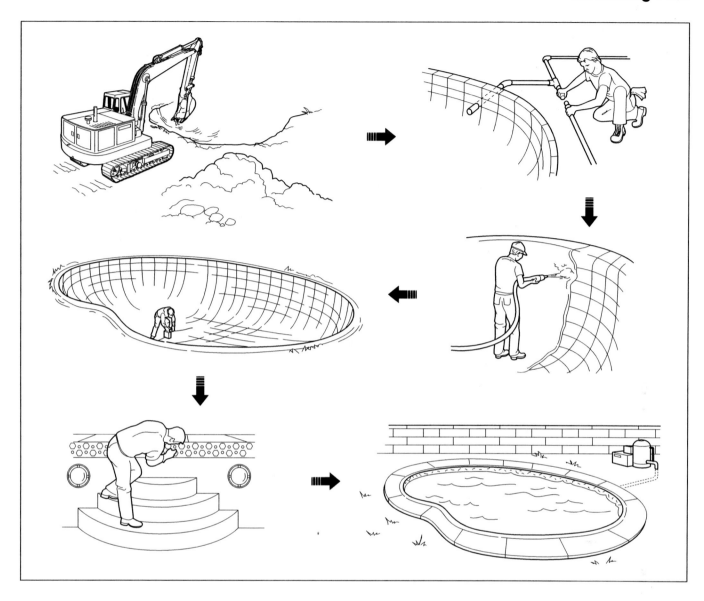

square of concrete is poured in the middle to secure the drain in place.

8 After installing the main drain the walls and corners are rechecked for level alignment and squareness. A concrete collar is poured around the perimeter of the pool at the base of the walls to secure the pool in place. The collar is made from a thin concrete mixture about 6 in (15.2 cm) deep around the pool wall. After the concrete collar is dry (about 24 hours) the bottom can be finished.

9 The concrete bottom is poured, either pumped from a truck or by a method called gunite. Gunite pumps dry cement and sand through a hose adding water at the nozzle.

10 Vermiculite and cement or grout provide a permanent bottom and is installed by an experienced mason. Plumbing is installed before backfilling. Black poly (coil pipe) or PVC schedule 40 pipe is placed along the top of the concrete collar.

11 The pipe is run from each fitting (skimmer, main drain, return) back to the filter. Pipe compound or non-hardening permatex is placed on the thread of the adapter.

12 Two pipe clamps are slipped over the end of the pipe and the pipe is pushed over the end of the fitting. Tightening clamps will set in approximately 15 minutes.

Depending on customer specifications, swimming pools vary in size, shape, and design.

13 After the vacuum has cleaned the area, water is added. If large rocks are used for the backfill, the pipes are covered with sand to prevent damage.

14 Once the pool has been filled with water all the equipment should be turned on. Chemicals such as chlorine, muratic acid, and stabilizer will need to be added. The filter system should be run continuously the first 24 hours until the water reaches the desired level of clarity. This typically represents the time required for 99% of the pool water to pass through the filter. Usually, once this level is reached the pool can run as little as six hours a day to maintain a healthy environment. The surface of the water usually contains the most pollutants (i.e., body oils, grease, sweat, and skin debris). To keep the pool clean, a skimmer (filtering device) should draw at least 70% of the pool water from the surface for filtration and treatment.

15 Water temperature has to be carefully monitored between 80.6–86°F (27–30°C). Temperatures that are too high may cause cracks in the pool structure or cause vinyl liners to expand and lose their elasticity. The higher temperatures may also destroy the polymers leaving the liner dry and brittle.

16 The maximum velocity in any suction pipe must not exceed 5 ft (1.52 m) per second. The maximum velocity in any pressure pipe must not exceed 9 ft (2.74 m) per second. The Filtration Rate is the speed or velocity of the water through the filtration media. The slower the Filtration Rate, the more effective the filtration.

Quality Control

As soon as the ditches are dug and the piping in place, the pipes are capped and filled with pressurized water to check for stress (leaks). Electrical code inspectors check the electrical wiring systems for safety. Concrete decks and any surfaces using gunite are checked for smoothness and integrity. Since the gunite is customized to the design specifications the concrete is inspected visu-

ally on each site. The steel reinforcing material is checked after installation looking for any stress or weakness in the material.

Byproducts/Waste

The variety of materials used in pool construction create a large proportion of recyclable materials. Galvanized steel reinforcing grids are manufactured at the factory with scrap metal being melted and reformed for future projects. Concrete byproducts from trimming the foundations are disposed of into large dumpsters onsite and extreme care is taken to diminish waste. City water systems are used to dispose of the backwashed pool water and plastics used in the liners are carefully fabricated with excess materials being recycled.

The Future

Improvements are constantly being made in the manufacturing and construction of swimming pools. Above ground pools are using lighter weight construction and are easier to assemble. Newer materials such as a laminate of woven polypropylene mesh fabric will not burst but cause water to slowly leak from the pool. Fiber optics are being used in underwater lighting systems and provide safe alternatives to electrical installations. Automated pool systems allow in-house operation and pre-programmed timing of pool maintenance activities.

Where to Learn More

Other

Above Ground Pools Web Page. December 2001. <http://www.above-ground-pools.com>.

Pleasure Pools Web Page. December 2001. <http://www.pleasurepoolinc.com>.

SwimCSI Web Page. December 2001. <http://www.swimcsi.com>.

The Swimming Pool Buyer's Guide Web Page. December 2001. <http://www.swimmingpoolbuyersguide.com>.

Bonny P. McClain

Swimsuit

Background

A swimsuit is an article of clothing used for swimming and sunbathing. For women, the swimsuit is either a two-piece bra and panty ensemble or a one-piece maillot style. Men's swimsuits are either a bikini-style brief or the longer and fuller swim trunk.

History

Although swimming is not a natural human ability, people have been drawn to water since ancient times. The Romans built the first swimming pools and by the first century B.C. had even created a heated pool. In Japan during this period swimming events were common. Europeans were slower to come to the sport because of a widespread fear of infections carried from other bathers through the water. However, by the mid-nineteenth century, a number of swimming organizations were founded, particularly in London.

The swimsuit as a particular article of clothing did not appear until the early twentieth century. In the late nineteenth century, bathers wore a bathing costume consisting of billowy bloomers and overblouses, stockings, and shoes. Suntans were considered a sign of low class, so many women covered their heads and faces. To many, even these body-obscuring outfits were considered shocking. Over the next several decades, the style and acceptance of bathing wear changed significantly.

At the beginning of the 1902, three young men in Oregon (John Zehntbauer, Roy Zehntbauer, and Carl Jantzen) owned a clothing company called the Portland Knitting Company. They were also avid members of a rowing club. Their financial futures were secured when one of their teammates asked them to create a wool rib-knit rowing suit that would retain body heat. Although the garment they created was not particularly suited for swimming (when wet it could weight up to 8 lb [3.6 kg]), an idea was born. The form-fitting knit suit, made by the company that would become Jantzen, featured a sleeveless shift over long shorts.

In the 1930s, sunbathing became a popular pastime. Women's styles began to feature lower-cut backs and armholes to allow more exposure to the sun. Jantzen introduced the Shouldaire model with a drawstring sewn above the bustline that allowed the wearer to lower the shoulder straps for better suntan coverage. It was also in the 1930s that women's midriffs were exposed for the first time; in cutouts and eventually in two-piece swimsuits

Manmade fabrics were introduced during this period. The evolution of the swimsuit as a form-fitting garment called for a flexible and elasticized material. Rayon was the first fabric used, and then American Rubber Company developed Lastex, an extruded rubber surrounded by fiber. Lastex's success was short-lived because it was not colorfast and did not retain shape when stretched. The fabric's flexibility was also affected by body oils.

In 1939, E. I. duPont de Nemours & Company developed a nylon called 6.6 polymer that revolutionized the manmade materials industry. In the years to come, other synthetic materials such as Dacron, Orlon, Lycra, and Spandex were invented and were used alone or blended to make swimsuits. During this period, textile manufacturers also began to experiment with woven patterns and bright colors.

In the late nineteenth century, bathers wore a bathing costume consisting of billowy bloomers and overblouses, stockings, and shoes.

365

The next major highlight in the history of the swimsuit occurred in 1946 when the bikini, a two-piece suit for women, was introduced in Paris. Supposedly named for the Pacific atoll where atomic bomb experiments were conducted, the bikini caused a furor. Although immediately popular on European beaches, the bikini was not worn in the United States until the 1960s.

In the 1970s, the use of materials such as Lycra became popular as a means to manipulate physical attributes. Soft one-piece maillots were popular among women and remain so today. In 1977, designer Rudi Geinrich's thong bikini, which features a mere strap on the rear portion of the suit bottom, hit the beaches of Brazil. It also remains a popular style around the world, but is often considered controversial on American beaches.

Innovations of the 1980s included suits of material that allowed the wearer to tan through the fabric. This style faded in popularity as the public became more aware of sun-related skin cancers. The French cut, leg openings high on the hip, also appeared during this decade.

In competitive swimming, the design of suits built for speed is an on-going challenge. At the 2000 Olympics in Sydney, Australia, the use of a full-body suit by many swimmers caused a uproar. The Fastkin suit, made by Speedo, is constructed of a sharkskin-like material and is marketed as a performance-enhancing suit. A number of teams objected to the use of the suit and sought to have it banned. However the Olympic governing committee allowed the suits.

Raw Materials

Fabric is the primary material. Some companies manufacture their own fabric while others purchase it from outside supplies. Synthetic dyes are used to color the fabric. Until the mid-nineteenth century, dyes were extracted from animal, vegetable, and mineral sources. In 1856, a young chemist in London named William H. Perkin accidentally discovered how to make mauveine, a purple dye, while he was attempting to synthesize quinine. Dyes are applied in a variety of ways depending on the type of dye, the type of fabric, and the desired effect. In the simplest process, cloth is dipped into a solution of water and dye. Sometimes an oxygen reagent is added to make the color more uniform. In mordant dying, a wet metallic solution of tin, chromium, iron or aluminum is applied directly to the fabric. Then a dye is applied on top of that and the color is formed within the cloth. Some dyes can be applied directly to the cloth. In this process, the fabric is immersed in a hot solution of the dye. Patterns of color are created by dying fabrics that have been woven with different types of yarn such as nylon and polyester. The yarns react differently, or not at all, to different dyes. In this manner, a pattern appears on the fabric.

The bra of a bikini or two-piece woman's swimsuit may have metal or plastic fasteners. Men's swim trunks often include a drawstring in the waistband. Lengths of elastic are used for straps, leg openings, and waistbands.

Design

Design is a crucial step in the manufacturing of swimsuits. As in any aspect of the apparel industry, designers pay close attention to what is being worn and what is new in fabric and color. Swimsuit designers are also concerned with comfort, colorfastness and elasticity of the fabric.

Designers use a combination of hand-sketches and computer-assisted-design software (CAD) to create new styles. Hand-sketches are enlarged to create paper patterns and a sample is cut from a material such as muslin. The garment is then fitted onto a mannequin and adjusted until the designer achieves the desired look. Colors and fabric are chosen and a sample is made and tried on by a human model. The designer again makes adjustments.

Designers using CAD draw with a stylus onto a digitizing pad that is connected to a computer. As the designer draws, the garment's image appears on the computer screen. Colors and fabrics can also be chosen and viewed on the computer screen. Templates are created and sent to the factory to be cut into pattern pieces.

The Manufacturing Process

The manufacture of swimsuits is largely a computerized and mechanized process with

Lady's bathing suit 1886   Modern bathing suits

factory workers running the machinery and occasionally guiding the fabric.

1 Spools of cotton and synthetic thread are loaded onto knitting machines that weave the threads into rolls of fabric. The rolls are fed into large tanks fitted with agitators. Pre-measured amounts of bleach and color-dyes are released into the tanks. After the fabric has been cleaned and dyed to the desired color, it is then placed into drying machines. The fabric is re-rolled and stored until it is needed.

2 Workers bring the bolts of fabric to spreading and measuring tables. The bolts are attached to one end of the table and the fabric is drawn across the table and wrapped around an empty bolt on the other side. The worker turns the empty bolt to take up the slack until the fabric is pulled taut across the table. The worker enters pre-determined length measurements into an encoder. The encoder then relays the information to electronic blades that cut the fabric.

3 After the entire bolt of fabric has been cut into lengths, the worker stacks them in heights up to 6 in (15 cm). He or she then takes the stacks to the piece-cutting machine. Here, another worker operates the computerized machinery that cuts the swimsuit pieces from the lengths of fabric. In smaller factories, pattern-marking may be done by hand before the fabric is cut. In larger companies, the pattern dimensions are fed into a computer that relays the information to the cutting machinery.

The number of pieces is determined by the style of swimsuit. A woman's one-piece maillot is usually made from two pieces. A bikini would have two sections for the brief and four to six pieces for the bra. Cups for the bra and for the top of the one-piece are also cut. Lining panels are cut for the crotch and for the bodice. Men's swim trunks are constructed from two to four panels.

4 Each piece is stitched to another at separate sewing stations. Depending on the size of the factory, the sewing is done by individual seamstresses working at industrial sewing machines, or by computerized stitching machines operated by workers. For a bikini, the bra cup is placed between the lining and the front bra panels and the three pieces are stitched together. A side panel is then stitched to each of the front panels.

If the design calls for straps, lengths of elastic are placed between two strap pieces and the three pieces are sewn together. The straps are then sewn onto the front and side panels of the bra. If hooks are used to close the back of the bra, a metal or plastic hook is sewn or ironed into a facing on the end of one side panel. A loop is made on the other side panel by folding the end piece over and stitching it to the panel.

Briefs, whether for a woman's two-piece suit or for a man's swim trunks, are pieced together in similar fashions. Lining is stitched into the front panel or panels. The front panel(s) are joined to the back panel(s). Lengths of elastic are inserted into waistbands and leg openings. The outer material is folded over to make a facing and the facing is then stitched to the garment.

5 The completed garments are pressed and labels are stitched onto the inside. The swimsuits are packaged in plastic bags and loaded into cartons for shipment to retail outlets.

Quality Control

Swimsuits are subjected to a number of tests in the factory before they are sold to the public. Tests for fabric and color changes include repeated washings in fresh, salt, and chlorinated water, as well as exposure to simulated sunlight. The suits are stretched and weighted down before and after washings to determine if they retain their original shapes. Samples suits are also given to volunteer testers who report back to the companies on comfort and wearability.

Byproducts/Waste

The primary waste products result from fabric dyeing and from the manufacture of synthetic fabrics. An entire industry dedicated to the recycling of dyes and synthetic materials now exists to serve the textile industry. Dye solutions are generally purified and reused. The residue from the manufacture of synthetic fabric are used to make other products, especially plastic bottles.

Very few waste materials exists after the sewing of the swimsuits. The computerized processes allow for precise measurement and cutting so that little excess remains. What extraneous bits of thread and fabric do exist are discarded.

The Future

Although no significant design changes have occurred in the swimsuit industry in the last 20 years, the popularity of the suit and the popularity of swimming is expected to continue. Innovations are likely to focus on figure enhancement. The use of computer-assisted-design and computer-assisted-manufacture is expected to increase. Industry associations are currently working to devise a universal standard for the computer language used.

Where to Learn More

Books

I Want to be a Fashion Designer. San Diego: Harcourt Brace, 1999.

Other

Good Housekeeping Magazine Web Page. December 2001. <http://goodhousekeeping.women.com>.

Jantzen Web Page. December 2001. <http://www.jantzen.com>.

"The History of the Bathing Suit." *Retro Web Page*. December 2001. <http://www.retroactive.com/mar98/swimsuit.html>.

Mary McNulty

Table

Background

The table is a basic piece of household furniture. It generally consists of a flat top that is supported by either a set of legs, pillars, or trestles. The top may be made of stone, metal, wood, or a synthetic material such as a plastic. Tables may be subdivided by any one of a number of criteria, the most basic of which is whether the table is a fixed table or a mechanical table. A fixed table has a top that does not move in any way to expand or reduce in size for storage. The tops on fixed tables can be quite sizable and may be supported by a single column or pedestal. Mechanical tables have tops or legs that move, fold, drop, or in some way may be reconfigured in order to save space or make them more flexible. Mechanical tables include drop leaf tables, tilting tables, or those with legs that fold up or collapse if a mechanism is unlocked.

Tables are more commonly subdivided by other criteria such as the material from which they are made, the purpose for which they are constructed, the form they take, and the style of any added decoration. Style is an extremely important part of a table. The look of the table may vary as a result of many factors. These include changing stylistic preferences, advances in technology that make available different materials for the table or methods for its construction, and new table forms that are the result of new human activities or needs.

Tables purchased in this country are most frequently mass-produced from wood and can be made with minimal cabinetmaking skill. American-made tables may be made from native hardwoods such as maple, oak, or alder, or soft woods such as pine. American tables may be manufactured unfinished meaning without any stain or sealer or may be purchased ready-to-use in standard or custom finishes. Some table manufacture takes place in the home; these are considered custom or specially made pieces of furniture that must be constructed by a cabinet maker.

History

Until about the sixteenth century, when decorative and stylistically distinctive furniture became very important, tables were less frequently found than either the chair or the chest (which held clothing as a chest of drawers does today). However, there were tables in the ancient world. Different cultures made them of different materials. Egyptian tables were of wood or stone and resembled pedestals. It is said the Assyrians made them of metal. Pompeii and Herculaneum populaces had tables made with supporting members of marble.

Cathedrals in the Middle Ages used communion tables that stood on masonry or on a base of stone. Castles often included large, rectangular plank tables with the master of the castle in the center and the less important inhabitants or guests at right angles to him. More ordinary medieval tables that survive include simple wooden tables supported by plain side members. Early seventeenth century American tables were generally of the trestle type, with a plank top and vertical planks on the side. Some could be dismantled if more room was needed; many were just moved against the wall to provide space when the table was not in use.

Decoration became very important to the wealthy about the sixteenth century as well.

Pompeii and Herculaneum populaces had tables made with supporting members of marble.

Stylish furniture was ornately carved and included turnings made on foot-pedal lathes. Until the mid-seventeenth century most furniture was constructed by *joiners* who made furniture much as they made houses, with pegs, mortise and tenon construction, and massive members for supporting the slab tops. In the later seventeenth and eighteenth century the cabinetmaker began making fine furniture, creating sculptural pieces that were veneered, carved, and expertly joined including the use of interlocked dovetailing for strength.

In the early nineteenth century the machine enabled manufacturers to provide attractive furniture far less expensively. Wood was cut by water, steam, or electrical saws, machine sanded, machine incised and decorated, turned on machine lathes, and so forth. By 1890, all but the very poorest Americans could afford to purchase an inexpensive table and chairs. In the early twentieth century the table changed again, this time because new, unconventional materials were used in its construction such as laminate, plastic, and chipboard, making tables truly affordable for all. As new activities were enjoyed and embraced, tables changed form, too. Table forms that were invented in the past 200 years include the card tables, gaming tables, tea tables, dressing tables, diapering tables, and computer tables.

Raw Materials

Raw materials vary greatly according to the type of table under production. Unfinished pine table made in quantity in this country include pine planks that are called one-by-fours or one-by-sixes. (These are boards that were once truly 1 in [2.5 cm] thick by 4 in [10.1 cm] wide or 6 in [15 cm] wide but are now cut slightly smaller than that size today.) Other materials include water-resistant glue formulated from polyvinyl acetate. Hardware, including screws, vary according to the price point of the piece but are often steel. Most American table manufacturers are careful to obtain woods that are certified, meaning the manufacturer can prove that the trees were harvested legally from controlled forests grown specifically for the manufacture of furniture. Furthermore, furniture-grade wood is especially important in the construction of unfinished tables, in which the grain may not be covered with paint. Furniture-grade wood is virtually knot-free or clear; when there are small knots the company must be sure they can use the wood in a hidden area of the piece such as the back or inside a drawer. Drawer bottoms or sides may be of a plywood, engineered wood (pressed wood chips formed into sheet goods), or even masonite.

Design

The decoration and configuration of tables are fairly important in the unfinished furniture industry. Additive or incised decoration may be found on the table apron (a board which goes across the front of the table running from leg to leg and may hold the drawer front), or on the legs themselves. Painted decoration may be seen at any place on the table. The shape or form of table top, table legs, or the apron determines style and may be created by specialized machinery

While the high-end manufacturers of ready-to-use furniture spend a great deal of time and money on the design of their furniture, the unfinished furniture generally provides basic forms to the consumer. The unfinished table manufacturer surely cares about selling an attractive table, but it is not likely of the most stylish or innovative table shown at the important furnishings markets. Most larger furniture firms have a design director on staff whose job is to ferret out new designs for their market and work with the production managers to create these styles economically. These larger firms haunt malls, study the shelter and fashion magazines, and perform some audience assessment of taste and style preferences.

However, smaller firms, such as those who produce medium to low-price products, point out that unfinished goods may spend less money on the development of styles and decoration, preferring to offer basic tables and forms to the consumer. Some smaller firms may assign the task of developing new products and styles to the production manager. This manager works with staff designers to craft tables that can be manufactured using the equipment used in-house. Interestingly, some prefer to design tables for which parts can easily be interchanged, resulting in a wide array of products with little re-design. For example, a console table may have the same front and back apron and drawers as the coffee table but have a narrower top, sides, and longer legs. A Queen Anne-style coffee table may have cabriole (curved) legs while a Shaker-style coffee table may be identical except the legs are rectilinear and slightly tapered.

Designers or production directors generally keep their eye on current styles, assessing what is leading the market and what trends are infiltrating the target market. Generally, when a new style or form is suggested for production, a team of directors, including the director of sales and marketing, the director of manufacturing, and in-house designers assess the viability of the new design. If the design is approved, the director of manufacturing and the designer works with an operator who uses a computer-based design and drafting system such as AutoCAD. This operator works with the design on a computer and then inputs that information into the computer in order to produce that product on computer-driven machines. All staff members work together to devise the best way to get the new table form through the system, especially vigilant that the costs of the new table will not exceed the price point of the intended market and that no new machinery or manufacturing expertise will be necessary to produce the new product.

The new table must be made in prototype in order to evaluate how the product will go through the established system. In addition, the staff must physically examine the proposed new table for aesthetics and durability. The prototype is made using templates made on machines. Any changes to the prototype are made, the AutoCAD operator changes computer settings for templates, and the piece is ready for production once approved.

The Manufacturing Process

1 The pine boards come into the factory fairly rough but cut to pre-determined lengths that are then cut to shorter lengths using a power saw. The wood must stay in a carefully temperature and humidity-controlled room or the wood may swell (too much humidity) or shrink (very dry) and the piece will have cracks when finished. Temperatures must stay in the range of 50–85°F (10–29°C).

2 Workers gather up the pine lengths and spread water-resistant wood glue to the long edges of the planks where they join with other boards to make a table top. The planks are then clamped together with furniture clamps to ensure a tight bond and a sturdy top. Several dozens of the tops can be made at one time and must sit for at least one hour.

3 The raw table tops are then sent to wide belt-driven sanders that sand the planks down to take away extreme roughness and splinters.

4 Most tables have some sort of apron or vertical board just below the table top. The apron must be cut next. A computer numerically controlled (CNC) router cuts several aprons according to computer specifications. This router is especially suited to cutting sheet goods such as planks of wood. The shapes to be cut or routed are defined by the drawing programs such as AutoCAD. The information is transferred to a CAD/CAM program that allows the user to define the path of the router tool. When the tool path has been defined, the computer software allows a tap file to be made, which actually runs the tool over the wood or plank, cutting it into the desired shape. The router is able to create high quality routing and carving effects with uniform consistency and with very little wasted wood. The cut aprons are put into a cart and head to the assembly area.

5 The board tops, now glued and sanded, are sent to the CNC router and cut to the desired shape as well as described above.

6 The legs are shaped on a profiler. Before the legs are shaped an aluminum template must be cut on the CNC router. Once the template is cut, the metal template is put on the profiler, and a bearing guide follows the template in order to shape the table legs.

7 Legs with special stylish feet such as a pad foot are put into a chucker which acts like a giant pencil sharpener, shaping the stylish feet. The table tops and legs are put into carts and are sent to the assembly area as well.

8 If the table is to receive a drawer, then the front, sides, and bottom of the drawer are cut on the CNC router. The front is of

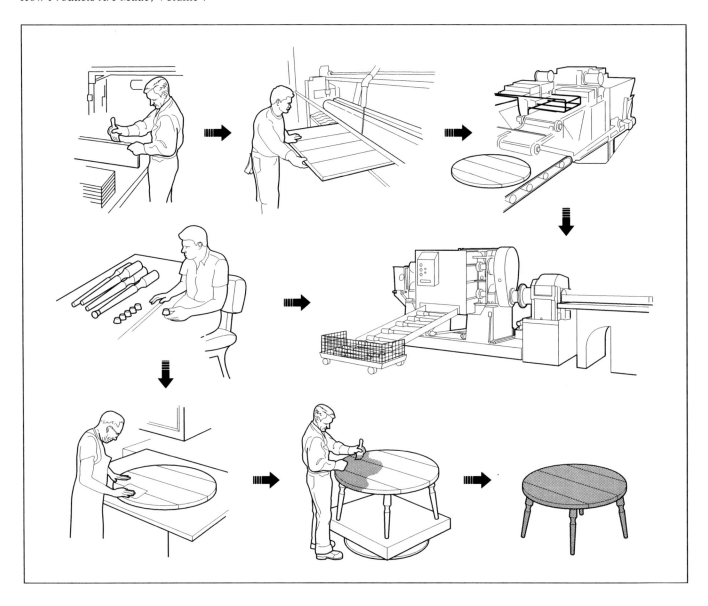

The manufacturing of a medium-sized round table.

pine, but the sides and the bottom of the drawer may be of inferior grade pine, plywood, or engineered wood. For the drawer front, a CNC router cuts a groove for the bottom and sides that will be inset into the back face of the drawer. The face is then flipped over, the grooves receive a coating of waterproof glue, and the bottom and sides are inserted into these tracks and may be nailed together using an automatic nailer for additional stability. The table is now ready for assembly.

9 Many unfinished furniture manufacturers do not ship their tables assembled as the shipping price increases wholesale costs significantly. Those that do not intend to assemble their products proceed to the final sanding before packaging. Companies that

do assemble their products employ workers who examine a work order and pull the appropriate parts from the bins, then ready them for assembly. The assembly of the table is done by hand. This is the preferred method in many shops as a manufacturer may produce several tables with very similar parts, thus making it difficult for a machine to discern which parts were used on specific product numbers. Jigs or templates mark where the screws must go. They are laid on the table tops so the table may be screwed in place precisely.

10 The tables are given a final sanding using a hand sander. While this is a fairly careful sanding it is not considered a fine sanding or finish sanding as the company presumes that the consumer will spend time

sanding the surface to close the grain, remove hand oils that prevent it from taking a good stain, and further reduce any mill marks.

Quality Control

Quality control is monitored at every step of production. Most storage rooms for wood raw materials stay between 50–85°F (10–29°C) and moderate humidity. The moisture content, known as MC in the trade, is the weight of water contained in the wood compared to the wood's oven-dry weight. This moisture content should never exceed 25% and ideally should stay around 12%. Most factories try to find one temperature and humidity and keep these constant so that the wood comes to an equilibrium moisture content. Temperature and humidity must be moderate to ensure that a product does not bend, crack, or warp after manufacture. Excessive humidity can result in the loosening and weakening of joints and even failure at the joint. In fact, most of the problems associated with wood in the manufacture of furniture are associated with dimensional changes or movement of the wood due to variation in humidity.

Everyone who assists with production is constantly performing visual checks of the wood. Wood is checked for cracks, knots, or discoloration that are unsightly or may weaken the piece structurally when boards are first chosen for the tops in the gluing process. Hand gluing and stapling of the drawer and hand assembly of the entire table helps ensure a strong, sturdy table. The jig is carefully placed on the table top and apron in order for the screws to be placed in the correct positions, further ensuring stability. Even after the tops are glued and sanded, they are checked again for flaws in the wood. Hand assemblers and hand sanders who come in at the end of the process give the table a visual examination as well. Finally, in packaging, the entire table is looked over, then sent to the warehouse for storage.

Byproducts/Waste

Wood waste generated from the routing and profiling is gathered up and sent in quantity to the factory's "hog" which chops the waste into fine chips. A variety of companies, including remanufactured furniture factories, paper product producers and manufacturers who make particle board, chip board, and so forth may arrange to take away these small, processed chips.

The Future

Currently, the availability of North American woods for the production of unfinished tables is not a problem. Increasingly there is interest from the consumer that the pine used in such tables is certified, meaning it has been legally and carefully harvested and is not a foreign rain forest product. Labor costs for the production of such pieces is not prohibitive and the abandonment of furniture production in this country is not likely in the immediate future. However, stylish, inexpensive, already finished tables of imported wood such parawood are proving to be challenges for some of these manufacturers. As Americans become more concerned about using these foreign woods it may be that these tables, particularly computer tables, will not sell well. However, their price points may be so competitive that the use of certified woods may be deemed unimportant.

Where to Learn More

Books

Krill, Rosemary Troy, and Pauline K. Eversmann. *Early American Decorative Arts, 1620–1860.* Walnut Creek, CA: 2000.

Other

"Quality Control in Furniture Manufacture: Moisture Content." December 2001. <http://www.mtc.com.my/publication/library/quality/content.html>.

Oral interview with Roger Shinn, Director of Production for Westview Products. Dallas, Oregon. July 2001.

Nancy EV Bryk

Tattoo

Background

A tattoo is a design that is permanently etched in the skin using needles and ink. The word tattoo is derived from the Tahitian term "tatua," which means "to mark." Tattoos have been displayed by people of all cultures for centuries, but they have only recently gained social acceptance in the United States.

History

Adding decorative illustrations to skin has been a popular practice since ancient times. Clay dolls have been found that indicate the Egyptians used tattoos as early as 4000 B.C. Over the centuries, different forms of tattoo art have been practiced by many different world cultures. For example, around 500 B.C., the Japanese began tattooing for both cosmetic and religious purposes. They even used tattoos to brand known criminals as part of their punishment. The Japanese method involved puncturing the skin with fine metal needles to create multicolor designs. Eskimos tribes developed their own technique using bone needles to pull soot-covered thread through the skin.

In the 1700s, Captain James Cook traveled to Tahitia and observed the natives' skin marking customs. In his book *The Voyage in H.M. Bark Endeavor*, Cook wrote, "they stain their bodies by indentings, or pricking the skin with small instruments made of bone, cut into short teeth; which indentings they fill up with dark-blue or black mixture prepared from the smoke of an oily nut. This operation, which is called by the natives 'tatua' leaves an indelible mark on the skin." In the years after Cook's voyages, sailors visiting the Polynesian islands spread the Tahitian ritual around the Pacific.

The popularity of tattoos continued to grow over the last 200 years. In the nineteenth century, tattoos became popular in England among the upper-class. For example, Lady Randolph Churchill, Winston Churchill's mother, had a snake tattooed around her wrist. In the United States, tattoos have been historically associated with sailors, motorcyclists, and prison inmates because tattoo shops were considered dangerous and socially unacceptable. However, since the 1980s this mindset has changed considerably and tattoos are becoming increasingly popular among men and women of all ages.

Design

A tattoo design is called "flash" and it can consist of any sort of artwork from simple symbols or letters to detailed sketches or caricature. Flash can be composed of one color or many. Tattoo parlors display a large assortment of flash on their walls with the larger ones having as many as 10,000 to choose from. In addition, clients may bring in their own design or they may work with the artist to develop custom flash.

When selecting a design it is important to consult with the artist to establish an appropriate size and location for the tattoo. The artist can also help decide on color schemes that will determine the price of the final art. Care should also be taken to identify a reputable tattoo parlor that follows the guidelines set forth by the Association of Professional Tattooists (APT). According to the APT, the tattooists should follow these precautionary measures: have the client fill out consent forms before beginning the procedure; wash and dry their hands immediately before and after working on the customer;

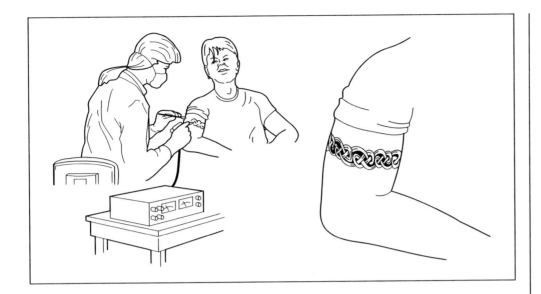

wear latex gloves at all times; only use instruments that have been sterilized in an autoclave; clean all surfaces with a disinfectant or biocidal cleanser; and dispose of used tissues and other waste material in a special leak-proof container to limit the transmittal of blood borne diseases.

Raw Materials and Equipment

Flash

The flash, or tattoo design, is simply a sketch or a piece of line art that can be used to create a tattoo. Flash may be shown in color or in black and white and they are displayed in the tattoo parlor either in books or along the walls.

Stencil

A stencil is a copy of a flash that is made on a special copying machine. The stencil allows the inked outline of the design to be transferred to the skin so it can be traced by the artist.

Ink

Tattoo supply houses sell special inks that are used to create tattoos. They are available in a variety of colors and are typically packaged in 4 oz plastic squeeze bottles so they are easily dispensed. These inks are liquid dispersions of pigments that, in the United States, are approved by the Food and Drug Administration. The ink consists of dyes derived from

metal components. For this reason, allergic reactions to the type of ink used is possible.

Tattoo Machine

The machine consists of a hand held needle gun connected to a power unit that provides pressure to move the needles. The needles may be of different sizes and shapes and are bundled together on a needle bar in different patterns depending on the requirements of the artwork. The unit is attached to a power supply that is activated by depressing a foot pedal on the floor beside the work station. When the pedal is depressed the tattoo needle bar moves up and down very quickly like the needle on a tiny sewing machine. It penetrates the skin to inject the dye 3,000 times per minute.

Miscellaneous supplies

During the course of the procedure the artist may use a variety of additional supplies including skin disinfectants, disposable razors, bandages, petroleum jelly, and biocidal cleaning supplies.

The Manufacturing Process

1 Before the process can begin, the artist should have the client sign a waiver that indicates they are over 18, understand the procedure is permanent, and realize it will create an open wound or abrasion. After the waiver is signed, the tattooist inspects the skin to ensure it is free from cuts and

The tattoo machine.

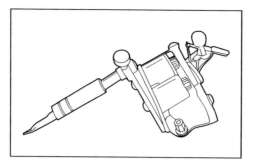

scrapes. He or she then sprays the skin with an antiseptic to kill germs and reduce the possibility of infection. The artist then shaves the area and disposes of the razor immediately afterward.

2 When the skin is ready, the artist prepares the design to be copied onto the skin. This is done using a copy machine to make a color copy of the flash that is the proper size. The copier uses a special carbon-type paper that allows the design to be transferred to skin. The client exposes the target area and the artist peels the design off the backing paper and applies it to the proper location. This creates an ink outline of the design on the skin that is used to guide the artist in creating the actual tattoo. The client must be careful not to touch the outline or to otherwise disturb the skin because the stencil can be accidentally smeared. At this point, the client can look at the design (in a mirror if necessary) to confirm it is correct. If there is something wrong with the stencil it can be washed off and applied again. Once the stencil is completed, the tattooist spreads a thin layer of ointment, such as petroleum jelly, over the area to be tattooed.

3 The artist directs the client to sit or lay in a position that exposes the skin to be tattooed. The client may recline in a dentist-type chair that can lay flat to aid in exposing the back or buttocks. The position must be comfortable for the client but must also provide a comfortable working position for the artist who typically sits on a stool next to the client's chair.

4 Next, the artist prepares the tattoo inks on a pallet, which is a plastic tray with a series of divots to hold the ink. The artist fills several of these wells with ink; black is typically used for the outline color. He or she then attaches a needle bar with three to five needles to the tattoo machine. The needles

are dipped into the ink well to suck up the colorant and the artist then activates the needle gun with the foot pedal and begins to trace the stencil. Because the needles are breaking the skin for the first time in this step, it is a very painful process. The needles deposit the dye in the second layer of the skin which is about 0.64–0.16 in (1.6–0.4 cm) deep. This process may take five minutes to an hour depending on the size and complexity of the design. The skin may be slightly numb by the time outlining is complete.

5 After the stencil is outlined the artist fills the ink wells with the colors to be used to finish the tattoo. He or she then connects a shader needle bar to the tattoo machine. This shader bar may contain five to thirteen flat needles or five to seven round needles. It is designed to apply color over a larger area to fill in the outline. The tattooist guides the tattooing machine over the skin coloring in all sections of the outline. He or she frequently stops the needle to wipe the blood and ink off the skin. The amount of bleeding caused by the needle penetration and the degree of pain experienced varies from person to person. The shading operation may take an hour or several hours depending on the complexity and size of the design. Shading is complete when the entire surface of the tattoo has colored in.

Quality Control

The key to ensuring a successful tattoo is taking care of it properly in the first few hours and days. Immediately after the tattoo is finished, the area is washed with a mild soap solution and then covered with an antibiotic ointment and a gauze bandage. After allowing the area to heal for about two hours, a bandage should be removed so dried blood can be washed away. For the first week, apply a vitamin lotion to the area daily. After that, apply a regular, mild skin lotion to keep the tattooed area moist until the wound has finished healing. The affected area should be kept out of hot tubs, swimming pools, and hot baths until the skin has healed fully. Submersion in water too soon can ruin the tattoo.

In general, taking care of the tattoo is like treating a minor burn. It must be kept clean and moist, and will experience the same sort of scabbing and crusting. The initial healing process usually takes about two weeks.

Tattoo Removal

It is not uncommon for someone to change their mind after a period of time and to want to have their tattoo removed. Tattoo removal is possible but the process is difficult, expensive, and not fully successful. In the past, a wire brush was used to sand the skin and destroy the first and second layers where the ink resided. Salt solutions were also used to leach out the ink or acid was used to burn the skin away.

All of these methods are painful and not very effective. Even if the tattoo can be removed, the affected area may lose its ability to produce normal skin pigment and some scarring may occur. Recently, lasers have been used to remove tattoos because they can destroy most of the ink pigments and cause very little scarring. Still, the process is expensive and the skin may never produce its normal pigmentation again.

The Future

Tattoos continue to grow in popularity as a method of self expression. It is also anticipated that tattoos will be used increasingly for medical and non-medical cosmetic applications. For example, tattoos can be used to obscure the reddish purple birthmarks known as "port-wine" stains. They may also be used to improve the skin color of patients with vitaligo, a disorder that causes the melanocytes in the skin to shut down and stop producing normal skin color. Tattooing is also being used to create permanent makeup, such as eye liner or blush, for burned or disfigured victims.

Where to Learn More

Books

Graves, Bonnie. *Tattooing and Body Piercing.* Capstone Press, 2000.

Wilkonson, Beth. *The Dangers of Tattooing, Body Piercing, and Branding.* Rosen Publishing Group, 1998.

Other

Tat2studio.com Web Page. December 2001. <http://www.tat2studios.com>.

Randy Schueller

Teflon

A fiberglass fabric with Teflon coating serves to protect the roofs of airports and stadiums.

Background

Teflon is the registered trade name of the highly useful plastic material polytetrafluoroethylene (PTFE). PTFE is one of a class of plastics known as fluoropolymers. A polymer is a compound formed by a chemical reaction which combines particles into groups of repeating large molecules. Many common synthetic fibers are polymers, such as polyester and nylon. PTFE is the polymerized form of tetrafluoroethylene. PTFE has many unique properties, which make it valuable in scores of applications. It has a very high melting point, and is also stable at very low temperatures. It can be dissolved by nothing but hot fluorine gas or certain molten metals, so it is extremely resistant to corrosion. It is also very slick and slippery. This makes it an excellent material for coating machine parts which are subjected to heat, wear, and friction, for laboratory equipment which must resist corrosive chemicals, and as a coating for cookware and utensils. PTFE is used to impart stain-resistance to fabrics, carpets, and wall coverings, and as weatherproofing on outdoor signs. PTFE has low electrical conductivity, so it makes a good electrical insulator. It is used to insulate much data communication cable, and it is essential to the manufacture of semi-conductors. PTFE is also found in a variety of medical applications, such as in vascular grafts. A fiberglass fabric with PTFE coating serves to protect the roofs of airports and stadiums. PTFE can even be incorporated into fiber for weaving socks. The low friction of the PTFE makes the socks exceptionally smooth, protecting feet from blisters.

History

PTFE was discovered accidentally in 1938 by a young scientist looking for something else. Roy Plunkett was a chemist for E.I. du Pont de Nemours and Company (Du Pont). He had earned a PhD from Ohio State University in 1936, and in 1938 when he stumbled upon Teflon, he was still only 27 years old. Plunkett's area was refrigerants. Many chemicals that were used as refrigerants before the 1930s were dangerously explosive. Du Pont and General Motors had developed a new type of non-flammable refrigerant, a form of Freon called refrigerant 114. Refrigerant 114 was tied up in an exclusive arrangement with General Motor's Frigidaire division, and at the time could not be marketed to other manufacturers. Plunkett endeavored to come up with a different form of refrigerant 114 that would get around Frigidaire's patent control. The technical name for refrigerant 114 was tetrafluorodichloroethane. Plunkett hoped to make a similar refrigerant by reacting hydrochloric acid with a compound called tetrafluoroethylene, or TFE. TFE itself was a little known substance, and Plunkett decided his first task was to make a large amount of this gas. The chemist thought he might as well make a hundred pounds of the gas, to be sure to have enough for all his chemical tests, and for toxicological tests as well. He stored the gas in metal cans with a valve release, much like the cans used commercially today for pressurized sprays like hair spray. Plunkett kept the cans on dry ice, to cool and liquefy the TFE gas. His refrigerant experiment required Plunkett and his assistant to release the TFE gas from the cans into a heated chamber. On the morning of April 6, 1938, Plunkett found he could not get the gas out of the can. To Plunkett and his assistant's mystification, the gas had transformed overnight into a white, flaky powder. The TFE had polymerized.

Polymerization is a chemical process in which molecules combine into long strings. One of the best known polymers is nylon, which was also discovered by researchers at Du Pont. Polymer science was still in its infancy in the 1930s. Plunkett believed that TFE could not polymerize, and yet it had somehow done so. He sent the strange white flakes to Du Pont's Central Research Department, where teams of chemists analyzed the stuff. The polymerized TFE was curiously inert. It did not react with any other chemicals, it resisted electric currents, and it was extremely smooth and slick. Plunkett was able to figure out how the TFE gas had accidentally polymerized, and he took out a patent for the polymerized substance, polytetrafluoroethylene, or PTFE.

PTFE was initially expensive to produce, and its value was not clear to Plunkett or the other scientists at Du Pont. But it came into use in World War II, during the development of the atomic bomb. Making the bomb required scientists to handle large amounts of the caustic and toxic substance uranium hexafluoride. Du Pont provided PTFE-coated gaskets and liners that resisted the extreme corrosive action of uranium hexafluoride. Du Pont also used PTFE during the war for making nose cones of certain other bombs. Du Pont registered the trademark name Teflon for its patented substance in 1944, and continued to work after the war on cheaper and more effective manufacturing techniques. Du Pont built its first plant for the production of Teflon in Parkersburg, West Virginia in 1950. The company marketed Teflon after the war's end as a coating for machined metal parts. In the 1960s, Du Pont began marketing cookware coated with Teflon. The slick Teflon coating resisted the stickiness of even scorched food, so cleaning the pans was easy. The company marketed Teflon for a variety of other uses as well. Other related fluoropolymers were developed and marketed in ensuing decades, some of which were easier to process than PTFE. Du Pont registered another variant of Teflon in 1985, Teflon AF, which is soluble in special solvents.

Raw Materials

PTFE is polymerized from the chemical compound tetrafluoroethylene, or TFE.

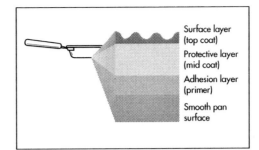

Surface layer (top coat)
Protective layer (mid coat)
Adhesion layer (primer)
Smooth pan surface

A non-stick pan is composed of varying non-stick layers.

TFE is synthesized from fluorspar, hydrofluoric acid, and chloroform. These ingredients are combined under high heat, an action known as pyrolosis. TFE is a colorless, odorless, nontoxic gas which is, however, extremely flammable. It is stored as a liquid, at low temperature and pressure. Because of the difficulty of transporting the flammable TFE, PTFE manufacturers also manufacture their own TFE on site. The polymerization process uses a very small amount of other chemicals as initiators. Various initiators can be used, including ammonium persulfate or disuccinic acid peroxide. The other essential ingredient of the polymerization process is water.

The Manufacturing Process

PTFE can be produced in a number of ways, depending on the particular traits desired for the end product. Many specifics of the process are proprietary secrets of the manufacturers. There are two main methods of producing PTFE. One is suspension polymerization. In this method, the TFE is polymerized in water, resulting in grains of PTFE. The grains can be further processed into pellets which can be molded. In the dispersion method, the resulting PTFE is a milky paste which can be processed into a fine powder. Both the paste and powder are used in coating applications.

Making the TFE

1 Manufacturers of PTFE begin by synthesizing TFE. The three ingredients of TFE, fluorspar, hydrofluoric acid, and chloroform are combined in a chemical reaction chamber heated to between 1094–1652°F (590–900°C). The resultant gas is then cooled, and distilled to remove any impurities.

Teflon can be used on a wide variety of cookware.

Suspension Polymerization

2 The reaction chamber is filled with purified water and a reaction agent or initiator, a chemical that will set off the formation of the polymer. The liquid TFE is piped into the reaction chamber. As the TFE meets the initiator, it begins to polymerize. The resulting PTFE forms solid grains that float to the surface of the water. As this is happening, the reaction chamber is mechanically shaken. The chemical reaction inside the chamber gives off heat, so the chamber is cooled by the circulation of cold water or another coolant in a jacket around its outsides. Controls automatically shut off the supply of TFE after a certain weight inside the chamber is reached. The water is drained out of the chamber, leaving a mess of stringy PTFE which looks somewhat like grated coconut.

3 Next, the PTFE is dried and fed into a mill. The mill pulverizes the PTFE with rotating blades, producing a material with the consistency of wheat flour. This fine powder is difficult to mold. It has "poor flow," meaning it cannot be processed easily in automatic equipment. Like unsifted wheat flour, it might have both lumps and air pockets. So manufacturers convert this fine powder into larger granules by a process called agglomeration. This can be done in several ways. One method is to mix the PTFE powder with a solvent such as acetone and tumble it in a rotating drum. The PTFE grains stick together, forming small pellets. The pellets are then dried in an oven.

4 The PTFE pellets can be molded into parts using a variety of techniques. However, PTFE may be sold in bulk already pre-molded into so-called billets,

which are solid cylinders of PTFE. The billets may be 5 ft (1.5 m) tall. These can be cut into sheets or smaller blocks, for further molding. To form the billet, PTFE pellets are poured into a cylindrical stainless steel mold. The mold is loaded onto a hydraulic press, which is something like a large cabinet equipped with weighted ram. The ram drops down into the mold and exerts force on the PTFE. After a certain time period, the mold is removed from the press and the PTFE is unmolded. It is allowed to rest, then placed in an oven for a final step called sintering.

5 The molded PTFE is heated in the sintering oven for several hours, until it gradually reaches a temperature of around 680°F (360°C). This is above the melting point of PTFE. The PTFE particles coalesce and the material becomes gel-like. Then the PTFE is gradually cooled. The finished billet can be shipped to customers, who will slice or shave it into smaller pieces, for further processing.

Dispersion polymerization

6 Polymerization of PTFE by the dispersion method leads to either fine powder or a paste-like substance, which is more useful for coatings and finishes. TFE is introduced into a water-filled reactor along with the initiating chemical. Instead of being vigorously shaken, as in the suspension process, the reaction chamber is only agitated gently. The PTFE forms into tiny beads. Some of the water is removed, by filtering or by adding chemicals which cause the PTFE beads to settle. The result is a milky substance called PTFE dispersion. It can be used as a liquid, especially in applications like fabric finishes. Or it may be dried into a fine powder used to coat metal.

Nonstick cookware

7 One of the most common and visible uses of PTFE is coating for nonstick pots and pans. The pan must be made of aluminum or an aluminum alloy. The pan surface has to be specially prepared to receive the PTFE. First, the pan is washed with detergent and rinsed with water, to remove all grease. Then the pan is dipped in a warm bath of hydrochloric acid in a process called etching. Etching roughens the surface of the

metal. Then the pan is rinsed with water and dipped again in nitric acid. Finally it is washed again with deionized water and thoroughly dried.

8 Now the pan is ready for coating with PTFE dispersion. The liquid coating may be sprayed or rolled on. The coating is usually applied in several layers, and may begin with a primer. The exact makeup of the primer is a proprietary secret held by the manufacturers. After the primer is applied, the pan is dried for a few minutes, usually in a convection oven. Then the next two layers are applied, without a drying period in between. After all the coating is applied, the pan is dried in an oven and then sintered. Sintering is the slow heating that is also used to finish the billet. So typically, the oven has two zones. In the first zone, the pan is heated slowly to a temperature that will evaporate the water in the coating. After the water has evaporated, the pan moves into a hotter zone, which sinters the pan at around 800°F (425°C) for about five minutes. This gels the PTFE. Then the pan is allowed to cool. After cooling, it is ready for any final assembly steps, and packaging and shipping.

Quality Control

Quality control measures take place both at the primary PTFE manufacturing facility and at plants where further processing steps, such as coatings, are done. In the primary manufacturing facility, standard industrial procedures are followed to determine purity of ingredients, accuracy of temperatures, etc. End products are tested for conformance to standards. For dispersion PTFE, this means the viscosity and specific gravity of the dispersion is tested. Other tests may be performed as well. Because Teflon is a trademarked product, manufacturers who wish to use the brand name for parts or products made with Teflon PTFE must follow quality control guidelines laid down by Du Pont. In the case of nonstick cookware manufacturers, for example, the cookware makers adhere to Du Pont's Quality Certification Program, which requires that they monitor the thickness of the PTFE coating and the baking temperature, and carry out adhesion tests several times during each shift.

Byproducts/Waste

Though PTFE itself is non-toxic, its manufacture produces toxic byproducts. These include hydrofluoric acid and carbon dioxide. Work areas must be adequately ventilated to prevent exposure to gases while PTFE is being heated, or when it cools after sintering. Doctors have documented a particular illness called polymer fume fever suffered by workers who have inhaled the gaseous byproducts of PTFE manufacturing. Workers must also be protected from breathing in PTFE dust when PTFE parts are tooled.

Some waste created during the manufacturing process can be reused. Because PTFE was at first very expensive to produce, manufacturers had high incentive to find ways to use scrap material. Waste or debris generated in the manufacturing process can be cleaned and made into fine powder. This powder can be used for molding, or as an additive to certain lubricants, oils, and inks.

Used PTFE parts should be buried in landfills, not incinerated, because burning at high temperatures will release hydrogen chloride and other toxic substances. One study released in 2001 claimed that PTFE also degrades in the environment into one substance that is toxic to plants. This is trifluoroacetate, or TFA. While current levels of TFA in the environment are low, the substance persists for a long time. So TFA pollution is possibly a concern for the future.

Where to Learn More

Books

Ebnesajjad, Sina. *Fluoroplastics.* Norwich, NY: Plastics Design Library, 2000.

Periodicals

Friedel, Robert, and Alan Pilon. "The Accidental Inventor." *Discover* (October 1996): 58.

Gorman, J. "Environment's Stuck with Nonstick Coatings." *Science News* (21 July 2001): 36.

Angela Woodward

Telephone Booth

The early telephone booths were manufactured from wood with ornate trim and design and typically had an attendant.

Background

Although Alexander Graham Bell is credited with the invention of the telephone, the "telephon" (made from a hollowed out beer barrel, a sausage skin, and a knitting needle) was an original prototype being researched in 1860 by Philipp Reis. The mechanism of the phone was uncovered in 1874 and focused on musical reproduction, but the actual resolution of electricity and voice transmission was actually invented in 1876 by Bell.

The early telephone booths were manufactured from wood with ornate trim and design. A heavy solid wood door allowed the attendants to lock the customer into the booth until the completion of the phone call. This prevented the customer from leaving the premises without making payment. The earlier model slowly evolved into the coin-operated phones of today.

History

It may be of little surprise that the telephone booth has been around for more than 100 years. Inventor William Gray invented the booth after realizing the difficulty of placing a phone call from outside the home. Early wooden telephone booths were primarily located in railroad stations, fancy hotels, or banks. They were located in heavy traffic areas to ensure that the attendant's salary could be paid by the earnings of the booth.

Gray was encouraged that a coin-operated phone booth without the need of an attendant would be more of a convenience then the more costly attended booths. The Hartford Bank in Connecticut became the site of the first coin-operated telephone booth. In the days where Western Electric manufactured thousand of these telephone booths, a phone call was only a nickel.

Raw Materials

The original telephone booths were constructed from hard woods such as mahogany and had plush carpets on the floors. The floor disappeared over the years evolving from reinforced steel or metal to just disappearing altogether. The current enclosures of hard plastic or 14 gauge steel have aluminum anodized or powder coat to protect against corrosion.

Design

The traditional wooden phone booth is still available but typically as a touch of nostalgia in restaurants or private offices. Telephone booths historically have had accordion doors but as these limit handicap access recent designs are usually partial enclosures with phones attached at lower heights to accommodate users in wheelchairs. Telephone enclosures designed for institutional use such as prisons, universities, or other high traffic areas subject to abuse and vandalism are constructed from heavy duty (14 gauge) steel housing.

There are many different mounting arrangements for modern telephone booths, depending on whether the user will be sitting, standing, or placing a phone call from their automobile. In the manufacturing process these methods strive to be as standardized as possible to keep the numbers of parts required to a minimum.

The Manufacturing Process

1 The steel arrives at the manufacturing plant in thick sheets. Its is loaded onto a conveyor belt and cut to the desired length and width.

2 Side panels are die cast. Molten steel is poured into the die and left to harden. When cooled, the frame panels are removed.

3 The framework is then taken into a room and sprayed with an aluminum anodized coat to protect them from wear and corrosion.

4 The framework is then welded to the frame. Each framework is molded to hold several panels inserted at a later stage in the assembly process.

5 Panels can be made from wood, particle board, plastic, or any other material that can be formed into sheets. The panels may be customized to reflect the business establishment providing the public phone access. Many are constructed from vacuum formed plastic and bearing the logo of the business. Display panels can also be made of white translucent butyrate.

6 The panels are manufactured with grooves along each of the outer surfaces. Correspondingly, moldings are provided with projections extending along at least a portion of the molding.

7 The panels are secured by moldings allowing the housing to be manufactured and assembled without nuts and bolts. The moldings are made of rubber, plastic, or similar materials and are manufactured by an extrusion process.

8 The phone is then fitted to the finished booth.

9 The booths are transported either to a dealer or directly to the customer. They are then manually installed.

Quality Control

Through the entire process, the quality of the materials is checked. Any defective molds, steel, or panels are discarded. The frame is checked for stability strength. The projections on the molding must be of equal

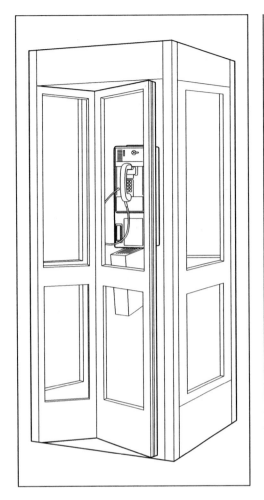

One version of a modern telephone booth.

or less than the depth of the grooves to be received securely. Design and Engineering manufacturing uses AutoCAD and VersaCAD computerized technology that constantly provides feedback and analysis of the manufacturing process providing a consistent and uniform product.

Byproducts/Waste

The steel and/or aluminum from the manufacturing process can be retrieved and recycled. Waste is kept to a minimum as the manufacturing tools run on compressed air keeping the work area free of debris.

The Future

It will be interesting to see where the telephone booth finds itself in the future. The full length booths have been replaced by smaller modular models and the convenience of cellular technology seems to leave fewer and fewer of us scrambling for spare change.

Kiosk systems are being developed that provide a multitude of communication options. Coinless phone options such as Internet capabilities and phone and fax services are all rendered possible from one multi-use communication enclosure.

Where to Learn More

Other

PBG, Telephone Enclosures Web Page. December 2001. <http://www.PBGinc.com>.

Redy Ref Web Page. December 2001. <http://www.redyref.com>.

Bonny P. McClain

Tiara

Background

Tiaras are marks of distinction and style worn by women of royalty and for special events such as pageants, proms, and weddings. Revivals of interest in romantic ensembles, like those seen in movies and in period costumes, make tiaras fascinating headdresses.

History

Egyptian artifacts show the first known tiaras. The huge headdresses crowning most Egyptian royalty and priests may be the most memorable, but the court's princesses wore tiaras shaped from wire and bearing images taken from nature.

Tiaras from both the Greek and Roman Empires began as simple bands of cloth worn around the top of the head or forehead and tied in the back. Powerful men and women wore these bands, and later versions were decorated with pearls and gems. Another kind of tiara gained popularity too. Wreaths of leaves were awarded to winners in sports, but officials also wore them for ceremonies. The natural wreaths were replaced when craftsmen used gold, silver, and plated metal to make imitation leaves embossed with natural patterns and shaped like wreaths.

Celtic tribes in the eighth to sixth centuries B.C. created elaborate headwear like the bronze diadem excavated from Cavenham Heath in Suffolk, England. This Celtic diadem was probably made in the period from the second century to the fourth century A.D. The front upper edge is cut into three semicircles, and the back has three points. The front and back stand higher than the sides, which are plain bronze strips. The front and back panels can slide on the bronze strips so the diadem can be adjusted to fit the wearer.

Terms for majestic headwear are often confusing and sometimes interchangeable. The Celtic diadem is unique because it is adjustable. Most other ornaments called diadems are fixed in size. Tiaras are also called diadems, but they are not full circles and have limited flexibility. Tiaras must either be sized specifically for the wearer or positioned differently on the heads of others. A tiara called the bandeau was popular in the 1920s. It was worn around the forehead with the ends tucked in the hair. These bandeaus have been revived today, but wearers sit them on the crowns of their heads like conventional tiaras. Yet another variation was designed that can also be worn as a small crown-like circlet or a necklace.

Nineteenth and twentieth century British monarchs benefited from their extensive Empire; South Africa, the world's largest producer of diamonds, and the fabulously wealthy maharajahs of India presented lavish jewelry, tiaras, and gemstones (later mounted in tiaras) for every occasion. Incidentally, although the Royal Family of England owns private collections of jewelry, many tiaras and other items are the property of the nation and are "on loan" to the Royal Family. The Queen has stated that even the private collections will someday be given to the state as part of Great Britain's history.

Raw Materials

The raw materials for a tiara begin with metal wire. Silver and gold are the leading choices. Square wire (box shaped when cut) is also preferred for forming the tiara band

Tiaras from both the Greek and Roman Empires began as simple bands of cloth worn around the top of the head or forehead and tied in the back.

Princess Diana.

Diana Frances Spencer was born July 1, 1961, in Norfolk, England. Tutored at home until she was nine, Diana then attended Riddlesworth Hall. At 12, she attended West Heath School, leaving at 16 for a Swiss finishing school. Moving to London, Diana worked as a kindergarten teacher's aide.

Prince Charles knew Diana most of her life, but they were reintroduced in 1977. In 1981, Charles proposed, and on July 29, 1981 they were married in St. Paul's Cathedral. A congregation of 2,500 and a television audience of 750 million watched the ceremony. Another two million spectators jammed the processional route.

On June 21, 1982, Diana gave birth to son William. Henry arrived September 15, 1984. On December 9, 1992, a Palace spokesperson announced the royal couple was separating, and on February 29, 1996, they divorced. Diana was barred from succeeding to the throne and had to drop the prefix Her Royal Highness. She shared custody of her sons with Charles and was involved in all decisions regarding their lives. Diana also received a lump-sum alimony payment of more than $20 million and was able to keep jewelry acquired during the marriage.

Tragedy struck August 31, 1997. The paparazzi's pursuit of Diana and Emad "Dodi" al Fayed combined with their driver speeding while legally drunk, caused an accident in Paris. Fayed and the driver were killed instantly; a bodyguard sustained critical injuries. Doctors worked to save Diana, but a few hours later she was pronounced dead.

On September 6, one million people gathered along the funeral route from Kensington Palace to Westminster Abbey. Walking behind Diana's casket were her sons, brother, ex-husband, former father-in-law, and five representatives from each of the 110 charities with which Diana had been associated.

and its upright designs, although round, half-round, flat, beaded, and other forms of wire may be preferred (or combined) for different designs. The wire may also be of different gauges (thickness). Tiaras typically call for 12–14 gauge wire. Gauges with larger numbers are the thinnest and smaller gauge numbers show thicker material. Binding wire holds pieces together until they can be soldered. Iron wire must undergo a heating process called annealing to make it malleable (soft) and flexible for binding small jewelry parts. Even after cooling, the annealed iron can be easily, smoothly, and tightly wrapped. Precious metal solder will not stick to the iron so the binding wire can be removed.

Solder is metal or a metal alloy that is melted to connect pieces of metal. In tiara manufacture, solder of the same precious metal as the wire is chosen. Not only should it be gold or silver, but it should also be the same shade and hardness so it blends seamlessly with the wire. Silver solder melts at a low, medium, or high temperature depending on the composition of the alloy. Alloys melt at lower temperatures than their component metals, so the combination of another metal and silver will produce an alloy that can be used as solder to join pure silver. High-temperature silver solder melts at just below the melting point of silver. Low-temperature solder (made of silver alloyed with a soft metal like lead or tin) can be applied over high-temperature solder (still made of silver but with a smaller amount of tin or a harder metal like zinc, as examples) so uneven surfaces can be leveled.

Stones that give tiaras their shimmer range from artificial to precious materials. Fake or artificial stones are made of glass (containing lead to add sparkle) or paste. Rhinestones are also artificial gems that are colorless and made of paste, glass, or gem-quality quartz. Despite their classification as artificial, rhinestones made of well-cut, gem-quality quartz can be expensive and beautiful. Austrian crystal is clear, high quality manufactured glass with a high lead content for brilliance; it is fabricated in many colors.

Chemicals, compounds, and polishes are needed to clean and finish tiaras. A solution of acid and water is used to pickle (clean) tarnish from metal. Typically, one part sulfuric acid is added to 10 parts water to make a pickling acid, and the solution is heated. Metal is polished in three steps using polishing compounds that are increasingly gentle to add more shine with each step. There are over 500 buffing and polishing compounds in liquid, cream, stick, and cake forms. Jewelers prefer a white polish known as Bobbing compound (also called Chrome Rouge) that will add a final finish to many good quality metal products, but it is used as the first polish for jewelry. Tripoli, a brown compound, is an excellent polish for silver that cleans and removes tiny scratches. To produce a mirror finish on precious metals (or to polish soft metals), a pink compound called Jewelers' Rouge gives tiaras their final polish.

Finally, rolls made of padded velvet cushion the wearers' heads against the weight of tiaras and skin-to-metal contact. Adjustable, elastic bands can be attached to the ends of the velvet headbands to make tiaras complete circles and to tighten them slightly. Hair combs (usually of tortoiseshell or plastic) are sometimes added to the cushions so the tiaras can be anchored to hair more securely.

Design

The tiara has two major and several minor components. The head band that sits on the wearer's head is called the band or circlet. It is usually silver or another precious metal with stones ranging from rhinestones to gemstones. The delicate, filigree portion (and the second major component) is called the top ornament or crown; it may rise from the band in a matching pattern or be dramatically different. The tiara band is typically small and lightweight so it is partially concealed in the wearer's hair and makes the top ornament radiate from her face. Wider bands are sometimes preferred to display more sparkle or carry out a theme.

Usually, the ornament has a center spike with identical sides bracketing the spike. The spike is often considered distinct from the ornament because its size is proportional to the ornament and the entire circlet can change the appearance from understated to excessive. Its proportions also influence the weight of the tiara, its balance and comfort on the wearer's head. In general, the spike should only be one third of the length of the band higher than the band. This is particu-larly effective in preventing delicate designs from being overpowered by the spike.

Other smaller components of the tiara include the interior headband padding and an elastic band or haircomb to help secure the tiara. The elastic band connects the two ends of the circlet at the back of the wearer's head, is adjustable, and can be hidden under the hair. Haircombs are less common but may be stitched in place near each end of the circlet. The combs point downward and are pushed into the hair. They are used most often on heavier tiaras.

The designer makes a drawing or rendering of the finished piece as it would appear if laid out flat. The drawing can be simple, perhaps for a new band with an existing top ornament. The drawing can also be a pencil or brush (painted) rendering or a computer-aided design. Pencil and brush illustrations are usually highly detailed and shaded, so the depth (thickness) of the piece, embossed patterns, and stone settings (bezels) are clear. Like designers of other types of jewelry, they keep notebooks of design ideas.

Using the design drawing, the artist makes a template on cardboard. The template for a tiara consists of two parts, the lower band and the top ornament, also called a crown. The band is sized to fit the wearer (based on direct measurement or a sizing chart), and, even though the tiara is not a full circlet, the measurement is increased to allow for the padding on the band. The upper and lower edges of the band will be straight lines on the template if the crown stands upright on the band, or the lines (and the flat shape of the template) will curve if the ornament slants out. Sometimes the sizing of the template is adjusted to fit the wearer before any metalworking is done. The template for the crown is a simple outline; details of the ornament are shown on the drawing. Instead, the crown template is used to confirm (before the two major parts are joined) that the centerlines of the crown and band match and that the lengths of both are compatible.

The Manufacturing Process

1 Jewelers and artists may work for firms that produce a variety of designs of tiaras and a significant volume or independently in

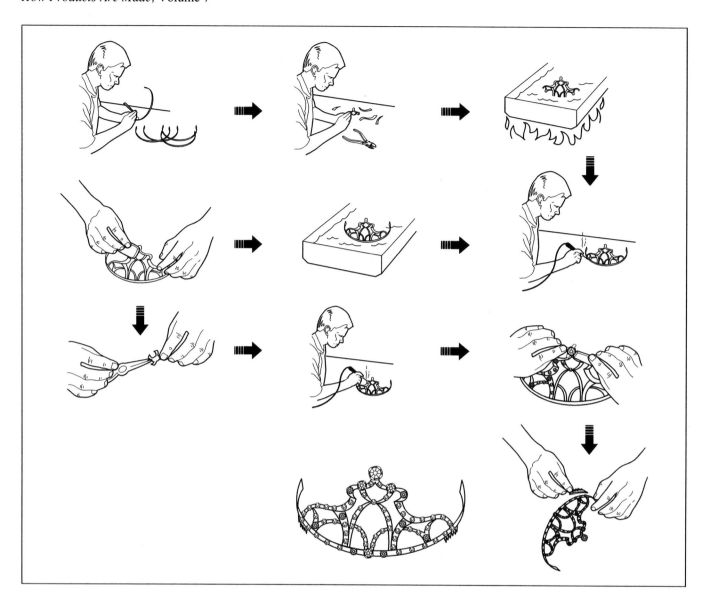

The tiara manufacturer follows a template to create a tiara according to the customer's specifications.

their own studios. Assuming the maker has chosen a design, the jeweler/artist begins by confirming that appropriate supplies are on hand. Frequently used stones may be stored in jewelers' boxes or safes in their studios. Precious gems or rare cuts are purchased from diamond and gem merchants, stone cutters, or other specialists. The tiara maker may also cut and polish gems.

2 Small bands are usually made from wire that is simply wound together. More complex windings can be crafted in any of a variety of patterns, called unit arrangements, rather like chains. Wire can be coiled around a tool called an arbor that is a specific shape in cross section, such as round, oval, or rectangular. The unit arrangements are repeated along the length of the band and sometimes

stacked for wider bands. These are soldered together at points specified on the design drawing. Fine iron binding wire is used as needed to hold pieces together while they are soldered. The band now looks like the template. The ends of the band are hammered flat and are squared with a file.

3 The processes of applying the binding wire, soldering the unit arrangements and binding, and shaping the band work the metal and tend to make it brittle. As the band is formed, the metal is softened and restored to its characteristic malleability so it can be worked again through a heating and cooling process called annealing. A borax paste called flux is applied to the solder and the metal being joined before annealing or soldering.

Flux serves two purposes. It prevents the silver or other metal from oxidizing as heat is applied during annealing or soldering; solder will roll into balls if it oxidizes and will fail to stick to the metal. Flux will flow under high heat and so it serves as a warning to the tiara maker to watch the temperature of the silver. Soldering also tarnishes metal; to clean it, the band is pickled in a weak solution of sulfuric acid and water while it is mildly heated. Annealing and pickling may be done several times while the band is being made. The band is still only metal without any gems as metal working to shape the circlet is completed.

4 Bezels are made for each of the stones that will be mounted on the tiara later. A bezel is a setting or mount that angles inward under the stone so more of the stone and less of the setting shows. A bezel is made of wire that matches the silver or gold of the tiara and is shaped by bending wire with needle-nosed pliers. There are several basic shapes and sizes for bezels.

Bezels are grouped in unit arrangements. They are sized to their stones, and soldered closed and numbered. The numbered bezels are assembled in their unit arrangements, held together with binding wire, and soldered together. They are annealed and pickled as needed.

5 The symmetrical patterns in the crown or top ornament are shaped with wire. Wire may be wound, coiled, bent with pliers, or formed to suit the design, as it was for the band. The crown is also made up of unit arrangements because there are matching or mirrored designs on both sides of the tiara. Periodically, the bezel units are test-fitted to the crown units. Each unit arrangement is soldered and then soldered again to a larger unit until the crown is complete. The crown should match the template.

6 The band and crown are checked several times during construction to make sure they match the design and complement each other. Next, they are shaped into rounds by gently wrapping them around an arbor and gently working them into circles. The band and crown are held together with binding wire, soldered to each other, and pickled.

7 Bezels in their unit arrangements are then soldered to the crown (and the band if the design requires). Low-temperature solder is used so that it will not melt other solder already in place. The tiara is pinned to a charcoal block or rotating base so the unit can be positioned from the outside but soldered on the inside. Apart from bezels, other ornaments like solid metal beads may be part of the design, and these are also soldered in place one at a time. Once ornaments and bezels are in place, the tiara is pickled again. It is also checked for smooth edges and true shape.

8 The tiara surfaces are finished in three polishing steps. Bobbing compound (also called Chrome Rouge) is a relatively coarse polish and removes more metal than the other polishes. Tripoli compound is finer but intermediate step for cleaning and removing fine scratches. The ultimate shine is produced with Jewelers' Rouge, the most delicate polish. Very small buffing pads are used on a flexible shaft buffer. The small pads and the flexible shaft let the artist reach every surface without applying too much pressure.

9 Next, the gems or stones are set in their bezels. The bezels may not fit snuggly, so a square-edged tool called a pusher is used to press the top edges against the stones. The padding on the inner, lower edge of the band is made of stuffed velvet. A Velcro strip is glued with white fabric or crafters' glue to the band and the velvet headband so it can be cleaned.

10 After a final inspection, the tiara is packaged in a padded hat box or velvet case and delivered to its new owner.

Quality Control

The jeweler, artist, or other skilled specialist performs every task and observes each minute detail of tiara production. The reputations of the artist, his or her studio, or the manufacturing firm rest on producing heirlooms and recommendations from clients. Even when jewelers produce custom tiaras for royalty, each royal member may have several preferred jewelers who compete for such honored work.

Byproducts/Waste

Tiara making does not result in byproducts. Both artistic design and jewelry making are involved, so it is common for tiara manufacturers to produce other types of jewelry.

Tiara production creates few wastes despite the use of metals, polishing compounds, and pickling acids. Manufacturers use only small amounts of pickling liquid that they replenish as needed. Tiaras are pickled in the container in which the acid bath liquid is stored. Evaporation is the only form of loss; no pickling liquid remains to require disposal. Polishing compounds and powdered metals can be hazardous, but workers are protected with filtered respirators. In the artist's studio, air filters and polishing machines suck up these powders and control dust in work areas. The dust collected is stored in a closed drum along with used air and respirator filters. When full, the drum is disposed in a landfill that is classified to receive this kind of waste.

Metal wastes consist of very fine trimmings and particles. These are swept up daily and stored in bins. They are sent to a metal refinery where the precious metals are extracted and recycled. Apart from environmental law and responsibility, artists benefit from metal recycling because refunds for metals recovered at the foundry may pay more than the artists' annual power bills.

The Future

Some manufacturers claim that tiaras are suddenly back in fashion, but, in fact, interest has been steady. They have long been treasured bridal accessories. If interests have changed, it is in the popularity of theme weddings; tiaras, period gowns, and other heirlooms are special in themselves but also contribute to unique wedding photographs. The artist is an expert in heraldry (symbols on shields, armor, and family crests) and jewelry from medieval and Renaissance times.

Real brides and princesses may have to step aside for Hollywood, however. Crown and coronet specialist Carl W. Lemke has made jeweled reproductions for many films and television programs including modern ornaments and orders as well as historic headwear. Although a few brides may be inspired by the wedding styles of their

mothers and friends, many more will want to copy the tiaras and fabulous jewels from *Shakespeare in Love, Titanic, The Princess Diaries,* and other motion pictures. Imagination and beautiful craftsmanship insure a long and healthy interest in tiaras.

Where to Learn More

Books

Ebbetts, Lesley. *Royal Style Wars.* New York: Crescent Books, 1990.

Latham, Caroline, and Jeannie Sakol. *The Royals.* New York: Congdon & Weed, Inc., 1987.

Menkes, Suzy. *The Royal Jewels.* London: Grafton Books, 1988.

Morton, Andrew. *Theirs is the Kingdom: The Wealth of the Windsors.* London: Michael O'Mara Books Limited, 1989.

Rose, Augustus F., and Antonio Cirino. *Jewelry Making and Design.* New York: Dover Publications, 1967.

Other

Anderson, Louise. "Why Wear a Tiara on Your Wedding Day?" *Roses By Design Web Page.* December 2001. <http://www.webview.co.nz/roses/wedding-tiaras/why-a-tiara.htm>.

Lemke, Carl W. "Making a Hollow-ware Coronet." *Carl W. Lemke Unique Jewelry Web Page.* December 2001. <http://www.signetring.com/Coronet/Coronet_how_to_make/coronet_how_to_make.htm>.

Minor, Margaret R. "Diadem." *The Power of Women in Celtic Society.* December 2001. <http://www.unc.edu/courses/art111/celtic/catalogue/femdruids/diadem.html>.

Palady, Dion. *Tiara Town Web Page.* December 2001. <http://www.tiaratown,com/history.htm>.

Tiara Enterprises, Inc. Web Page. 1998. December 2001. <http://www.tiaras.com>.

Gillian S. Holmes

Titanium

Background

Titanium is known as a transition metal on the periodic table of elements denoted by the symbol Ti. It is a lightweight, silver-gray material with an atomic number of 22 and an atomic weight of 47.90. It has a density of 4510 kg/m3, which is somewhere between the densities of aluminum and stainless steel. It has a melting point of roughly 3,032°F (1,667°C) and a boiling point of 5,948°F (3,287 C). It behaves chemically similar to zirconium and silicon. It has excellent corrosion resistance and a high strength to weight ratio.

Titanium is the fourth most abundant metal making up about 0.62% of the earth's crust. Rarely found in its pure form, titanium typically exists in minerals such as anatase, brookite, ilmenite, leucoxene, perovskite, rutile, and sphene. While titanium is relatively abundant, it continues to be expensive because it is difficult to isolate. The leading producers of titanium concentrates include Australia, Canada, China, India, Norway, South Africa, and Ukraine. In the United States, the primary titanium producing states are Florida, Idaho, New Jersey, New York, and Virginia.

Thousands of titanium alloys have been developed and these can be grouped into four main categories. Their properties depend on their basic chemical structure and the way they are manipulated during manufacture. Some elements used for making alloys include aluminum, molybdenum, cobalt, zirconium, tin, and vanadium. Alpha phase alloys have the lowest strength but are formable and weldable. Alpha plus beta alloys have high strength. Near alpha alloys have medium strength but have good creep resistance. Beta phase alloys have the highest strength of any titanium alloys but they also lack ductility.

The applications of titanium and its alloys are numerous. The aerospace industry is the largest user of titanium products. It is useful for this industry because of its high strength to weight ratio and high temperature properties. It is typically used for airplane parts and fasteners. These same properties make titanium useful for the production of gas turbine engines. It is used for parts such as the compressor blades, casings, engine cowlings, and heat shields.

Since titanium has good corrosion resistance, it is an important material for the metal finishing industry. Here it is used for making heat exchanger coils, jigs, and linings. Titanium's resistance to chlorine and acid makes it an important material in chemical processing. It is used for the various pumps, valves, and heat exchangers on the chemical production line. The oil refining industry employs titanium materials for condenser tubes because of corrosion resistance. This property also makes it useful for equipment used in the desalinization process.

Titanium is used in the production of human implants because it has good compatibility with the human body. One of the most notable recent uses of titanium is in artificial hearts first implanted in a human in 2001. Other uses of titanium are in hip replacements, pacemakers, defibrillators, and elbow and hip joints.

Finally, titanium materials are used in the production of numerous consumer products. It is used in the manufacture of such things as shoes, jewelry, computers, sporting equipment, watches, and sculptures. As tita-

Titanium is the fourth most abundant metal making up about 0.62% of the earth's crust.

nium dioxide, it is used as a white pigment in plastic, paper, and paint. It is even used as a white food coloring and as a sunscreen in cosmetic products.

History

Most historians credit William Gregor for the discovery of titanium. In 1791, he was working with menachanite (a mineral found in England) when he recognized the new element and published his results. The element was rediscovered a few years later in the ore rutile by M. H. Klaproth, a German chemist. Klaproth named the element titanium after the mythological giants, the Titans.

Both Gregor and Klaproth worked with titanium compounds. The first significant isolation of nearly pure titanium was accomplished in 1875 by Kirillov in Russia. Isolation of the pure metal was not demonstrated until 1910 when Matthew Hunter and his associates reacted titanium tetrachloride with sodium in a heated steel bomb. This process produced individual pieces of pure titanium. In the mid 1920s, a group of Dutch scientists created small wires of pure titanium by conducting a dissociation reaction on titanium tetraiodide.

These demonstrations prompted William Kroll to begin experimenting with different methods for efficiently isolating titanium. These early experiments led to the development of a process for isolating titanium by reduction with magnesium in 1937. This process, now called the Kroll process, is still the primary process for producing titanium. The first products made from titanium were introduced around the 1940s and included such things as wires, sheets, and rods.

While Kroll's work demonstrated a method for titanium production on a laboratory scale, it took nearly a decade more before it could be adapted for large-scale production. This work was conducted by the United States Bureau of Mines from 1938 to 1947 under the direction of R. S. Dean. By 1947, they had made various modifications to Kroll's process and produced nearly 2 tons of titanium metal. In 1948, DuPont opened the first large scale manufacturing operation.

This large scale manufacturing method allowed for the use of titanium as a structural material. In the 1950s, it was used primarily by the aerospace industry in the construction of aircraft. Since titanium was superior to steel for many applications, the industry grew rapidly. By 1953, annual production had reached 2 million lb (907,200 kg) and the primary customer for titanium was the United States military. In 1958, demand for titanium dropped off significantly because the military shifted its focus from manned aircraft to missiles for which steel was more appropriate. Since then, the titanium industry has had various cycles of high and low demand. Numerous new applications and industries for titanium and its alloys have been discovered over the years. Today, about 80% of titanium is used by the aerospace industry and 20% by non-aerospace industries.

Raw Materials

Titanium is obtained from various ores that occur naturally on the earth. The primary ores used for titanium production include ilmenite, leucoxene, and rutile. Other notable sources include anatase, perovskite, and sphene.

Ilmenite and leucoxene are titaniferous ores. Ilmenite ($FeTiO_3$) contains approximately 53% titanium dioxide. Leucoxene has a similar composition but has about 90% titanium dioxide. They are found associated with hard rock deposits or in beaches and alluvial sands. Rutile is relatively pure titanium dioxide (TiO_2). Anatase is another form of crystalline titanium dioxide and has just recently become a significant commercial source of titanium. They are both found primarily in beach and sand deposits.

Perovskite ($CaTiO_3$) and sphene ($CaTiSiO_5$) are calcium and titanium ores. Neither of these materials are used in the commercial production of titanium because of the difficulty in removing the calcium. In the future, it is likely that perovskite may be used commercially because it contains nearly 60% titanium dioxide and only has calcium as an impurity. Sphene has silicon as a second impurity that makes it even more difficult to isolate the titanium.

In addition to the ores, other compounds used in titanium production include chlorine gas, carbon, and magnesium.

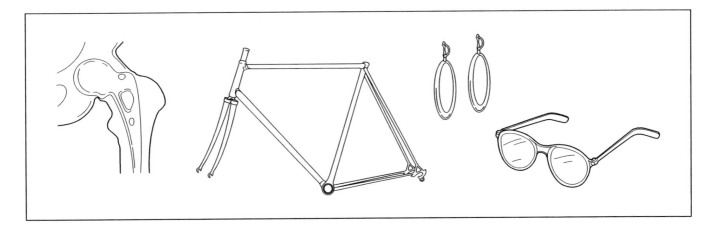

The Manufacturing Process

Titanium is produced using the Kroll process. The steps involved include extraction, purification, sponge production, alloy creation, and forming and shaping. In the United States, many manufacturers specialize in different phases of this production. For example, there are manufacturers that just make the sponge, others that only melt and create the alloy, and still others that produce the final products. Currently, no single manufacturer completes all of these steps.

Extraction

1 At the start of production, the manufacturer receives titanium concentrates from mines. While rutile can be used in its natural form, ilmenite is processed to remove the iron so that it contains at least 85% titanium dioxide. These materials are put in a fluidized-bed reactor along with chlorine gas and carbon. The material is heated to 1,652°F (900°C) and the subsequent chemical reaction results in the creation of impure titanium tetrachloride (TiCl4) and carbon monoxide. Impurities are a result of the fact that pure titanium dioxide is not used at the start. Therefore the various unwanted metal chlorides that are produced must be removed.

Purification

2 The reacted metal is put into large distillation tanks and heated. During this step, the impurities are separated using fractional distillation and precipitation. This action removes metal chlorides including those of iron, vanadium, zirconium, silicon, and magnesium.

Production of the sponge

3 Next, the purified titanium tetrachloride is transferred as a liquid to a stainless steel reactor vessel. Magnesium is then added and the container is heated to about 2,012°F (1,100°C). Argon is pumped into the container so that air will be removed and contamination with oxygen or nitrogen is prevented. The magnesium reacts with the chlorine producing liquid magnesium chloride. This leaves pure titanium solid since the melting point of titanium is higher than that of the reaction.

4 The titanium solid is removed from the reactor by boring and then treated with water and hydrochloric acid to remove excess magnesium and magnesium chloride. The resulting solid is a porous metal called a sponge.

Alloy creation

5 The pure titanium sponge can then be converted into a usable alloy via a consumable-electrode arc furnace. At this point, the sponge is mixed with the various alloy additions and scrap metal. The exact proportion of sponge to alloy material is formulated in a lab prior to production. This mass is then pressed into compacts and welded together, forming a sponge electrode.

6 The sponge electrode is then placed in a vacuum arc furnace for melting. In this water-cooled, copper container, an electric arc is used to melt the sponge electrode to form an ingot. All of the air in the container is either removed (forming a vacuum) or the atmosphere is filled with argon to prevent contamination. Typically, the ingot is

Titanium is used for a wide variety of items, such as bike frames, hip implants, eyeglass frames, and earrings.

remelted one or two more times to produce a commercially acceptable ingot. In the United States, most ingots produced by this method weigh about 9,000 lb (4,082 kg) and are 30 in (76.2 cm) in diameter.

7 After an ingot is made, it is removed from the furnace and inspected for defects. The surface can be conditioned as required for the customer. The ingot can then be shipped to a finished goods manufacturer where it can be milled and fabricated into various products.

Byproducts/Waste

During the production of pure titanium a significant amount of magnesium chloride is produced. This material is recycled in a recycling cell immediately after it is produced. The recycling cell first separates out the magnesium metal then the chlorine gas is collected. Both of these components are reused in the production of titanium.

The Future

Future advances in titanium manufacture are likely to be found in the area of improved ingot production, the development of new alloys, the reduction in production costs, and the application to new industries. Currently, there is a need for larger ingots than can be produced by the available furnaces. Research is ongoing to develop larger furnaces that can meet these needs. Work is also being done on finding the optimal composition of various titanium alloys. Ulti-

mately, researchers hope that specialized materials with controlled microstructures will be readily produced. Finally, researchers have been investigating different methods for titanium purification. Recently, scientists at Cambridge University announced a method for producing pure titanium directly from titanium dioxide. This could substantially reduce production costs and increase availability.

Where to Learn More

Books

Othmer, K. *Encyclopedia of Chemical Technology.* New York: Marcel Dekker, 1998.

U.S. Department of the Interior U.S Geological Survey. Minerals Yearbook Volume 1. Washington, DC: U.S. Government Printing Office, 1998.

Periodicals

Freemantle, M. "Titanium Extracted Directly from TiO2." *Chemical and Engineering News* (25 September 2000).

Eylon D. "Titanium for Energy and Industrial Applications." *Metallurgical Society AIME* (1987).

Other

WebElements Web Page. December 2001. <http://www.webelements.com>.

Perry Romanowski

Toaster

Background

A toaster is a small appliance that uses heat to brown and harden bread.

History

The practice of browning bread is an ancient one. Early civilizations placed bread over an open fire in order to keep the bread from growing mold. The Romans brought the idea back from Egypt in 500 B.C. and then took it to Great Britain when they invaded in A.D. 44.

In the eighteenth century, a hinged fork was used to hold the bread and prevent it from falling into the fire. With the appearance of wood and coal stoves in the 1880s, a new toasting method was needed. This led to a tin and wire pyramid-shaped device. The bread was placed inside and the device was heated on the stove.

Fire was the source of heat for toasting bread until 1905 when Albert Marsh, an engineer, created an alloy of nickel and chromium, called Nichrome. Marsh's invention was easily shaped into wires or strips and was low in electrical conductivity. Within months, other inventors were using Nichrome to produce electric toasters.

In 1905, Westinghouse introduced a toaster stove that featured a heating element on a raised base. The bread was placed in a wire mesh tray that lay across the heating element. Five years later, General Electric received the first patent for an electric toaster. This first model was a bare wire skeleton with a rack for the bread. A single exposed heating element (nickel wire woven through sheets of mica) toasted the bread. Tempera-ture could not be controlled, and the bread had to be turned manually.

At first, the electric toaster was primarily used in restaurants because the majority of homes had limited access to electrical power. In most cases, electricity was not available during the daylight hours. However, as public demand for labor-saving appliances in the home increased, electric companies began to offer 24-hour service.

In the early twentieth century, a flip-top toaster appeared. It had a side panel that lowered for insertion and turning of the bread slices. The Estate Stove Company developed a square box that could toast four slices at the same time. Still, these early toasters did not have any type of sensor, so the bread had to be constantly monitored. During World War I, an employee in a Minnesota manufacturing company named Charles Strite began developing a toaster that would work automatically. Strite first built an adjustable time and attached it to a spring. The timer turned off the heating element and released the spring, causing the toast to pop up.

Strite patented his toaster in 1919 and began selling it to restaurants under the brand name Toastmaster. Because each toaster was made by hand, production was extremely slow. In 1925, Murray Ireland joined the company and redesigned Strite's original plans so that the toasters could be mass-produced.

The first toaster with a piece of bimetal in the timer circuit appeared in the late 1920s. In 1930, the Proctor Electric Company produced an automatic toaster with a timing strip. By the end of the 1940s, the Sunbeam Corporation had improved the bimetal sen-

Early civilizations placed bread over an open fire in order to keep the bread from growing mold.

sor control so that it was triggered by the heat from the bread instead of the heating element. When the bread reached 310°F (154°C) the sensor automatically turned off the heating element. In 1955, General Electric created its Toast-R-Oven.

In the late 1950s, two men in the United States invented the microchip. Jack Kilby and Robert Noyce, working independently, discovered that they could create an integrated circuit on a thin layer of silicon. The microchip transformed many industries. For toaster manufacturers it meant a sophisticated means for gauging the moisture in the bread and determining the precise heating time for different types of breads.

In the late 1970s, the development of heat-resistant plastics offered more options for toaster design. Models with rounded sides in a variety of colors became popular. Wider slots for bagels and thick slices of bread also appeared, as well as models with up to six slots for multiple toasting.

By the twenty-first century, the term smart toasters was a frequent marketing term. Using microchip technology, these toasters are programmed to toast a variety of baked goods from bagels to English muffins to frozen pastries. Dual versions were also introduced. Hamilton Beach introduced a combination toaster/oven/broiler that can toast bread and bake coffee cake. Black & Decker's Versa-Toast 4-Slice model allows consumers to toast two separate types of bread, such as an English muffin and a slice of raisin bread, at the same time.

Raw Materials

Toasters and toaster ovens are constructed from a long list of intricate parts that includes a heating element, spring, bread rack, heat sensor, trip plate, level, timing mechanism, electromagnet, catch, and browning control. The various parts are constructed on site from a variety of metals and molded plastics. Screws, nuts, bolts, and washers are used to join the parts together.

To make heat-resistant plastic cases, flame retardants and smoke suppressants are added to the petroleum and coal compounds of plastic. If the toaster case is to be metal, aluminum is generally used. Mica is also used because this fire proof, flat sheet made of aluminosilicate minerals holds the nichrome wire.

Design

Design is an integral step in the manufacture of toasters. Toasters are made with a variety of features and are in a constant state of re-design. Design innovations of the last two decades include wider slots for larger slices of bread and bagels, heat resistant plastics for toaster bodies, and microchip controls. The construction process is a complex one; therefore, detailed designs are needed to insure the proper assembly of the toaster's countless parts. The manufacturer's design staff creates prototypes of updated innovations. The prototypes are then subjected to a battery of tests. Some of the prototypes may be test-marketed for consumer approval.

The type of toaster dictates the number and types of parts. Manual toasters, those with turnable toast racks, are of the simplest construction. Semi-automatic toasters incorporate some type of signal features, such as a buzzer, to announce the end of the toasting cycle. These toasters may also have a thermostat that is controlled by the surface temperature of the bread and a regulation dial to adjust the level of brownness. Semi-automatic toasters do not have a pop-up feature.

Automatic toasters use an electrical current to brown the bread. The pop-up feature appears in all automatic toasters in which the heating element is connected to a thermostatic switch or timing device.

The Manufacturing Process

Making the case

1 Today, toaster cases are generally made out of pressure molded plastic. Plastic pellets are fed into a hopper and heated to 350°F (177°C). As the compound heats, it becomes semi-fluid. It is then poured into a toaster case mold and cooled. After the plastic cools and hardens, it is removed from the mold. If the toaster case is aluminum, a sheet of aluminum is placed on a conveyor belt and taken to a punch press. The punch press has a die that quickly exerts pressure

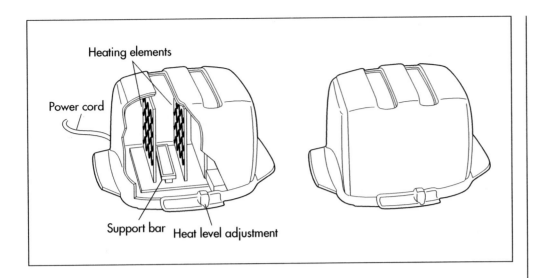

Heating elements

Power cord

Support bar   Heat level adjustment

on the aluminum sheet, punching out the specific size needed.

2 The punched aluminum sheet is then conveyed to a backward extrusion press. The sheet is placed inside of a toaster-shaped die and the press forces a punch against the sheet. The metal flows backwards around the punch forming a hollow, toaster-shaped shell.

3 Batches of Nichrome wire are shipped from a manufacture to the toaster factory. These wires are loaded into a machine that automatically wraps the wire around a mica sheet.

4 The mica sheets wrapped with Nichrome are then placed back on the assembly line and two are bolted to the toaster base.

5 In between the mica sheets is the spring-loaded tray. This tray is manually set and screwed into place.

6 A single bi-metallic element is connected to a support bracket, which is then mounted onto a second bracket that allows for a pivoting motion. A spring is installed between the two brackets. A stop is mounted on a mounting bracket to lock the support bracket when it is activated by the toaster carriage assembly. A release assembly is attached to the bi-metallic element.

Assembling the terminal board

7 Two rectangular sheet-metal conductors are layered above and below a sheet of insulation. The terminal board is then screwed or welded to the toast rack. The ter-

minal ends of the temperature-sensing device are attached to the ends of the conductors by screws.

Installing the base

8 The electrical cord is threaded into the molded plastic base. The various toaster parts such as the wire brackets and terminal board are attached to the base. The toaster body is then fitted over the base.

Byproducts/Waste

Very few byproducts result from the manufacture of toasters. However, there is frequently excess waste from the plastic molding or metal stamping processes. Plastic waste is collected and reprocessed for use in other plastic products. Scrap metal such as aluminum can be recycled into a new metal form as well for reuse.

Quality Control

Toasters are regularly tested on the assembly line. They are tested for durability and functionality. Companies typically generate toasting reports consisting of how well the bread was toasted, at what temperature the heating element turned off, and how quickly the bread toasted. This can result in large amounts of toasted bread. One company donates the toast to ranchers who use it as livestock feed.

The Future

The toaster continues to be modified according to new technology. Designers create new, versatile looks that fit into any kitchen. Size and capabilities range from simplistic to extreme. In 1996, two University of Washington engineering students designed and created a talking toaster. This toaster allows the user to specify verbally what setting they would like (light, medium, or dark) and the toaster verifies this aloud. Although the design specifications are available, there is no sign of its mass production.

The toaster itself has become an icon. Toasters can be found on clothing and on CD covers, as salt-and-pepper shakers, and as artwork. The most outrageous example of the latter is a mosaic made with 3,053 pieces of toast for an Italian art gallery.

Where to Learn More

Books

Alphin, Elaine Marie. *Toasters.* Minneapolis: Carolrhoda Books, 1998.

Travers, Bridget, ed. *World of Invention: History's Most Significant Inventions and the People Behind Them.* Detroit: Gale Research Inc., 1996.

Other

Anderson, Corin. *The Talking Toaster Home Page.* 8 September 1996. December 2001. <http://www.the4cs.com/~corin/cse477/toaster/>.

Huggler; Peter E. "Temperature Sensitive Timing Assembly for Toaster Appliance." *United States Patent and Trademark Office Web Page.* Patent no. 59,01,639. 19 December 1996. December 2001. <http://www.uspto.gov/patft/index.html>.

The Toaster Museum Foundation Web Page. September 2001. <http://www.toaster.org/museumintro.html>.

Mary McNulty

Tuxedo

Background

The tuxedo is a man's tailored suit used for semi-formal or formal wear. It may be sewn from a wide variety of colors and fabrics; increasingly, brighter colors and unconventional designs are pervasive in tuxedo styling. Nevertheless, most tuxedos are produced in black. While tuxedos are available for purchase, most men rent these fancy suits for special occasions since they are infrequently worn and seen as an unwise investment.

Tuxedo jackets often include satin on the lapels that are attached to the collars. Tuxedo pants resemble men's tailored trousers except that they generally have a satin or ribbon stripe sewn over the outside seam of the leg. Most tuxedos are worn with specific accessories that include the slightly stiffened, sometimes fancy, white pleated shirt that closes with old-fashioned shirt studs rather than buttons. Another important accessory is the cummerbund or fabric belt that encircles the waistband of the trousers and secures in the back.

The tuxedo is, essentially, a ready-to-wear garment made in specific, standard sizes. They may be purchased or rented from an apparel store at a moment's notice. Custom or couture tuxedos are available through a personal tailor and are made to fit the wearer to his specifications. Tuxedos are constructed just as a man's tailored, pattern-graded ready-to-wear suit would be produced except the fabric is a bit dressier, the lapel includes satin and a decorative stripe is sewn onto the trousers. Companies that make men's suits may also be involved in tuxedo production.

History

Interestingly, the tuxedo did not begin as formal wear. Rather, it was seen as a less formal alternative to men's formal wear. Until the early twentieth century, gentlemen wore *frock coats* for formal wear, choosing a black frock coat with tails and gray-striped trousers for formal wear during the day. A black frock coat with tails, a white waistcoat (sometimes referred to as a vest), white shirt with stiffened bosom, and black trousers were worn with a black silk top hat and was the typical formal evening wear for gentlemen.

About the turn-of-the-century, legend suggests that American gentlemen in and around Tuxedo Park in New York, an enclave of the wealthy, chose to simplify formal wear and drop the fancy tail coats preferred for evening wear. They chose instead to wear a black coat styled much like their work suitcoats. The gentleman thought they could then wear these simple black trousers for semi-formal occasions. The jackets, known as tuxedo jackets, were often decorated with rich black silk satin on the lapels and that detail persists in many tuxedos today. The ribbon stripe on the outside edge of conventional tuxedo trousers may be reminiscent of the gray-striped trousers popular for day formal wear in the nineteenth century. By the second decade of the twentieth century, the black tuxedo had supplanted the formal black tailcoat as acceptable formal and semi-formal wear.

The wealthy had their fine tuxedo jackets and matching trousers made by a personal tailor in the early twentieth century. However, with the development and refinement of the American ready-to-wear industry, tuxedos were available in standard sizes by the

About the turn-of-the-century, legend suggests that American gentlemen in and around Tuxedo Park in New York, an enclave of the wealthy, chose to simplify formal wear and drop the fancy tail coats preferred for evening wear. They chose instead to wear a black coat styled much like their work suitcoats.

early twentieth century. Today, few men own such suits, instead they are frequently rented for special events. There is no question that today we see these suits as quite formal and do not consider them semi-formal. Colors and styles are varied today, including bright colors, patterns, double-breasted styles, even long coats are popular again. The design of the tuxedo is only as limited as the imagination can create and the market can bear.

Raw Materials

Tuxedos may be made from a great variety of fabrics today. These include wool, polyester, and rayon. Fancy detailing is generally an imitation silk satin such as polyester or rayon. Linings may be acetate or polyester. Stiffeners are an important part of the tuxedo as they help the shoulders, collar and lapel retain their shape. These stiffeners may be felt (underneath the collar) and buckram, a coarsely-woven fabric used in more structured ready-to-wear outfits. Fasteners typically include synthetic component buttons that can hold up to the chemical bombardment they receive during endless dry cleanings, and metal-toothed zippers in the trousers.

Design

The design of the tuxedo may be the most important part of a successful manufacturing process. Popular trends in men's clothing help set the style for tuxedos. A group of designers study men's fashion and suggest what tuxedo styles will appeal to a broad group of consumers. This group finds illustrations and may create illustrations of the styling they hope to reproduce within the factory. Fabrics, new colors, interesting lapel shapes, length of coat, or flare of the trousers may be among the new styling features the designers manipulate to produce new products.

Pattern makers provide the tools that will enable the manufacturer to produce these new tuxedos—the patterns. The process for this is fairly straight-forward; the pattern parts are sketched on paper and once there is consensus that these parts will create the targeted design, the pieces are digitized into a Computer-Aided Design (CAD) system. All men's fashions are drafted in prototype pattern form in one size referred to as *40 regu-*

lar, which includes a jacket with a 40-inch chest, a 32–33 inch sleeve length, and a pair of trousers with a 33–34 inch waist. (Generally, in standard sizing for men's suits, the waist is 6 in less than the chest size of the jacket; thus, a 48 regular jacket would be accompanied by a pair of trousers with a 42-inch waist.) All subsequent patterns are then graded from this standard 40 regular pattern.

The prototype pattern is used to cut out a size 40 regular tuxedo. The company then assesses the styling and decides whether the tuxedo will indeed be marketable as well as the complexity and expense involved in production. Upon approval, the pattern is graded—proportionally scaled, up or down off of size 40 regular, lengthening or broadening the pattern as necessary. The variety of pattern sizes produced is significant since many tuxedo manufacturers offer the product in sizes from 36 extra short to 60 XXL. Specifications for cutting patterns is fed into the CAD system so that the pattern pieces are devised on a computer-generated system that produces all subsequent sizes of the 40 regular prototype.

The designers and other members of the manufacturing team suggest the appropriate fabrics for production of the tuxedo. Some tuxedos are produced in dozens of fabrics and colors and utilize a variety of linings, buttons and other notions. The designers and the pattern-makers are keenly aware that each fabric type utilized affects other aspects of production including how the fabric is cut, the lining and tapes that must be used to reinforce the fabric type, the kind of needle that most cleanly pierces the fabric, the type of thread that will ensure the fabric will not be pulled, etc. Once these specifications for production are established, production is ready to proceed.

The Manufacturing Process

Most tuxedos are worked on over a period of many days, even several weeks. There are so many small parts or tasks to be completed before the tuxedo is finished that much time is spent in production. If the time it took to cut, sew and finish a single tuxedo was condensed into one single day, it is estimated that it would take eight to 12 hours to produce one unit.

1 Fabric pieces may be cut out in one of three ways depending on the manufacturer. All of the methods described enable multiple layers of fabric to be cut out at one time, cutting approximately 25 layers at once (this amount varies according to the thickness of the fabric). The fabric pieces may be cut by hand using manual shears or very sharp, heavy tailor's scissors. A second method employs an electric round wheel much like a circular saw that is held in the hand. A third method entails cutting fabric using a motorized machine that is run from a computer program.

2 Each piece is tagged with special identification indicating the specific bolt of fabric from which the piece was cut because all tuxedo cloth must be cut from the same bolt and dye lot (or the parts may not match precisely in color) and the size of the tuxedo for which it is intended. Also, the tag may indicate which tuxedo style the piece is intended if more than one style is in production at the same time. The pieces are either carried to the operators at sewing machines for assembly, or are stored until they are needed.

3 Operators sitting at individual stations generally sew the pieces together using industrial grade sewing machines (these machines are able to handle the heavy fabrics and linings used in men's suits and tuxedos). At one company, the construction of a tuxedo has been divided into 150 different sewing operations, meaning that many different operators actually work on a single garment. The coat generally consists of 110 operations and the trousers 40 different operations.

Assembling the coat

The sequence of operations includes the following general steps, each with many subcomponents.

1 First, the two front panels are sewn together, which generally includes some stiffening in parts of the bosom. The stiffening fabric is sewn to each panel so they become a single unit. The fabrics are sewn inside out in order to hide the stitching upon reversal.

2 Pockets are sewn in by the operator next. If they are patch pockets, like a breast pocket, they are sewn on the outside of the panel. Pockets in the seams have a lining that is sewn to the inside of the panel along the seam opening. The pocket edges are finished off by tucking excess fabric and stitching the seam edges to smooth and secure the seam at its openings.

3 The back of the coat is constructed by sewing the two back panels together down the center. The front panels are connected to the back at the shoulder seams but not the side seams. Stiffening or padding may be sewn in at this point if needed.

4 If the sleeves are to be lined they are juxtaposed with thin lining and sewn down the inner arm on a sewing machine. Again, the fabric is sewn from the interior in order to hide the seams and produce a more polished looking garment.

5 The remaining lining is added to the coat body at this point. A thin layer of satin-like fabric is usually used for the lining and is cut to the dimensions of the front and back panels. The lining is sewn with both finished sides facing each other, and then flipped right side out. The sleeves, already sewn together, are attached to the coat at the armhole.

6 Finally, the collar, including the lapel, is assembled. This has a shell or top of the lapel of satin (characteristic of a tuxedo) and an interfacing consisting of felt with a piece of canvas built into it and buckram to give it strength. The interfacing is cut to the shape of the collar and sewn into a "sleeve" of the outer fabric. Contrasting fabrics, such as satin striping along the edge of the collar are sewn onto the outer fabric as well prior to jacket attachment. The lapel is constructed using the same process as the collar, but in different shapes and styling. The lapel is sewn along the front opening of the front panels. After assembling and attaching both the lapel and collar, the coat is complete.

Assembling the trousers

Trousers are not generally sewn to a specific length. Instead, the end is often left with a pinked edge so the store can hem each leg up or down as needed.

1 If the trouser legs are to be lined, the liner fabric is cut to match the size and shape of the trousers. The thin lining, usually a

The fabrication of a tuxedo.

satin-like fabric, is juxtaposed on the interior of the leg before the legs are put together. Once the lining has been sewn in, the trousers are sewn together along the back inseam and along the outer side of each leg.

2 The characteristic satin stripe is applied along the outside of each trouser leg with topstitching. The legs are then sewn together at the interior curved seat seam and interior leg seams as well.

3 The waistband, which is generally folded at the top and stiffened within with buckram or some other interfacing, is sewn all around the upper edge of the raw edge of the trousers. The belt loops are constructed of small, machine-sewn strips of self-fabric and are attached at regular intervals onto the waistband.

4 The zipper is sewn to the interior of the trousers so that the overlapping fly fabric covers the metal teeth of the fastener.

5 When the coat and trousers are completely assembled, the parts must be finished. Finishing refers to closing off raw edges with closely stitched thread, such as that seen around buttonholes. It also includes sewing buttons onto the coat and pressing both the trousers and the coat. The hem of the trousers may remain a raw edge. The tuxedo is now complete.

Quality Control

All fabric is carefully inspected upon arrival for any flaws or irregularities that could produce an inferior suit with imperfections. The industry examines a length of material in a 100-yard piece and has determined that an acceptable bolt of yard goods can only have a specified number of flaws per piece. Dye lots, in which yard goods are colored in the same dye vat at the same time, are carefully marked so that the tuxedo is not sewn from bolts colored at different times. These dyes vary widely even when the same recipe is used for their formulation. Seamstresses and tailors are vigilant in using fabrics from the same dye lot. Requirements are determined for each of the sewing operations performed on the tuxedo; thus each job is evaluated against that specific criteria. Also, since so much of the construction of the tuxedo is completed by human operators at sewing machines they easily and quickly perform visual checks at each stage of production. Garments are fully inspected after finishing as well, especially along seams for durability and closure.

An important part of quality control is prototyping each new design and ironing out all design flaws carefully before production begins. Armholes that are too small, lapels that have no body, trousers with improper flare, all can be avoided with thoughtful feedback on the prototyped tuxedo.

Byproducts/Waste

There is a considerable amount of wasted fabric resulting from cutting out the tuxedo parts. One manufacturer estimated that perhaps as much as 12% of the fabric is unusable after pattern pieces are cut. Most garment-makers try to recoup losses related to this unusable fabric by selling this scrap to companies that make reconstituted fibers. These fibers are used in everything from other garments to floor coverings.

The Future

Tuxedo manufactures need to keep up with changing men's fashions; men's styles change almost as frequently as women's fashions. Couturiers with great cache greatly affect the design of higher-style tuxedos. New styles by well-known designers seen at very public events, such as the Academy Awards presentation, certainly have resonance in the manufacture of tuxedos. New colors, and occasionally new fabrics creep into tuxedo use but the days of outrageous tuxedos are largely over. In fact, the conservative black tuxedo with white shirt used for middle-class weddings rarely varies from year to year. The challenges that face tuxedo manufacturers primarily revolve around their ability to construct tuxedos competitively.

Where to Learn More

Books

Constantino, Maria. *Men's Fashion in the Twentieth Century.* New York: Fashion Press, 1997.

Hollander, Ann. *Sex and Suits.* New York: Alfred A. Knopf, 1994.

Other

Oral interview with Barry Cohen, Vice-President of Manufacturing for Hartz and Company. Frederick, MD. September 2001.

Nancy E.V. Bryk

Typewriter

The first manufactured typewriter appeared in 1870 and was the invention of Malling Hansen. It was called the Hansen Writing Ball and used part of a sphere studded with keys mounted over a piece of paper on the body of the machine.

Background

Typewriters fall into five classifications. The standard typewriter was the first kind manufactured. It was too heavy (15–25 lb or 5.6–9.3 kg) to move often, so it was kept on a desk or typing table. The standard typewriter had a wider platen (a rubber-covered, steel cylinder for absorbing typing impact) in the carriage (the part that moved the paper into place) that could hold oversized forms. The portable manual typewriter was smaller in size, lighter in weight, and equipped with a carrying case for easier movement and storage. Portable typewriters were popular for home and school use.

Electric typewriters were heavier than standard machines because of their motors and electrical parts. Electric machines made typing easier because less effort was needed to strike the keys. Electric portables were smaller and lighter than desktop machines, and they had carrying cases with storage for the power cord.

The most recent kind of typewriter to be produced—the electronic typewriter—eliminated many of the disadvantages of both standard and electric machines. Circuit boards made the electronic typewriter much lighter (about 10 lb or 3.7 kg) than other models. Personal word processors (PWPs) were closely related to computers.

History

Writing machines were built as early as the fourteenth century. The first patented writing machine was made in England in 1714 but never built. The first manufactured typewriter appeared in 1870 and was the invention of Malling Hansen. It was called the

Hansen Writing Ball and used part of a sphere studded with keys mounted over a piece of paper on the body of the machine.

Christopher L. Sholes and Carlos Glidden developed a machine with a keyboard, a platen made of vulcanized rubber, and a wooden space bar. E. Remington & Sons purchased the rights and manufacture began in 1874. To avoid jamming typebars with adjacent and commonly used pairs of letters, Sholes and Glidden arranged the keyboard with these first six letters on the left of the top row and other letters distributed based on frequency of use. Their "QWERTY" system is still the standard for arranging letters.

The first Remington typewriter only printed capital letters, but a model made in 1878 used a shift key to raise and lower typebars. The shift key and double-character typeface produced twice as many characters without changing the number of typebars. By 1901, John Underwood was producing a machine that had a backspace, tab, and ribbon selector for raising and lowering the ribbon.

George Blickensderfer produced the first electric typewriter in 1902, but practical electric typewriters were not manufactured until about 1925. In 1961, International Business Machines (IBM) introduced the Selectric electric typewriter. From about 1960 to 1980, the standard typewriter industry in the United States withered away. The IBM Selectric II debuted in 1984, but IBM stopped making electric models in favor of the electronic Wheelwriter in the early 1990s. By this time personal computers were becoming more popular.

By the late 1990s, most of the manual typewriters supplied to the United States came from three firms. Olympia in Germany makes standard portables, Olivetti in Italy makes a standard office typewriter and two portable models, and the Indian firm Godrej & Boyce Manufacturing Company is the largest producer of manual typewriters.

Raw Materials

Carrying cases can be made of wood, steel, or plastic. Steel is the material used for most of the parts in standard models. Typewriters use hundreds to thousands of moving parts, and cold-rolled steel is one of the most reliable materials.

The platen is a steel tube covered with a rubber sleeve. The rubber sleeves are made of a special form of rubber from the "buna-N" family. Glue is used to adhere the rubber sleeve to the platen tube.

The keys were molded of plastic in a two-shot, injection-molding process that made white characters with the surrounding key tops in other colors. From the 1970s forward, a pad printing process has been used to apply the characters in ink and coat the keys with a durable "clearcoat" finish.

Mylar (plastic) ribbons with ink on one side are used to transfer the typeface. These ribbons are contained in plastic cartridges that could be thrown away.

Miscellaneous materials are also used. These include glue, paint, chemical solvents and other fluids, zinc and chromium for plating some components, and acetic acid for building protective coatings on some parts.

Design

Typewriters have several parts that allow them to produce typed papers; the keyboard being the most obvious. Each key is connected to a typebar that lifted a typeface to strike the paper. Each typeface has upper and lower case forms of a letter or numbers and symbols. The assemblage of typebars and typefaces is called the typebasket.

Mylar (a plastic produced in very thin ribbons and coated with ink on the platen side) typewriter ribbon uses ink to transfer images on the typeface to the paper. Its alignment parallels the platen and the paper, and ribbon guides raise the ribbon to print and then lowers it.

The platen stops the typeface but allows enough force to the paper for the image to print. The carriage is a box-like container in the upper, rear part of the typewriter that carries the platen, the lever for carriage returns and line spacing, guides to help direct and grip the paper, and the paper itself. The paper is inserted in a feed rack (paper support) in the back of the carriage, supported and curved up toward the typing surface in a paper table or paper trough, and held against the underside of the platen by two feed rollers.

An escapement (a device that allows motion in only one direction and in precise steps) controls the motion of the carriage to the left after each character was typed. A mainspring in the escapement transmits energy to move the carriage on ball bearings.

To move the paper up after a line of typing is complete, a line-spacing lever rotates the platen toward the rear of the typewriter. The lever is also the carriage-return that disengages the escapement and pushes the carriage back to the right for the new line. Knobs on the ends of the platen are turned so the paper can be removed.

The Manufacturing Process

1 Metal (primarily pre-tempered steel) for typewriter parts arrives as round stock. Round stock is supplied in 10–12-ft-long (3–3.75-m-long) rods of steel, brass, or other metals and in a range of diameters for making screws, bolts, and rivets.

2 Rods of round stock are distributed to machines where fabricators mark and cut them to length for rivets, bolts, or screws. Screw machines (lathe-like devices) turn round stock into screws by cutting the threads, points, and heads. Hobs (another type of cutting tool) are often used to cut other fasteners to length and shape.

3 The parts are taken to plating or finishing stations where they are treated for protection from wear and rust. Zinc or chromium plating is applied by treating the metal parts in baths of non-metallic solutions that

conduct electricity. The parts are subjected to slight electrical charges that cause atoms from small pieces of zinc or chromium to be attracted to them when the baths are given opposite charges. Electrically bonded coatings made up of thin layers of atoms of zinc or chromium protect all surfaces of the metal parts.

4 Parts of the typewriter on the inside of each machine are treated in a series of baths of acetic acid to color the metal black. This process of creating the black layer (called black oxide) is something like dyeing clothing; the general term for the process is bluing. After the acetic baths, the metal parts are bathed again in a dip tank containing a type of light oil. The hot oil dries and leaves a protective coating over the black oxide. These treatments protect the parts against rust.

5 At finishing stations, exterior parts are polished. Operators apply buffing compounds to buffing wheels on machines and hold the typewriter parts against them. The rotating wheels coat the parts with the compounds and shine the typewriter components. Workers polish very small parts by hand, also using polishing compounds and hand-held buffers.

6 Pieces are then riveted or brazed to form complete parts for assembly. Brazing is similar to a soldering process that uses alloys with lower melting temperatures than the metal pieces being joined to avoid melting or warping those pieces. Both brazing and riveting create rigid joins, although rivets are also used when parts have to be free to move. Screws, bolts, and other fasteners also make moveable connections.

7 The platen is a specialized subassembly because it requires precision grinding with heavy machinery and the process produces rubber dust. The internal steel tube (sometimes called the axle or shaft) is cut from hollow round stock. It is finished on the outside for easier addition of the rubber and on the ends for smoothness. Similarly, the internal metal rod is also cut from round steel stock. The centers are stamped from steel in sheet form.

8 A rubber sleeve is then heated slightly to fit over the platen, and an air press pushes the sleeve over the tube coated with glue. A rod and the two platen centers are added to the steel tube, and fittings are added to hold the rod and centers tightly.

9 To make the typefaces, blank pieces of metal called "type slugs" are formed in the machines by vibrating the slugs into die sets bearing the letters and other characters. As the slugs are worked into the dies and hardened, the typefaces are spit out of the machine. Then transferred to the subassembly section where they are soldered on the typebars.

10 The rail system uses ball bearings to glide the carriage from left to right. Subassembly of the carriage consists of mounting the rail to the base of the carriage, installing the ball bearings, and attaching the spring and linkages.

11 The carriage-return lever extends over the top of the typewriter. Although the it is attached to the carriage to move it, it also has several linkages to the platen, paper handling system, and escapement. The lever and one set of the ends of its linkages are connected to the carriage. The parts of the metal feed rack (also called a paper support) that hold the paper as it is put in the typewriter are assembled, and the rack is attached to the back of the carriage.

12 The paper-handling system is another subassembly. It includes the paper trough (also called a paper table), two feed rollers (like miniature platens) that holds the paper against the underside of the platen, the paper-release lever, and a paper-alignment scale (paper bail). The paper-handling system allows the paper to be inserted in the typewriter, held firmly during typing, and rolled out when the page is complete. The paper trough is a U-shaped piece of steel stamped out of sheet stock, curved, and plated.

13 The escapement's subassembly is a system of gears, small gears called pinions, springs, chains, pawls, and fasteners. A pawl is a small bar with a tooth at each end that drops into the teeth of a gear, ratchet, or pinion. The pawls move the gear system forward, and the gears advance the escapement rack that pulls the carriage of the typewriter to each space needed for a new typed image. The escapement is assembled

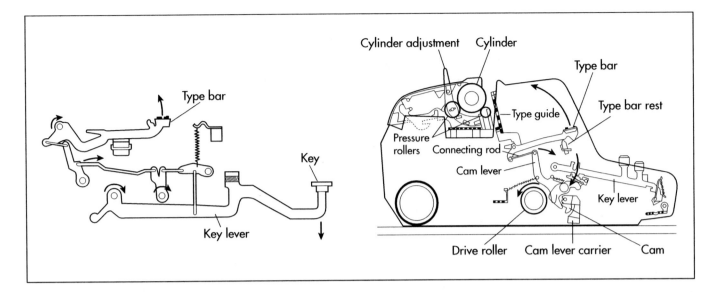

in a fitted, tray-like frame that will be set into the inner face of the strong underside of the typewriter jacket. This heavy underside and the arrangement of portions of other subassemblies that would be attached over the escapement protects the sensitive works.

14 The subassembly for the typebasket contains many of the 3,200–3,500 parts in the typewriter. The typebasket subassembly holds the typebars with typefaces on their ends as well as the spring system that connects the typebars to the keys. Each typeface is soldered to the end of its typebar. Each typebar has a unique angular bend so its typeface will strike flat against the platen. Like those in the carriage, sets of ball bearings are added to help move the typebars from upper to lower case and back. The assembler inserts the typebars in their positions in the typebasket and attaches the ends at the bottom of the basket to the appropriate springs. The springs will be connected to the keys when the keyboard and typebasket subassemblies are linked to each other during main assembly.

15 To begin the keyboard subassembly, the cap of each key is soldered to the correct key lever. The key levers are connected to springs that allow the keys to be depressed. The levers are put in appropriate slots in an internal keyboard frame. The spring system is also mounted to the keyboard frame to be connected with the springs for the typebasket subassembly during main assembly.

Main assembly

16 The five key subassemblies of the standard typewriter (the carriage, paper handling system, escapement, typebasket, and keyboard) are put on trucks and moved to the main assembly line where they are added to the typewriter frame.

17 Inside the body, the tray-like frame of the escapement is bolted into the inner face of the underside jacket of the typewriter.

18 The rail on the underside of the carriage is fixed to its matching half on the upper part of the body frame. The platen is set into place in openings in the carriage frame. A knob is added to the extruding end of the center rod on the right side of the platen; on the left end, a fitting holding the carriage-return and line-spacing lever is fitted on the rod, and is finished with another knob.

19 The keyboard and typebasket are inserted, their frames bolted to the body frame. A steel, V-notched typeguide is attached across the half-moon of the typebasket facing the platen; the V-notch provides an opening for the typefaces to strike the platen. The springs for each key and its typebar are linked together.

20 The typebars are also connected to the escapement and carriage linkages. To align typebars with the opening in the typeguide and strike the platen at the correct angles, the workers use three-pronged pliers to bend each typebar gently.

21 When the jacket of the typewriter is made from steel, it is attached to the main frame. The strong underside of the jacket had been installed on the main frame earlier because it also serves as a support for the escapement subassembly. Two pieces of steel forming lower sides of the jacket around the carriage are attached to the carriage frame. Two upper sides are also mounted on the carriage frame. These match the lower sides to provide round openings for the inner ends of the platen knobs so they can be used to turn the platen. The back and top L-shaped sections of the carriage jacket are attached to the body frame. The sides and top jacket of the keyboard are fastened in place over the keyboard. All of the sidepieces of the jacket are attached to the underside to strengthen the frame and jacket; the firm fit also seals the underside to limit the amount of dust that could enter the interior of the typewriter.

Quality Control

When raw materials are delivered to the typewriter fabrication plant, the receivers log in the materials and compare them to blueprints and specifications provided by design and manufacturing engineers. The quality control engineers also use a number of instruments for determining that parts and materials are acceptable such as verniers (short sliding rulers), micrometers (also called micrometer calipers) that are vice-like gauges for measuring thickness precisely, and height gauges to confirm dimensions.

When the typewriters are complete, a final quality control check is done by actually using each machine to test its performance. Each typewriter is checked for binding keys, print quality, advance of the ribbon, and movement of the carriage, among many other performance characteristics. Its appearance is careful examined for any flaws that might lead to rusting.

Byproducts/Waste

Most of the waste is generated during fabrication. Steel wastes such as the "skeletons" left after stamping or punch pressing and turnings and bushings (fragments) from screw-machine production of rivets and other parts are sold to salvage dealers, or melted and reused.

Plastic parts are used increasingly, plastic runners and rejected parts are also recycled. In the fabrication plant, they are reground, and these plastics were added to new batches of plastic. The percentage of reground plastic in a batch varied depends on the criticality of the part and the decision of the manufacturing engineers.

A large volume of rubber dust was produced when platens were ground round. The dust was carefully controlled and placed in collection boxes. The cooled dust was taken in the collection boxes to landfills. Machine exhaust was hooded to the outside. Minor quantities of other materials were disposed or recycled. Inked ribbons and cassettes containing Mylar ribbons were sometimes rejected and were also disposed in landfills.

The Future

Typewriters have a minor future in the Western World because computers have replaced them almost completely. Some businesses still need typewriters for limited uses, and many people find typewriters more convenient for single or small tasks.

Standard, electric, and electronic typewriters do have some future remaining in developing countries, and manufacturers in Asia and Europe supply this market. Brother makes typewriters in Japan, China has two or three factories, and Godrej & Boyce Manufacturing Company in India is the largest typewriter producer in the world. The Hermes, Olivetti, Olympia, and Royal brands are made in one or two factories in Europe. At the peak of standard typewriter manufacture, Smith Corona dominated production with a 54% market share; the company no longer makes its own typewriters, but, as a small supplier, it purchases them from a factory in Korea.

Rare use of typewriters today and their distinction as truly magnificent machines has made them popular and given them a respected future as collectibles. Antique dealers and other specialists buy and sell rare models on the Internet, and collectors exchange information using newsletters and web sites.

Where to Learn More

Books

Bryant, Carl. *All About Typewriters and Adding Machines.* New York: Hawthorn Books, Inc., 1973.

Davies, Margery. *Woman's Place is at the Typewriter: Office Work and Office Workers 1870–1930.* Philadelphia: Temple University Press, 1982.

Linoff, Victor M., ed. *The Typewriter: An Illustrated History.* Dover Publications, 2000.

Periodicals

Frazier, Ian. "Typewriter Man." *The Atlantic Monthly* Vol. 280, no. 5 (November 1997): 81–92.

Groer, Annie. "True to Type." *The Washington Post* (3 May 2001): H01.

Other

"Typewriter History at a Glance." *My Typewritter.com Web Page.* December 2001. <http://www.mytypewriter.com>.

Gillian S. Holmes

Unicycle

While standard unicycles have a height of about 3 ft (0.91 m), Giraffe unicycles can be 15–20 ft (4.6–6.1 m) tall. The record height for a Giraffe unicycle is about 100 ft (30.5 m).

Background

A unicycle is a single-wheeled vehicle traditionally used during circus performances. It consists of a spoke wheel, pedals, and a tube shaped body attached to a seat. Unicycles are made like bicycles; individual parts are produced separately then pieced together by the manufacturer.

History

The unicycle's history began with the invention of the bicycle. Comte De Sivrac first developed bicycles during the late eighteenth century. His device, called a celerifere, was a wooden horse that had two wheels joined by a wooden beam. In 1816, wooden, riding horses like these had become improved by the addition of a steering mechanism. In 1840, Kirkpatrick Macmillan introduced a mechanism for powering the hobbyhorse with his feet. During the 1860s significant progress was made with the introduction of rubber tires, metal spoke wheels, and ball bearing hubs. During 1866, James Stanley invented a unique bicycle called the Penny Farthing. It is this vehicle that is thought to be the inspiration for the unicycle.

During the late nineteenth century the Penny Farthing was a popular bicycle. It had a large front wheel and a small rear wheel. Since its pedal cranks were connected directly to the front axle, the rear wheel would go up in the air and the rider would be moved slightly forward. This likely prompted riders to see how long they could ride with the back wheel in the air and the unicycle was born. Evidence for this theory of development can be found in pictures from the late eighteenth century that show unicycles with large wheels.

Since the unicycle requires a greater degree of skill to ride than a bicycle, many people that could ride them became entertainers. Over the years, unicycle enthusiasts have inspired manufacturers to create new designs such as seatless and tall, giraffe unicycles. During the late 1980s some extreme sportsmen took an interest in the unicycle and outdoor unicycling on rugged terrain was born. Today, the unicycle remains a relatively obscure vehicle however there are more people riding unicycles now then ever before.

Design

A unicycle is a single-wheel vehicle. Typically, it consists of a seat attached to a frame that is attached to the wheel hub. Seats are available in a variety of sizes and shapes. Unlike a bicycle seat, they are often curved slightly downward and symmetrical. They are padded and may have "bumpers" on each end to protect them during one of the rider's inevitable falls. Some seats are designed with handles on the front to enable the rider to do various tricks.

The seat is attached to the frame via the seat post. The various types of posts that are available can differentiated by the way they allow for height adjustment. The most basic seat post is a metal tube with holes drilled at regular intervals. There is also a hole in the main body through which a bolt is placed to hold the seat post. To adjust the seat higher or lower, the bolt is removed, the seat is adjusted to the next hole and the bolt is replaced. A more sophisticated seat post has no holes but is attached to the frame through a clamp. The clamp is loosened or tightened with a blot or Allen wrench screw to adjust the height of the seat post. This design allows for a finer adjustment of the seat height. In addition to these differences some seat posts allow for angle adjustments.

The unicycle frame is a metal structure with two forks that attach to the wheel and a hollow tube that connects to the seat post. Different designs are available. The simplest has flat forks bolted together above the wheel and attached to the seat post. More sophisticated versions of this design have curved forks for better rigidity. A better design is the tubular one-piece body with a squared or rounded fork crown. These structures have less points of weakness and are more durable.

The frame is attached to the center of the wheel along with the bearings, pedal cranks and the spoke hub. The bearings are sealed balls of steel that reduce friction. The pedal cranks are attached to the bearings, and at the end of the cranks are the pedals. The spoke hub is the area where each of the wheel spokes are attached. The hub can have straight or angled flanges that attach to the spokes and have a cottered or cotterless shaft.

The spokes are thin metal tubes measuring in thickness from 0.08 to 0.125 in (0.2–0.32 cm). The number of spokes on a unicycle varies depending on the design and can be from 28 to 48. In general, more spokes are better. The spokes can be arranged in different patterns on the wheel. For example they can have a three or four cross pattern. Additionally, they can have an interleaved pattern that adds to strength and stability. The spokes are attached to the tire rim through a series of holes. These holes can be straight or angled to match the angle of the spoke.

The final component of a unicycle is the tire. A standard unicycle tire has a round cross-section and a flat or smooth tread. This design is ideal for riding on flat surfaces. Some outdoor unicycle tires have thicker treads that are better for rugged terrain. The size of a standard unicycle tire is 26 in (66 cm).

In addition to the standard unicycle just described, other designs are available. The Giraffe unicycle is a taller vehicle that is chain-driven like a bicycle. While standard unicycles have a height of about 3 ft (0.91 m), Giraffe unicycles can be 15–20 ft (4.6–6.1 m) tall. The record height for a Giraffe unicycle is about 100 ft (30.5 m). Another type of unicycle is called the Ultimate Wheel. This version has no seat or body consisting of only a wheel and pedals. The spokes are typically replaced by a plywood disk to reduce injury to the rider's ankles. The Impossible Wheel consists of a wheel and side posts that the rider stands on. The challenge to this unicycle is figuring out how to propel oneself. Finally, a rare type of unicycle is the monocycle. This design consists of a large wheel with the seat on the inside.

Raw Materials

Numerous raw materials are used to create a unicycle. Since the main body must provide strength and rigidity but also remain lightweight, it is typically made of a steel alloy, aluminum, or titanium. Steel is a material made up of primarily iron. Other metals that can be incorporated include aluminum, manganese, titanium, tungsten, vanadium and zirconium. During the 1990s, composite materials such as carbon fibers were introduced and are now sometimes used to make unicycle frames. Depending on the frame material it can be protected with various coatings including a baked-on enamel, powder coating or chrome plating.

Metal alloys are also used in the construction of the rim, spokes, pedal cranks, hub, and seat post. The most basic rim is composed of chromed steel. More sophisticated structures use a steel alloy or chromed steel alloy. A typical spoke is made of zinc plated steel. The may be also made of stainless steel or chrome-plated steel.

Other components of the unicycle are made using various materials. Pedals are made out of rubber or plastic. A basic pedal has a solid rubber, block construction with no spindle adjustment. Better pedals are made of a solid plastic with an adjusting device to fit different sized riders. The seats can be made from various materials such as leather or vinyl. They can also have a polystyrene or polypropylene padding.

The Manufacturing Process

The production of a unicycle is typically done in two phases. In the first phase the in-

A unicycle.

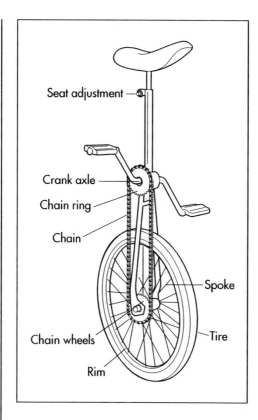

Seat adjustment

Crank axle

Chain ring

Chain

Spoke

Chain wheels

Tire

Rim

dividual components are made. Separate manufacturers that specialize in a specific component typically do this. In the second phase, the unicycle manufacturer buys the components and assembles them.

Creating the frame

1 Seamless tubes for the frame are made from solid blocks of steel. The steel is heated and molded into a cylinder shape. This cylinder is put in a furnace making it white-hot. It is then rolled under high pressure. The rolling stretches or draws out the steel and a hole forms in the center. A bullet shaped piercer point is pushed through the middle of the steel as its being rolled. After the piercing stage, the tube is passed through a series of rolling mills to correct any irregular shape or thickness.

2 To fashion the tubes into a unicycle frame, they are first heated until softened. This allows the tubes to be bent and shaped into the appropriate structure. In one type of unicycle design, the forks are flattened and crimped at the end to seal them. This end is also welded to provide greater stability and strength. A hole is drilled at the crimped end to provide a place for the frame to be attached to the wheel axle. This

process can be done on a machine or by hand. The piece of the frame that attaches to the seat post is drawn out to the appropriate thickness using rollers and then cut. The three frame pieces are then joined together in a metal coupler that has two holes that attach to the forks and one hole on the opposite side in the middle that attaches to the third piece. These links are welded closed.

3 The frame is put on a jig and checked to ensure that it is aligned and shaped properly. Since it is still hot, subtle adjustments can be made. Excess flux and brazing metal is cleaned off and the welds are ground smooth. After the metal is cooled, other adjustments may be made.

4 To protect the metal frame and give it a more appealing appearance, it is typically coated. Coating may be done by hand spray painting or by passing the frame through an automatic electrostatic spraying room. In this room, frames are given a negative charge and sprayed with positively charged paint. The frame is rotated to get full coverage. A finishing lacquer is then applied. Some unicycles are coated with chrome plating instead of paint.

5 The seat is typically made by an outside contractor and shipped to the plant. The seat post is bolted to the seat's bottom. The seat post is then inserted into the unicycle frame and attached with a bolt or clamp.

Wheels, rims, and spokes

6 Rims are produced on machines that roll steel strips into circular hoops. The hoops are then welded and further shaped. Holes are drilled in the rim at set intervals and angles. A larger hole is also drilled in the rim to allow the tire's air nozzle to be accessible.

7 Spokes are attached to the rim in these holes. On one end of the spoke is a nipple that screws into the rim holes. The other end of the spoke is attached to the wheel hub, which is a metal disc shaped device. The spokes are placed in the rim holes and hubs in a specific pattern. The spokes are tightened in a uniform direction to straighten the wheel. A liner is placed on the inside of the rim to protect the inner tube from damage by the spokes.

8 A rubber innertube is placed around the rim and then the outer tire is attached. The air nozzle is pulled through the rim frame and the innertube is inflated to an appropriate air pressure.

Final assembly

9 To complete the unicycle, the various pieces are connected. The frame is attached to the wheel via a solid metal tube. The bearings are attached along to the same area, along with the pedal cranks. The entire assembly is locked in place with nuts and bolts. The unicycle tire is inflated and final adjustments are made.

Quality Control

A variety of tests are done to ensure that each part of the unicycle meets specifications. The first phase of quality control is provided by the suppliers of the unicycle parts. Visual inspections are done during most manufacturing processes. For the makers of plastic parts visual inspection can find things such as deformed pieces and improperly fitted parts. Additionally, other quality control measures are taken. For example, the steel tube manufacturers are required to form tubes at a specific thickness. To do this, they utilize a device known as an X-ray gauge. This instrument is attached directly to the production line and controls the rollers to resize the steel tube if a change in thickness is determined.

The second phase of quality control is done at the unicycle manufacturing plant. The incoming component pieces are physically checked to ensure they meet specifications. For example, the diameter of the tire is measured or the color of the frame is checked. After the unicycle is assembled a quality control specialist tests it for obvious defects. For instance, the wheel is spun to ensure that it is straight or the bolts are checked for tightness.

The Future

While the unicycle designs have changed little over the years, inventors have not stopped trying to produce a better cycle. Most of these attempts have been related to making the unicycle safer and easier to ride. For example, a patent issued in the United States during 1994 describes a unicycle that is designed to have a limited tilt. This invention uses a ground contact attached to the pedals to limit the amount of tilt that the rider experiences. Another patent issued in 1999 describes a unicycle that is equipped with handlebars that can aid beginning riders. In addition to these new designs, other improvements in unicycles will likely be in the form of new composite materials that make the vehicles stronger, more durable, and lighter.

Where to Learn More

Books

Kirk-Othmer Encyclopedia of Chemical Technology. New York: John Wiley & Sons, 1994.

Periodicals

Johnson, R. C. "Unicycles and Bifurcations." *American Journal of Physics* (July 1998).

Martin, S. "Miyata Unicycle." *Bicycling* (April 1993).

Other

The Unicycle Web Page. December 2001. <http://www.unicycling.org>.

Perry Romanowski

Vending Machine

Vending operations in the United States have evolved into a $36.6 billion industry.

Background

From humble single-cent beginnings, vending operations in the United States have evolved into a $36.6 billion industry. Canned cold drinks were the industry's top sellers in 1999, posting $15.7 billion in sales and accounting for 42.9% of the industry's gross sales volume. Packaged candy and snacks ranked second, with 19.7% of the industry and $7.2 billion in sales. More than 857,000 coin-operated vending machines were produced and shipped in 1999. More than half (477,102) were refrigerated units vending canned and bottled soft drinks.

Vending accounts for a significant portion of the sales and profits of beverage bottlers and snack makers. In 2000, vending machines generated 14% of total foodservice sales in venues such as college campuses, factories, businesses, hospitals, and schools. Bottlers pay colleges, schools, and other institutions millions of dollars for exclusive rights to place vending machines on campuses. Vended soft drink sales may represent only as much as 15% of a bottler's total yearly sales, but that same volume could easily account for half of their annual profits.

History

The first documented vending machine dates from about 215 B.C., when the mathematician Hero invented a device that accepted bronze coins and dispensed holy water in the temples of Alexandria, Egypt. In A.D. 1076, Chinese inventors developed a coin-operated pencil vendor. Coin-activated tobacco boxes appeared in English taverns during the 1700s.

The United States government began granting patents for coin-operated vendors in 1896. However, it was not until 1888 that vending became a viable market in the United States. In that year, the Adams Gum Company developed gum machines that were placed on elevated train platforms throughout New York City. The machines dispensed a piece of tutti-frutti gum for a single penny.

In 1926, William Rowe invented a cigarette vending machine that started a trend toward higher priced merchandise, including soft drink and nickel-candy machines that evolved throughout the late 1920s and 1930s. Coffee vendors were developed in 1946, and refrigerated sandwich vendors followed in 1950. In 1984, Automatic Products International, Ltd. (APi) introduced a vending machine that ground and brewed fresh coffee beans.

Practically anything that can be vended, has been at one time or another. The first beverage vendor, dated to 1890 in Paris, France, offered beer, wine, and liquor. Items that have been found in vending machines include clothing, flowers, milk, cigars and cigarettes, postage stamps, condoms, cologne, baseball cards, books, live bait for fishermen, comic books, cassettes and CDs, lottery tickets, and cameras and film. Some modern vending machines dispense hot foods such as pizza, popcorn, and even french fries.

Raw Materials

Vending machines are constructed primarily from four major raw materials: galvanized steel, Lexan or other plastic, acrylic powder coatings, and polyurethane insulation.

The bulk of the machine is constructed from galvanized steel ranging from 10 gauge to

22 gauge in thickness. The thicker gauges are used for the outside cabinet, external doors, and internal tank. Thinner gauges are used for internal doors and plates, can stacks, and mechanisms such as coin validators and product trays.

Lexan, a tough polycarbonate plastic, is used in the front panels of the vending machine. Sheets of Lexan in vending machines usually range from 0.13 in (3.18 mm) to 0.25 in (6.35 mm) in thickness. Lexan is very difficult to break, flame retardant, relatively easy to shape, and can be treated to restrict UV rays, light, and heat transmission. Product logos, names, and illustrations are silk-screened on Lexan sheets, which are installed in channels in the doors of the vending machines.

Acrylic powder coatings are colored powders used to "paint" the surfaces of vending machines. The powder is applied in a uniform layer and baked on during the manufacturing process. Acrylic coatings withstand the rigors of weather and abuse better than paints that are applied wet. In addition, acrylic powders more readily meet governmental environmental standards.

Polyurethane foam provides the insulation for the inside of the vending machine. The foam is blown between the outer cabinet and internal tank of the machine, where it cures into a very tough, rigid material. In addition to thermal insulation, the stiff foam adds structural stability to both the cabinet and tank of the machine.

Some manufacturers, such as Dixie-Narco, also make the complicated electronic devices used in vending machines, while others purchase them pre-made and install them as part of the manufacturing process. These components include bill and coin validators, computer control boards, refrigeration units, and lighting.

Design

The basic design of a vending machine begins with the cabinet, the steel outer shell that holds all internal components and which determines the machine's overall size and shape. Inside the cabinet is a steel inner lining called the tank. The tank and the cabinet fit closely together, leaving enough room in-between for a layer of polyurethane

foam insulation. In combination, the tank and the foam insulation help keep internal temperatures stable and protect products against temperature extremes outside the cabinet. Although all products and dispensing mechanisms are contained in the cabinet, in the strictest sense, they are actually installed within the tank.

The outer surfaces of the cabinet are coated with an acrylic powder finish that is baked into place. Powder coatings enable the machine to withstand extreme temperatures, salt or sand, abuse by customers, and other conditions requiring high surface durability.

To store and dispense products, can feeder stack columns or feeder trays are installed inside the machine. Each tray is equipped with a large rotating wire spiral that holds the products. Feeder trays slide in and out of the machine for easy maintenance and restocking of merchandise. The feeder stacks and trays also contain the motor controls that physically push the products forward until they are released from the stack and fall to the access area. When a customer selects a product, a rotor turns and advances a single item, dropping cans or bottles one at a time. In the same way, spirals on snack food trays rotate and push products forward until they fall off the tray.

Some vending machines, especially cold drink vendors, have two doors. The internal door seals the inside of the machine and provides additional insulation. The outer door contains the electronic controls that allow customers to purchase and receive goods. The outer door also includes signage and illustrations, generally silk-screened onto a panel of Lexan that fits into the front panel of the door. Lighting for the front panel is generally installed behind the Lexan panels. The outer door includes heavy-duty hasps, locks, and hinges to deter theft and vandalism.

Electronic components, such as coin and bill validators, test coins and scan dollar bills that have been inserted to ensure that the cash is genuine and in the proper amount. A panel of control buttons lets customers make their selections. These buttons are connected to the motor controls of the feeder stacks and trays, activating the rotors that release products to the bins. Change-makers hold quantities of coins and release the correct change after a selection has been made. More recent machines may also include card validators for accepting debit and credit cards, LCD panels with pricing details and machine status information, and speech chips that give transaction details to customers by voice.

Design changes occur most frequently in the mechanisms for handling and dispensing the vast number of different types of bottles, cans, boxes, bags, and other packages available on the market. When 20 oz (592 ml) plastic soda bottles were first introduced to vending, they tended to jam in the machine. Designers had to re-work the way those bottles were stored in the machine and delivered to the customer. Constant changes in product packaging have ensured that designers must always look for practical and more efficient ways to vend products.

The Manufacturing Process

The most popular type of vendor in the United States is the cold canned soft drink machine, which vends the traditional 12 oz (355 ml) aluminum can of sodas and soft drinks. The manufacture of a cold can drink vending machine is often accomplished on several automated, concurrently running assembly lines that make all the components simultaneously. Manufacturing processes for vending machines can be as varied as the products dispensed in them.

1 The cabinet is made from a roll or coil of galvanized steel. At the start of the assembly line, the raw steel passes through automated presses that flatten it and cut it into sheets. Cabinets are frequently made of two or more separate pieces.

2 Other presses then punch and notch the sheets. The punching process creates holes in the cabinet for bolts and fasteners, openings for electrical cords, slits for vents, and other necessary openings. Corners and edges of the sheets of steel are notched where necessary to accommodate the fitting of components onto the finished cabinet.

3 Sheets of steel automatically enter and exit heavy-duty air and hydraulic presses. Each press exerts 200–400 tons of force or more. The steel sheets lie flat as the presses crimp the edges, create bends in the metal, and form the steel into the cabinet's basic shape.

4 The seams are secured using resistance welding, also known as spot welding, a process that uses a high-voltage charge through two contacts that melts metal surfaces together.

5 Finally, the cabinet is unloaded from the line and taken to the finishing area to undergo powder finishing and await installation of the tank.

6 While the cabinet is being made, another line forms the tank. The tank is created by a process similar to that used to make the cabinet. Galvanized steel is cut from rolls of raw material and enters the automated line. Openings for bolts, feeder stacks, motors, and other mechanisms are punched in the sheets, and the corners are notched so that the tank will fit inside the cabinet more easily and securely once it is welded.

7 Presses provide any necessary forming and bending, and separate pieces are welded together to create a whole unit.

8 The tank is then taken to the finishing area, where it is fitted to the correct cabinet.

9 Before powder finishing, the cabinet undergoes an eight-stage pre-treatment. First, the cabinet is attached to an overhead conveyor that runs the cabinet 210 ft (64 m) through the pre-treatment system. Nozzles mounted on both sides of the conveyor thoroughly spray the cabinet at each stage, beginning with an alkaline wash to remove heavy surface soil.

10 A second alkaline bath cleans the surface even further.

11 The cabinet is then rinsed, coated with zinc phosphate, and rinsed again.

12 Chromic acid, a sealer, is then applied and the surface is rinsed once more.

13 Finally, the entire surface is thoroughly rinsed with deionized (DI) water. Units then go into the drying oven for approximately 30 minutes to an hour at temperatures ranging from 350°F to 400°F

(177°C to 204°C) to ensure that no moisture is left on the surface.

14 Upon leaving the drying ovens, the cabinet remains on the overhead conveyor system where it begins the powder finishing process. This process "paints" the cabinet in various colors by applying a coating of acrylic powder to the surfaces of the cabinet and baking it into place. Cabinets, tanks, internal mechanisms, and other parts may all be powder finished during the manufacture of a cold can soda vending machine.

15 Cabinets travel along the convey into environmentally controlled powder booths. The booths are constructed of polypropylene, which is believed to attract less excess powder, improving spray efficiency and reducing the need for clean-up. Each booth applies a single color with an array of 18 to 22 spray guns, all of which are designed to move in order to provide better coverage, reach into cavities, and track along with the cabinet. The guns apply an acrylic powder in a uniform 0.0015–0.002 in (1.5–2 mm) thick layer on the cabinet. The positively charged powder adheres easily to the grounded cabinet surface. Oversprayed powder is collected, mixed with virgin powder, and resprayed, resulting in very little waste of coating material. As much as 95% of oversprayed powder can be recovered.

16 When the powder application is finished, coated cabinets exit the booth and go into the cure oven for 20–30 minutes at 370°F (188°C). The powder finish is permanently baked on.

17 Finished cabinets and tanks are then united for the process of foaming, the application of polyurethane foam insulation to the interior of the machine. The tank is fitted into the cabinet, and both are pre-heated to approximately 120–150°F (49–66°C) while awaiting foaming. When the correct temperature is reached, high-pressure foaming fixtures blow the insulation between the cabinet and the tank using carefully metered shots of material. The foam solidifies, providing not only effective insulation but also a degree of additional structural stability to the cabinet and tank.

At one time, the resin additives used to create the structure of foam insulation, called blowing agents, contained large amounts of CFCs, or chloroflurocarbons. However, environmental concerns over the use of CFCs led to the use of alternative blowing agents, including halocarbon (HCFC) substitutes and water.

18 When foaming is completed, refrigeration units are installed in the bottom of the cabinet. These units are often acquired pre-manufactured and are slid into place, bolted securely, and wired into the cabinet by a single operator on the assembly line.

19 Can feeder stack columns are installed inside the cabinet. These columns are manufactured and finished using steel forming, punching, notching, and

welding processes similar to those used to make the cabinet and tank. The process is monitored carefully to make sure the correct size stack is matched to the proper unit.

20 In the final stage of manufacturing, doors are installed on the cabinets. The exterior door units hold most of the highly sophisticated electronics of the cold drink vending machine, including coin and bill validators, selection buttons, control panels, change tubes, signs, and lighting. Interior doors act as additional seals for the inside of the machine. The majority of doors on can vending machines do not have clear windows for displaying products. Instead, the door fronts often contain artwork, logos, and graphics that advertise the machine's contents. However, machines that vend bottles, candy, and snacks often have clear windows made of Lexan or some other tough, break-resistant material.

21 The basic door shell is created and finished in much the same way as cabinets and tanks, with additional punching necessary to accommodate the controls on exterior doors. Components of doors are often assembled as completely as possible in one area rather than on an assembly line. It is essential that the correct door is fitted to the correct cabinet, so door assembly is centralized in order to reduce the possibility of errors in matching doors with units. During manufacture, doors are placed on revolving, indexible carousels that allow operators to turn and move them.

22 All assembly items are placed on the door while it is in this carousel. Operators fit the coin and bill validators, card acceptors, selection buttons, LCD displays, and other external controls in the appropriate spots and attach them with bolts or screws. Selection buttons are wired to the motors of the proper can stacks. Front panels of silk-screened Lexan are installed in channels on the doors. Lighting fixtures are bolted in place and wired into the power supply. Finished doors are carried to the cabinets and automatically attached. The completed can soda vending machine is then ready for quality checking and testing.

Quality Control

Finished machines are tested after assembly. Checks are made to ensure that all electronic components function properly, that the can stacks are fitted correctly, and that cans are dispensed accurately and safely. If problems are found, the machines are sent to stations capable of holding multiple units. A rework technician makes the needed repairs, and the machine is tested again. If problems remain, the unit is rejected and flagged.

Additional tests include refrigeration pull-down testing, in which the interior temperature of the unit is "pulled down" to 30–31°F (-1 – -0.5°C). This test ensures that the machine controls and temperature controls are working properly.

Tested units are given a final visual check, then cleaned and wrapped in clear plastic shrink-wrap. Finished vending machines are transported by conveyor to the warehouse where they are packed and readied for shipping to customers.

The Future

In 2000, the Coca-Cola Company announced plans to invest more than $100 million in online "Dial-a-Coke" vending technology from Atlanta's Marconi Online. For vending companies, the technology will improve efficiency, data collection, and maintenance. Machines will employ remote diagnostics to alert staff at headquarters when machines need refilling or servicing. Detailed stock status will be transmitted to centralized locations so that route drivers and technicians will know exactly what products are needed to re-stock their machines before they even begin their routes. Sophisticated data collection will provide more immediate feedback on what products are selling and what items need to be replaced with more popular merchandise. Machines will transmit real-time data on transactions, allowing companies tighter control and more precise data on cash and stock accountability. For customers, the Dial-a-Coke technology will allow cashless purchasing. Consumers will be able to use their cell phones to dial up a particular vending machine to select and pay for soft drinks.

Data can also be transmitted to the machines. Changing prices, for example, usually requires a technician to physically visit a machine to make the changes by hand. Remote vending will allow changes to be made remotely, permitting vendors the flexibility to accommodate conditions such as promotional pricing or lowered prices during non-peak hours to increase overall sales volume. Cashless systems will also reduce theft and vandalism in machines by reducing the amount of cash stored in a vending machine at any given time. APi has already began production on a vending machine that will accept credit or debit cards.

Technological improvements in wireless machine monitoring systems in early 2001 have allowed companies to utilize the Internet as a type of wide-area network for monitoring and maintaining remote vending sites. Handheld computers have also become increasingly popular as a way to capture sales and stock data directly from machines.

Where to Learn More

Periodicals

Babyak, Richard J. "New Era for Insulation (Change is in the Wind for Blowing Agents)." *Appliance Manufacturer* 41, no. 8 (August 1993): 47–48.

Bailey, Jane M. "Vending Machines Take a Beating." *Industrial Finishing* 67, no. 4 (April 1991): 36–37.

"Coca-Cola Customers to Buy Vending Machine Drinks Using Marconi's GSM Dial-a-Coke Solution." *Wireless Internet* 3, no. 5 (May 2001): 7.

Marcus, David L., Leslie Roberts, and Jeffery L. Sheler. "A Hot Idea From Those Cold-Drink Folks." *U.S. News and World Report* 127, no. 18 (8 November 1999): 10.

Prince, Greg W. "100 Years of Vending Innovation." *Beverage World* 117, no. 1651 (January 1998): 214–216.

Simpson, David. "A Peak in the Heart of Dixie (Dixie-Narco Inc.'s Use of Powder Coatings)." *Appliance* 46, no. 8 (August 1989): 56–57.

Somheil, Timothy. "Vending Innovation." *Appliance* 55, no. 1 (January 1998): 87–89.

Stevens, James R. "The Dixie-Narco Story." *Appliance* 47, no. 6 (June 1990): 31–4.

Sutej, Joseph M. "Evaluating Low-CFC Foam Insulation." *Machine Design* 62, no. 10 (24 May 1990): 108–109.

Other

National Automatic Merchandising Association Page. 8 July 2001. <http://www.vending.org>.

Vending Times Web Page. 8 July 2001. <http://vendingtimes.net>.

Jeffrey W. Roberts

Videotape

Background

Videotape is an integral component of the video technology that has profoundly impacted the media and home entertainment industries. First controlled by the television industry, videotape and video technology are now widely available to the private sector and have led to significant changes in the way that information is distributed and entertainment is created.

Videotape is all about magnetic recording. First introduced commercially in 1956, magnetic recording was a relatively new technology. Videotape and all other forms of recording tape are the same in that they are magnetic. Videotape, in fact, is very similar in composition to audiotape. Most videotapes consist of a layer of tiny magnetic particles applied to Mylar, a strong, flexible plastic material. About a billion magnetic particles cover a square inch of tape and function like microscopic bar magnets. When the tape passes over an electromagnet, information is recorded and played back.

The magnetic particles are the most important part of the tape, as they are responsible for picking up and carrying the video signal. Particle size, composition, density, and distribution determine the quality of a tape. During the manufacturing process, the particles are arranged in the tape's coating. During the recording process, video heads arrange the particles into patterns dictated by the changing voltage of the video signal. When the tape is played back, the patterns are picked up by a playback head and become the video image.

History

The first video recording tapes were rust covered with paper backing. The first video recording machines recorded signals on a thin metal wire. When it was discovered that magnetic videotape produced better results, magnetic recorders were built. Some of the early machines used small electromagnets that magnetized iron alloy wire as it passed between spools while crossing over the electromagnet.

Magnetic tape proved easier to work with because it does not curl or bend like wire, At first, metal oxides like iron were powdered and applied to the tape. Magnetic particles in early videotapes were relatively large. While this made the manufacturing process easier, the size of the particles limited the effectiveness of the tapes. Later, cobalt was added to particles to improve their magnetic properties.

When researchers explored ways to reduce the size of the particles, they discovered that smaller particles resulted in a better tape. However, smaller particles proved more difficult to disperse in the binding material during the manufacturing process. Binding material is a liquid mix of ingredients that later harden and give structure to the magnetic layer of the tape. Researchers later focused on better binder formulations and application techniques and significantly improved videotape quality.

In 1951, Bing Crosby Enterprises conducted one of the first demonstrations of magnetic videotape recording. However, the poor speed of the first videotape made it commercially impractical. Still, the benefits of videotapes were immediately recognized. Potential advantages included improved broadcast quality, reusable tapes and less expensive production costs.

In 1956, the Ampex company introduced the first practical videotape machine. This first model was a large reel-to-reel machine that used four record heads and two-inch wide tape. Obviously, this invention attracted the interest of the television broadcasting industry and, on November 30, 1956, CBS became the first network to broadcast a program using videotape.

A major innovation was introduced in 1969, when Sony presented its EIAJ-standard, three-quarter-inch U-Matic series, the first videocassette system to become widely accepted. The videocassette was a vast improvement on the reel-to-reel format and had a profound effect on the video field.

In the early 1980s, Sony became the first company to establish a consumer market for the videocassette system with its Betamax format. Other manufactures soon followed, and the VHS system introduced by JVC, with its 0.5-in (1.3 cm) tape, soon dominated and continues to dominate the market. In 1984, Kodak and General Electric introduced the eight-millimeter video recorder, or camcorder. In 1995 digital videotape was introduced. Digital recording resulted in less background noise and less degradation of picture and sound quality.

The introduction of digital tape coupled with the emergence of the Digital Video Disc (DVD), led many to believe the days of the standard videocassettes were numbered. However, in 2001, it was estimated that 90% of households had VCRs while only 10% had DVD players. With new advances in tape manufacturing, as well as a consumer preference for videotapes, it seems it will take a long time for videotape to go the way of the dinosaur.

Raw Materials

Today's magnetic tape is composed of three main layers: the base film, the magnetic layer, and the back coating. The base film provides the physical support and main strength of the tape. Base film consists mostly of Polyethylene terephthalate (PET) because it is durable and resistant to stretching. The magnetic layer determines the tape's magnetic properties. It is composed of magnetic powders, ranging from lower strength iron oxides to high-energy metal particles, that determine the tape's magnetic properties; binders that provide structure to the layer, including polymers, adhesives, lubricants, cleaners, solvents, dispersion agents and static controlling compounds; additives, such as carbon black, that enhance tape properties; and lubricants, that decrease friction and wear. The back coating, which is made up of lubricants, enhances the tape's durability and performance.

Design

No two tape manufacturers employ the exact same videotape manufacturing process. Each manufacturer uses its own materials and variations on the process to gain an edge over competitors. However, the basic manufacturing process is the same for all companies. Essentially, manufacturing videotape involves taking a roll of clear plastic sheeting, painting it with a mixture of binding material and magnetic particles, baking it in an oven until this "paint" dries, and cutting the wide plastic roll into thin strips that are wound onto reels. The magnetic particles that coat a videotape are needle-shaped. This allows them to be tightly packed onto the tape surface. The greater the magnetic density, the better the tape. The best tapes use smaller particles packed in greater concentration.

The Manufacturing Process

The manufacturing process, according to Sony, one of the major videotape producers, is a multi-stage process that includes mixing, coating, calendaring, slitting, and finishing.

1 The first step, mixing, involves creating the magnetic layer, or "paint," that will coat a tape. The magnetic powder is first pre-mixed with solvents and dispersants, typically polyurethane, that keep the powder from floating on top of the solvent. The mixing process starts with the powder being dispersed by large planetary mixers in large, metal tanks. When the magnetic material is dispersed, the binders, additives, and lubricants are added and mixed. The paint is then passed automatically onto the milling stage, when it is milled, or rolled, by large rollers. This creates a shearing action that prevents agglomerations from forming in the paint.

2 In the second step, the coating stage, the magnetic paint is applied to the base film in a continuous process that starts when large rolls of the base film are fed into a machine called a coater. The coater is 120 ft (36.6 m) long and uses 400 kw of power. The film is drawn by six motors through the coater at low tension. At the start of the process, the tape enters the head end of the coater, which is called the un-wind end. New rolls are automatically spliced on when the coater senses the end of a roll. A mechanical device called a flying splice attaches the new roll of base film, cutting off the end of the old, while the rolls are in motion. (The device is called a flying splice because the splice can be made "on the fly" while the process continues operating.)The coater heads stops painting the film until the new splice goes through.

3 The coater head applies the magnetic layer to the base film. Computers monitor and control the process, so that a thin, even layer of magnetic paint can be applied without defects. The thin layers are measured in submicrons, or millionths of a meter. The coating method most often used is called extrusion. In this process, the paint is applied to the film as it is blown out, by pressure, through a small opening. As the base film passes over the extruded paint, the paint sticks to the surface of the base film in a smooth layer.

4 After coating, the still-wet tape is automatically passed through a strong magnetic field which physically orients the magnetic material in a newly coated layer. At first, the particles are randomly arranged on the binder. As the physical alignment is important for the most effective magnetic recording, the particles are then oriented in the same direction. This is accomplished by passing the tape through the magnetic field as the binder hardens. The more uniform the dispersion and orientation of the particles, the better the tape will perform. The magnetic layer is now ready to be dried.

5 Drying fixes the now-oriented particles before they can change position. During drying, solvents used for mixing are evaporated and recovered. The tape's magnetic layer is stabilized. When the drying is complete, the tape passes through an X-ray scanner that checks the evenness of the newly applied layer.

6 After the drying and scanning, the tape is wound back onto large rolls measuring up to 4 ft (1.2 m) in width and over a 1,000 ft (305 m) long. Like the flying splice that started the process, a slit is made in the tape and the new roll automatically starts wind-up. The tension of the tape is held steady during coating. As the newly wound roll gradually increases in size, the speed of the wind is gradually decreased to keep the tension even from start to finish.

7 The third step involves a process called calendaring. At this point, the space between the magnetic particles has not completely set. Therefore, it is possible to minimize this space by the calendaring process, which involves compacting the layer that has just been formed between steel rollers and elastic rollers. This increases the packing density of the magnetic particles to maximize the tape's magnetic density, and it smoothes the surface of the magnetic layer, which provides better tape-to-head contact. During the process the tape is automatically fed through a series of the rollers. The steel rollers make contact with the tape's magnetic side and the elastic rollers make contact with the back of the tape, a method that creates an even pressure across the film. After this process, the tape's final surface characteristics and thickness are set. The tape is now ready to be slit.

8 Before being loaded into cassettes, the tapes are slit to the width determined by its format. Commonly used tape formats include 0.5 in (1.3 cm), 0.75 in (1.9 cm), or 1 in (2.5 cm). The tape is spliced onto a slitter comprised of two sets of blades. As the tape is drawn through the slitter, the blades cut the tape into the desired width. The slitting is a very precise process and it is critical to the quality of the tapes. The process is usually monitored by lasers to detect defects such as folds or pinholes. The tape is slit within microns of the desired width to insure smooth operation in a VCR or camcorder. A poorly slit edge can shed oxide, base film and back-coated particles. After the tape is slit, it passes over a cleaning wipe that removes any debris accumulated during the process. The tape is then wound onto long "pancake" rolls, which resemble rolls of movie film. The tape is now ready for the finishing stage, when it will be loaded into cassettes.

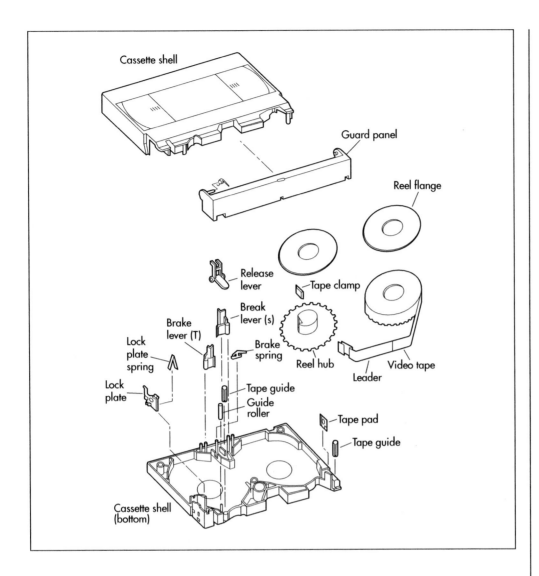

Cassette shell

Guard panel

Reel flange

Release lever

Tape clamp

Break lever (s)

Brake lever (T)

Brake spring

Lock plate spring

Reel hub

Leader

Video tape

Lock plate

Tape guide

Guide roller

Tape pad

Tape guide

Cassette shell (bottom)

9 In the finishing stage, the tape cassettes are produced. The cassette shells themselves are created from melted plastics that are placed in a metal cavity, or mold, and formed into the top and bottom halves of the shells. These halves are held together by five screws. Components of the videocassette shells include two spools that hold the tape itself; a moving, spring-loaded door that houses the tape safely inside the shell; stainless steel pins and rollers that protect the tape itself against scratches; low-friction rollers that guide the tape during recording or playback; anti-static leader that prevents buildup of dust; and two, spring-loaded locks that prevent the tape from rolling around inside the shell. The empty cassette shells have hubs with leader attached. The leader is automatically spliced to the pancake rolls, which are placed on an in-cassette loading machine that spools a measured amount of tape into the cassette. Next, the loaded cassettes are assembled with packing inserts and placed in protective sleeves. Finally, the cassettes are packaged and boxed for shipment.

Byproducts/Waste

At some manufacturing plants, during production, the solvent used in the coating process is recovered and purified and then used again in the manufacture of more tapes. The solvents are evaporated in dryers that use air currents. This creates a mixture of air and solvent that is carried through pipes to a solvent recovery station. At this station, the solvent and air are separated. The solvent is then distilled and stored for use.

Quality Control

Generally, quality control is continuous throughout the manufacturing process.

Computers, x rays, and lasers are used to monitor various stages. Ingredients are also tested. At Sony, before the manufacturing begins, ingredients are checked by the quality control lab against specifications in the tape's formulation. Oxide and metal particles are checked for evenness and size. Magnetic "footprints" are tested to make sure they conform to magnetic characteristics. Binders and lubricants are checked for purity. The polyester base film is checked for consistency and strength.

The Future

Recent advances in technology and manufacturing are making even better tapes possible and are pointing the way to the future. By the turn of the new century, most of the tapes people have been using were oxide tapes. The active magnetic coating has been some form of oxidized metal. Tapes made with coatings of pure metal, or metal evaporated tape, have proven superior. In making metal evaporated tape, manufactures employ a different process to deposit magnetic particles. Instead of the magnetic particles being carried in a binder and painted onto the tape, they are vaporized from a solid and deposited onto base film. An electronic beam heats metal to thousands of degrees inside a vacuum chamber. The metal then vaporizes and adheres to a specially prepared base film. A protective coating is applied to this magnetic layer. The result is a smooth, thin, densely packed film of pure magnetic particles. Because no binder is used, the particles mesh with a density that approaches solid metal. The Sony corporation led the way with this new process with its Advanced Metal Evaporated, or AME, process. This kind of tape represents a relatively new technology, and its full potentials await to be tapped.

Where to Learn More

Books

Alldrin, L., et al. *The Computer Videomaker Handbook.* 2nd ed. Boston: Focal Press, 2001.

Nmungwun, A. *Video Recording Technology: Its Impact on Media and Home Entertainment.* New Jersey: Lawrence Erlbaum Associates, 1989.

Heller, N., and T. Bentz. *The Great Tape Debate: Evolution of the New Video Format.* New York: Knowledge Industry Publications, 1987.

Other

Stoffel, T. *Videotape Systems Theory Web Page.* December 2001. <http://www.lionlmb.org/quad/theory.html>.

MTC-Open.net Web Page. December 2001. <http://www.mtc-open.net>.

Dan Harvey

Vinegar

Background

Vinegar is an alcoholic liquid that has been allowed to sour. It is primarily used to flavor and preserve foods and as an ingredient in salad dressings and marinades. Vinegar is also used as a cleaning agent. The word is from the French *vin* (wine) and *aigre* (sour).

History

The use of vinegar to flavor food is centuries old. It has also been used as a medicine, a corrosive agent, and as a preservative. In the Middle Ages, alchemists poured vinegar onto lead in order to create lead acetate. Called "sugar of lead," it was added to sour cider until it became clear that ingesting the sweetened cider proved deadly.

By the Renaissance era, vinegar-making was a lucrative business in France. Flavored with pepper, clovers, roses, fennel, and raspberries, the country was producing close to 150 scented and flavored vinegars. Production of vinegar was also burgeoning in Great Britain. It became so profitable that a 1673 Act of Parliament established a tax on so-called vinegar-beer. In the early days of the United States, the production of cider vinegar was a cornerstone of farm and domestic economy, bringing three times the price of traditional hard cider.

The transformation of wine or fruit juice to vinegar is a chemical process in which ethyl alcohol undergoes partial oxidation that results in the formation of acetaldehyde. In the third stage, the acetaldehyde is converted into acetic acid. The chemical reaction is as follows: $CH_3CH_2OH=2HCH_3CHO=CH_3COOH$.

Historically, several processes have been employed to make vinegar. In the slow, or natural, process, vats of cider are allowed to sit open at room temperature. During a period of several months, the fruit juices ferment into alcohol and then oxidize into acetic acid.

The French Orléans process is also called the continuous method. Fruit juice is periodically added to small batches of vinegar and stored in wooden barrels. As the fresh juice sours, it is skimmed off the top.

Both the slow and continuous methods require several months to produce vinegar. In the modern commercial production of vinegar, the generator method and the submerged fermentation method are employed. These methods are based on the goal of infusing as much oxygen as possible into the alcohol product.

Raw Materials

Vinegar is made from a variety of diluted alcohol products, the most common being wine, beer, and rice. Balsamic vinegar is made from the Trebbiano and Lambrusco grapes of Italy's Emilia-Romagna region. Some distilled vinegars are made from wood products such as beech.

Acetobacters are microscopic bacteria that live on oxygen bubbles. Whereas the fermentation of grapes or hops to make wine or beer occurs in the absence of oxygen, the process of making vinegars relies on its presence. In the natural processes, the acetobacters are allowed to grow over time. In the vinegar factory, this process is induced by feeding acetozym nutrients into the tanks of alcohol.

Mother of vinegar is the gooey film that appears on the surface of the alcohol product

By the Renaissance era, vinegar-making was a lucrative business in France. Flavored with pepper, clovers, roses, fennel, and raspberries, the country was producing close to 150 scented and flavored vinegars.

as it is converted to vinegar. It is a natural carbohydrate called cellulose. This film holds the highest concentration of acetobacters. It is skimmed off the top and added to subsequent batches of alcohol to speed the formation of vinegar. Acetozym nutrients are manmade mother of vinegar in a powdered form.

Herbs and fruits are often used to flavor vinegar. Commonly used herbs include tarragon, garlic, and basil. Popular fruits include raspberries, cherries, and lemons.

Design

The design step of making vinegar is essentially a recipe. Depending on the type of vinegar to be bottled at the production plant—wine vinegar, cider vinegar, or distilled vinegar—food scientists in the test kitchens and laboratories create recipes for the various vinegars. Specifications include the amount of mother of vinegar and/or acetozym nutrients added per gallon of alcohol product. For flavored vinegars, ingredients such as herbs and fruits are macerated in vinegar for varying periods to determine the best taste results.

The Manufacturing Process

The Orléans method

1 Wooden barrels are laid on their sides. Bungholes are drilled into the top side and plugged with stoppers. Holes are also drilled into the ends of the barrels.

2 The alcohol is poured into the barrel via long-necked funnels inserted into the bungholes. Mother of vinegar is added at this point. The barrel is filled to a level just below the holes on the ends. Netting or screens are placed over the holes to prevent insects from getting into the barrels.

3 The filled barrels are allowed to sit for several months. The room temperature is kept at approximately 85°F (29°C). Samples are taken periodically by inserting a spigot into the side holes and drawing liquid off. When the alcohol has converted to vinegar, it is drawn off through the spigot. About 15% of the liquid is left in the barrel to blend with the next batch.

The submerged fermentation method

1 The submerged fermentation method is commonly used in the production of wine vinegars. Production plants are filled with large stainless steel tanks called acetators. The acetators are fitted with centrifugal pumps in the bottom that pump air bubbles into the tank in much the same way that an aquarium pump does.

2 As the pump stirs the alcohol, acetozym nutrients are piped into the tank. The nutrients spur the growth of acetobacters on the oxygen bubbles. A heater in the tank keeps the temperature between 80 and 100°F (26–38°C).

3 Within a matter of hours, the alcohol product has been converted into vinegar. The vinegar is piped from the acetators to a plate-and-frame filtering machine. The stainless steel plates press the alcohol through paper filters to remove any sediment, usually about 3% of the total product. The sediment is flushed into a drain while the filtered vinegar moves to the dilution station.

The generator method

1 Distilled and industrial vinegars are often produced via the generator method. Tall oak vats are filled with vinegar-moistened beechwood shavings, charcoal, or grape pulp. The alcohol product is poured into the top of the vat and slowly drips down through the fillings.

2 Oxygen is allowed into the vats in two ways. One is through bungholes that have been punched into the sides of the vats. The second is through the perforated bottoms of the vats. An air compressor blows air through the holes.

3 When the alcohol product reaches the bottom of the vat, usually within in a span of several days to several weeks, it has converted to vinegar. It is poured off from the bottom of the vat into storage tanks. The vinegar produced in this method has a very high acetic acid content, often as high as 14%, and must be diluted with water to bring its acetic acid content to a range of 5–6%.

4 To produce distilled vinegar, the diluted liquid is poured into a boiler and

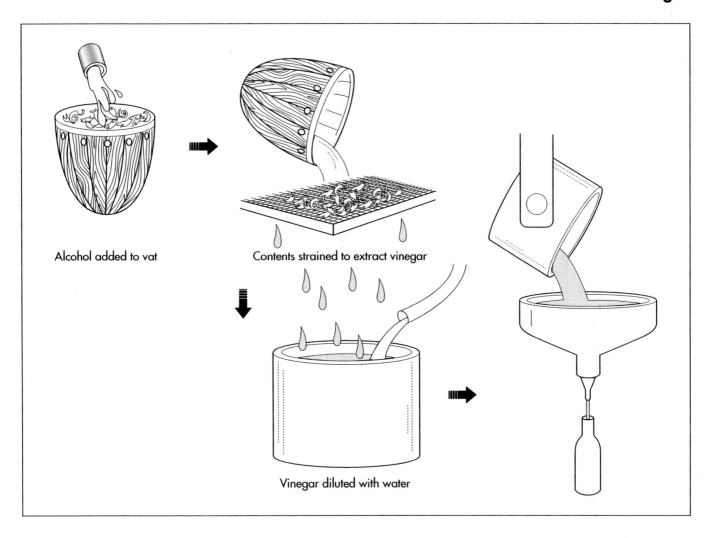

Alcohol added to vat

Contents strained to extract vinegar

Vinegar diluted with water

The production of vinegar.

brought to its boiling point. A vapor rises from the liquid and is collected in a condenser. It then cools and becomes liquid again. This liquid is then bottled as distilled vinegar.

Balsamic vinegar

1 The production of balsamic vinegar most closely resembles the production of fine wine. In order to bear the name balsamic, the vinegar must be made from the juices of the Trebbiano and Lambrusco grapes. The juice is blended and boiled over a fire. It is then poured into barrels of oak, chestnut, cherry, mulberry, and ash.

2 The juice is allowed to age, ferment, and condense for five years. At the beginning of each year, the aging liquid is mixed with younger vinegars and placed in a series of smaller barrels. The finished product absorbs aroma from the oak and color from the chestnut.

Quality Control

The growing of acetobacters, the bacteria that creates vinegar, requires vigilance. In the Orléans Method, bungholes must be checked routinely to ensure that insects have not penetrated the netting. In the generator method, great care is taken to keep the temperature inside the tanks in the 80–100°F range (26–38°C). Workers routinely check the thermostats on the tanks. Because a loss of electricity could kill the acetobacters within seconds, many vinegar plants have backup systems to produce electrical power in the event of a blackout.

Byproducts/Waste

Vinegar production results in very little by-products or waste. In fact, the alcohol product is often the by-product of other processes such as winemaking and baker's yeast.

Some sediment will result from the submerged fermentation method. This sediment is biodegradable and can be flushed down a drain for disposal.

The Future

By the end of the twentieth century, grocery stores in the United States were posting $200 million in vinegar sales. White distilled vinegar garners the largest percentage of the market, followed in order by cider, red wine, balsamic, and rice. Balsamic vinegar is the fastest growing type. In addition to its continued popularity as a condiment, vinegar is also widely used as a cleaning agent.

Where to Learn More

Books

Lang, Jenifer Harvey, ed. *Larousse Gastronomique.* New York: Crown, 1984.

Proulx, Annie, and Lew Nichols. *Cider: Making, Using and Enjoying Sweet and Hard Cider.* Pownat, VT: Storey Communications, 1997.

Watson, Ben. *Cider Hard and Sweet: History, Traditions, and Making Your Own.* Woodstock, Vermont: Countryman Press, 1999.

Other

Alcoholic Drinks of the Middle Ages: Vinegar. December 2001. <http://www.geocities.com/Paris/1265/cvinegar.html>.

Sonoma Vinegar Works Web Page. December 2001. <http://www.sonomavinegar.com>.

Mary McNulty

Windmill

Background

A windmill is a structure or machine that converts wind into usable energy through the rotation of a wheel made up of adjustable blades. Traditionally, the energy generated by a windmill has been used to grind grain into flour. Windmills are designed by skilled craftsmen and can be constructed on site using hand tools. Windmills developed steadily over the centuries and achieved their most prominence in Europe during the eighteenth century. They were largely replaced as a power generating structure when steam power was harnessed during the nineteenth century. Today, windmill technology is experiencing a renaissance and the wind turbine promises to be an important alternative to fossil fuels in the future.

History

Man has used wind to power machines for centuries. The earliest use was most likely as a power source for sail boats, propelling them across the water. The exact date that people constructed windmills specifically for doing work is unknown, but the first recorded windmill design originated in Persia around A.D. 500–900. This machine was originally used for pumping water then it was adapted for grinding grain. It had vertical sails made from bundles of lightweight wood attached to a vertical shaft by horizontal struts. The design, known as the panemone, is one of the least efficient windmill structures invented. It should be noted that windmills may have been used in China over 2,000 years ago making it the actual birthplace for vertical-axis windmills. However, the earliest recorded use found by archeologists in China is A.D. 1219.

The concept of the windmill spread to Europe after the Crusades. The earliest European designs, documented in A.D. 1270, had horizontal axes instead of vertical ones. The reason for this discrepancy is unknown, but it is likely a result of two factors. First, the European windmills may have been patterned after water wheels that had a horizontal axis. The water wheel had been known in Europe for long before this. Second, the horizontal axis design was more efficient and worked better. In general, these mills had four blades mounted on a central post. They had a cog and ring gear that translated the horizontal motion of the central shaft into vertical motion for the grindstone or wheel which would then be used for pumping water or grinding grain.

The European millwrights improved windmill technology immensely over the centuries. Most of the innovation came from the Dutch and the English. One of the most important improvements was the introduction of the tower mill. This design allowed for the mill's blades to be moved into the wind as required and the main body to be permanently fixed in place. The Dutch created multi-story towers where mill operators could work and also live. The English introduced a number of automatic controls that made windmills more efficient.

During the pre-industrial world, windmills were the electric motors of Europe. In addition to water pumping and grain grinding, they were used for powering saw mills and processing spices, dyes, and tobacco. However, the development of steam power during the nineteenth century, and the uncertain nature of windmill power resulted in a steady decline of the use of large windmill structures. Today, only a small fraction of

The first recorded windmill design originated in Persia around A.D. 500–900. This machine was originally used for pumping water then it was adapted for grinding grain.

the windmills that used to power the world are still standing.

Even as larger windmills were abandoned, smaller fan-type windmills were thriving. These windmills were used primarily for pumping water on farms. In America, these designs were perfected during the nineteenth century. The Halladay windmill was introduced in 1854 followed by the Aermotor and Dempster designs. The later two designs are still in use today. In fact, between 1850 and 1970 in the United States over six million were constructed.

Design

There are two classes of windmill, horizontal axis and vertical axis. The vertical axis design was popular during the early development of the windmill. However, its inefficiency of operation led to the development of the numerous horizontal axis designs.

Of the horizontal axes versions, there are a variety of these including the post mill, smock mill, tower mill, and the fan mill. The earliest design is the post mill. It is named for the large, upright post to which the body of the mill is balanced. This design gives flexibility to the mill operator because the windmill can be turned to catch the most wind depending on the direction it is blowing. To keep the post stable a support structure is built around it. Typically, this structure is elevated off the ground with brick or stone to prevent rotting.

The post mill has four blades mounted on a central post. The horizontal shaft of the blades is connected to a large break wheel. The break wheel interacts with a gear system, called the wallower, which rotates a central, vertical shaft. This motion can then be used to power water pumping or grain grinding activities.

The smock mill is similar to the post mill but has included some significant improvements. The name is derived from the fact that the body looks vaguely like a dress or smock as they were called. One advantage is the fact that only the top of the mill is moveable. This allows the main body structure to be more permanent while the rest could be adjusted to collect wind no matter what direction it is blowing. Since it does not move, the main body can be made larger and taller.

This means that more equipment can be housed in the mill, and that taller sails can be used to collect even more wind. Most smock mills are eight sided although this can vary from six to 12.

Tower mills are further improvements on smock mills. They have a rotating cap and permanent body, but this body is made of brick or stone. This fact makes it possible for the towers to be rounded. A round structure allows for even larger and taller towers. Additionally, brick and stone make the tower windmills the most weather resistant design.

While the previous windmill designs are for larger structures that could service entire towns, the fan-type windmill is made specifically for individuals. It is much smaller and used primarily for pumping water. It consists of a fixed tower (mast), a wheel and tail assembly (fan), a head assembly, and a pump. The masts can be 10–15 ft (3–15 m) high. The number of blades can range from four to 20 and have a diameter between 6 and 16 ft (1.8–4.9 m).

Raw Materials

Windmills can be made with a variety of materials. Post mills are made almost entirely of wood. A lightweight wood, like balsa wood, is used for the fan blades and a stronger, heavier wood is used for the rest of the structure. The wood is coated with paint or a resin to protect it from the outside environment. The smock and tower mills, built by the Dutch and British prior to the twentieth century, use many of the same materials used for the construction of houses including wood, bricks and stones.

The main body of the fan-type mills is made with galvanized steel. This process of treating steel makes it weather resistant and strong. The blades of the fan are made with a lightweight, galvanized steel or aluminum. The pump is made of bronze and brass that inhibits freezing. Leather or synthetic polymers are used for washers and o-rings.

The Manufacturing Process

Windmills are always erected on site using pre-made parts. The following description relates to the fan-type windmill. The basic

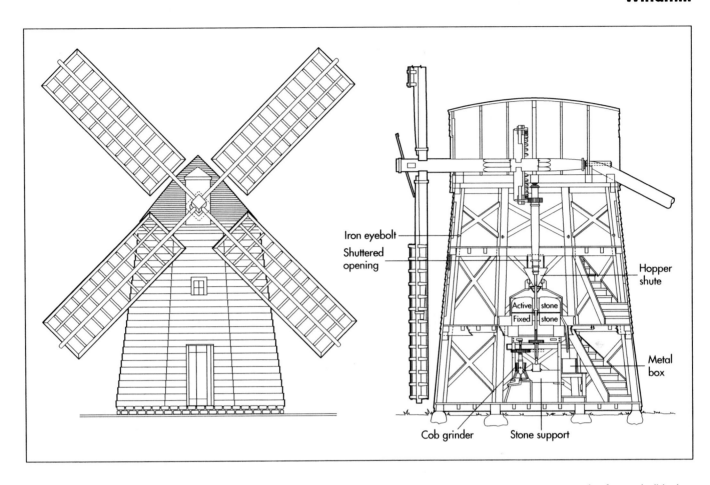

Iron eyebolt
Shuttered opening
Hopper shute
Active stone
Fixed stone
Metal box
Cob grinder
Stone support

An example of a windmill built in 1797.

steps include making the parts and then assembling the structure.

Making the tower parts

1 The tower parts are made from galvanized steel. This process begins with a roll of coiled sheet metal. The coils are put on a de-spooling device and fed to the production line. They are run under a straightener to remove any kinks or twists. The pieces are cut to the appropriate size and shape. In some cases, pieces may be put on a machine that rolls them and welds the seam. The ends are passed under a crimping machine and the pieces are moved to the finishing station.

2 At the finishing station, holes are drilled in the metal parts at specific places as required by the windmill design. The parts may also be painted or coated before being arranged in the final windmill kit.

Making the gearbox

3 The gearbox is an intricate assembly made up of various gears, axles, rotors, and wheels. The parts are die cast and assembled by hand. The are placed in an weather resistant housing that is designed to accommodate the gearbox parts and the attached wheel and tail assembly.

Making the fan

4 The fan is made up of a metal rim with slightly curved blades attached. The rim is produced on a machine that rolls steel strips into circular hoops. A hole is drilled in both ends, and they are connected with a small clamp and screw after the fan blades are attached. A center axle is then connected to the rim and attached with small steel spokes. A typical design will have five pairs of spokes attached a evenly spaced intervals along the rim.

5 The fan blades and tail are cut from pieces of sheet metal. The blades are then run through a machine that gives them a slight curve. They are attached to the metal rim with small bolts and metal clamps. They are attached in such a way that they can be raised or lowered depending on the wind conditions.

A modern steel windmill.

clamped and bolted to the top of the tower. The main shaft is then inserted into the bottom of the gearbox. Next, the fan and its attached axle are connected to the gearbox. Finally, the tail section is attached to the gearbox. The pump is then hooked up to the main shaft and the windmill is operational.

Quality Control

Various tests may be done to ensure that each part of the windmill meets the specifications laid out in the design phase. The most basic of these are simple visual inspections. These will catch most of the obvious production flaws. Since windmills are erected by hand, the quality of each part goes through an additional visual inspection. The quality of workmanship that goes into construction of the windmill will be primarily responsible for the quality of the finished product. To ensure that it remains efficient during operation, regular maintenance checks are necessary.

The Future

Windmills have changed little over the last hundred years. In fact, one basic design conceived in the 1870s is still sold today. The major improvements have come in the types of materials used in construction. This trend will likely continue in future windmill products. However, the future of harnessing wind power is not in traditional windmills at all. The United States government has spent millions of dollars researching and developing wind turbines for electricity generation. In California, numerous wind farms are already in operation. Various other states and cities have plans for creating similar wind farms. In the future, wind power promises to be an environmentally friendly substitute for fossil fuels.

Preparing the site

6 Finding and preparing the construction site is a crucial step in creating a functional windmill. First, an area with a prevailing wind of at least 15 mph (24 km/hr) is needed. Then the area needs to be cleared of trees and other structures that may block wind. In some cases, a dirt mound or concrete base is erected to raise the windmill off the surface to catch more wind.

Final assembly

7 The parts of the main body are connected first. They are bolted together on the ground and then raised up vertically. The outer poles are joined with the connecting rods. Clamps are bolted at each joint for stability. After the tower is raised it is loosely bolted to the solid base. Next stay wires are strung from the frame down to the ground and attached to tensioners and ground anchors. When the structure is level, the bolts are tightened and the structure integrity is tested. In some cases a ladder is built into the frame design to allow access to the fan on top which makes cleaning an maintenance easier.

8 The fan wheel, gearbox, and main shaft are next attached. The gearbox is first

Where to Learn More
Books

Baker, T. Lindsay. *North American Windmill Manufacturers' Trade Literature.* University of Oklahoma Press, 1998.

Clegg, Alan John. *Windmills.* Horseshoe Publications, 1995.

Hills, Richard L. *Power from Wind: A History of Windmill Technology.* Cambridge University Press, 1994.

Hooker, Jeremy. *In Praise Of Windmills.* Circle Press Pubns, 1990.

Watts, Martin. *Water and Wind Power.* Shire Publications, 2000.

Perry Romanowski

Windshield Wiper

Background

Windshield wipers are used to clean the windshield of a car so that the driver has an unobstructed view of the road. A typical wipe angle for a passenger car is about 67 degrees. The blades are 12–30 in (30–76 cm) long with lengths increasing in 2-in (5-cm) increments.

History

The history of the windshield wiper began with the invention of the automobile. Most transportation vehicles did not have wipers. Horse-drawn carriages and trucks moved at slow speeds, and glass was not needed to protect the driver or passengers or to act as a windbreak.

The first windshield wipers were brushes. Inventor J. H. Apjohn came up with a method of moving two brushes up and down on a vertical plate glass windshield in 1903. In the same year, Mary Anderson devised a swinging arm that swept rain off the windshield when the driver moved a lever located inside the car. Anderson patented her invention of the mechanical windshield wiper in 1905, and it became standard equipment by 1913. Electric motors were not used yet to power automobile essentials or accessories, and Anderson's device had a drawback. Without another power source, a driver had to use one hand to move the lever. The driver's other hand steered the car (with either a wheel or steering tiller) and worked the stick-mounted gear shift and brake grips standing on the floor of the car or outside the driver's side on the running board.

Rubber strips replaced brushes as the cleaning tools on wipers in 1905. Unfortunately, the hazardous need for drivers to wipe windshields while driving was not eliminated until 1917. The solution was to use an electric motor to move a single wiper with a long rubber blade back and forth. Hawaiian dentist Dr. Ormand Wall invented the automatic wiper by placing an electric motor in the top center of the windshield so the wiper arced down over the hood of the car in a semi-circular or rainbow shape. Wipers were one of the first electrical devices in automobiles after the electric starter was developed in 1912. Most wipers on cars before 1930 were paired and hung down from the top of the windshield. They were moved to the base of the windshield as electrical systems became more complicated.

Windshield washers were added to the wiper on/off levers, and these required spray nozzles in front of the windshield, a tank for washer fluid in the engine compartment, and electrical connections to coordinate these operations. In 1962, Bob Kearns invented the intermittent wiper with intervals and speeds that the driver could change. The advent of electronic systems with fuses and circuit breakers to operate, regulate, and coordinate electrical components expanded the possibilities for more diverse wipers. Wipers were added to headlights in the 1980s, requiring connections between the lighting and wiper systems. In the 1990s, microsensors were built into windshields to detect rain on the windshield, activate the wipers, and adjust speed and intermittent use for the amount of rain.

Raw Materials

The manufacturer purchases all of the parts from companies that specialize in fabricating parts from aluminum and steel, rubber blades, plastic bushings for the linkages, and the motors. Windshield wipers and windshield wiper systems (with motors) are different assemblies; some manufacturers make both, and others produce wipers only.

The connecting and drive links and the pivots that move the wipers are made of galvanized steel. Galvanization is the process of applying zinc coating to steel to protect it from corrosion. Drive arms for boats and vehicles used in the marine industry are made of stainless steel that resists damage from salt water. The wiper suspension and claws are also galvanized steel. The galvanizing zinc coating is easier to paint than uncoated steel. Steel is also the material in the small parts of wipers, such as washers, screws, nuts, springs, and brackets.

The blade frame is made from aluminum. The blades are made of natural rubber or synthetic compounds. Some rubber blades are composites of soft rubber on the wiping edge (the squeegee surface) and firm rubber that supports the wiping edge in the rest of the blade.

Other materials that comprise parts of windshield wipers are rubber for washers in the pivots and plastic bushings that line holes for connecting parts of the linkage. The wiper suspension is typically painted black. If the wiper manufacturer also builds wiper systems, motors are purchased from subcontractors. The motors are contained in steel housings and include permanent magnet motors wound with copper wire. Each housing has connections for the electrical wires that are part of the vehicle and wiring harnesses are furnished specific to operating the wipers. Each motor also contains one or more electronic circuits depending on the sophistication of the system that the motor controls.

Design

Windshield wipers are designed and made to clear water from a windshield. Most cars have two wipers on the windshield, and they may have one on the rear window and one on each headlight. The wiper parts visible from outside the car are the rubber blade, the wiper arm holding the blade, a spring linkage, and parts of the wiper pivots. The wiper itself has up to six parts called pressure points or claws that are small arms

under the wiper. The claws distribute pressure from the wiper along the back of the blade. This is described as a balance beam with a suspension system, where the wiper is the beam and the claws are the suspension components. The claws keep the blade flexed against the windshield to distribute even pressure to clean the glass all along the blade. More claws usually distribute the pressure better and are suited to large or highly curved windshields.

Although the rubber is the familiar part of the blade, the blade actually includes a metal strip called a blade frame with a slot along the length of the frame and replacement holes in the frame. The replacement holes provide access for replacing the rubber blade with a refill. The blade on its aluminum frame can also be changed as a unit.

The standard two windshield wipers are usually operated as a single-motor, tandem scheme with one wiper on the driver's side and one positioned near the middle of the windshield that moves across the passenger's view. The wipers are secured to pivots. A wiper and pivot are mounted on brackets at both ends of a long rod called the connecting link, and, as the force from the motor pushes on the driver's end of the connecting link, it in turn moves the other wiper. The connecting link is attached to another long rod called the drive link near the wiper motor. A slender spring linkage ties the pivot to the drive link to return the wiper to its resting or park position, hug the wiper close to the windshield, and keep it attached to the car if the links are damaged.

Between the motor and the drive link, a linkage system consisting of a cam (another short rod) and pivot, a gear output shaft, and a worm gear controls the force of the motor delivered to the drive arm. The worm gear slows the speed of the motor while multiplying its torque (force). The gear allows a small motor to produce enough force to move the blades across the glass. This description is based on using a single motor to drive both wipers. If one motor powers each wiper, more links are needed to move the two wipers together in a so-called unitized motor system.

This multiplied force is required to accelerate the blades from being stopped at both

Windshield wiper systems.

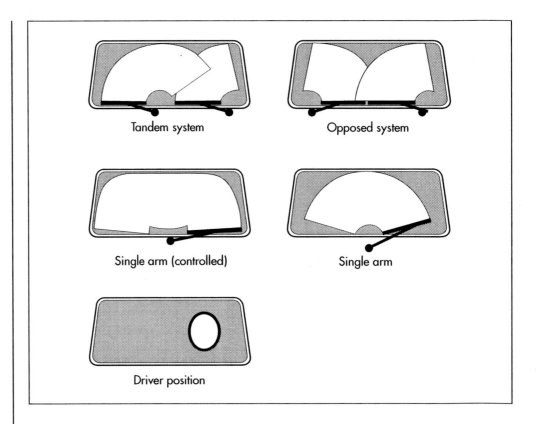

Tandem system

Opposed system

Single arm (controlled)

Single arm

Driver position

ends of their movement, to resist the friction of the rubber against the glass, to resist the friction of the rubber on dirt on the glass, and to oppose wind pressure on the windshield.

The tandem scheme is the most common because the blades produce overlapping cleared areas on the windshield with the greatest overlap in front of the driver. An opposed scheme with two blades begins with both blades on the windshield toward the sides of the car, and the blades overlap as they both pivot toward the center of the windshield. A single wiper that swings in an arc from the center of the windshield is also used. The single-arm controlled wiper is the most complex; as it sweeps over the glass, the wiper arm lengthens toward the car sides and retracts again as it points straight up at the middle of the windshield. Each of the two wipers in the tandem and opposed operating schemes and the one wiper in the single-wiper scheme make an arc with a single radius and so are called radial arm wipers. The single-arm-controlled wiper produces a multiple-radius arc.

The electric motor, worm gear, gear shaft, cam, drive link, and pivots are built into the underside of the dash. The connecting link and wiper pivots are located below the windshield and behind the trim molding. Wipers called depressed wipers also rest behind the molding when they are not being used. Non-depressed wipers are above the windshield trim molding even at rest and are visible from outside the car and from the passenger compartment. In the passenger compartment, the wiper's on/off lever is usually attached to the steering column. When the wipers are turned on, an electronic circuit inside the wiper motor starts it. When the wipers are turned off the circuit stops the power to the wiper motor. Intermittent operation of the wipers is basically short on-and-off periods for the wiper motor that the circuit also regulates.

The Manufacturing Process

1 Wiper manufacturers carry large stocks of materials provided by subcontractors. As the materials are received, the receiving inspectors confirm that the types and quantities of parts are correct, compile an inventory, and store the parts.

2 The worker begins by putting together the pivot shaft for each wiper. The pivot shaft is made of a set of fasteners and spac-

ers that hold the wiper arm securely while allowing it to pivot and sweep the design wipe angle. The shaft assembly includes the pivot shaft itself and (from the end near the small connecting link to the tip of the shaft) a rubber washer, metal washer, nut, nut cap, knurled driver, washer, and acorn nut. The knurled driver is a type of nut with ridges on the sides that grip any attachment. The wiper arm will sit on the knurled driver, which keeps it from shifting out of position on the shaft, and the washer and acorn nut hold the arm on the shaft. The pivot shaft is then attached to the small connecting link with a washer and spring clip. A pin on the pivot shaft can be inserted in any one of three pin positions when the shaft is attached to the link, depending on the design for the pivot and link.

3 For a single-arm wiper scheme, a U-shaped, galvanized steel bracket is fixed to the small connecting link on the only pivot shaft with two shaft screws. The other end of the bracket will be attached to the drive link later. For a scheme with two wipers, the small connecting link for the wiper on the passenger's side is joined with a bracket to the end of the longer connecting link with shaft screws. Similarly, a bracket is put on the small connecting link for the driver's side wiper, and it is attached to the opposite end of the longer connecting link. Later, this end will also be attached to the drive link.

4 The drive link will be attached to the motor in the next step. The motor with the worm gear reduction and other linkage is a stock item provided by a vendor, and the wiper system manufacturer does not make any changes to it. The drive link must be secured precisely on the cam (drive arm) on the end of the gear shaft so the wiper will sweep correctly but also so it can be parked in the right position under the car molding. The connection between the cam and the drive link will be fixed by using another bracket called the mirror bracket.

5 To set the angle between the drive link and the cam and motor, the motor, cam, mirror bracket, and drive link are put inside a die set. The die set is an outline-like pattern made of steel with areas fitted for the four parts. Wiper system makers have a collection of die sets with various angles for mounting

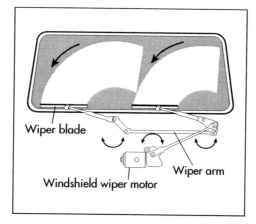

A tandem system motor.

the drive link to the motor. The bracket is put on the drive link with a set of screws. The bracket is then attached to the cam.

6 With the angle established, the cam is checked for fit with the drive shaft. Spacer washers are added (if needed) between the cam and drive shaft, and the two are connected with a set of motor nuts and screws. For a single-wiper scheme, the bracket with the single small link and pivot is screwed on the drive link. For a scheme with two wipers, the bracket on the end of the long connecting link that also supports the driver's side pivot and small connecting link is fastened to the drive link.

7 In the final steps in assembling the windshield wiper system, linkages made of springs are added to connect each pivot shaft to the drive link. The wiper arms and blades are connected to the pivots. The starter on the motor is also moved into the park position, and the wipers are placed in their park positions.

8 If the customer is purchasing windshield wiper systems, accessories may be included. A system of washers with water bottles, tubing, and controls for the dash is the most common accessory set. A wiring harness with the washer controls and the other electrical connections for the wipers is provided with the motor.

9 The completed windshield wipers are given a final quality control inspection as described below and transferred to the packing area. Depending on the items ordered, each set consisting of wipers, a motor system, and accessories is boxed with operating, maintenance, and return information.

The separate boxes are bundled together and packed in shipping cartons if the customer ordered several items.

Quality Control

During assembly, the workers observe the conditions of the parts during their work, but their only specific quality control activity is to check the operation of the motors by turning them on to make sure they start and by listening to the sounds they make as indications of performance.

The last inspection is performed when the assemblies are complete and before the wipers and systems are packed. The manufacturing director or final quality control inspectors look at the general appearance of the assemblies, confirm that the wipers have been sized and angled correctly for their sweep, and check that the assemblies are in the park position. The director or inspectors also check to see that the correct accessories are ready to be packed with the assemblies.

Byproducts/Waste

Small quantities of steel and aluminum scraps from trimmings or rejected or damaged parts are collected in bins and sold to salvage dealers who, in turn, sell them to metal manufacturers who melt the scrap down for recycling. Packaging from received parts is also collected and recycled.

The Future

As of 2002, windshield wipers and wiper systems are evolving because of changes in automobiles and other vehicles, technical improvements, and consumer demand. Wiper blades are as much as 30 in (76 cm) long, creating more resistance as they clean the windshield. Night-vision screens for windshields are in development, and these also increase resistance and change the dimensions needed for wipers. Blades are being improved with increasingly flexible rubber, so-called "boots" that fit around the blades to keep out ice and snow, and non-stick coatings on the squeegee edges of the blades to keep oil and wax from adhering and aging them.

Motor systems are also being increased in voltage to power longer wipers and more accessories. Engineers are investigating fully automated systems that do not require any actions by drivers to start and stop wiper systems. Inventors expect the capabilities of the rain-detecting sensors available in the late 1990s to widen to prompt the wipers to clean dirty windshields with no rain, for example. Windshield wipers are among most reliable automotive devices—the design life of a wiper system is 1.5 million wipes.

Where to Learn More

Books

Billiet, Walter E., and Leslie F. Goings. *Automotive Electrical Systems*. Alsip, Illinois: American Technical Publishers, Inc., 1970.

Clymer, Floyd. *Those Wonderful Old Automobiles*. New York: Bonanza Books, 1953.

Day, John. *The Bosch Book of The Motor-Car*. New York: St. Martin's Press, 1976.

Halderman, James D. *Automotive Electrical and Electronic Systems*. Englewood Cliffs, NJ: Prentice Hall, 1988.

Setright, L. J. K., and Ian Ward, ed. *Anatomy of the Automobile*. New York: Crescent Books, 1977.

The World of Automobiles: An Illustrated Encyclopedia of the Motor Car. Vol. 22. New York: Columbia House, 1974.

Other

Anco Web Page. December 2001. <http://www.federal-mogul.com/anco/wiper_anco.html>.

Cleveland Ignition Co. Web Page. December 2001. <http://www.windshieldwipersystems.com>.

Gillian S. Holmes

X-Ray Glasses

Background

X-ray glasses are a novelty product designed to create the illusion that the user can see through solid objects. They are plastic framed eyeglasses with special lenses made of cardboard. The manufacturing process involved in their production includes plastic stamping, papermaking, printing and gluing. Inspired by the real discovery of x rays by nineteenth century scientists, x-ray glasses were introduced during the 1940s as a gag product. Today, they have become timeless icons of American pop culture.

X-ray glasses do not actually allow the user to see through objects. They are constructed of a plastic frame with lenses of heavy cardboard. In the center of each lens, small holes about 0.25 in (0.64 cm) in diameter are punched out. Feathers that create a polarizing effect cover the holes. When looking through the hole, objects appear to have a transparent outline and a solid center. This effect creates the illusion of x-ray vision. The prankster then uses the glasses and pretends to be able to see through things like walls, clothes, and other objects.

History

The idea to create x-ray glasses could not have been conceived until real x rays were discovered. This was done by Wilhelm Roentgen around 1895. In the late nineteenth century, Roentgen and other scientists were working with electrons. He wanted to visually capture their movement so he wrapped a Crookes tube in black photographic paper. During the experiment Roentgen found that a plate coated with fluorescent material began to glow. He knew this should only happen if the material was exposed to visible light. Roentgen reasoned that there must be some kind of invisible light and upon further investigation he characterized this newly discovered light.

Roentgen found that the invisible light could penetrate materials such as aluminum, wood and human skin. He called the new discovery x rays and was first to publish an x-ray image. Over the next few decades, the science of x rays was further understood and the technology was applied to medicine. While scientists had a basic understanding of x rays, the general public developed some erroneous notions. Many people mistakenly believed that x rays could penetrate solid materials, that they could use them to see through things like walls, boxes, and clothes. This myth was reinforced with comic book heroes like Superman who could use x-ray vision to see through solid objects.

An early precursor to x-ray glasses was the Wonder tube, introduced during the 1940s. This product, produced by the S. S. Adams Company, was a tube with a single hole that had a small feather across it. The feather was hidden inside the tube so the user was unaware of its existence. When the user looked at their hand through the tube, it appeared as though they could see their bones. The fine spacing between the strands that make up the feather created this illusion.

X-ray specs were later introduced by the Adams Company. At the time, they were relatively popular. This was probably because the notion of x rays was new and not well understood. Today, the public is more knowledgeable and people do not really believe there are glasses that can allow them to see through objects. However, x-ray glasses con-

Inspired by the real discovery of x rays by nineteenth century scientists, x-ray glasses were introduced during the 1940s as a gag product.

tinue to amuse the novelty buying public and they remain an icon of American pop culture.

Raw Materials

Various raw materials are used in the construction of x-ray glasses. These include plastics, cardboard, feathers, and finishing materials.

Plastics are high molecular weight polymers made through various chemical reactions. Plastics suitable for x-ray glasses manufacture need to be easily colored, heat stable, and durable. Plastics that may be used are thermoplastic materials. Various materials are added to the plastic to make it more useful. Colorants are added to the plastic in order to make the material match the design requirements. Plasticizers such as glycerol are added to improve the workability and flexibility of the material and reinforcement materials such as fiberglass may be added. Finally, stabilizers and antioxidants are also added to improve the durability of the plastic.

The plastic is usually delivered to the manufacture in a semi-prepared state such as pellets that can be melted down or sheets that can be punched out from a plastics supplier. Some of the additives may need to be mixed in at the point when the plastic is prepared for molding and formation of the x-ray glasses frame.

The lenses of x-ray glasses are made out of cardboard. Cardboard is a type of paper made from cellulose fibers extracted from wood or recycled paper. There are several different grades of cardboard available, differing in their thickness and exterior coating. For the lenses, a thin, compact cardboard material should be used that will resist bending. The cardboard should also be treated with a coating that will allow paint or silkscreen ink to adhere to the surface.

Various finishing materials are needed to complete the x-ray glasses. Small, thin white feathers are used to help create the illusion. Glue is used to hold the lenses in their frames and the feathers in place. Inks are used to coat the lenses and provide a visual appearance that enhances the trick. This is typically a spiral pattern that confuses the eyes.

Design

Design of the x-ray glasses focuses primarily on the frame size, shape, and color, and the construction of the lenses. The frame may be produced in different sizes, one to fit adults and one for children. The shape usually resembles thick black plastic frames of the 1950s, but may vary.

The lenses are carefully designed in conjunction with the feather in order to create a realistic x-ray image when looking through them. Other synthetic objects may be used to produce a similar effect as well. The exterior design of the lenses varies the most. It is decorated with either inks or an adhesive covering (similar to a sticker). Most commonly, a spiral design is used, but human or animal eyes may be painted on, as well as holograms for a more haunting visual effect.

The Manufacturing Process

The manufacture of x-ray glasses can be broken down into three distinct phases, frame forming, lens creation, and decoration and packaging.

Frame Forming

1 The glasses frame is made from plastic and typically formed using injection molding. In the first step of this process, a hollow metal mold is prepared that is the exact size and shape of the desired frame. The frame may be produced as a single unit, or separate parts that must be assembled. Most often, they are created as a single unit.

2 Next, pellets of the plastic polymer are released from a holding bin into a machine that liquefies them. The pellets are melted at very high temperatures, usually around 300°F (149°C). Once the pellets have been melted into a liquid plastic, they are ready for injection into the mold.

3 The molten plastic, called melt, is forced into a prepared mold at a very high speed and under intense pressure, approximately 300–700 psi. The melt fills the mold, then a bit more melt is added in order to compensate for the contraction due to cooling and solidification of the plastic.

4 Once the polymer has cooled, it is separated from the mold. This solidified frame is automatically removed from the mold and moved to the next step either by conveyer belt or manually. The melt and in-

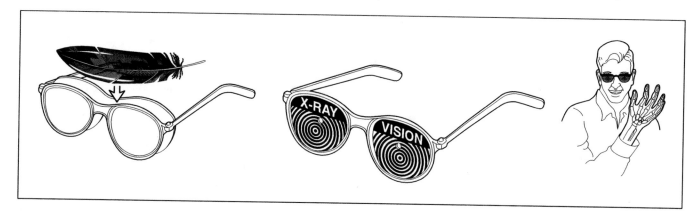

jection cycle is repeated every 10-100 seconds, depending on the time it takes the plastic to cool and set.

5 The frames may then be smoothed to remove any rough edges and excess plastic. This is typically done by an abrasive machine, which has rotating discs with sandpaper-like consistencies to smooth out the mold seams. A worker guides the frame over the rotating disc until the desired texture is achieved. The frames are then transported to the fitting station so they can have lenses attached.

Lens creation

6 X-ray glasses have lenses made from cardboard. The cardboard is purchased from a supplier in large sheets of the thickness and grade specified by the designers. Prior to cutting the lenses out, the cardboard sheets are sent to printing presses. The sheets are drawn into a machine where they pass through large rollers. The rollers transfer ink onto the lenses creating the outward appearance. This design is typically a black and white spiral. Numerous lenses are printed onto each sheet.

7 Once the decoration has been applied and set, the lenses are cut out of the cardboard sheets using blankinging machines. The sheet of cardboard is put into a machine and rolled under stampers. The are attached to the roller and are essentially a series of metal cookie cutters the exact size and shape of the lens. As the cardboard rolls through, the stampers come down on the cardboard and cut out the exact shape for the lenses. The cutting is done such that the decoration on the front of the lens is perfectly positioned. The shape for

the lens is actually cut so it can be folded over to make a double layered lens.

Finishing

8 Holes are next stamped through the double layered lenses using a punching machine. The cut-outs are placed so that a metal rod with the same diameter as the desired holes can punch through the center of each lens.

9 The lens is then opened and a feather is glued over each hole. The lens is then refolded and glued together.

10 The lenses are then attached to the plastic frame. A groove on the frame inside each lens area allows the cardboard lenses to snap in. The x-ray glasses are then put into a plastic bag packaging and a cardboard card is stapled to the top to close the bag. The card is decorated with illustrations to show how the glasses are used.

Quality Control

Quality control begins with the incoming plastic and cardboard which are used to make the x-ray glasses. The manufacturer checks to ensure the cardboard and plastic measure up to specifications related to physical appearance, dimensions, consistency, and other characteristics. For materials that are supplied by outside vendors, the x-ray glasses manufacturer typically requires quality control inspection by their suppliers. During production, the sheets are randomly checked for consistency and printing errors. Defective sheets are removed prior to production. Line inspectors are also stationed at various points on the production line to ensure only quality parts are used for produc-

A feather between the lenses has a polarizing effect, creating the illusion that objects have a transparent outline and solid center.

tion. Production speeds are verified by computers as are pressure, blanking, and resistance in rolling. Computerized sensors also help monitor each stage of production. Other things that are tested for include color uniformity, water resistance and flexibility of the plastic frame.

Byproducts/Waste

The waste products resulting from the manufacture of x-ray glasses includes excess plastic run-off from the injection molding process and scrap cardboard left over from the blanking process. Most of the run-off plastic can be melted down and reused again if it has not been contaminated and still meets the necessary specifications. If it does not, the plastic may be recycled for use in other products. Cardboard can be recycled, so excess scraps are collected for this purpose.

The Future

X-ray glasses are not nearly as popular as they once were. Even the novelty stores that carry this type of gag product are difficult to find. Therefore, there is likely to be little change in the design of the current novelty x-ray glasses. In the future, technological improvements in paper and plastic production will be incorporated into x-ray glasses production.

While the production of truly working x-ray glasses is impossible, the idea of creating glasses that can see through some materials looks promising. One product that has been introduced is a pair of goggles that lets the user see through certain types of clothes. They are not based on the principle of x rays but rather on light rays. Like x rays, certain visible and infrared light rays can pass through some materials, like clothing fibers. The object then reflects this light and it is this reflected light that can be seen. Typically, the light reflected from the clothing surface overwhelms a person's ability to see the reflected light from the body. These goggles are equipped with a sensitive camera that filters out the light from the clothing and captures the reflected infrared light from the body. This makes the covering invisible and allows the object to be visible.

Where to Learn More

Books

"Paper." *Kirk-Othmer Encyclopedia of Chemical Technology*. Vol. 18. R. E. Kirk and D. F. Othmer, eds. New York: John Wiley & Sons, 1996.

"Plastics Processing." *Kirk-Othmer Encyclopedia of Chemical Technology*. Vol. 19. R. E. Kirk and D. F. Othmer, eds. New York: John Wiley & Sons, 1996.

Periodicals

Munden, M. "Mike Harden: Prank Items Reflect Changing Sense of Humor." *The Columbus Dispatch* (16 February 2001).

Others

McCarron, Brett. "The Real Secret Behind X-Ray Specs." *How It's Done Web Page*. December 2001. <http://www.olywa.net/blame/how/how4.htm>.

The S. S. Adams Company Web Page. December 2001. <http://www.ssadams.com>.

Perry Romanowski

Index

A

Abbott Laboratories, 7:310

Above-ground swimming pools, 7:361

Absorbable sutures, 7:356, 7:357

Accomplished Cook (May), 7:69

Accordions, **3:1–5**

Acetic acid (Vinegar), **7:425–428**

Acetobacters, 7:425, 7:427

Acetozym nutrients, 7:426

Acetylene, **4:1–5**

Acid-based drain cleaners, 7:87–90

Acrylic fingernails, **3:6–10**

Acrylic plastic, **2:1–5**
    diving bells, 7:84
    stereoptic viewers, 7:337, 7:338

Acrylic powder coatings, 7:415, 7:416–417

Acrylonitrile butadiene styrene (ABS), 7:292

Action figures, **6:1–4**

Activated carbon, 7:1–2, 7:4

Active night scopes, 7:244

Adams Gum Company, 7:414

Adams, W. V., 7:177

Additives, in hairspray, 7:173

Adhesives
    cellophane tape, **1:105–108**
    contact glue, 7:126, 7:127
    duct tape, **6:150–153**
    glues, **5:234–237**
    rubber cement, **6:322–324**
    super glues, **1:444–447**
    surgical glue, 7:360

Adrian, Edgar Douglas, 7:96

Advanced Metal Evaporated process, 7:424

Aermotor windmills, 7:430

Aerosol delivery systems
    hairspray, 7:172–173, 7:174–175
    spray paint, 7:329

Agriculture. *See* Crops; Farm machinery

Air bags, **1:1–7**

Air conditioners, **3:11–15**

Air fresheners, **6:5–8**

Air purifiers, **7:1–5**

Air tests, for lawn sprinklers, 7:213

Aircraft
    airships, **3:16–20**
    business jets, **2:80–85**
    helicopters, **1:223–229**
    hot air balloons, **3:220–224**
    jet engines, **1:230–235**

Airships, **3:16–20**

Alcoholic beverages. *See* Beverages

Alcometer breath tester, 7:47

Alginate, for molding, 7:158

Alloys, of titanium, 7:391, 7:393–394

Aluminum, **5:1–5**
    beverage cans, **2:6–10**
    in diving bells, 7:84
    foil, **1:8–13**
    in lawn sprinklers, 7:212
    in microphones, 7:231
    in movie projectors, 7:242
    in oxygen tanks, 7:250–251
    in revolving doors, 7:282–283
    in scales, 7:296
    in Stirling cycle engines, 7:343
    in toasters, 7:396–397

Aluminum beverage cans, **2:6–10**

Aluminum foil, **1:8–13**, 7:199, 7:200

Amber, **7:6–10**

Ambulances, **5:6–10**

American Academy of Pediatrics, 7:201

American Rubber Company, 7:365

American Society for Testing and Materials (ASTM), 7:4, 7:18

American Society of Mechanical Engineers (ASME), 7:84, 7:85

Ammunition, **2:11–15**
    bullets, 7:51–56
    grenades, **7:160–162**

Ampex Company, 7:421

Amplifiers, for EEG machines, 7:96–97, 7:98

Amusement park rides
carousels, **4:88–93**
ferris wheels, **6:171–175**
roller coasters, **6:316–321**

Analog insulin, 7:190–191

Anatase, 7:392

Anderson, Mary, 7:434

Aneroid barometers, **7:11–14**

Angioplasty balloons, **6:9–12**

Animation, **3:21–25**

Annealing, of sutures, 7:359

Antibacterial soaps, **4:6–11**

Antibiotics, **4:12–16**

Antihistamines, in sleep aids, 7:310

Antilock brake systems, **2:16–20**

Antiperspirant/deodorant sticks, **5:11–14**

Antishoplifting tags, **3:26–29**

Apjohn, J. H., 7:434

Appliances
air conditioners, **3:11–15**
automatic drip coffee makers, **3:50–52**
cast iron stoves, **4:94–99**
dishwashers, **6:136–139**
furnaces, **7:144–146**
microwave ovens, **1:286–290**
refrigerators, **1:355–360**
surge suppressors, **3:424–426**
telephones, **5:447–451**
televisions, **3:438–444**
toasters, **7:395–398**
toilets, **5:458–463**
vacuum cleaners, **6:439–443**
washing machines, **1:473–477**

See also Household items and tools

Arkwright, Richard, 7:321

Armored trucks, **4:17–21**

Armstrong, Edwin, 7:275

Aronson, Louis V., 7:215

Artificial blood, **5:15–20**

Artificial eyes, **3:30–34**

Artificial flowers, **5:21–24**

Artificial heart valves, **6:18–21**

Artificial hearts, **6:13–17**

Artificial limbs, **1:14–18**

Artificial skin, **3:35–38**

Artificial snow, **4:22–25**

Artificial turf, **7:15–19**

Artillery shells, 7:304–308

Asbestos, **4:26–30**

Aseptic containers, 7:198–199, 7:201

Aseptic Packaging Council, 7:198–199

Aspartame, **3:39–43**

Asphalt cement, **2:21–26**

Asphalt pavers, **3:44–49**

Aspirin, **1:19–23**

Association of Professional Tattooists (APT),
7:374

Astatic Corporation, 7:229

Astrodome (Houston, Texas), 7:15, 7:16

AstroTurf, 7:15

Atmospheric pressure, 7:11

Atraumatic needles, 7:357

Audio assemblies, in movie projectors, 7:239,
7:240

Auer von Welsbach, Carl, 7:147, 7:215

Automated external defibrillators, 7:116–117

Automatic drip coffee makers, **3:50–52**

Automatic Products International, Ltd., 7:414,
7:419

Automobile windshields, **1:31–34**

Automotive products
air bags, **1:1–7**
antilock brake systems (ABS), **2:16–20**
spark plugs, **1:420–423**
speedometers, **7:317–320**
tires, **1:453–457**
windshield wipers, **7:434–438**
windshields, **1:31–34**

Averill, D. R., 7:329

B

Baby carriers, **6:22–26**

Baby formulas, **4:31–35**

Baby strollers, **4:36–38**

Baby wipes, **6:27–30**

Backhoes, **6:31–36**

Baeyer, Adolf von, 7:310

Bagels, **4:39–43**

Bagpipes, **6:37–42**

Baking, of pizza, 7:262

Baking powder, **6:43–46**

Baking soda, **1:35–38**

Baleen, 7:74

Ball bearings, **1:39–43**

Balloons, **2:27–31**

Ballpoint pens, **3:53–57**

Balsamic vinegar, 7:427

Bands, for tiaras, 7:387, 7:388–389

Bank pins, 7:257

Bank vaults, **7:20–24**

Banting, Frederick G., 7:188

Bar code scanners, **1:44–49**

Barbed wires, **2:32–36**

Barbiturates, 7:310

Barometers, aneroid, **7:11–14**

Barrels, of pens, 7:141–142

Barton, Otis, 7:84

Bascules, for guillotines, 7:165, 7:166

Baseball bats, **2:37–40**

Baseball caps, **4:44–47**

Baseball gloves, **1:55–59**

Baseballs, **1:50–54**

Basketballs, **6:47–53**

Batch cooking/mixing
artificial turf, 7:17
birdseed, 7:38
candy canes, 7:58, 7:59–60
condensed soups, 7:70–72
drain cleaners, 7:88–89
fabric softeners, 7:122–123
fluoride treatment, 7:137–138
hairspray, 7:174, 7:175
Pyrex, 7:272, 7:273
sleeping pills, 7:311
spray paint, 7:330–331

Batelli, Frederic, 7:115

Bath towels, **4:53–57**

Bathgate, Andy, 7:156

Bathing suits, **7:365–368**

Bathroom scales, 7:296–297

Bathtubs, **2:41–46**

Bathyscaphs, 7:84

Bathyspheres, 7:83–84

Batteries, **1:60–64**
external defibrillators, 7:116, 7:117, 7:119
laser pointers, 7:208, 7:209

Batts, of felt, 7:131–133

Beam, E. D., 7:178

Bean bag plush toys, **5:25–30**

Beans, in maracas, 7:225, 7:226

Beauchamp, George, 7:100

Beauty products. *See* Health and beauty
products

Beaver felt, 7:130–131

Beck, Claude, 7:115

Bed sheets, **5:31–35**

Beebe, William, 7:84

Beecher, Catherine, 7:194

Beef jerky, **4:58–61**

Beepers, **2:47–51**

Beer, **2:52–56**

Behrend, Otto, 7:309

Bekker, Yvonne, 7:289

Bell, Alexander Graham, 7:229, 7:333, 7:382

Bell Laboratories, 7:207, 7:229, 7:276

Bells, **2:57–60**

Bending machines, 7:283

Benedict, Clint, 7:156

Benzodiazepines, 7:309, 7:310

Berger, Hans, 7:95

Bernard, Claude, 7:66

Bernstein, Leonard, 7:225

Best, Charles H., 7:188

Betamax videotapes, 7:421

Beverages
beer, **2:52–56**
brandy, **7:41–45**
champagne, **3:82–86**
cider, **4:119–123**
coffee, **1:134–138**
cognac, **6:107–110**
green tea, **5:242–247**
instant coffees, **3:235–241**
juice boxes, **7:198–202**
orange juice, **4:348–352**
soft drinks, **2:418–422**
soy milk, **5:406–409**
tea bags, **2:443–447**
vodka, **5:484–488**
water, **4:457–461**
whiskey, **2:481–486**
wine, **1:488–493**

Bezels, in tiaras, 7:389

BIC lighters, 7:215

Bicycle seats, **7:25–29**

Bicycle shorts, **1:65–68**

Bicycles, **2:61–66**

Bikini swimsuits, 7:366, 7:367

Billboards, **5:36–41**

Bimetallic strips
aneroid barometers, 7:12, 7:13
hair dryers, 7:170
toasters, 7:395–396, 7:397

Bing Crosby Enterprises, 7:420

Binoculars, **7:30–32**

Bioceramics, **5:42–45**

Biodegradable plastic substitutes, 7:328

Biologic decompression (Diving), 7:86

Biological drain cleaners, 7:87, 7:90

Bird cages, **7:33–36**

Birdseed, **7:37–40**

Birth control pills, **4:62–65**

Bisque porcelain figurines, **5:46–50**

Black & Decker, 7:396

Black boxes, **3:58–61**

Blades, for guillotines, 7:164, 7:165

Blaisdell, George G., 7:215

Blanks, for pens, 7:141–142

Bleach, **2:67–70**

Bleichroeder, Frizt, 7:66

Blickensderfer, George, 7:404

Blood alcohol content, 7:46, 7:50

Blood glucose, 7:151, 7:187

Blood pressure monitors, **1:69–73**

Blood pumps (Artificial hearts), **6:13–17**

Bloodless glucometers, 7:155

Blowers, in furnaces, 7:144, 7:145–146

Blue jeans, **1:74–79**

Bobbing compound, 7:387, 7:389

Bobbins, in spinning wheels, 7:321, 7:322

Bodies, of electric guitars, 7:102, 7:104

Body fat measurement, 7:296

Boil and cool tests, 7:254, 7:256

Bond, George, 7:83

Books, **1:80–85**

Books (Pop up), **7:263–265**

Boomerangs, **6:54–58**

Boric acid, in Pyrex, 7:270–271, 7:272, 7:273

Borkenstein, Robert, 7:47

Boron, 7:270

Borosilicate glass, 7:270, 7:271, 7:272

Bottles, for model ships, 7:298, 7:299

Bowling balls, **4:66–69**

Bowling pins, **4:70–74**

Bows and arrows, **5:51–56**

Boxing, 7:266

Boxing gloves, **6:59–62**

Braided sutures, 7:356, 7:357, 7:358–359

Braille publications, **4:75–78**

Brakes
    antilock brake systems (ABS), **2:16–20**
    in revolving doors, 7:283

Brandy, **7:41–45**

Brass, **6:63–66**
    cigarette lighters, 7:215–216, 7:217
    shrapnel shells, 7:304, 7:306, 7:307

Brassieres, **5:57–60**, 7:367

Breads, **2:71–74**

Breakout mechanisms, in revolving doors, 7:283–284

Breath alcohol testers, **7:46–50**

Breath mints, **6:67–72**

Breathalyzer, 7:47

Brewster, Sir David, 7:336

Bricks, **1:86–90**

Brine solutions, for smoked ham, 7:314, 7:315

Bromides (Sleep aids), 7:309–310

Brooms, **6:73–77**

Brosnan, Cornelius, 7:258

Brother typewriters, 7:408

Broughton, Jack, 7:266

Buccal insulin, 7:192

Building and construction materials
    asbestos, **4:26–30**
    asphalt cement, **2:21–26**
    barbed wire, **2:32–36**
    bricks, **1:86–90**
    ceramic tiles, **1:109–113**
    concrete, **1:158–163**, 7:20, 7:22, 7:361, 7:362–364
    decorative plastic laminates, **2:157–160**
    drywall, **2:171–174**
    dynamite, **2:175–179**
    fiberboard, **3:143–147**
    foam rubber, **5:201–204**
    hammers, **4:251–255**
    nails, **2:307–311**
    paintbrushes, **5:348–352**
    paints, **1:310–314**
    pavers (asphalt), **3:44–49**
    saws, **3:359–362**
    screws, **3:372–374**
    shingles, **3:386–399**
    vermiculite, **6:444–447**
    vinyl floorcoverings, **4:452–456**
    wallpapers, **3:467–471**

Bulldozers, **3:62–66**

Bulletproof vests, **1:91–95**

Bullets, **7:51–56**

Bungee cords, **2:75–79**

Burchmore, Bill, 7:156–157

Burial traditions, 7:180

Burners, in furnaces, 7:144, 7:145

Burning bars, 7:24

Business jets, **2:80–85**

Business supplies. *See* Office and school supplies

Butter and margarine, **2:86–91**

Buttons, **2:92–97**

Byproducts, waste, and recycling
    juice boxes, 7:201
    precious metals, 7:390
    typewriters, 7:408

C

Cabinetmaking (Tables), 7:370, 7:371–373

Cabinets, for vending machines, 7:415, 7:416–418

Cable-tooled oil drilling, 7:204

CAD. *See* Computer-aided design

Cages (Bird), **7:33–36**

Cages, for goalie masks, 7:157, 7:159

Calabash gourds, 7:224, 7:225

Calendaring, of magnetic tape, 7:422

Calibration, of speedometers, 7:319

California Air Resources Board (CARB), 7:329, 7:330

Cambridge University (Britain), 7:394

Camcorders, 7:421

Camera lenses, **2:98–102**

Cameras, **3:67–71**

Campbell's Soup Company, 7:69–70, 7:72

Cams
    cigarette lighters, 7:217
    lawn sprinklers, 7:211, 7:212, 7:213

Canals and locks, **6:78–83**

Candies and confections
    breath mints, **6:67–72**
    candy canes, **7:57–61**
    candy corn, **6:83–87**
    caramel, **6:88–91**
    chewing gum, **1:124–128**
    chocolates, **1:129–133**
    cotton candy, **4:157–161**
    gummy candies, **3:186–189**
    ice cream, **3:225–230**
    ice cream cones, **6:224–229**
    jawbreakers, **6:239–242**
    jelly beans, **2:262–266**
    licorice, **4:314–318**
    lollipops, **6:253–256**
    M&M candies, **3:257–260**
    popsicles, **6:307–310**

Candles, **1:96–99**

Candy canes, **7:57–61**

Candy corn, **6:83–87**

Canning, of soup, 7:72

Capacitors, in radios, 7:276

Capillary attraction, in fountain pens, 7:139

Capsules, in aneroid barometers, 7:12–13

Caramel, **6:88–91**

Carbon, activated, 7:1–2

Carbon dioxide, as dry ice, 7:91–94

Carbon fibers, **4:79–83**, 7:157

Carbon microphones, 7:229, 7:230

Carbon monoxide detectors, **4:84–87**

Carbon papers, **1:100–104**

Cardano, Girolamo Carolano, 7:290

Cardboard
    corrugated, **1:169–173**
    in x-ray glasses, 7:440, 7:441

Carding, of felt, 7:131, 7:132

Carousels, **4:88–93**

Carpets, **2:103–108**

Carriages, in typewriters, 7:405, 7:406, 7:407

Carton blanks, for juice boxes, 7:199–200

Cases, for cigarette lighters, 7:215, 7:216, 7:217–218

Cash registers, **7:62–65**

Cassettes, for videotapes, 7:422–423

Cast iron stoves, **4:94–99**

Castanets, **5:61–64**

Casting
    bullets, 7:53–54
    grenades, 7:161–162
    manhole covers, 7:220–222
    Stirling cycle engines, 7:345

Cat litter, **2:109–113**

CAT scanners, **3:72–76**

Catagenesis, 7:203

Catgut, 7:356–357

Catheters, **7:66–68**

Cathode ray tubes, **2:114–118**

Caustic drain cleaners, 7:87–90

CBS Corporation, 7:421

Celerifere, 7:410

Cellophane tape, **1:105–108**

Cellulose, 7:425–426

Celtic diadems, 7:385

Centers for Disease Control, 7:353

Ceramic filters, **5:65–69**

Ceramic microphones, 7:229, 7:230

Ceramic rolling pins, 7:286, 7:287

Ceramic tiles, **1:109–113**

Cereals, **3:77–81**

Certified wood, 7:370, 7:373

Chalk, **1:114–118**

Chalkboards, **2:119–122**

Champagne, **3:82–86**

Change machines, **4:100–104**

Charcoal briquettes, **4:105–109**

Chase Company, 7:329

Cheek plates, in handcuffs, 7:178

Cheese, **1:119–123**

Cheese curls, **5:70–73**

Cheevers, Gerry, 7:158

Chemgrass, 7:15

Chemical drain cleaners, 7:87–90

Chemicals and base materials
    acetylene, **4:1–5**
    acrylic plastic, **2:1–5**
    amber, **7:6–10**

brass, **6:63–66**
compost, **5:86–90**
copper, **4:142–146**
cork, **5:103–106**
cubic zirconia, **5:128–132**
cushioning laminates, **4:172–176**
decorative plastic laminates, **2:157–160**
diamonds, **2:165–170**
dry ice (carbon dioxide), **7:91–94**
fertilizers, **3:138–142**
fiberglass, **2:190–194**
fibers, **5:308–309**
gasoline, **2:219–222**
glues, **5:234–237**
gold, **1:204–208**
helium, **4:256–261**
indigo, **6:230–233**
iron, **2:257–261**
krypton, **4:297–301**
latex, **3:245–250**
litmus paper, **6:249–252**
lubricating oils, **1:277–280**
lumber, **3:251–256**
mercury, **4:324–328**
natural gas, **6:270–272**
oxygen, **4:353–357**
plywood, **4:379–383**
polyurethane, **6:298–301**
propane, **3:335–338**
Pyrex, **7:270–274**
salt, **2:389–393**
sand, **3:354–358**
shellac, **4:418–422**
silicon, **6:357–359**
silver, **3:390–393**
sodium chlorite, **6:380–383**
stainless steel, **1:424–429**
Teflon, **7:378–381**
tin, **4:437–442**
titanium, **7:391–394**
zinc, **2:487–492**
zirconium, **1:505–507**
Chemstrand Company, 7:15
Cherries, **6:92–96**
Cherry wood, for spinning wheels, 7:322
Chess, **6:97–101**
Chewing gum, **1:124–128**
Chicken, **5:74–78**
Child care products
    baby carriers, **6:22–26**
    baby formulas, **4:31–35**
    baby strollers, **4:36–38**
    baby wipes, **6:27–30**
    child safety seats, **5:80–85**
    clothing, **4:110–114**
    disposable diapers, **3:118–122**
    pacifiers, **7:253–256**
Child safety seats, **5:80–85**
Child, Thomas, 7:329

Children's clothing, **4:110–114**
Chilton seed-cleaning system, 7:38, 7:39
Chimenti, Jacopo, 7:336
Chip logs, 7:317
Chloral hydrate, 7:309
Chlorofluorocarbon (CFC) propellants
    hairspray, 7:173
    spray paint, 7:329, 7:330
Chocolates, **1:129–133**
Choke holes, in pacifiers, 7:253, 7:256
Chopsticks, **4:115–118**
Chorpening, C. M., 7:229
Chrome Rouge, 7:387, 7:389
Chromium, in electroplating, 7:350
Churchill, Lady Randolph, 7:374
Cider, **4:119–123**
Cigarette lighters, **7:215–219**
Cigarettes, **2:123–128**
Cigars, **3:87–91**
CinemaScope lenses, 7:239
Cinematographes, 7:238
Circlets, for tiaras, 7:387, 7:388–389
Circuit boards. *See* Printed circuit boards
Clarification, of amber, 7:8
Clarinets, **3:92–96**
Clavellinas, 7:224
Clement, Saint, 7:130
Closed-cell foam, 7:26, 7:27
Cloth. *See* Fabrics and textiles
Clothes irons, **6:102–106**
Clothing and accessories
    baseball caps, **4:44–47**
    bicycle shorts, **1:65–68**
    blue jeans, **1:74–79**
    brassieres, **5:57–60**
    children's clothing, **4:110–114**
    corsets, **7:73–76**
    cowboy boots, **2:148–152**
    galoshes, **5:220–224**
    high heels, **5:258–262**
    iron-on decals, **3:240–244**
    leather jackets, **2:283–287**
    neckties, **1:301–304**
    pantyhose, **1:315–319**
    raincoats, **6:311–315**
    shoelaces, **6:349–352**
    Stetson hats, **3:414–418**
    sunglasses, **3:419–423**
    swimsuits, **7:365–368**
    t-shirts, **2:439–442**
    tiaras, **7:385–390**
    tuxedos, **7:399–403**
    umbrellas, **1:469–472**
    watches, **1:478–481**

wooden clogs, **4:471–473**

CNC machine tools, **2:129–134**
    for spinning wheels, 7:322–323
    for tables, 7:371–372

Coating, of sutures, 7:359

Coca-Cola Company, 7:418

Coffees, **1:134–138**

Coffins, **4:128–131**

Cognac (Spirits), **6:107–110**

Coins, **2:135–138**

Coir, **6:111–115**

Cold rolling, of bird cages, 7:34

Coleman Arc Lantern, 7:147, 7:148

Coleman, W. C., 7:147

Collars, for pet containment, 7:279–281

Collimating optics, in laser pointers, 7:208, 7:209

Coloring, in candy canes, 7:59

Column stills, 7:43–44

Combination locks, **1:139–142**, 7:20, 7:22

Combustion chambers, of Stirling cycle engines, 7:346

Comic books, **6:116–121**

Communication kiosks, 7:384

Compact disc players, **1:153–157**

Compact discs, **1:148–152**

Compost, **5:86–90**

Compressed Gas Association, 7:252

Compression, of dry ice, 7:92, 7:93

Computer-aided design (CAD)
    electric guitars, 7:105
    manhole covers, 7:223
    swimsuits, 7:366
    tables, 7:371
    tuxedos, 7:400

Computer control, by EEG machine, 7:99

Computer mouse, **5:91–96**

Computer numerically controlled (CNC) machines, **2:129–134**
    for spinning wheels, 7:322–323
    for tables, 7:371–372

Computerized movie projectors, 7:242

Concrete, **1:158–163**
    reinforced, in bank vaults, 7:20, 7:22
    swimming pools, 7:361, 7:362–364

Concrete beam bridges, **4:137–141**

Concrete blocks, **3:97–102**

Concrete dams, **5:97–102**

Condensed and evaporated milk, **6:162–164**

Condensed soups, **7:69–72**

Condenser microphones, 7:229, 7:230, 7:231

Conditioning agents, for fabric softeners, 7:121, 7:122

Condoms, **2:139–143**

Conductive gels, for external defibrillators, 7:116

Cones, in speakers, 7:333, 7:334

Construction equipment and features
    bulldozers, **3:62–66**
    canals and locks, **6:78–83**
    concrete beam bridges, **4:137–141**
    concrete blocks, **3:97–102**
    concrete dams, **5:97–102**
    cranes, **5:116–121**
    draw bridges, **6:145–149**
    geodesic domes, **6:184–189**
    hard hats, **6:200–204**
    skyscrapers, **6:365–371**
    suspension bridges, **5:434–440**
    tunnels, **6:429–434**

Consumer Products Safety Commission (CPSC)
    hair dryer standards, 7:168, 7:170
    pacifier regulations, 7:254–256

Contact glue, for feather dusters, 7:126, 7:127

Contact lenses, **2:143–147**

Continuous vinegar process, 7:425

Cook, James, 7:374

Cooking. *See* Batch cooking/mixing

Cooking oils, **1:164–168**

Copper, **4:142–146**

Coral baby necklaces, 7:253–254

Cork, **5:103–106**

Corkscrews, **6:122–125**

Corn, in birdseed, 7:37–38

Corn syrup, **4:147–151**, 7:59

Corning Glass Works, 7:270, 7:271

Coronary catheters, 7:66, 7:68

Correction fluid, **4:152–156**

Corrugated cardboard, **1:169–173**

Corsets, **7:73–76**

Cotton, **6:126–130**
    softeners for, 7:120
    swabs, 4:162–167

Cotton candy, **4:157–161**

Cotton swabs, **4:162–167**

Cough drops, **5:107–110**

Cowboy boots, **2:148–152**

CPSC. *See* Consumer Products Safety Commission

Crafts and craft supplies
    ships in a bottle, **7:298–303**
    spinning wheels, **7:321–324**
    yarn, **3:478–482**, 7:233–234

Cranberries, **5:111–115**

Cranes, **5:116–121**

Crankshafts, in Stirling cycle engines, 7:345, 7:346

Crash test dummies, **5:122–127**

Crayons, **2:153–156**

Credit cards, **4:168–171**

Crest with Fluoristan toothpaste, 7:136

Critelli, Tom, 7:268

Crops
    in birdseed, 7:37–38
    cherries, **6:92–96**
    cotton, **6:126–130**
    industrial hemp, **6:234–238**
    macadamia nuts, **5:312–315**
    olives, **5:343–347**
    raisins, **4:399–403**
    rice, **5:383–387**
    seedless fruits and vegetables, **6:340–343**
    sunflower seeds, **5:430–433**

Cross-linking
    amber, 7:7
    grenade polymers, 7:161–162

Crotonic acid, in hairspray, 7:172

Crowns, for tiaras, 7:387

Crude oil recovery, 7:203–204

Cubic zirconia, **5:128–132**

Cuckoo clocks, **6:131–135**

Cultivation, of nori, 7:352–353

Cultured pearls, **3:103–108**

Cumberland, Duke of, 7:266

Cushioning laminates, **4:172–176**

Cut-meters, 7:317

Cutlery, **1:174–179**

Cutting torches, in bank robberies, 7:21

Cyclotrons, **7:77–81**

D

Dean, R. S., 7:392

Decapitation, by guillotine, 7:163, 7:166

Déclics, for guillotines, 7:165, 7:166

Decompression (Diving), 7:82, 7:83, 7:86

Decorative plastic laminates, **2:157–160**

Deep Diving Systems (DDS), 7:82, 7:83

Deep drawing, of metal, 7:216

Defibrillators (External), **7:114–119**

Dempster windmills, 7:430

Dental caries, 7:135

Dental crowns, **4:177–181**

Dental drills, **3:109–113**

Dental floss, **2:161–164**

Deodorant sticks, **5:11–14**

Dephlogisticated air, 7:249

DHTDMAC (Dihydrogenated tallow dimethyl ammonium chloride), 7:121

Diabetes mellitus, 7:151, 7:187–188

Diadems, 7:385

Diagenesis, 7:203

Dial-a-Coke, 7:418

Diamonds, **2:165–170**

Diana, Princess of Wales, 7:386

Diapers, disposable, **3:118–122**

Diaphragms, in microphones, 7:230, 7:232

Dice, **4:182–186**

Dies, for pop up books, 7:264–265

Diesel fuel, 7:206

Digital barometers, 7:14

Digital movie projectors, 7:242

Digital radios, 7:278

Digital Video Discs (DVDs), 7:421

Digital videotapes, 7:421

Dihydrogenated tallow dimethyl ammonium chloride (DHTDMAC), 7:121

Dillinger, John, 7:21

Dimethyl sulfoxide kerosene extraction, 7:205–206

Diodes, in lasers, 7:207–209

Diphenhydramine antihistamines, 7:310

Disc-operating epilators, 7:111–112

Discrete Epilady model, 7:111

Dishwashers, **6:136–139**

Disinfecting, for surgery, 7:356–357

Dispersion polymerization, 7:380

Disposable diapers, **3:118–122**

Distillation
    brandy, 7:41, 7:42–44
    crude oil, 7:204, 7:205
    titanium, 7:393

Diving bells, **7:82–86**

DNA synthesis, **6:140–144**, 7:188–191

Dog biscuits, **5:133–136**

Doorknobs, **5:137–140**

Doors (Revolving), **7:282–285**

Dopyera, John, 7:100

Dorrance, John T., 7:69–70

DOT-3AL oxygen tank marking, 7:252

Double-lock handcuffs, 7:177–178

Doughnuts, **5:141–145**

Douglas, John Sholto (Marquees of Queensberry), 7:266

Dowels, for rolling pins, 7:287–288

Downy Ball, 7:123

Doxylamine succinate, 7:310

Dragout, in electroplating, 7:350

Drain cleaners, **7:87–90**

Draw bridges, **6:145–149**

Drawing, of glass, 7:246

Drayes Wetting Test, 7:123

The Dream Beam, 7:155

Drew-Baker, Kathleen, 7:352–353

Drilling, for granite, 7:181

Drinking straws, **4:186–190**, 7:199, 7:200

Driving wheels, in spinning wheels, 7:322

Drop spindles, 7:321

Droplet-in-matrix phase structure, 7:272

Drug delivery systems, for insulin, 7:192

Drums, **4:191–195**

Drunkometer breath tester, 7:47

Dry ice, **7:91–94**

Dry mops, 7:233, 7:235–236

Dryer fabric softener sheets, 7:120

Drying, of sleep aid mixtures, 7:311

Drywall, **2:171–174**

Duct tape, **6:150–153**

Ductile cast iron
      manhole covers, 7:220, 7:222–223
      Stirling cycle engines, 7:343

Ductility tests, for cast iron, 7:222–223

Dulcimers, **4:196–201**

Dump tanks, for ice resurfacers, 7:184

DuPont, 7:365, 7:378–379, 7:381, 7:392

Dust mops, 7:233, 7:235–236

Dusters (Feather), **7:125–129**

DVD players, **4:202–206**

Dyeing, of fabric, 7:366, 7:368

Dynamic microphones, 7:229, 7:230, 7:231–232

Dynamic speakers, 7:333–335

Dynamite, **2:175–179**

E

E. coli bacteria, 7:189–190, 7:191

E. I. du Pont de Nemours & Company
      (DuPont), 7:365, 7:378–379, 7:381, 7:392

E. Remington & Sons, 7:404

Ebco Technologies, Inc., 7:78

Eckert, Jacob, 7:63

Edison, Thomas Alva, 7:229, 7:237, 7:238, 7:333

Edward I (King of England), 7:295

Edward Miller Company, 7:147

EEG machines, **7:95–99**

Efficient Lamp, 7:147

Eggs, **5:146–149**

EKG machines, **3:123–127**

Elastic, in corsets, 7:74, 7:75

Electric blankets, **6:154–157**

Electric guitars, **7:100–105**

Electric motors, for windshield wipers, 7:435, 7:436

Electric shavers, 7:110

Electric shocks
      from hair dryers, 7:168–169
      for pet containment, 7:279, 7:280
      to restart heart, 7:114, 7:115

Electric tea kettles, **7:106–109**

Electric typewriters, 7:404

Electrocution, from hair dryers, 7:168, 7:170

Electrodes
      EEG machines, 7:95, 7:96–97
      external defibrillators, 7:114, 7:116, 7:118, 7:119

Electroencephalogram (EEG) machines, **7:95–99**

Electron beam welding, 7:13

Electronic cash registers, 7:63

Electronic ink, **6:158–161**

Electronic typewriters, 7:404

Electronics
      bar code scanners, **1:44–49**
      beepers, **2:47–51**
      black boxes, **3:58–61**
      breath alcohol testers, 7:47, 7:48
      cash registers, 7:63
      cathode ray tubes, **2:114–118**
      compact disc players, **1:153–157**
      compact discs, **1:148–152**
      cyclotrons, **7:77–81**
      DVD players, **4:202–206**
      EEG machines, 7:97–98, 7:97–98
      electronic ink, **6:158–161**
      external defibrillators, 7:117
      hearing aids, **2:228–231**
      integrated circuits, **2:251–256**, 7:276, 7:396
      laser pointers, **7:207–210**
      Liquid Crystal Displays (LCDs), **1:272–276**
      microwave ovens, **1:286–290**
      mouse, computer, **5:91–96**
      player pianos, **2:356–360**
      printed circuit boards, **2:365–370**, 7:47, 7:48, 7:97–98, 7:117–118, 7:208, 7:209, 7:276, 7:277, 7:278
      radios, **7:275–278**
      robots, industrial, **2:245–250**
      satellite dishes, **1:390–394**
      speedometers, 7:318
      surge suppressors, **3:424–426**

telephones, **5:447–451**

televisions, **3:438–444**

typewriters, 7:404

Electroplating

    cigarette lighters, 7:216, 7:217

    straight pins, 7:348, 7:350

Electrostatic precipitator air filters, 7:1–3, 7:4–5

Elevators, **2:180–185**

Eli Lilly Corporation, 7:188–189, 7:310

Emulsifiers, in fabric softeners, 7:121, 7:122

Emulsion polymers, 7:121

Enameling, of gas lanterns, 7:149

Energy efficiency, with revolving doors, 7:282

Engines, types of, 7:342

Engler Sr., Bill, 7:37

Engraving, of headstones, 7:183

Entertainment

    compact disc players, **1:153–157**

    compact discs, **1:148–152**

    DVD players, **4:202–206**

    LP records, **5:301–306**

    movie projectors, **7:237–243**

    radios, **7:275–278**

    televisions, **3:438–444**

    video games, **5:478–483**

    videotapes, **7:420–424**

Envelopes, **5:157–161**

Environmental concerns

    drain cleaners, 7:89–90

    electroplating, 7:350

    hairspray propellants, 7:173, 7:175

    hog farm waste, 7:316

    juice boxes, 7:198–199

    lead, in bullets, 7:55

    manhole cover casting, 7:223

    mercury barometers, 7:14

    Pyrex, 7:273–274

    seafood for sushi, 7:355

    spray paint, 7:329, 7:330

    Teflon, 7:381

    whale population decline, 7:76

See also Health hazards and safety concerns

Epilady hair removal device, 7:110, 7:111

Epilation devices, **7:110–113**

Epoxy resins, in goalie masks, 7:157, 7:159

Erasers, **5:162–166**

Escalators, **3:128–132**

Escapements, in typewriters, 7:405, 7:406–407, 7:408

Estate Stove Company, 7:395

Etruscan civilization, 7:286

Euclid, 7:336

Eurydamus, 7:266

Evans, Sir Arthur, 7:73

Evaporated and condensed milk, **6:162–164**

Excavation, for swimming pools, 7:362

Executioners, 7:166

Expanded Polystyrene Foams (EPFs), **1:180–184**

Expanding bullets, 7:52, 7:53, 7:55

Explosives

    in bank robberies, 7:21

    dynamite, **2:175–179**

    grenades, **7:160–162**

    in shrapnel shells, 7:305, 7:307

External combustion engines, 7:344, 7:346

External defibrillators, **7:114–119**

Extraction

    kerosene, 7:203–206

    titanium, 7:393

Extrusion

    candy canes, 7:60

    magnetic tape, 7:422

    oxygen tanks, 7:250, 7:251

    revolving doors, 7:283

    sporks, 7:326

    sutures, 7:358

    toasters, 7:397

ExxonMobil, 7:206

Eyeglass frames, **5:167–172**

Eyeglass lenses, **1:185–189**

F

Fabric softeners, **7:120–124**

Fabrics and textiles

    coir fibers, **6:111–115**

    cotton, **6:126–130**

    fake furs, **3:133–137**

    felt, **7:130–134**

    lace curtains, **4:302–305**

    linen, **4:319–323**

    lyocell, **5:308–311**

    patent leather, **6:282–287**

    polyester, **2:361–364**

    polyester fleece, **4:384–388**

    rayon, **1:350–354**

    ribbons, **3:349–353**

    silk, **2:398–402**

    softeners for, 7:120–124

    spandex, **4:423–427**

    spinning wheels, **7:321–324**

    straight pins, **7:348–351**

    for swimsuits, 7:365, 7:366

    thread, **5:452–457**

    for tuxedos, 7:400, 7:401–403

    wool, **1:494–499**

    yarn, **3:478–482**

Fake furs, **3:133–137**

Fan-type windmills, **7:430–432**

Farm and gardening equipment

combines, **1:143–147**
hay balers, **2:223–227**
lawn mowers, **1:252–255**
lawn sprinklers, **7:211–214**
milking machines, **2:303–306**
wheelbarrows, **5:494–497**

Fastkin swimsuits, 7:366

Fat substitutes, **2:186–189**

Fats and oils
butter and margarine, **2:86–91**
cooking oils, **1:164–168**
fat substitutes, **2:186–189**
lubricating oils, **1:277–280**
olive oils, **3:304–308**

Faucets, **6:165–170**

FDA. *See* U. S. Food and Drug Administration

Feather dusters, **7:125–129**

Feathers, in x-ray glasses, 7:439, 7:440, 7:441

Federal Meat Inspection Act of 1967, 7:316

Feeder stacks, for vending machines, 7:415, 7:417–418

Felt, **7:130–134**, 7:216, 7:217

Fender, Leo, 7:101

Fermentation
brandy, 7:41, 7:42
rice, 7:352
vinegar, 7:425–427

Ferris wheels, **6:171–175**

Fertilizers, **3:138–142**

Fessenden, Reginald Aubrey, 7:275

Fiberboard, **3:143–147**

Fiberglass, **2:190–194**
goalie masks, 7:156–157, 7:159
swimming pools, 7:361, 7:362

Fiberglass Canada, 7:156

Fibers, **5:308–309**

See also Fabrics and textiles

Fick, Adolph, 7:66

15 Puzzle, 7:290

Figg, James, 7:266

File cabinets, **1:190–193**

Filigree, in tiaras, 7:387

Fill dams, **5:173–179**

Filling processes
hairspray, 7:174, 7:175
spray paint, 7:331

Fills, for artificial turf, 7:18

Filters (Ceramic), **5:65–69**

Filters (Swimming pools), 7:364

Fingerboards, of electric guitars, 7:102–103, 7:104

Fiorello, Frank A., 7:260

Firearms
ammunition, **2:11–15**
bulletproof vests, **1:91–95**
bullets, **7:51–56**
Thompson submachine guns, **6:410–414**

See also Military equipment

Fireworks, **2:201–205**

Fisher-Price, Inc., 7:336

Fixed tables, 7:369

Flash heating and cooling, 7:199

Flash (Tattoo designs), 7:374, 7:375, 7:376

Flashlights, **6:176–179**

Flat movie projector lenses, 7:239

Flavorings, in candy canes, 7:57, 7:59

Flea collars, **4:217–220**

Flint glass, 7:271

Flint, in cigarette lighters, 7:217

Floppy disks, **1:199–203**

Flour, **3:153–158**, 7:260–261

Flowers (Artificial), **5:21–24**

Fluorescent anodes, 7:245, 7:246, 7:247

Fluoride treatment, **7:135–138**

Flutes, **5:196–200**

Flux, in jewelry-making, 7:388–389

Fluxes, in Pyrex, 7:272, 7:273, 7:387

Foam
closed-cell, 7:26, 7:27
expanded polystyrene, **1:180–184**

Foam rubber, **5:201–204**

Foley catheters, 7:66–67

Foley, Frederick E. B., 7:66–67

Food
aspartame, **3:39–43**
baby formulas, **4:31–35**
bagels, **4:39–43**
baking powder, **6:43–46**
baking soda, **1:35–38**
beef jerky, **4:58–61**
bread, **2:71–74**
butter and margarine, **2:86–91**
cereals, **3:77–81**
cheese, **1:119–123**
cheese curls, **5:70–73**
cherries, **6:92–96**
chicken, **5:74–78**
condensed soups, **7:69–72**
cooking oils, **1:164–168**
corn syrup, **4:147–151**
cranberries, **5:111–115**
dog biscuits, **5:133–136**
doughnuts, **5:141–145**
eggs, **5:146–149**
evaporated and condensed milk, **6:162–164**
fat substitutes, **2:186–189**

flour, **3:153–158**
fortune cookies, **2:206–208**
freeze-dried food, **2:209–213**
frozen vegetables, **5:210–215**
frozen yogurt, **2:214–218**
fruit leathers, **5:216–219**
fruitcakes, **4:221–225**
gelatin, **5:225–228**
graham crackers, **3:181–185**
honey, **5:269–273**
hot dogs, **4:272–276**
ice cream, **3:225–230**
imitation crab meat, **3:230–234**
jams and jellies, **5:284–287**
ketchup, **2:272–276**
macadamia nuts, **5:312–315**
maple syrup, **3:266–272**
marshmallows, **3:276–280**
mayonnaise, **6:262–264**
milk, **4:329–333**
molasses, **5:316–320**
oatmeal, **5:340–342**
olive oils, **3:304–308**
olives, **5:343–347**
pasta, **2:331–335**
peanut butter, **1:320–325**
pepper, **5:359–362**
pet food, **2:341–345**
pickles, **4:369–373**
pita bread, **5:368–371**
pizza, **7:260–262**
popcorn, **5:377–382**
potato chips, **3:331–334**
pretzels, **4:394–398**
raisins, **4:399–403**
rice, **5:383–387**
rice cakes, **4:404–407**
salad dressings, **6:332–334**
salsa, **1:381–384**
salt, **2:389–393**
seedless fruits and vegetables, **6:340–343**
shortbread, **6:353–356**
smoked ham, **7:314–316**
soy sauce, **3:408–413**, 7:353
Spam, **6:388–391**
sugar, **1:439–443**
sunflower seeds, **5:430–433**
sushi rolls, **7:352–355**
tofu, **2:453–457**
tortilla chips, **1:458–463**
TV dinners, **5:469–473**
vegetarian burgers, **5:474–477**
vinegar, **7:425–428**
yogurt, **4:474–478**
See also Beverages; Candies and confections
Food and Drug Administration (FDA). *See* U. S. Food and Drug Administration
Food-grade dry ice, 7:91
Foot-and-mouth disease, 7:316
Footbags, **6:180–183**

Football helmets, **3:162–164**
Footballs, **3:159–161**
Foote-Taylor, Daniel, 7:348
Ford Foundation, 7:15
Forging
    shrapnel shells, 7:305
    Stirling cycle engines, 7:345
Forks, history of, 7:325
Formers, in Pyrex, 7:271–272
Formex kerosene extraction, 7:206
Forrester, Glenn, 7:47
Fortune cookies, **2:206–208**
Fossils, in amber, 7:6
Foundries, 7:221–222
Fountain pens, **7:139–143**
Fragmentation shells, 7:307
Frames, for unicycles, 7:411, 7:412
Frazier, Joe, 7:267
Free, Helen, 7:152
Freeze-dried foods, **2:209–213**
Frequency amplifiers, in radios, 7:276
Frigidaire, 7:378
Frisbees, **5:205–209**
Frit, 7:246
Frock coats, 7:399
Frozen vegetables, **5:210–215**
Frozen yogurt, **2:214–218**
Fruit juice, in boxes, 7:198
Fruit leathers, **5:216–219**
Fruitcakes, **4:221–225**
Fruits and vegetables. *See* Crops
Fuel cell breath alcohol testers, 7:46–49, 7:50
Fuels
    gasoline, **2:219–222**
    natural gas, **6:270–272**
Full-range speakers, 7:333
Fuller, Walter, 7:100
Fulling, of felt, 7:132–133
Furnaces, **7:144–146**, 7:221–222
Furniture polishes, **4:226–230**
Fuses, in shrapnel shells, 7:305, 7:306–307

G

Galilei, Galileo, 7:30
Gallium arsenide, in night scopes, 7:245
Galoshes, **5:220–224**
Games. *See* Toys and games
Gapachos, 7:224

Gardening equipment. *See* Farm and gardening equipment

Gas lanterns, **7:147–150**

Gas masks, **3:171–174**

Gasoline, **2:219–222**

Gasoline pumps, **4:231–235**

Gears
    lawn sprinklers, 7:211, 7:212, 7:213
    windshield wipers, 7:435, 7:436, 7:437

Geinrich, Rudi, 7:366

Gelatin, **5:225–228**

Gem paper clips, 7:257, 7:258, 7:259

Gems and precious metals
    amber, **7:7–8**
    cubic zirconia, **5:128–132**
    diamonds, **2:165–170**
    gold, **1:204–208**
    silver, **3:390–393**
    synthetic rubies, **4:428–432**
    in tiaras, 7:385–386

General Aniline & Film Corporation, 7:336

General Electric, 7:395, 7:396, 7:421

Generator vinegar process, 7:426–427

Genetically engineered insulin, 7:188–191

Geodesic domes, **6:184–189**

Gesner, Abraham, 7:203

Gibson, Orville, 7:100, 7:101

Gift wrap, **6:456–460**

Gilding metal, 7:304

Gillette, King, 7:110

Giraffe unicycles, 7:411

Glass
    in binoculars, 7:30, 7:31
    characteristics of, 7:270, 7:271
    in gas lanterns, 7:148, 7:149
    in movie projectors, 7:241
    in night scopes, 7:245–246
    ornaments from, **5:229–233**
    Pyrex, **7:270–274**
    in rolling pins, 7:286, 7:287
    for ships in a bottle, 7:298

Glass fibers, in HEPA filters, 7:1, 7:2, 7:3–4

Glass forming, 7:272, 7:273

Glidden, Carlos, 7:404

Globes, **4:236–240**

Globes (Lanterns), 7:148, 7:149

Glucometer test kits, **7:151–155**

Glues, **5:234–237**

Glues (Surgical), 7:360

Goalie masks, **7:156–159**

Goatskin, for punching bags, 7:267

Godrej & Boyce Manufacturing Company, 7:405, 7:408

Gold, **1:204–208**, 7:385–386

Golden Age of Radio, 7:275–276

Golf carts, **1:209–213**

Golf clubs, **4:241–245**

Golf tees, **5:238–241**

Gothic paper clips, 7:257

Gourds, for maracas, 7:224–225, 7:226

Graham crackers, **3:181–185**

Granite, in headstones, 7:181–183

Graves, Harold, 7:336

Gravestones, 7:180–183

Gray cast iron, 7:220, 7:223

Gray, William, 7:382

Green sand, for casting, 7:220–221, 7:222, 7:223

Green tea, **5:242–247**

Greenberg, Leon, 7:47

Greeting cards, **5:248–252**

Gregor, William, 7:392

Grenades, **7:160–162**

Grinding wheels, **1:214–218**

Ground Fault Circuit Interrupters (GFCI), 7:169, 7:170

Gruber, William, 7:336

Guillotin, Joseph Ignace, 7:163

Guillotines, **7:163–167**

Guitars, **1:219–222**

Guitars (Electric), **7:100–105**

Gummy candies, **3:186–189**

Gunite, 7:363, 7:364

Gunpowder, 7:305

Guns. *See* Firearms

Gutta-percha, 7:32

Gyroscopes, **6:190–193**

H

Hair dryers, **7:168–171**

Hair dyes, **3:190–194**

Hair removers, **4:246–250**, **7:110–113**

Hairspray, **7:172–176**

Halladay windmills, 7:430

Halley, Edmund, 7:83

Halogen lamps, **6:194–199**

Ham (Smoked), **7:314–316**

Hamilton Beach, 7:396

Hamilton, Sir William, 7:290

Hammers, **4:251–255**

Hand grenades, 7:161

Handcuffs, **7:177–179**

Hang gliders, **5:253–257**

Hangiri, 7:352

Hansen, Malling, 7:404

Hansen Writing Ball, 7:404

Hard hats, **6:200–204**

Harger, R. N., 7:47

Harmonicas, **3:195–187**

Harps, **3:198–201**

Harpsichords, **6:205–211**

Hats, from felt, 7:130–131

Hay balers, **2:223–227**

Headstones, **7:180–183**

Health and beauty products
    acrylic fingernails, **3:6–10**
    antibacterial soaps, **4:6–11**
    antibiotics, **4:12–16**
    antiperspirant/deodorant sticks, **5:11–14**
    aspirin, **1:19–23**
    birth control pills, **4:62–65**
    condoms, **2:139–142**
    cotton swabs, **4:162–167**
    cough drops, **5:107–110**
    dental floss, **2:161–164**
    deodorant sticks, **5:11–14**
    diapers, disposable, **3:118–122**
    epilation devices, **7:110–113**
    fluoride treatment, **7:135–138**
    hair dryers, **7:168–171**
    hair dyes, **3:190–194**
    hair removers, **4:246–250**
    hairspray, **7:172–176**
    home pregnancy tests, **4:268–271**
    lipsticks, **1:267–271**
    mascaras, **3:281–283**
    mouthwashes, **6:265–268**
    nail polishes, **1:297–300**
    nicotine patches, **3:300–303**
    perfume, **2:336–340**
    safety razors, **5:388–392**
    shampoos, **3:381–385**
    shaving cream, **1:406–409**
    soap, **2:414–417**
    sunscreen, **2:429–433**
    tattoos, **7:374–377**
    teeth whiteners, **6:406–409**
    toothbrushes, **2:458–461**
    toothpastes, **3:455–458**
    vitamins, **3:462–466**
    wigs, **3:472–477**

Health hazards and safety concerns
    airborne pollutants, 7:1
    corsets, 7:74
    fluoridation, 7:136
    hair dryers, 7:168–169, 7:170
    mercury poisoning, 7:131

Teflon, 7:381
    zinc toxosis, 7:35
See also Environmental concerns

Hearing aids, **2:228–231**

Heart physiology, 7:114

Heart valves (Artificial), **6:18–21**

Heat exchangers, in furnaces, 7:144–145

Heat pumps, **3:202–206**

Heat rejection, 7:342

Heat treating, of oxygen tanks, 7:250, 7:251

Heating elements
    hair dryers, 7:168, 7:169
    toasters, 7:395, 7:397

Heavy duty trucks, **3:207–213**

Heavy machinery. *See* Industrial and heavy machinery

Helicopters, **1:223–229**

Helium, **4:256–261**

Hemp, industrial, **6:234–238**

Hendrix, Jimi (James Marshall), 7:101

HEPA filters, 7:1–5

Herbal sleep aids, 7:309

Hermes typewriters, 7:408

Hero (Mathematician), 7:414

HFC (Hydrofluorocarbon) propellants, 7:173, 7:330

Hibbard, Susan, 7:125

Hides (Animal). *See* Leather

High Efficiency Particulate Air (HEPA) filters, 7:1–5

High heels, **5:258–262**

Hobbies. *See* Crafts and craft supplies; Sporting goods and equipment; Toys and games

Hobbyhorses, 7:410

Hockey masks, **7:156–159**

Hockey pucks, **6:212–216**

Hockey sticks, **4:262–267**

Hoe, Robert, 7:348

Holding agents, in hairspray, 7:172

Holiday lights, **5:263–268**

Holograms, **3:214–219**

Home pregnancy tests, **4:268–271**

Honey, **5:269–273**

Hooker, D. R., 7:115

Hoop rim spinning wheels, 7:321

Horizontal-axis windmills, 7:429, 7:430

Horner, William George, 7:237

Horseshoes, **6:217–219**

Hot air balloons, **3:220–224**

Hot dogs, **4:272–276**

Hourglasses, **5:274–278**

Household items and tools
air fresheners, **6:5–8**
air purifiers, **7:1–5**
aluminum foil, **1:8–13**
antibacterial soaps, **4:6–11**
artificial flowers, **5:21–24**
bath towels, **4:53–57**
bathtubs, **2:41–46**
bed sheets, **5:31–35**
bleach, **2:67–70**
brooms, **6:73–77**
bungee cords, **2:75–79**
buttons, **2:92–97**
candles, **1:96–99**
carpet, **2:103–108**
cat litter, **2:109–113**
cellophane tape, **1:105–108**
charcoal briquettes, **4:105–109**
chopsticks, **4:115–118**
clothes irons, **6:102–106**, 7:194
combination locks, **1:139–142**
corkscrews, **6:122–125**
cuckoo clocks, **6:131–135**
cutlery, **1:174–179**
diapers, disposable, **3:118–122**
doorknobs, **5:137–140**
drain cleaners, **7:87–90**
drinking straws, **4:186–190**
duct tape, **6:150–153**
electric blankets, **6:154–157**
electric tea kettles, **7:106–109**
fabric softeners, **7:120–124**
faucets, **6:165–170**
feather dusters, **7:125–129**
flashlights, **6:176–179**
flea collars, **4:217–220**
fountain pens, **7:139–143**
furniture polishes, **4:226–230**
gas lanterns, **7:147–150**
glass ornaments, **5:229–233**
greeting cards, **5:248–252**
halogen lamps, **6:194–199**
holiday lights, **5:263–268**
incense sticks, **5:279–283**
ironing boards, **7:194–197**
lace curtains, **4:302–305**
lava lamps, **4:306–309**
lawn sprinklers, **7:211–214**
lead crystal, **4:310–313**
light bulbs, **1:256–260**
LP records, **5:301–306**
magnets, **2:293–298**
matches, **3:284–288**
mattresses, **1:281–285**
mirrors, **1:291–296**
mops, **7:233–236**
mosquito repellents, **3:295–299**
mousetraps, **5:321–324**
nutcrackers, **5:335–339**
paintbrushes, **5:348–352**
pencils, **1:326–329**
pillows, **6:288–291**
plastic wrap, **2:351–355**
recliners, **3:345–348**
rolling pins, **7:286–289**
rubber bands, **1:367–370**
safety pins, **2:385–388**
sandpaper, **1:385–389**
scales, **7:295–297**
screwdrivers, **1:395–399**
sofas, **3:398–403**
sponges, **5:417–420**
sporks, **7:325–328**
staplers, **1:430–433**
steel wool, **6:397–400**
super glue, **1:444–447**
tables, **7:369–373**
thread, **5:452–457**
toilet paper, **6:415–418**
vacuum cleaners, **6:439–443**
vinyl floorcoverings, **4:452–456**
wallpaper, **3:467–471**
wheelbarrows, **5:494–497**
wood stains, **6:452–455**
wrapping paper, **6:456–460**
yarn, **3:478–482**
zippers, **1:500–504**

See also Appliances

Howe, John Ireland, 7:348

Howe Manufacturing Company, 7:348

Hughes Aircraft, 7:207

Hula Hoops, **6:220–223**

Hulls, for ships in a bottle, 7:300

Humane Slaughter Act of 1958, 7:316

Humbucking electric guitar pickup, 7:101

Hunt, Seth, 7:348

Hunt, Waldo, 7:263

Hunter, Matthew, 7:392

Hunting knives, **2:232–236**

Hybrid artificial/biological turf, 7:19

Hydrapulping, of juice boxes, 7:201

Hydrocarbon propellants, 7:173, 7:330

Hydrofluorocarbon (HFC) propellants, 7:173, 7:330

Hydrometers, 7:44

Hydrostatic testing, of oxygen tanks, 7:250–251, 7:252

Hypocausts, 7:144

I

IBM (International Business Machines), 7:404

Ice cream, **3:225–230**

Ice cream cones, **6:224–229**

Ice resurfacing machines, **7:184–186**

Ice skates, **2:237–240**

Ideal Toy Company, 7:291, 7:336

Idiophones, 7:224

Ilmenite, 7:392, 7:393

Image intensifier tubes, 7:245–247

Imgard, August, 7:58

Imitation crab meat, **3:230–234**

Implantable blood glucose sensors, 7:154–155

Impossible Wheel unicycles, 7:411

Improved Conventional Munition (ICM),
    7:307–308

In-ground swimming pools, 7:361

In-line skates, **2:241–244**

Incandescence, 7:147

Incense sticks, **5:279–283**

Incorruptible Cashier, 7:62, 7:63

Indigo, **6:230–233**

Industrial and heavy machinery
    asphalt pavers, **3:44–49**
    backhoes, **6:31–36**
    bulldozers, **3:62–66**
    CNC machine tools, **2:129–134**, 7:322–323,
        7:371–372
    forklifts, rough terrain, **2:380–384**
    grinding wheels, **1:214–218**
    hard hats, **6:200–204**
    heavy duty trucks, **3:207–213**
    laboratory incubators, **1:236–241**
    robots, industrial, **2:245–250**
    Stirling cycle engines, **7:342–347**
    windmills, **7:429–433**

Industrial hemp, **6:234–238**

Industrial robots, **2:245–250**

Infant care products
    baby carriers, **6:22–26**
    baby formulas, **4:31–35**
    baby strollers, **4:36–38**
    baby wipes, **6:27–30**
    child safety seats, **5:80–85**
    disposable diapers, **3:118–122**
    pacifiers, **7:253–256**

Inflation, of punching bags, 7:268

Infrared breath alcohol testers, 7:46

Infrared night scopes, 7:244

Ingots, of titanium, 7:393–394

Inhaled insulin, 7:192

Injection molding
    air purifier cases, 7:3
    bicycle seat shells, 7:26–27, 7:28
    external defibrillator casings, 7:117
    glucometers, 7:153–154
    hair dryers, 7:169–170
    lawn sprinklers, 7:212
    pacifiers, 7:255, 7:256

Rubik's cubes, 7:291, 7:292
    stereoptic viewers, 7:338
    tea kettle components, 7:107, 7:108
    x-ray glasses, 7:440–441

Ink
    electronic, **6:158–161**
    for pens, 7:140
    for tattoos, 7:375, 7:376

Inspection. *See* Quality control and regulation;
    specific government agencies

Instant coffees, **3:235–241**

Instant Insanity puzzle, 7:290

Instant lottery tickets, **4:277–280**

Insulated bottles, **4:281–284**

Insulin, 7:151, **7:187–193**

Integrated Circuits (ICs), **2:251–256**
    radios, 7:276
    toasters, 7:396

Interference fitting, 7:339

Intermediate-acting insulin, 7:187

International Business Machines (IBM), 7:404

International Commission for the Conservation
    of Atlantic Tuna (ICCAT), 7:355

International Organization for Standardization
    (ISO), 7:319

Intoximeter breath tester, 7:47

Ionic hair dryers, 7:171

Ionizing units (Air purifiers), 7:2–3, 7:4

Ireland, Murray, 7:395

Iron, **2:257–261**
    manhole covers, 7:220–222
    tea kettles, 7:106

Iron-on decals, **3:240–244**

Ironing boards, **7:194–197**

Irons (Clothes), **6:102–106**, 7:194

Ishige, Terutoshi, 7:291

ITT Corporation, 7:244, 7:245

J

J. R. Clark Company, 7:194

Jacketing, for bullets, 7:52, 7:54

Jams and jellies, **5:284–287**

Jantzen, Carl, 7:365

Jantzen Company, 7:365

Japanese Agriculture Standard (Japan), 7:353

Jawbreakers, **6:239–242**

Jelly beans, **2:262–265**

Jeremiah Symphony (Bernstein), 7:225

Jet engines, **1:230–235**

Jet fuel, 7:206

Jet injectors, for insulin, 7:192

Jet piercing, for granite, 7:181

Jets (Business), **2:80–85**

Jewelers' Rouge, 7:387, 7:389

Jewelry-making
   amber, 7:8–9
   tiaras, 7:385–389

Juice boxes, **7:198–202**

Jukeboxes, **5:288–291**

Julien, Jean Baptiste Gilbert Payplat dis, 7:69

JVC, 7:421

K

Kaleidoscopes, **6:243–248**

Kayaks, **2:267–271**

Kazoos, **4:289–292**

Kearns, Bob, 7:434

Kehoe, Simon D., 7:266

Keller, Gregory, 7:58

Keller, Hannes, 7:83

Kerosene, **7:203–206**

Ketchup, **2:272–276**

Kevlar, in goalie masks, 7:157

Keyboards, in typewriters, 7:405, 7:407, 7:408

Kilby, Jack, 7:276, 7:396

Kilns, for smoked ham, 7:314

Kinetoscopes, 7:237–238

Kitty litter, **2:109–113**

Klaproth, M. H., 7:392

Knives, history of, 7:325

Knots (Nautical speed), 7:317

Kodak, 7:421

Konaclip paper clips, 7:258

Kouwenhoven, William B., 7:115

Kremer, Tom, 7:291

Kroll process, 7:392, 7:393–394

Kroll, William, 7:392

Krylon Company, 7:332

Krypton, **4:297–301**

L

Lace curtains, **4:302–305**

Lamilux, 7:337

Laminates
   cushioning, **4:172–176**
   decorative plastic, **2:157–160**

Lamps, in movie projectors, 7:239, 7:240–241

Lancing devices, 7:151–152, 7:154

Langworthy, Orthello, 7:115

Laser-guided missiles, **1:242–246**

Laser pointers, **7:207–210**

Lasers
   for blood glucose testing, 7:155
   for hair removal, 7:113
   for headstone etching, 7:183
   semiconductor, **6:344–348**, 7:207, 7:208
   solid state, **6:384–387**

Lastex, 7:365

Latex, **3:245–250**, 7:254

Lathes (Wood), 7:225, 7:226

Laundry detergents, **1:242–246**

Lava lamps, **4:306–309**

Lavoisier, Antoine-Laurent, 7:249

Lawn mowers, **1:252–255**

Lawn sprinklers, **7:211–214**

Lawrence, Ernest O., 7:77, 7:78

Lead, **2:277–282**
   bullets, 7:51, 7:55
   shrapnel balls, 7:304, 7:306

Lead crystal, **4:310–313**

Leather
   jackets, **2:283–287**
   patent leather, **6:282–287**
   in punching bags, 7:266–268

Leather jackets, **2:283–287**

Leibniz, Gottfried, 7:12

Lemke, Carl W., 7:390

Lenses
   binoculars, 7:30, 7:31
   camera, **2:98–102**
   movie projectors, 7:239, 7:241
   night scopes, 7:245
   stereoptic viewers, 7:337, 7:338
   x-ray glasses, 7:439, 7:440, 7:441

Leucoxene, 7:392

Lever Brothers, 7:120, 7:123

Levers, in scales, 7:296

Lexan, in vending machines, 7:415, 7:418

License plates, **5:292–295**

Licorice, **4:314–318**

Liebig, Justus von, 7:309

Life vests, **2:288–292**

Light bulbs, **1:256–260**

Light-Emitting Diodes (LEDs), **1:261–266**

Light ray goggles, 7:442

Lighters (Cigarette), **7:215–219**

Limbs (Artificial), **1:14–18**

Linen, **4:319–323**, 6:288

Linkages, in windshield wipers, 7:435, 7:436, 7:437

Lippershey, Jan, 7:30

Lipsticks, **1:267–271**

Liquid air, 7:250, 7:251

Liquid breathing (Diving), 7:86

Liquid carbon dioxide, 7:92, 7:93

Liquid Crystal Displays (LCDs), **1:272–276**

Liquid paper fluid, **4:152–156**

Lister, Joseph, 7:356–357

Litmus paper, **6:249–252**

Littleton, Jesse T., 7:271

Lloyd, Sam, 7:290

Loar, Lloyd, 7:100

Locks, **5:296–300**, 7:20

Locks and canals (Water), **6:78–83**

Locock, Sir Charles, 7:309

Loeb, W., 7:66

Lollipops, **6:253–256**

Lombardi, Gennaro, 7:260

London Prize boxing rules, 7:266

Long-acting insulin, 7:187

Lorena, Guglielmo de, 7:83

Lottery tickets, **4:277–280**

Louis, Antoine, 7:163

Louis XVI (King of France), 7:163, 7:164

LP records, **5:301–306**

Lubricating oils, **1:277–270**, 7:51, 7:54–55

Lucas, Edouard, 7:290

Lumber, **3:251–256**

Lumière, Auguste Marie Louis, 7:238

Lumière, Louis Jean, 7:238

Lunettes, for guillotines, 7:165–166

Lurgi Arosolvan kerosene extraction, 7:205

Lyocell, **5:308–311**

M

M&M candies, **3:257–260**

Macadamia nuts, **5:312–315**

Machining
    oxygen tanks, 7:250
    Stirling cycle engines, 7:345–346

MacMahon, Percy, 7:290

Macmillan, Kirkpatrick, 7:410

Macro-emulsions, 7:121

Magnetic Resonance Imaging (MRI), **3:261–265**

Magnetic sound systems, for films, 7:239

Magnetic tapes, 7:420–424

Magnets, **2:293–298**
    cyclotrons, 7:77, 7:79
    microphones, 7:229–230, 7:231–232
    speedometers, 7:318, 7:319
    stereo speakers, 7:333, 7:334

Maillot swimsuits, 7:365, 7:366, 7:367

Makisu, 7:352, 7:353

Manhole covers, **7:220–223**

Mantles, in lanterns, 7:147–148, 7:150

Manufacturing processes
    annealing, 7:359
    baking, 7:262
    batch cooking/mixing, 7:17, 7:38, 7:58, 7:59–60, 7:88–89, 7:122–123, 7:137–138, 7:174, 7:175, 7:272, 7:273, 7:311, 7:330–331
    bending, 7:283
    brining, 7:315
    calendaring, 7:422
    canning, of soup, 7:72
    carding, 7:131, 7:132
    casting, 7:53–54, 7:161–162, 7:220–222, 7:345
    cold rolling, 7:34
    compression, of dry ice, 7:92, 7:93
    condensed soups, 7:70–72
    deep drawing, 7:216
    drawing, of glass, 7:246
    drying, 7:311
    electron beam welding, 7:13
    electroplating, 7:216, 7:217, 7:348, 7:350
    engraving, 7:183
    excavation, 7:362
    extraction, 7:203–206, 7:393
    extrusion, 7:60, 7:250, 7:251, 7:283, 7:326, 7:358, 7:397, 7:422
    fermentation, 7:41, 7:42, 7:352, 7:425–427
    filling, 7:174, 7:175, 7:331
    flash heating and cooling, 7:199
    forging, 7:305, 7:345
    fulling, of felt, 7:132–133
    glass forming, 7:272, 7:273
    heat treating, 7:250, 7:251
    injection molding, 7:3, 7:26–27, 7:28, 7:107, 7:108, 7:117, 7:153–154, 7:169–170, 7:212, 7:255, 7:256, 7:291, 7:292, 7:338, 7:440–441
    Kroll process, 7:392, 7:393–394
    machining, 7:250, 7:345–346
    metalworking, 7:112, 7:145–146, 7:148–149, 7:178, 7:195–196, 7:216–217, 7:241–242, 7:296, 7:405–408, 7:412–413, 7:416–418, 7:431
    molding, 7:22, 7:112, 7:158–159, 7:162
    nonstick coating, 7:380–381
    painting, 7:149
    polishing, 7:181–183, 7:386–387, 7:389
    polymerization, 7:379–380
    punching, 7:34, 7:54, 7:107–108, 7:212, 7:216, 7:231, 7:240, 7:241, 7:396–397
    quarrying, 7:181
    sewing, 7:75–76, 7:234, 7:268, 7:367–368, 7:401–402
    smoking, of ham, 7:315
    soldering, 7:98, 7:208, 7:209, 7:386, 7:389
    stamping, 7:349–350

sterilizing, 7:200, 7:359
swaging, 7:54, 7:250
tabletting, 7:312
tanning, 7:267
thermoforming, 7:326–327
tufting, 7:15, 7:17
vinegar processes, 7:426–427
welding, 7:84, 7:195, 7:216, 7:241–242
wire forming, 7:258–259
woodworking, 7:102–104, 7:126,
    7:141–142, 7:164–166, 7:226,
    7:287–288, 7:300, 7:322–323, 7:370,
    7:371–373

Maple syrup, **3:266–272**

Maple wood, for spinning wheels, 7:322

Maracas, **7:224–228**

Marbles, **2:299–302**

Marconi, Guglielmo, 7:275

Marconi Online, 7:418

Marie-Antoinette (Queen of France), 7:163

Markers, **3:272–275**

Marsh, Albert, 7:395

Marshmallows, **3:276–280**

Mascaras, **3:281–283**

Matches, **3:284–288**

Matryoshka dolls, **6:257–261**

Mattel, Inc., 7:336

Matthews, B. C. H., 7:96

Mattresses, **1:281–285**

May, Robert, 7:69

Mayblox puzzle, 7:290

Mayonnaise, **6:262–264**

McConnell, Howard Mack, 7:268

McKay, Frederick S., 7:135

McLoughlin Brothers, 7:263

Mechanical speedometers, 7:318

Mechanical tables, 7:369

Medical devices and equipment
    angioplasty balloons, **6:9–12**
    artificial blood, **5:15–20**
    artificial eyes, **3:30–34**
    artificial heart valves, **6:18–21**
    artificial hearts, **6:13–17**
    artificial limbs, **1:14–18**
    artificial skin, **3:35–38**
    blood pressure monitors, **1:69–73**
    CAT scanners, **3:72–76**
    catheters, **7:66–68**
    contact lenses, **2:143–147**
    cyclotrons, **7:77–81**
    dental crowns, **4:177–181**
    dental drills, **3:109–113**
    DNA synthesis, **6:140–144**
    EEG machines, **7:95–99**
    EKG machines, **3:123–127**
    external defibrillators, **7:114–119**

    eyeglass frames, **5:167–172**
    eyeglass lenses, **1:185–189**
    glucometer test kits, **7:151–155**
    hearing aids, **2:228–231**
    insulin, 7:151, **7:187–193**
    Magnetic Resonance Imaging (MRI),
        3:261–265
    microscopes, **3:289–294**
    needle-free injection systems, **6:274–277**
    oxygen tanks, **7:249–252**
    pacemakers, **3:309–313**
    sleeping pills, **7:309–313**
    stethoscopes, **1:434–438**
    sutures, **7:356–360**
    syringes, **3:427–431**
    thermometers, **1:448–452**
    vaccines, **2:466–470**

Meffert, Uwe, 7:291

Mephitic air, 7:249

Mepro Company, 7:111, 7:113

Mercury, **4:324–328**, 7:131

Mercury barometers, 7:11–12, 7:14

Metal evaporated tapes, 7:424

Metals
    aluminum, **5:1–5**
    brass, **6:63–66**
    copper, **4:142–146**
    gold, **1:204–208**
    iron, **2:257–261**
    lead, **2:277–282**
    mercury, **4:324–328**
    silver, **3:390–393**
    stainless steel, **1:424–429**
    tin, **4:437–442**
    titanium, **7:391–394**
    zinc, **2:487–492**
    zirconium, **1:505–507**

See also Gems and precious metals

Metalworking
    cigarette lighters, 7:216–217
    epilation devices, 7:112
    furnaces, 7:145–146
    gas lanterns, 7:148–149
    handcuffs, 7:178
    ironing boards, 7:195–196
    movie projectors, 7:241–242
    scales, 7:296
    typewriters, 7:405–408
    unicycles, 7:412–413
    vending machines, 7:416–418
    windmills, 7:431

Metric system, 7:295

Mica
    hair dryers, 7:169
    toasters, 7:396, 7:397

Micro-emulsions, 7:121

Microchannel plates, 7:245–247

Microchips. *See* Integrated Circuits

Microphones, **7:229–232**

Microscopes, **3:289–294**

Microwave ovens, **1:286–290**

Middlebrook, William, 7:258

Midrange speakers, 7:334

Milady Décolletée razor, 7:110

Military equipment
    grenades, **7:160–162**
    night scopes, **7:244–248**
    shrapnel shells, **7:304–308**

Milk, **4:329–333**

Milk cartons, **4:334–338**

Milk (Evaporated and condensed), **6:162–164**

Milking machines, **2:303–306**

Millet, in birdseed, 7:38

Milling. *See* Machining

Milnor, William, 7:115

Mini rolling pins, 7:287

Mining, of amber, 7:7–8, 7:9

Mirrors, **1:291–296**, 7:207

Mitterand, François, 7:164

Model kits (Toys), **6:419–424**

Model ships, in bottles, 7:298–303

Model trains, **4:339–342**

Moisture content, in wood, 7:373

Molasses, **5:316–320**

Molding processes
    bank vaults, 7:22
    epilator housings, 7:112
    goalie masks, 7:158–159
    grenade shells, 7:162

See also Injection molding

Moly (Molybdenum disulfide), 7:51, 7:54–55

Monofilament sutures, 7:356, 7:358

Mops, **7:233–236**

Morita, Akio, 7:334

Morrison, Jim (James Douglas), 7:230

Mosgrove, Roy, 7:156

Mosquito repellents, **3:295–299**

Mother of vinegar, 7:425–426

Motorcycles, **4:343–347**

Motors
    lawn sprinklers, 7:211, 7:212, 7:213
    windshield wipers, 7:435, 7:436

Mousetraps, **5:321–324**

Mouth guards, in pacifiers, 7:253, 7:256

Mouthwashes, **6:265–268**

Moutons (Guillotine weights), 7:164, 7:165

Moveable books, **7:263–265**

Movie projectors, **7:237–243**

Moving coil microphones, 7:229, 7:230, 7:231–232

Mozzarella cheese, 7:261

Musical instruments and equipment
    accordions, **3:1–5**
    bagpipes, **6:37–42**
    castanets, **5:61–64**
    clarinets, **3:92–96**
    drums, **4:191–195**
    dulcimers, **4:196–201**
    electric guitars, **7:100–105**
    flutes, **5:196–200**
    guitars, **1:219–222**
    harmonicas, **3:195–187**
    harps, **3:198–201**
    harpsichords, **6:205–211**
    kazoos, **4:289–292**
    maracas, **7:224–228**
    microphones, **7:229–232**
    pianos, **3:320–326**
    pipe organs, **5:363–367**
    player pianos, **2:356–360**
    radios, **7:275–278**
    saxophones, **6:335–339**
    sheet music, **5:393–397**
    stereo speakers, **7:333–335**
    trumpets, **1:464–468**
    tubas, **5:464–468**
    ukuleles, **6:435–438**
    violin bows, **2:476–480**
    violins, **2:471–475**
    xylophones, **6:461–465**

Mylar typewriter ribbons, 7:405

N

Nail polishes, **1:297–300**

Nails, **2:307–311**

National Cash Register Company, 7:63

National Hockey League (NHL), 7:157

National Institute of Justice Standards, 7:178

Natural gas, **6:269–273**

Natural vinegar process, 7:425

Necks, of electric guitars, 7:103, 7:104

Neckties, **1:305–309**

Needle-free injection systems, **6:274–277**

Needles
    surgical, 7:357, 7:359
    for tattoos, 7:375, 7:376

Negative ion cyclotrons, 7:78

Neodymium iron boron magnets, 7:229–230, 7:231–232

Neon signs, **2:312–316**

Newspapers, **2:317–322**

NHL (National Hockey League), 7:157

Nibs, for pens, 7:139, 7:140, 7:142

Nichrome heating elements, 7:169, 7:395, 7:397

Nickel plating, of straight pins, 7:349, 7:350

Nickerson, William, 7:110

Nicotine patches, **3:300–303**

Night scopes, **7:244–248**

Nipples, for pacifiers, 7:253, 7:254

Non-absorbable sutures, 7:356, 7:357

Non-aerosol hairsprays, 7:174, 7:175–176

Nonex glass, 7:271

Nonstick cookware, 7:380–381

Nori, in sushi, 7:352–354

Notes (Self-adhesive), **2:394–397**

Noyce, Robert, 7:276, 7:396

Nuclear submarines, **5:329–334**

Nutcrackers, **5:335–339**

Nylon
    artificial turf, 7:16
    dog collars, 7:279–280
    speedometers, 7:318

O

O-rings, in oxygen tanks, 7:250

Oatmeal, **5:340–342**

Obturation, 7:304

Oddzon Company, 7:291

Odor absorption, with activated carbon, 7:1–2, 7:4

Off shore oil drilling, 7:204

Office and school supplies
    ballpoint pens, **3:53–57**
    bar code scanners, **1:44–49**
    cash registers, **7:62–65**
    chalkboards, **2:119–122**
    correction fluid, **4:152–156**
    crayons, **2:153–156**
    envelopes, **5:157–161**
    erasers, **5:162–166**
    file cabinets, **1:190–193**
    floppy disks, **1:199–203**
    fountain pens, **7:139–143**
    globes, **4:236–240**
    liquid paper fluid, **4:152–156**
    markers, **3:272–275**
    paintbrushes, **5:348–352**
    paper clips, **7:257–259**
    pencils, **1:326–329**
    rubber cement, **6:322–324**
    rubber stamps, **4:408–412**
    scissors, **3:363–367**
    self-adhesive notes, **2:394–397**
    staplers, **1:430–433**
    typewriters, **7:404–409**

Oldsmobile Curved Dash Runabout, 7:317

Olive oils, **3:304–308**

Olives, **5:343–347**

Olivetti typewriters, 7:405, 7:408

Olympia typewriters, 7:405, 7:408

Online vending technology, 7:418–419

Onomastos, 7:266

Optic glass, in night scopes, 7:245

Optical equipment
    binoculars, **7:30–32**
    camera lenses, **2:98–102**
    cameras, **3:67–71**
    microscopes, **3:289–294**
    photographic film, **2:346–350**
    photographs and photography, **4:363–368**
    telescopes, **2:448–452**

Optical fibers, **1:305–309**

Optical sound systems, for films, 7:239

Oral insulin, 7:192

Orange juice, **4:348–352**

Orléans vinegar process, 7:425, 7:426

Ornaments (Glass), **5:229–233**

Oscillating sprinkler arms, 7:211, 7:212, 7:213

Ostrich feathers, for dusters, 7:125–128

Otto cycle engines, 7:342, 7:343

Over-the-counter sleep aids, 7:309, 7:310–312

Oxygen, **4:353–357**

Oxygen tanks, **7:249–252**

P

Pacemakers, **3:309–313**

Pacifiers, **7:253–256**

Packaging, for birdseed, 7:39

Padding, for bicycle seats, 7:26, 7:27

Paddles, **2:323–326**

Paintball, **6:278–281**

Paintbrushes, **5:348–352**

Paints and painting, **1:310–314**
    cigarette lighters, 7:217–218
    with feather dusters, 7:129
    gas lanterns, 7:149
    magnetic tape, 7:421–422
    maracas, 7:225, 7:226
    spray paint, **7:329–332**

Panemone windmills, 7:429

Pantyhose, **1:315–319**

Paper, **2:327–330**
    carbon papers, **1:100–104**
    currencies, **3:314–319**
    in juice boxes, 7:199–200
    litmus paper, **6:249–252**
    toilet paper, **6:415–418**
    wrapping paper, **6:456–460**

Paper clips, **7:257–259**

Paper currencies, **3:314–319**

Paper engineers, 7:263, 7:264

Paper-handling systems, in typewriters, 7:406, 7:407

Parachutes, **5:353–358**

Parade floats, **4:358–362**

Paraffin, from kerosene, 7:206

Parents, Bernie, 7:157

Paris, John Ayrton, 7:237

Parker Brothers, 7:290

Parker Pen Company, 7:140, 7:142

Parks, William, 7:69

Particle filtration, by air filters, 7:1, 7:4, 7:5

Passenger vehicles, **1:24–30**

Passive night scopes, 7:244–247

Pastas, **2:331–335**

Patches
    insulin, 7:192
    nicotine, **3:300–303**

Patent leather, **6:282–287**

Pattern-making, for manhole covers, 7:221

Patterson, John H., 7:63

Paul, Les, 7:100–101

Pavers (Asphalt), **3:44–49**

Pawls, in handcuffs, 7:178

PDMS (Polydimethylsiloxane), 7:121

Peanuts, in birdseed, 7:38

Pelleted bird food, 7:40

Pelletier, Nicolas, 7:163

Pencils, **1:326–329**

Penny Farthing bicycles, 7:410

Pens
    ballpoint, **3:53–57**
    fountain, **7:139–143**
    insulin, 7:192

Pepper, **5:359–362**

Peptide p1025, 7:138

Père-Lachaise Cemetery (France), 7:180

Perfumes, **2:336–340**

Perkin, William H., 7:366

Permanent depilatory devices, 7:113

Perovskite, 7:392

Perret, Jean Jacques, 7:110

Persistence of vision, 7:237

Personal Transfer Capsules (PTC), 7:82–86

Pesticides, **1:330–334**

Pet foods, **2:341–345**

Pet supplies
    bird cages, **7:33–36**
    birdseed, **7:37–40**
    cat litter, **2:109–113**
    dog biscuits, **5:133–136**
    flea collars, **4:217–220**
    pet food, **2:341–345**
    radio collars, **7:279–281**

Petroleum byproducts, 7:203

Phelps, Orson C., 7:177

Phenakistiscopes, 7:237

Philadelphia Cooking School, 7:271

Phlogiston, 7:249

Phosphor screens, 7:245, 7:246, 7:247

Photo cathodes, 7:245, 7:245–246, 7:247

Photographic film, **2:346–350**

Photographs and photography, **4:363–368**

Pianos, **3:320–326**

Piccard, Auguste, 7:84

Piccard, Jacques, 7:84

Pickles, **4:369–373**

Pickling, of metal, 7:386–387, 7:390

Pickups, for electric guitars, 7:100, 7:101, 7:102, 7:105

Piezo, in electric guitars, 7:105

Piezoelectric crystal microphones, 7:229

Pigments, in spray paint, 7:329–331

Pillows, **6:288–291**

Pinball, **6:292–297**

Pins (Straight), **7:348–351**

Pipe organs, **5:363–367**

Pistons, in Stirling cycle engines, 7:344–345, 7:346

Pita bread, **5:368–371**

Pizza, **7:260–262**

Plante, Jacques, 7:156–157

Plastic
    air purifier cases, 7:2, 7:3
    bicycle seat shells, 7:26–27
    dolls, **5:382–376**
    epilator housings, 7:112
    external defibrillator casings, 7:117
    feather dusters, 7:126, 7:128
    gas lanterns, 7:149
    hair dryers, 7:169–170
    ice resurfacers, 7:185, 7:186
    lawn sprinklers, 7:212
    maracas, 7:225
    microphones, 7:231
    mops, 7:234
    Rubik's cubes, 7:291, 7:292
    speedometers, 7:318
    sporks, 7:325, 7:326–327
    stereoptic viewers, 7:337, 7:338–339
    tea kettles, 7:107
    toasters, 7:396

wraps, **2:351–355**
x-ray glasses, 7:440

See also specific plastics

Plastic dolls, **5:382–376**

Plastic wraps, **2:351–355**

Plasticene clay, 7:299, 7:300, 7:302

Plate-hardening, of felt, 7:132

Plateau, Joseph Antoine Ferdinand, 7:237

Platens, in typewriters, 7:404, 7:405, 7:406

Player pianos, **2:356–360**

Playing cards, **4:374–378**

Plucking, of feathers for dusters, 7:126–127

Plumbing fixtures
bathtubs, **2:41–46**
faucets, **6:165–170**
toilets, **5:458–463**

Plunkett, Roy, 7:378–379

Plywood, **4:379–383**

Point of sale (POS) terminals, 7:62, 7:63, 7:64

Polishing
granite, 7:181–183
tiaras, 7:386–387, 7:389

Pollutants. *See* Byproducts, waste, and
recycling; Health hazards and safety concerns

Polybutylene terephthalate (PBT) polyester,
7:318

Polydimethylsiloxane (PDMS), 7:121

Polyester fleece, **4:384–388**

Polyesters, **2:361–364**

Polyethylene, in juice boxes, 7:199, 7:200

Polyethylene terephthalate (PET), 7:421

Polymerization, of Teflon, 7:379–380

Polymers
amber resins, 7:7
grenades, 7:160, 7:161–162
hairspray, 7:172, 7:173
sutures, 7:357–359

See also Plastic; *specific polymers*

Polypropylene (PP)
artificial turf, 7:16
Rubik's cubes, 7:292
sporks, 7:325, 7:326
tea kettles, 7:107

Polystyrene (PS)
sporks, 7:325, 7:326
stereoptic viewers, 7:337, 7:338

Polytetrafluoroethylene (PTFE/Teflon),
7:378–381

Polyurethane, **6:298–301**

Polyurethane foam insulation, 7:415, 7:417

Polyvinylchloride (PVC) piping, for pools,
7:362, 7:363

Polyvinylpyrrolidone vinyl acetate (PVPVA),
7:172

Pool tables, **6:302–306**

Pop up books, **7:263–265**

Popcorn, **5:377–382**

Popsicles, **6:307–310**

Porcelain, **1:335–339**

Pork
hot dogs, **4:272–276**
smoked ham, **7:314–316**
Spam, **6:388–391**

Porro, Ignatio, 7:30

Porrongo gourds, 7:224–225

Portable toilets, **3:327–330**

Portable typewriters, 7:404

Portland Knitting Company, 7:365

Positron emission tomography (PET), 7:77

Post windmills, 7:430

Postage stamps, **1:340–344**

Potato chips, **3:331–334**

Pottery, **4:389–393**

Poughkeepsie pins, 7:348

Powder coatings, for vending machines, 7:415,
7:416–417

Precious metals. *See* Gems and precious metals

Pressure gauges, **1:345–349**

Pressure mantles, in lanterns, 7:147–148, 7:150

Pressure Vessels for Human Occupancy
(PVHO), 7:84

Prest-Air Devices Company, 7:92

Pretzels, **4:394–398**

Prevost, Jean-Louis, 7:115

Priestley, Joseph, 7:249, 7:250

Printed circuit boards, **2:365–370**
breath alcohol testers, 7:47, 7:48
EEG machines, 7:97–98
external defibrillators, 7:117–118
laser pointers, 7:208, 7:209
radios, 7:276, 7:277, 7:278

Printing
electronic ink, **6:158–161**
pop up books, 7:264–265

Prisms, in binoculars, 7:30, 7:32

Procter and Gamble, 7:120, 7:123, 7:136

Proctor Electric Company, 7:395

Prohibition (1920-1933), 7:42

Proinsulin, 7:190

Projection, of speedometer readouts, 7:319–320

Prokofiev, Sergey Sergeyevich, 7:225

Proof test, for diving bells, 7:84–85

Propane, **3:335–338**

Propellants
hairspray, 7:173, 7:174
spray paint, 7:329, 7:330

Pseudomonofilament sutures, 7:356

PTFE (Polytetrafluoroethylene/Teflon),
7:378–381

Pull tabs, for juice boxes, 7:199, 7:200

Pull tests, 7:256

Pumps, for insulin, 7:192

Punching
bird cage panels, 7:34
bullets, 7:54
cigarette lighters, 7:216
lawn sprinklers, 7:212
microphones, 7:231
movie projectors, 7:240, 7:241
tea kettles, 7:107–108
toasters, 7:396–397

Punching bags, **7:266–269**

Purification, of kerosene, 7:204–206

Putty, for ships in a bottle, 7:299–300, 7:302

PVC piping, for pools, 7:362, 7:363

Pyraminx puzzle, 7:290–291

Pyrex, **7:270–274**

Q

Quality control and regulation
breath alcohol testers, 7:49–50
cigarette lighters, 7:219
diving bells, 7:84, 7:85
electroplating, 7:350
fabric softeners, 7:123
felt density, 7:133
fluoride standards, 7:136, 7:137
hair dryers, 7:168
handcuffs, 7:178–179
lawn sprinklers, 7:213
manhole covers, 7:222–223
oxygen tanks, 7:250–251, 7:252
pacifiers, 7:253, 7:254–256
revolving doors, 7:283
shrapnel shells, 7:307
sleeping pills, 7:309, 7:310, 7:312–313
smoked ham, 7:315–316
sporks, 7:325–326
spray paint, 7:331–332
stereoptic viewers, 7:340
sushi rolls, 7:353
sutures, 7:356, 7:357, 7:359
tattoos, 7:374–375

See also specific government agencies

Quarrying, of granite, 7:181, 7:183

Quartz, in movie projector bulbs, 7:239, 7:240

Quaternary ammonium compounds, 7:121

Queensberry boxing rules, 7:266

Quill pens, 7:139

R

Rabbeting (Electric guitars), 7:102

Radio frequency receivers, 7:279, 7:280

Radio pet collars, **7:279–281**

Radios, **7:275–278**

Raincoats, **6:311–315**

Raisins, **4:399–403**

Rammed earth constructions, **3:339–344**

Randolph, Mary, 7:69

Rankine cycle, 7:342, 7:344

Rapid-acting insulin, 7:187

Ratchets, in handcuffs, 7:177, 7:178

Rattles, for babies, 7:253–254

Rausing, Ruben, 7:198

Rayon, **1:350–354**

Razors (Safety), **5:388–392**, 7:110

RCA Corporation, 7:229

Reagent strips, for blood glucose, 7:152, 7:153

Recliners, **3:345–348**

Recombinant DNA, 6:140–144, 7:188–191

Reconditioning, of goalie masks, 7:159

Records (LP), **5:301–306**

Recycling. *See* Byproducts, waste, and
recycling

Red laser pointers, 7:208–210

Redox (Recycle Extract Dual Extraction)
process, 7:206

Reed, J. W., 7:287

Reels, for stereoptic viewers, 7:336, 7:337–338,
7:340

Refining, of crude oil, 7:204–206

Refrigerant 114, 7:378

Refrigerators, **1:355–360**

Regenerative heat exchangers, 7:344, 7:345,
7:346

Reinforced concrete, in bank vaults, 7:20, 7:22

See also headings beginning with Concrete

Reis, Johann Philipp, 7:382

Remington Company, 7:110, 7:404

Remote vending technology, 7:418–419

Removal, of tattoos, 7:377

Renal catheters, 7:66, 7:68

Resch, Glenn Chico, 7:158

Resins, of amber, 7:6–7

Resonators, in lasers, 7:207

Revolving doors, **7:282–285**

Rey, Alvino, 7:100

Rheaume, Manon, 7:157

Ribbon microphones, 7:229, 7:230, 7:231

Ribbons, **3:349–353**

Rice, **5:383–387**, 7:352

Rice cakes, **4:404–407**

Rickenbacker, Adolph, 7:100, 7:101

Rimless spinning wheels, 7:321

Ritty, James, 7:62, 7:63

Road signs, **2:371–375**

Roentgen, Wilhelm, 7:439

Roget, Peter Mark, 7:237

Roller coasters, **6:316–321**

Rolling pins, **7:286–289**

Romeo and Juliet (Prokofiev), 7:225

Ropes, **2:376–379**

Rosin, in shrapnel shells, 7:306

Rotary oil drilling, 7:204

Rotary tweezing epilators, 7:111–112

Rotundas, in revolving doors, 7:282, 7:283

Rough terrain forklifts, **2:380–384**

Rowe, William, 7:414

Royal typewriters, 7:408

Rubber
    artificial turf, 7:17
    bands, **1:367–370**
    catheters, 7:67
    cement, **6:322–324**
    latex, **3:245–250**
    pacifiers, 7:254
    stamps, **4:408–412**
    windshield wipers, 7:435

Rubber bands, **1:367–370**

Rubber cement, **6:322–324**

Rubber stamps, **4:408–412**

Rubies (Synthetic), **4:428–432**

Rubik, Erno, 7:290, 7:291

Rubik's cubes, **7:290–294**

Runcible spoons, 7:325–328

Running shoes, **1:371–375**

Rutile, 7:392, 7:393

S

S. S. Adams Company, 7:439

Saddle bicycle seats, 7:25, 7:26

Safes (Bank), 7:20

Safety and protective equipment
    air bags, **1:1–7**
    antilock brake systems (ABS), **2:16–20**
    antishoplifting tags, **3:26–29**
    bulletproof vests, **1:91–95**
    carbon monoxide detectors, **4:84–87**
    child safety seats, **5:80–85**
    combination locks, **1:139–142**
    crash test dummies, **5:122–127**
    fire engines, **2:195–200**
    fire extinguishers, **1:194–198**
    gas masks, **3:171–174**
    life vests, **2:288–292**
    locks, **5:296–300**
    smoke detectors, **2:410–413**

Safety concerns. *See* Health hazards and safety concerns

Safety pins, **2:385–388**

Safety razors, **5:388–392**, 7:110

Sailboats, **6:325–331**

Sails, for ships in a bottle, 7:301–302

Salad dressings, **6:332–334**

Salmon farming concerns, 7:355

Salsa, **1:381–384**

Salt, **2:389–393**

Salter, Richard, 7:295

Samovars, 7:106

Sand, **3:354–358**

Sandblasting, of granite, 7:183

Sandpaper, **1:385–389**

Sanson, Charles-Henri, 7:163

Sanson, Henri, 7:163

Sargent, James, 7:20

Sas Maki Kenukesushi, 7:352

Satellite dishes, **1:390–394**

Satin trimming, on tuxedos, 7:399

Saturation diving, 7:82, 7:85

Saws, **3:359–362**

Sawyer's, Inc., 7:336

Saxophones, **6:335–339**

Scales, **7:295–297**

Scheele, Carl Wilhelm, 7:249–250

Schick, Jacob, 7:110

Schmidt, Tobias, 7:163

School buses, **4:413–417**

School supplies. *See* Office and school supplies

Schooley, Matthew, 7:257–258

Schrafft's (Ice cream), 7:92

Scissors, **3:363–367**

Scotch Brite Corporation, 7:233

Scratch and sniff, **3:368–371**

Screw conveyors, in ice resurfacers, 7:185, 7:186

Screwdrivers, **1:395–399**

Screws, **3:372–374**

Seats, for unicycles, 7:410–411

Seaweed, in sushi, 7:352–354

Seedless fruits and vegetables, **6:340–343**

Seeds, in maracas, 7:225, 7:226

Seismographs, **1:400–405**

Selectric typewriters, 7:404

Self-adhesive notes, **2:394–397**

Semiconductor lasers, **6:344–348**, 7:207, 7:208

Sensors, for windshield wipers, 7:434, 7:438

Separation, of crude oil, 7:204, 7:205

Seven Towns Ltd. (London), 7:291, 7:293

Sewell, Ike, 7:260

Sewers (Waste), 7:220

Sewing
  corsets, 7:75–76
  punching bags, 7:268
  swimsuits, 7:367–368
  tuxedos, 7:401–402
  yarn mops, 7:234

Shackle bracelets, in handcuffs, 7:178

Shamoji, 7:352

Shampoos, **3:381–385**

Shaving creams, **1:406–409**

Sheet music, **5:393–397**

Shellac, **4:418–422**

Shells, for shrapnel, 7:304, 7:305–306

Shelters (Storm), **6:401–405**

Shingles, **3:386–399**

Ships in a bottle, **7:298–303**

Shoelaces, **6:349–352**

Shoes. *See* Clothing and accessories

Shokunin, 7:352, 7:353

Sholes, Christopher L., 7:404

Short-acting insulin, 7:187

Shortbread, **6:353–356**

Shouldaire swimsuit, 7:365

Shrapnel, Henry, 7:304

Shrapnel shells, **7:304–308**

Shroud lines, for ships in a bottle, 7:301

Siemens, Ernst, 7:333

Silicon, **6:357–359**

Silicon dioxide, 7:271–272, 7:273

Silicone derivatives, in fabric softeners, 7:121, 7:122

Silicone rubber
  catheters, 7:67
  pacifiers, 7:254, 7:256

Silk, **2:398–402**

Silly Putty, **5:398–400**

Silver, **3:390–393**
  tea pots, 7:106
  tiaras, 7:385–386

Sivrac, Comte de, 7:410

Skateboards, **6:360–364**

Skis, **2:403–409**

Skyscrapers, **6:365–371**

Slater, Samuel, 7:321

Sled units, in ice resurfacers, 7:184, 7:185, 7:186

Sleeping pills, **7:309–313**

Slime, **6:372–375**

Slinky toys, **3:394–397**

Slocum, Samuel, 7:348

Slow vinegar process, 7:425

Smeaton, John, 7:83

Smith, Adam, 7:348

Smith Corona typewriters, 7:408

Smock windmills, 7:430

Smoke detectors, **2:410–413**

Smoked ham, **7:314–316**

Smoking equipment
  cigarettes, **2:123–128**
  cigars, **3:87–91**
  lighters, **7:215–219**

Snowshoes, **6:376–379**

Soaps, **2:414–417**

Soaps (Antibacterial), **4:6–11**

Soc-o-Mac punching bags, 7:268

Soccer balls, **5:401–405**

Soda bottles, **1:410–413**

Sodium chlorite, **6:380–383**

Sodium hypochloride, in drain cleaners, 7:87, 7:88–89

Sofas, **3:398–403**

Soft drinks, **2:418–422**

Soil analysis, for swimming pools, 7:362

Solar cells, **1:414–419**

Solar heating systems, **3:404–407**

Soldering
  EEG machines, 7:98
  laser pointers, 7:208, 7:209
  tiaras, 7:386, 7:389

Solid-body electric guitars, 7:100–101, 7:102

Solid state lasers, **6:384–387**

SoloSpar punching bags, 7:268

Solvents
  hairspray, 7:172–173
  spray paint, 7:330

Sonata (Zaleplon), 7:309, 7:313

Sony Corporation, 7:421, 7:424

Soups, condensed, **7:69–72**

Soy milk, **5:406–409**

Soy sauce, **3:408–413**, 7:353

Spacesuits, **5:410–416**

Spam, **6:388–391**

Spandex, **4:423–427**

Spark plugs, **1:420–423**

Spars, for ships in a bottle, 7:300

Speakers (Stereo), **7:333–335**

Speedo Company, 7:366

Speedometers, **7:317–320**

Sphene, 7:392

Spineela, Linda, 7:158

Spinning wheels, **7:321–324**

Spokes, for unicycles, 7:411, 7:412

Sponge mops, 7:233, 7:234, 7:235

Sponges, **5:417–420**

Sponges (Titanium production), 7:393

Spool assemblies, in movie projectors, 7:238–239, 7:241–242

Spoons, history of, 7:325

Sporks, **7:325–328**

Sporting goods and equipment
    ammunition, **2:11–15**
    artificial snow, **4:22–25**
    artificial turf, **7:15–19**
    baseball bats, **2:37–40**
    baseball caps, **4:44–47**
    baseball gloves, **1:55–59**
    baseballs, **1:50–54**
    basketballs, **6:47–53**
    bicycle seats, **7:25–29**
    bicycle shorts, **1:65–68**
    bicycles, **2:61–66**
    binoculars, **7:30–32**
    boomerangs, **6:54–58**
    bowling balls, **4:66–69**
    bowling pins, **4:70–74**
    bows and arrows, **5:51–56**
    boxing gloves, **6:59–62**
    bungee cords, **2:75–79**
    diving bells, **7:82–86**
    footbags, **6:180–183**
    football helmets, **3:162–164**
    footballs, **3:159–161**
    Frisbees, **5:205–209**
    goalie masks, **7:156–159**
    golf carts, **1:209–213**
    golf clubs, **4:241–245**
    golf tees, **5:238–241**
    hang gliders, **5:253–257**
    hockey pucks, **6:212–216**
    hockey sticks, **4:262–267**
    Hula Hoops, **6:220–223**
    hunting knives, **2:232–236**
    ice resurfacing machines, **7:184–186**
    kayaks, **2:267–271**
    life vests, **2:288–292**
    paddles, **2:323–326**
    parachutes, **5:353–358**
    punching bags, **7:266–269**

running shoes, **1:371–375**
saddles, **1:376–380**
skateboards, **6:360–364**
skates, ice, **2:237–240**
skates, in-line, **2:241–244**
skis, **2:403–409**
snowshoes, **6:376–379**
soccer balls, **5:401–405**
surfboards, **2:434–438**
swimming pools, **7:361–364**
swimsuits, **7:365–368**
tennis rackets, **3:445–449**
trampolines, **3:459–461**
unicycles, **7:410–413**
wet suits, **4:462–466**

See also Toys and games

Spray paint, **7:329–332**

Spring balance scales, 7:295, 7:296–297

Springs, **6:392–396**

Stabilizers, in Pyrex, 7:272, 7:273

Stained glass, **2:423–248**

Stainless steel, **1:424–429**
    cigarette lighters, 7:216, 7:217
    furnaces, 7:145
    tea kettles, 7:107

Stamping, of straight pins, 7:349–350

Standard typewriters, 7:404

Stanley, James, 7:410

Staplers, **1:430–434**

Staples (Surgical), 7:359

Starlight night vision systems, 7:244

Static electricity
    dusters, 7:125
    mops, 7:234

Statuary, **5:421–424**

Steam engines, 7:342–343, 7:344

Steel
    diving bells, 7:84
    galvanizing of, 7:35
    guillotines, 7:165
    handcuffs, 7:178
    ice resurfacers, 7:184–185
    ironing boards, 7:195–197
    lanterns, 7:148–149
    mops, 7:234, 7:235
    movie projectors, 7:240, 7:241–242
    shrapnel shells, 7:304, 7:305–306
    speedometers, 7:317, 7:318
    telephone booths, 7:382, 7:383
    typewriters, 7:405–408
    unicycles, 7:411
    vending machines, 7:414–415, 7:416, 7:417–418
    windmills, 7:430, 7:431
    windshield wipers, 7:435
    wool, **6:397–400**

See also Stainless steel

Steel pens, 7:139

Steel wire
  bird cages, 7:33, 7:34
  feather dusters, 7:126, 7:128
  paper clips, 7:258–259
  straight pins, 7:349

Steel wool, **6:397–400**

Stereo speakers, **7:333–335**

Stereoptic viewers, **7:336–341**

Sterilizing
  juice boxes, 7:200
  for surgery, 7:356–357
  sutures, 7:359

Stethoscopes, **1:434–438**

Stetson hats, **3:414–418**

Steward, Thomas W., 7:233

Stirling cycle engines, **7:342–347**

Stirling, Robert, 7:343

Stock, for soup, 7:70

Stone quarrying, 7:181, 7:183

Storm shelters, **6:401–405**

Straight pins, **7:348–351**

Straws, for drinking, **4:186–190**, 7:199, 7:200

Strength tests, for cast iron, 7:222–223

Stretching, of sutures, 7:358, 7:359

Striking bags, 7:266–268

Strite, Charles, 7:395

Sublimation, of dry ice, 7:91, 7:93

Submerged fermentation vinegar process, 7:426

Sugar, **1:439–443**, 7:59

Sulfolane kerosene extraction, 7:205

Sulfuric acid
  drain cleaners, 7:87, 7:88
  feltmaking, 7:132–133

Sullivan, Eugene G., 7:271

Sunbeam Corporation, 7:395

Sunflower seeds, **5:430–433**, 7:37

Sunglasses, **3:419–423**

Sunscreen, **2:429–433**

Super glues, **1:444–447**

Superhetrodyne radios, 7:275

Surface-oriented diving, 7:82

Surfboards, **2:434–438**

Surge suppressors, **3:424–426**

Surgical glue, 7:360

Surgical needles, 7:357, 7:359

Surgical staples, 7:359

Surgical zippers, 7:360

Sushi rolls, **7:352–355**

Suspension bridges, **5:434–440**

Sutures, **7:356–360**

Swaging
  bullets, 7:54
  oxygen tanks, 7:250

Swimming pools, **7:361–364**

Swimsuits, **7:365–368**

Swivels, in mops, 7:234–235

Swords, **5:441–446**

Synthetic Blood International, 7:155

Synthetic insulin, 7:188–191

Synthetic rubies, **4:428–432**

Synthetic sutures, 7:356, 7:357–359

Synthetic turf, 7:15–19

Syringes, **3:427–431**

T

Tables, **7:369–373**

Tabletting, of sleep aids, 7:312

Talking toasters, 7:398

Tanning, of leather, 7:267

Target assemblies, in cyclotrons, 7:78, 7:80

Taste testing
  brandy, 7:44
  candy canes, 7:60

Tattoos, **7:374–377**

Taylor, William C., 7:271

Tea, 7:106

Tea bags, **2:443–447**

Tea kettles (Electric), **7:106–109**

Teddy bears, **3:432–437**

Teeth whiteners, **6:406–409**

Teflon, **7:378–381**

Telecaster electric guitar, 7:101

Telephone booths, **7:382–384**

Telephones, **5:447–451**

Telescopes, **2:448–452**

Televisions, **3:438–444**

Temporary tattoos, **4:433–447**

Tennis rackets, **3:445–449**

Tesla, Nikola, 7:317

Test strips, for blood glucose, 7:152, 7:153

Tetra Brik containers, 7:198

Tetrafluoroethylene (TFE), 7:378–379

Textiles. *See* Fabrics and textiles

Thaumatropes, 7:237

Thermal efficiency, 7:342, 7:343, 7:344

Thermal fuses, in hair dryers, 7:170

Thermal night scopes, 7:244

Thermic torches, 7:24

Thermodynamic cycles, 7:342, 7:344

Thermoforming, of sporks, 7:326–327

Thermometers, **1:448–452**

Thompson submachine guns, **6:410–414**

Thong bikinis, 7:366

Thorium, in gas lanterns, 7:147, 7:148

Thread, **5:452–457**, 7:300–301, 7:302

Three-dimensional images, 7:336, 7:340

3M Company, 7:15

Thuras, A. C., 7:229

Tiaras, **7:385–390**

Tibor, Laczi, 7:291

Timed locks, 7:20, 7:22

Tin, **4:437–442**

Tires, **1:453–457**

Tires, for unicycles, 7:411, 7:413

Tissue ablation, with catheters, 7:68

Titanium, **7:391–394**

Toast-R-Oven, 7:396

Toasters, **7:395–398**

Toastmaster, 7:395

Tobacco equipment. *See* Smoking equipment

Tofu, **2:453–457**

Toilet paper, **6:415–418**

Toilets, **5:458–463**

Tomato soup, condensed, 7:70–72

Tombstones, 7:180–183

Tooth decay, 7:135

Toothbrushes, **2:458–461**

Toothpastes, **3:455–458**, 7:136

Topographic maps, **4:443–447**

Toposcopes, 7:96

Torricelli, Evangelista, 7:11

Tortilla chips, **1:458–463**

Towels, in ice resurfacers, 7:185, 7:186

Tower Company, 7:177

Tower, John, 7:177

Tower of Hanoi puzzle, 7:290

Tower windmills, 7:429, 7:430

Toy model kits, **6:419–424**

Toys and games
  action figures, **6:1–4**
  balloons, **2:27–31**
  bean bag plush toys, **5:25–30**
  chess, **6:97–101**
  crayons, **2:153–156**
  dice, **4:182–186**
  gyroscopes, **6:190–193**
  kaleidoscopes, **6:243–248**
  marbles, **2:299–302**
  matryoshka dolls, **6:257–261**
  model trains, **4:339–342**
  paintball, **6:278–281**
  pinball, **6:292–297**
  plastic dolls, **5:382–376**
  playing cards, **4:374–378**
  pool tables, **6:302–306**
  Rubik's cubes, **7:290–294**
  ships in a bottle, **7:298–303**
  Silly Putty, **5:398–400**
  slime, **6:372–375**
  Slinky toys, **3:394–397**
  stereoptic viewers, **7:336–341**
  teddy bears, **3:432–437**
  toy model kits, **6:419–424**
  video games, **5:478–483**
  water guns, **6:448–451**
  whistles, **4:467–470**
  x-ray glasses, **7:439–442**

See also Sporting goods and equipment

Traffic signals, **2:462–464**

Training bags, 7:268

Trampolines, **3:459–461**

Transistors, in radios, 7:276

Transmitters, for pet containment systems, 7:279, 7:280–281

Transparencies, for stereoptic viewers, 7:337

Transportation vehicles
  airships, **3:16–20**
  ambulances, **5:6–10**
  armored trucks, **4:17–21**
  business jets, **2:80–85**
  heavy duty trucks, **3:207–213**
  helicopters, **1:223–229**
  hot air balloons, **3:220–224**
  motorcycles, **4:343–347**
  nuclear submarines, **5:329–334**
  parade floats, **4:358–362**
  passenger vehicles, **1:24–30**
  sailboats, **6:325–331**

Trieste (Bathyscaph), 7:84

Tripoli compound, 7:387, 7:389

Trophies, **6:425–428**

Tru-Vue Company, 7:336

Trumpets, **1:464–468**

Tubas, **5:464–468**

Tubing (Metal)
  ironing boards, 7:195
  lawn sprinklers, 7:212

Tufting (Artificial turf), 7:15, 7:17

Tuna, in sushi, 7:353, 7:355

Tunnels, **6:429–434**

Tutmarc, Paul H., 7:100

Tuxedo Park (New York City), 7:399

Tuxedos, **7:399–403**

TV dinners, **5:469–473**

Tweeters (Speakers), 7:334

Twisted sutures, 7:356, 7:357, 7:358

Tyco Toys, 7:336

Typebaskets, 7:405, 7:407

Typewriters, **7:404–409**

U

Uchatius, Baron von, 7:237

Uchiwa, 7:352

Udex kerosene extraction, 7:205

Ukuleles, **6:435–438**

UL (Underwriters Laboratory, Inc.), 7:23

ULPA filters, 7:5

Ultimate Wheel unicycles, 7:411

Ultra-concentrated fabric softeners, 7:120, 7:123

Ultra Low Penetrating Air (ULPA) filters, 7:5

Umbrellas, **1:469–472**

Undergarments, for women, 7:73–74

Underwood, John, 7:404

Underwriters Laboratory, Inc. (UL), 7:23

Unger, E., 7:66

Unicycles, **7:410–413**

Uniform Building Code, 7:283

Union Carbide kerosene extraction, 7:206

United States Pharmacopeia, 7:356, 7:359

University of Washington, 7:398

Urinalysis, for glucose, 7:152

U. S. Bureau of Mines, 7:392

U. S. Department of Agriculture, 7:315

U. S. Department of Defense
    Night Vision Laboratory, 7:244, 7:245
    particle filtration standards, 7:4

U. S. Department of Energy, 7:119

U. S. Department of Transportation
    breath alcohol testers, 7:49–50
    oxygen tanks, 7:252

U. S. Environmental Protection Agency (EPA),
    7:350

U. S. Food and Drug Administration
    fluoride standards, 7:136, 7:137
    medical device standards, 7:118
    raw fish guidelines, 7:353
    sleeping pill regulation, 7:309, 7:310,
        7:312–313
    spork regulation, 7:326
    suture regulation, 7:356, 7:357

U. S. Navy, 7:84, 7:85

V

Vaaler, John, 7:258

Vaccines, **2:466–470**

Vacuum cleaners, **6:439–443**

Vacuums

barometers, 7:11
cyclotrons, 7:79

Van Brode Milling Company, 7:325

Vaults, for banks, **7:20–24**

Vegetable cooking oils, **1:164–168**

Vegetables. *See* Crops

Vegetarian burgers, **5:474–477**

Vending machines, **7:414–419**

Ventricular fibrillation, 7:114–115

Vermiculite, **6:444–447**

Veronal, 7:310

Versa-Toast 4-Slice toaster, 7:396

Vertical-axis windmills, 7:429, 7:430

VHS videotapes, 7:421

Video games, **5:478–483**

Videotapes, **7:420–424**

Vidie, Lucien, 7:12

View-Master viewers, 7:336, 7:340

Vinegar, **7:425–428**

Vinyl acetate, in hairspray, 7:172

Vinyl floorcoverings, **4:452–456**

Vinyl-lined swimming pools, 7:361, 7:362–364

Violin bows, **2:476–480**

Violins, **2:471–475**

Virginia House-Wife (Randolph), 7:69

Vitamins, **3:462–466**

Vitascopes, 7:238

VOCs. *See* Volatile organic compounds

Vodka, **5:484–488**

Voice coils
    microphones, 7:230, 7:231, 7:232
    stereo speakers, 7:333, 7:334

Volatile organic compounds (VOCs)
    hairspray, 7:175
    spray paint, 7:329

Vosbikian, Peter, 7:233

The Voyage in H. M. Bark Endeavor, 7:374

W

Wagner, Simon, 7:37

Wall, Ormand, 7:434

Wallpapers, **3:467–471**

Walnut wood, for spinning wheels, 7:322

Walsh, Donald, 7:84

Walter, W. Gray, 7:96

Warner, A. P., 7:317

Warner Instrument Company, 7:317

Wasabi, in sushi, 7:353

Washing machines, **1:473–477**

Waste. *See* Byproducts, waste, and recycling

Watches, **1:478–481**

Water, **4:457–461**

Water guns, **6:448–451**

Water jet piercing, for granite, 7:181

Watercraft
    nuclear submarines, **5:329–334**
    sailboats, **6:325–331**

Waterman, Lewis Edson, 7:139

Watt, James, 7:342–343

Wealth of Nations (Smith), 7:348

Weight standards, 7:295

Welding
    cigarette lighters, 7:216
    diving bells, 7:84
    ironing boards, 7:195
    movie projectors, 7:241–242

Wente, E. C., 7:229

Western Electric, 7:382

Western Lighting Company, 7:148

Westinghouse, 7:395

Wet mops, 7:233, 7:235–236

Wet suits, **4:462–466**

Whalebone corsets, 7:73, 7:74, 7:76

Wheat flour, **3:153–158**, 7:260–261

Wheatstone, Sir Charles, 7:336

Wheelbarrows, **5:494–497**

Wheels, for unicycles, 7:411, 7:412–413

Wheelwriter typewriters, 7:404

Whiskey, **2:481–486**

Whistles, **4:467–470**

White dummies (Pop up books), 7:264

Wicks, for cigarette lighters, 7:217

Wigs, **3:472–477**

William, Mac, 7:114

Wind chimes, **5:498–501**

Wind turbines, **1:482–487**

Windmills, **7:429–433**

Windshield wipers, **7:434–438**

Wine, **1:488–493**

Wire mesh, in bird cages, 7:33, 7:34

Wonder tubes, 7:439

Wood
    clogs, **4:471–473**
    electric guitars, 7:101–104
    feather duster handles, 7:126
    fountain pens, 7:141–142
    guillotines, 7:164–166
    maracas, 7:225, 7:226
    mops, 7:234, 7:235
    plywood, **4:379–383**
    rolling pins, 7:286–288, 7:289
    ships in a bottle, 7:298, 7:300
    smoked ham, 7:314
    spinning wheels, 7:322–323
    stains for, **6:452–455**
    tables, 7:369, 7:370, 7:371–373
    telephone booths, 7:382
    windmills, 7:430

Wood stains, **6:452–455**

Wooden clogs, **4:471–473**

Woodworth, F. H., 7:229

Woofers (Speakers), 7:334

Wool, **1:494–499**, 7:131

Working fluids, in engines, 7:343, 7:344

Wrapping paper, **6:456–460**

Wright, Lemuel, 7:348

Writing Instrument Manufacturers Association, 7:139

X

X-ray glasses, **7:439–442**

X-rays, 7:439

Xenon bulbs, in movie projectors, 7:239–241

Xylophones, **6:461–465**

Y

Yale Jr., Linus, 7:20

Yarn, **3:478–482**, 7:233–234

Yeast, in pizza, 7:260–261

Yogurt, **4:474–478**

Yogurt (Frozen), **2:214–218**

Yttrium, in gas lanterns, 7:148, 7:149

Z

Zahn, Johann, 7:30

Zaleplon, 7:309, 7:313

Zamboni, Frank J., 7:184

Zamboni machines, 7:184–186

Zehntbauer, John, 7:365

Zehntbauer, Roy, 7:365

Zeiss, Carl, 7:30

Zinc, **2:487–492**, 7:35

Zippers, **1:500–504**

Zippers (Surgical), 7:360

Zippo lighters, 7:215

Zirconium, **1:505–508**

Zoetropes, 7:237

Zoll, Paul, 7:115